The Science of Composting

European Commission
International Symposium

THE SCIENCE
OF COMPOSTING

Edited by

Marco de Bertoldi
Paolo Sequi
Bert Lemmes
Tiziano Papi

BLACKIE ACADEMIC & PROFESSIONAL
An Imprint of Chapman & Hall

London · Glasgow · Weinheim · New York · Tokyo · Melbourne · Madras

Published by
Blackie Academic & Professional, an imprint of Chapman & Hall,
Wester Cleddens Road, Bishopbriggs, Glasgow G64 2NZ

Chapman & Hall, 2–6 Boundary Row, London SE1 8HN, UK

Blackie Academic & Professional, Wester Cleddens Road, Bishopbriggs, Glasgow G64 2NZ, UK

Chapman & Hall GmbH, Pappelallee 3, 69469 Weinheim, Germany

Chapman & Hall USA, 115 Fifth Avenue, Fourth Floor, New York, NY 10003, USA

Chapman & Hall Japan, ITP-Japan, Kyowa Building, 3F, 2-2-1 Hirakawacho, Chiyoda-ku, Tokyo 102, Japan

DA Book (Aust.) Pty Ltd, 648 Whitehorse Road, Mitcham 3132, Victoria, Australia

Chapman & Hall India, R. Seshadri, 32 Second Main Road, CIT East, Madras 600 035, India

First edition 1996

© 1996 Chapman & Hall

Printed in England by Clays Ltd., St Ives plc

ISBN 0 7514 0383 0

A catalogue record for this book is available from the British Library.

Library of Congress Catalog Card Number has been applied for.

∞ Printed on acid-free text paper, manufactured in accordance with ANSI/NISO Z39.48-1992 (Permanence of Paper)

Introduction

The European Union initially demonstrated its interest in waste in the late 70s with the progamme on Waste Recycling Research and Development. At that time composting was only present as a coordination activity and it was only later that specific research programmes in the area were within Europe which was largely instrumental in setting up a series of European conferences, seminars and workshops. Some of these have resulted in publications which have made significant contributions to developments in the understanding of composting and the use of composts. In particular the outputs from meetings in Oxford (1984), Udine (1986), Neresheim (1988) and Angers (1991) are worthy of note.

Composting has seen significant changes since the 70s when the major thrust in Europe was using mixed municipal solid waste as a feed material. Many composting plants which were built to use this material were closed due to the poor quality of the compost which made it very difficult to market. As a result the main areas of interest, as far as the municipalities are concerned, are now with biowaste and source-separated organics. This interest is apparent from the many new plants which are being constructed across Europe, and the ready market which exists for the products. In parallel with the renewed interest of the municipalities other areas, such as agriculture and the wastewater treatment industries, are also developing their own schemes.

The EU produces in excess of 2500×106 tonnes of waste each year a high proportion of which is organic. The disposal options for this waste are fewer every year as we become aware of the potentially adverse environmental impacts of using them. A clear alternative for the organic fraction of this waste is composting which not only enables us to treat the waste but also to maximise its beneficial properties through re-use. This is particularly important in southern Europe where the soils are very poor in relation to their organic content, but it is not only these regions which need this attention to their soils. Many areas in our northern regions also need this type of input following many years of exploitation of inorganic fertilizers without attention to the long term implications for soil structure.

The many new composting plants opening in Europe confirm the interest in the utilisation of our organic wastes but we need to move forward with research and development to ensure that this resurgence does not die in a similar way to that of the 70s. Conferences, like the one in Bologna, are vital to this process bringing together researcher and practioners in a forum where their ideas can be presented, discussed and moved forwards.

The interest in this area is confirmed by the attendance of in excess of 500 delegates from 36 different countries. This is a recognition of the vision of the sponsors who have actively supported this initiative.

Marco de Bertoldi
Professor of Industrial Microbiology
University of Udine

September 1995

Publisher's note

Due to the short preparation time required to produce this book, some papers were not received in time to be included in the first setting. To ensure completeness of coverage, these papers have been included at the end of the appropriate volume.

Volume 1
Round Table on Legislation
W.A. Van Belle, ORCA

Legislation in the European Union on Compost Production and Use
A. Piavaux, European Commission

Volume 2
Biological Treatment, the Perfect Eco-efficient Tool in a Sustainable Integrated Waste Management
B. Lemmes, Managing Director, ORCA

ORCA Compostability Criteria: A Framework for the Evaluation of Feedstock for Source-Separated Composting and Biogasification
K. Mesuere, ORCA

Contents

PLENARY SECTION   1

Historical review of composting and its role in municipal waste management   3
C.G. GOLUEKE AND L.F. DIAZ

The Thermie programme and composting projects in the energy from biomass and waste sector   15
G.L. FERRERO

The role of composting in sustainable agriculture   23
P. SEQUI

I.S.W.A. policy in the regard of composting as an integrated system of waste management   30
J.H. SKINNER

Waste management and legislation   41
W.A. VAN BELLE

PART A1   COMPOSTING PROCESS   47

Composting control: principles and practice   49
E.I. STENTIFORD

Composting plant design and process management   60
R.T. HAUG

Odor emissions from composting plants   71
W. BIDLINGMAIER

Environmental impact of composting plants   81
K. FISHER

Production of functional compost which can suppress phytopathogenic fungi of lawn grass by inoculating Bacillus subtilis into grass clippings   87
K. NAKASAKI, M. KUBOL AND H. KUBOTA

Nutrient transformation of pig manure under pig-on-litter system   96
N.F.Y. TAM, S.M. TIQUIA AND L.L.VRIJMOED

Heat evolution during composting of sewage sludge   106
F.C. MILLER

Process control based on dynamic properties in composting:
moisture and compaction considerations   116
K. DAS AND H.M. KEENER

Compost facility operating guide   126
P.B. LEEGE

Glucose influence on the asymbiotic nitrogen fixation during
lignocellulosic waste composting   137
K. EZELIN, G. BRUN, M. KAEMMERER AND J.C. REVEL

Taxonomic and metabolic microbial diversity during composting   149
T. BEFFA, M. BLANC, L. MARILLEY, J.L. FISHER, P.F. LYON AND
M. ARAGNO

A new composting plant realized within the THERMIE program of
the European Commission   162
T. PAPI, G. MARANI, R. MANIRONI AND M. DE BERTOLDI

PART A2 THE QUALITY OF COMPOST   173

Evaluation of compost stability   175
R. BARBERIS AND P. NAPPI

Quality of composts: organic matter stabilization and trace metal
contamination   185
C. MASSIANI AND M. DOMEIZEL

Chemical and physico-chemical parameters for quality evaluation
of humic substances produced during composting   195
N. SENESI AND G. BRUNETTI

Heavy metals in compost and their effect on soil quality   213
G.A. PETRUZZELLI

Occurrence of microorganisms pathogenic for man and animals in
source-separated biowaste and compost: importance, control,
limits, epidemiology   224
D. STRAUCH

Phytohygienic aspects of composting 233
G.J. BOLLEN AND D. VOLKER

Canadian national compost standards 247
N. FOLLIET-HOYTE

**Ninhydrin reactive nitrogen of CHCl₃ fumigated and non-fumigated
compost extracts as a parameter to evaluate compost stability** 255
M. DE NOBILI, M.T. BACA, F. FORNASIER AND C. MONDINI

**Impact of composting type on compost organic matter
characteristics** 262
C. SERRA-WITTLING, E. BARRIUSO AND S. HOUOT

**Respirometric techniques in the context of compost stability
assessment: principles and practice** 274
K.E. LASARIDI AND E.I. STENTIFORD

**Different location of acid and alkaline phosphatases extracted from
a compost of urban refuse** 286
J.C. RAD MORADILLO AND S. GONZALES CARCEDO

**Biological parameters to estimate the effect of biogenic waste
composts on plant growth in pot-trials** 294
L. POPP AND P. FISCHER

**Parameters to estimate the nitrogen effect of biogenic waste
composts** 306
TH. EBERTSEDER, R. GUTSER AND N. CLAASSEN

**Biowaste compost and heavy metals: a danger for soil and
environment?** 314
F. AMLINGER

**Polychlorinated dibenzo-p-dioxins and dibenzofurans: level of
contamination and dynamic in bio and yard waste composting** 329
K. FRIKE, U. EINZMANN, A. BRUNETTI, H. FIEDLER AND
H. VOGTMANN

**The influence of composts and mineral fertilizers on the heavy metal
concentration and transfer in soil and plants** 346
H. VOGTMANN, G. BOURS AND W. FUCHSHOFEN

PART A3 LEGISLATION   355

USEPA regulations for compost production and use   357
J. WALKER

PART A4 USE OF COMPOST   371

Suppression of plant diseases by composts   373
H.A.J. HOITINK, A.G. STONE AND M.E. GREBUS

Formation and properties of humic substance originating from
composts   382
Y. CHEN, B. CHEFETZ AND Y. HADAR

Waste researches at the IRNAS, CSIC   394
R. LOPEZ, F. CABRERA, J.M. MURILLO AND A. TRONCOSO

Compost as a source of organic matter in Mediterranean soils   402
M.T. FELIPÒ

Utilizing composts in land management to recycle organics   413
W.H. SMITH

Effect of compost/fertilizer blends on crop growth   423
L.J. SIKORA

The influence of compost and sewage sludge on agricultural crops   431
G. BALDONI, L. CORTELLINI AND L. DAL RE

Compost from selected organic wastes as a substitute for
straw-bedded horse manure in *Agaricus bisporus* production   439
G. GOVI, G. INNOCENTI, G. FERRARI, C. GALLI AND G. SACCHINI

Use of composted societal organic wastes for sustainable crop
production   447
M.S. RODRIGUES, J.M. LOPEZ-REAL AND H.C. LEE

Effects on the content of organic matter, nitrogen, phosphorus and
heavy metals in soil and plants after application of compost and
sewage sludge   457
L. CORTELLINI, G. TODERI, G. BALDONI AND A. NASSISI

An evaluation of soil nutrient status following the application of
(i) co-composted MSW and sewage sludge and (ii) greenwaste to
maize   469
R. PARKINSON, M. FULLER, S. JURY AND A. GROENHOF

Development of compost products   477
H. HAUKE, H. STOPPLER-ZIMMER AND R. GOTTSCHALL

**Political implications for integrating composting into solid waste
management in West Virginia, USA**   495
R.G. DIENER, A. COLLINS, R. MENESES, M. MORRIS AND A. DAME

**Yard waste composting strategies: considering technical conditions
and organizing systems**   507
E. FAVOINO, M. CENTEMERO, M. CONSIGLIO, A. PANZIA OGLIETTI,
E. ACCOTTO AND G. NERI

**Experiences of compost use in agriculture and in land reclamation
projects**   517
F. PINAMONTI AND G. ZORZI

PART B1 STARTING MATERIAL   529

Importance of lignocellulosic compounds in composting   531
P.J. WHITNEY AND J.M. LYNCH

Composting of agricultural wastes   542
J.M. LOPEZ-REAL

Seven decades of sludge compost marketing   551
K. KELLOG JOHNSON

Utilizing scraps from blue crab and calico scallop processing plants 557
J.C. CATO

Dairy cattle slurry and rice hull co-composting   567
P.L. GENEVINI, F. ADANI AND C. VILLA

**Field scale study of the effect of pile size, turning regime and leaf to
grass mix ratio on the composting of yard trimmings**   577
F.C. MICHEL, J.F. HUANG, C.A. REDDY AND L.J. FORNEY

**Optimisation of anaerobically digested primary sludge as a
composting substrate**   585
D. WONG AND H. HOFSTEDE

**The production of compost from agricultural and municipal
solid waste**   593
P.F. BLOXHAM AND I.L. COLCLOUGH

Manufacture of artificial soil by composting coal fly ash and
bottom ash with poultry litter 603
H.L. BRODIE, L.E. CARR, G.A. CHRISTIANA AND J.R. UDINSKEY

Composting of a mixture of vegetable, fruit and garden waste and
used paper diapers 612
C. VERSCHUT, T.D. TNO-ME AND V. BRETHOUWER

Processes regulating grass straw composting 627
W.R. HORWATH, L.F. ELLIOTT, D.B. CHURCHILL AND H.F. MINSHEW

Olive-mill wastewater bioremediation: evolution of a composting
process and agronomic value of the end-product 637
U. TOMATI, E. GALLI, L. PASETTI AND E. VOLTERRA

Agricultural composting in the United States: trends and driving
forces 648
R.M. KASHMANIAN AND R.F. RYNK

PART B2 STATE OF THE ART OF COMPOSTING 661

Effectiveness of the Rutgers system in composting several different
wastes for agricultural uses 663
A. ROIG AND M.P. BERNAL

Composting in Finland: Experiences and perspectives 763
K. HANNINEN

Perspectives and state of the art of composting in France 684
J.M. MERILLOT

Composting of agricultural waste in Denmark 691
M. HEDEGAARD AND I. KRUGER

Composting in Italy: Current state and future outlook 698
G. ZORZI, S. SILVESTRI AND A. CRISTOFORETTI

State of the art and perspectives for composting in the United States
of America 714
J. GOLDSTEIN

The present situation and a new trend on composting in Japan 722
M. SHODA

Composting: experiences and perspectives in Brazil 729
J.T. PEREIRA NETO

Structural changes to a MSW composting plant in accordance with modern waste management concepts in Italy 736
D. SCHONAFINGER

The power of composting; the power of partnership 743
J. BEYEA

Composting plant in the city of Forlì: the public administration experience 748
T. GIUNCHI, G, VERONESI AND G. ZECCHI

Digestion by WAASA process of optically separated waste 758
R. WESTERGAARD

PART B3 COMPOSTING AS AN INTEGRATED SYSTEM OF WASTE MANAGEMENT 765

The co-treatment of municipal and industrial wastes 767
R.P. BARDOS, S. FORSYTHE AND K. WESTLAKE

The importance of waste characteristics and processing in the production of quality compost 784
G.M. SAVAGE

The role of biological treatment in integrated solid waste management 792
P. WHITE

Results of laboratory and field studies on wastepaper inclusion in biowaste in view of composting 803
B. DE WILDE, J. BOELENS AND L. DE BAERE

Fuel recovery: valorization of RDF and PDF 813
M. FRANKENHAEUSER AND H. MANNINEN

A database for I.W.M. covering recycling and composting 822
J. HUMMEL

Alternative utilization of MSW compost in landfills 831
R. COSSU AND A. MUNTONI

PART B4 BIOREMEDIATION   847

Stabilization of hazardous wastes through biotreatment   849
L.F. DIAZ, G.M. SAVAGE AND C.G. GOLUCKE

**Starch based biodegradable materials in the separate collection and
composting of organic waste**   863
C. BASTIOLI AND F. DEGLI INNOCENTI

**Degradation of naphthalene by microorganisms isolated from
compost**   870
M. CIVILINI AND N. SEBASTIANUTTO

**Composting and selected microorganisms for bioremediation of
contaminated materials**   884
M. CIVILINI, C. DOMENIS, M. DE BERTOLDI AND N. SEBASTIANUTTO

Bioremediation of PAH-contaminated soil   892
R. LILYA, J. UOTILA AND H. SILVENNOINEN

**Minimum effective compost addition for remediation of
pesticide-contaminated soil**   903
X. LIU AND M.A. COLE

**Enhancement of the biological degradation of contaminated soils by
compost addition**   913
K. HUPE, J.C. LÜTH, J. HEERENKLAGE AND R. STEGMANN

**Heavy metals removal by clinoptilolite in pepper cultivation using
compost**   924
E.G. KAPETANIOS

PART B5 COMPOSTING DESIGN   937

Composting technology in the United States: research and practice   939
R.J. TARDY AND R.W. BECK

Reconversion of traditional composting plants for a policy of quality 948
B. RANINGER

Basic processing technologies and composting plant design in Italy   958
F. CONTI, G. URBINI AND G. ZORZI

Design of passively aerated compost piles   973
N.J. LYNCH AND R.S. CHERRY

A review of features, benefits and costs of tunnel composting systems in Europe and in the USA   948
D. BORDER, R. DE GAMO AND K. PANTER

PART B6 MARKETING AND ECONOMY   987

Compost marketing trends in the United States   989
L.L. EGGERTH

The natural markets for compost   999
R.W. TYLER

Monitoring strategies and safeguarding of quality standards for compost   1011
J. BARTH

Minimizing the cost of compost production through facility design and process control   1020
H.M. KEENER, D.L. ELWELL, K.C. DAS AND R.C. HANSEN

FINAL REPORTS   1035

A1 Composting process   1037
E.I. STENTIFORD

B1 Starting materials   1039
J. LOPEZ-REAL AND J. MERILLOT

B2 The state of the art of composting and perspectives   1040
J. GOLDSTEIN

B3 Composting as an integrated system of waste management   1043
B. LEMMES

B4 Bioremediation   1048
L. DIAZ

B5 Composting design   1050
R.T. HAUG

B6 Marketing and economic   1053
L.L. EGGERTH

POSTERS 1055

Additional papers (see publisher's note, page vi)

Index of posters I1

Index of contributors I4

Steering Committee

Marco de Bertoldi, University of Udine, Chairman

Giovanni Brunelli, Ministero dell'Ambiente
Jurgen Busing, European Commission DG XII
Bert Lemmes, Orca, Bruxelles
Tiziano Papi, Caviro, Faenza (Ra)
Gianni Antonio Petruzzelli, Consiglio Nazionale delle Ricerche
Paolo Sequi, Ministero Risorse Agricole, Alimentari e Forestali
Giordano Urbini, University of Pavia
Gianni Zorzi, Istituto Agrario S. Michele dell'Adige (TN)
W.A. Van Belle, Procter & Gamble E.T.C.

International Advisory Board

Patrons

European Commission
Ministero dell'Ambiente
Ministero Risorse Agricole, Alimentari e Forestali
Ministero della Sanità
Regione Emilia Romagna, Assessorato Ambiente
Università degli Studi di Udine
Università degli Studi di Bologna
Consiglio Nazionale delle Ricerche
International Society of Soil Science: Working Group
Soil Organic Fertilizers & Amendments
International Solid Waste Association
Federambiente
Federchimica Assofertilizzanti
Confederazione Cooperative Italiane, Roma
Confederazione Generale Agricoltura Italiana
Confederazione Italiana Agricoltori
Confederazione Nazionale Agricoltori Diretti
Provincia di Forlì Cesena
Provincia di Ravenna
Comune di Faenza

Co-organizers

CAVIRO (Faenza RA)   ORCA (Bruxelles)

Sponsors

Caviro Faenza (Ra)
ORCA Bruxelles
European Commission
Procter & Gamble E.T.C.
Ecologia S.P.A. Milano
Tetra Pak Italiana S.p.A., Modena
Consorzio Italiano Compostatori
Igiene, Azienda Municipalizzata Igiene Urbana di Bologna
Saccecav Depurazioni Sacede S.p.A. di Milano
Assessorato Ambiente Regione Emilia Romagna
Federambiente
Conel S.n.c. Faenza
Ctf S.c.a.r.l. Faenza
Cmcf S.c.a.r.l. Faenza
Co El Me, Cesena
Vallicelli S.r.l. Forlì
Sit, Vicenza
Cassa di Risparmio Ravenna
Credito Romagnolo di Faenza
Banca Popolare dell'Emilia Romagna di Faenza
Cassa Rurale di Faenza
Banca Commerciale Italiana, Forlì
Servizi Ecologici Faenza
Comune di Faenza
Bureau Veritas, Milano
Pozzi, Santarcangelo di Romagna
Novamont, Novara
Gemos, Faenza

Preface

The scale of the problem of waste management and disposal continues to grow as landfilling and incineration face both environmental and economic concerns. The European Union alone has 2000 million tonnes of refuse a year to deal with and, in addition to the rest of the developed world, developing countries are having to confront the issue. With this background, increasing attention is now being given to integrated or alternative solutions involving composting and recycling. This book provides a state-of-the-art description of available and potential composting technologies, and all the associated scientific, economic and legislative issues. It indicates the best practical solutions for composting as an integrated method of waste management and discusses the adaptation of composing technologies to suit social, economic and geographic conditions in different countries.

Among its aims are the identification of the most profitable directions of future research in the field, and the examination of legislation and guidelines suitable for the practical implementation of composting policies.

Written for professionals and academics concerned with waste technology, science and management, this book is also an essential reference source for those with an interest in environmental technology including microbiologists, biochemists, chemists, biotechnologists and opinion and policy makers in government institutions.

Composting in Finland: Experiences and Perspectives

KARI HÄNNINEN – VTT Energy, P. O. Box 1603,
FIN-40101 Jyväskylä, Finland

Abstract

Approximately 1 million m^3 of compost is produced in Finland annually. Sewage sludges and liquid animal manures are the major waste groups composted in windrows or smaller heaps. Owing to the severe odour and leachate water problems in wintertime, however, in the major cities and towns will probably shift towards closed processing in the near future. At present there are about 100 drum composters in use on Finnish farmes. With the new environmental legislation and targets set by the Ministry of the Environment, source separation of biowaste, and centralized composting as the most effective method of treating it, can be expected to expand rapidly. Approximately 3% of detached houses employ small composters and growth is likely to be small. Although many industrial waste fractions could be composted, this is only marginal in Finland.

The composting process is an active area of research in Finland. Of particular interest are the compostability of different waste matrices and management of the process; quality aspects of composts, including the chemistry of biodegradation; the occupational hygiene of the process; and the composting of toxic waste fractions, particularly those containing organo chlorine aromatics.

Introduction

Being mainly rural before the 1960's and relatively sparsely populated Finland used to have few problems with waste management – always there was plenty of space for waste disposal. With the intensive urbanization beginning in the 1960's, and the considerable rise in living standards, the need to develop waste management systems for communities suddenly became pressing (FAMW 1967). Since the 1960's waste management in Finland has undergone rapid change, both in practical arrangements and in the minds of people. In the 1970's the main aim was to destroy waste, in the 1980's treatment became important; and then in the 1990's goals and principles resembling those of the United States and Western Europe were adopted, with emphasis on the reuse and recycling and particularly the reduction of wastes. The change in direction is clearly seen in the new environmental legislation, in literature (e.g. Kurki-Suonio and Heikkilä 1994) and concretely in the reduction in the number of public landfills: from a maximum of 1,799 to the present 762, of which 548 handle municipal wastes (Lettenmeier 1994).

According to statistics provided by the Ministry of the Environment, the annual generation of wastes in Finland today is 65–70 million tonnes: agricultural wastes 34%, municipal wastes 6% (comprising 1 million tonnes of sewage sludge, and 3.1 million tonnes of solid municipal wastes), industrial wastes 17%, construction and demolition wastes 10%, mining and concentration wastes 32%, and hazardous wastes 1% (Central Statistical Office 1994).

Treatment of organic wastes has been spearheading the development of waste management in Finland during recent decades, with composting the most important method of treatment (Haukioja et al. 1983, Paatero et al. 1984). Composting is a process that can accelerate biodegradation of organic wastes in a controlled manner, and so to avoid environmental and public health risks due to their uncontrolled degradation. Composting in urban areas began in the late 1960's with windrow composting. The reactor composting of mixed municipal wastes was tested too, with Dano composter, located in Helsinki and Turku.

Great amounts of wastes do not necessarily mean great problems for waste management. Political problems may nevertheless arise if the waste handling too closely affects people in their everyday lives, or if the risks associated with the wastes are thought to be great. So municipal wastes have been subject to heavy debate (Hukkinen 1994). Consequently it is natural that composting is currently mainly applied to agricultural manures, and sewage sludges, and large-scale composting of the source separated biowaste fraction of solid municipal wastes took a rapid leap forward in 1994.

The main producers of 'compost peat' are Vapo Oy (60,000 m³/a) and Kekkilä Oy (100,000–150,000 m³/a). Assuming in an average windrow compost to be roughly 40% sludge/liquid manure, 40% bark/wood chips, and 20% peat (by volume), we can estimate that about 300,000 – 400,000 m³ of sewage sludges and liquid animal manures (with dry matter content of 15–25%) are composted annually in Finland. Vapo also produces 210,000 m³ of peat annually for use as animal litter (Katainen 1995). After use, some of this material presumably is composted. These estimates suggest that about 1 million m³ of compost is produced in Finland annually (Vasara 1995). This amount is equivalent to roughly half the commercial market for soil (Mantsinen 1995).

State of the art of composting

Composting of municipal wastes

Composting of sewage sludges. When the amounts of sewage sludges in cities first began to increase, the digested sludges were applied as such, to public green areas.

In cities, however, the smell and other inevitable accidents in which people came in contact with the sludge have persuaded city officials to compost the sludges before spreading them (Ahmio 1987, Jyränkö 1986).

For its part, digested wastewater sludge, without further treatment, meets Finnish requirements even for agricultural use. Utilization of sludges in agriculture increased rapidly to 50% in the 1980's, while the deposition of sewage sludges on sanitary landfills decreased to 25%. In the 1980's, however, the farmers' union took a negative stance towards the use of all sludges in agriculture. Of the 1 million tonnes of sewage sludge currently generated annually (with a dry matter content of 150,000 – 200,000 tonnes), about 30% is used in agriculture, about 30% is landfilled, about 30% is used in public green areas, and about 10% is kept on storage areas. Since with composting the value of sludges improves to the point where they become acceptable in agriculture, the amount of sludges composted is growing all the time.

The City of Helsinki Sewage Works (CHSW) treats the sewage wastes generated by 750,000 people, and in 1993 produced digested sludge amounting to 79,800 m^3 (dry matter content 15–25%). Almost 70% of this is mixed with peat and lime (the final volume being 67,100 m^3) and used in agriculture. The rest is composted with chips, mixed with peat and sand (the final volume being 31,400 m^3) and used for public green areas. The direct utilization and composting of the digested sewage sludge is actually now a commercial business at CHSW, although the revenues from the product come indirectly, via the savings in landfill costs. CHSW has been able, with active marketing and reliable customers nearby, to maintain 100% utilization of its sludge for the last ten years. (CHSW 1993).

CHSW has been able to do the open air windrow composting in Viikki, just ten kilometres from the Helsinki market place. Now windrow composting of all sludges is carried out at the new modern composting field in Sipoo, 25 km east of Helsinki. Since the composting in windrows of relatively wet sludges during wintertime is difficult, the dry matter content of the sludge coming from the digestor has been increased to 30% with improved centrifugation. So now the sludge can be stored in heaps over hard winters and composted during the milder seasons. There are no plans to switch to reactor composting in the near future (Lundström 1995).

*Composting of source separated biowastes (SSSB).*Voluntarily adopting an environmentally sound way of thinking, in the late 1980's the Helsinki Metropolitan Area Council (HMAC) undertook its first experiments in collecting and composting biowastes. This led in April 1993 to the windrow composting, at the Ämmässuo sanitary landfill, of wastes from North Helsinki, an area of 100,000 people. The collection area is gradually increased to involve by the year 1997 the whole Helsinki Metropolitan Area (HMA) with its 850,000 inhabitants.

The Finnish winter creates serious problems for open air windrow composting of SSSB. Leachate waters in composting fields freeze, rendering the sewer system out of order and causing a terrible mess in the spring when the snow melts. Because of the cold, odorous compounds volatalize less readily than in summer. Moreover, in cold weather, air containing large amounts of odorous gases may remain intact for many hours, in really cold weather even for days. With a light wind, a layer like this may drift without dispersing, and if it moves into an inhabited area the result will be most unpleasant. The foul smell produced by anaerobicity in the biowaste compost is so unbearable as to drive people crazy. During the first one and a half years, with the exception of a few short periods, the main problem was the offensive odour, and procedures have been expressly directed to dealing with this problem. Recently HMAC adjusted the crushing intensity, reduced the size of windrows, and removed the anaerobic cores by also aerating the windrows mechanically through channels running along the bottom. The odour intensity has been reduced. (Hänninen et al. 1994). The suitability for agriculture of biowaste composted for periods of two to 18 months was investigated through physical and chemical analyses and with pot trials for barley in the summers of 1993 and 1994 (Mäkelä–Kurtto et al 1995).

The experiments carried out suggested that biowaste can be composted in open windrows with adequate hygiene and without unacceptable odour problems and with the production of a qualitatively good compost. However, composting in closed spaces is taken the preferred approach, the main advantages over the open air alternative being better hygiene, odourlessness, and more efficient water management. A closed composting plant for the HMA is currently being planned (Paavilainen 1995).

With the goal of utilizing both composting and incineration, VTT Energy is involved in the development of the Molok deep collection system of recent Finnish design. In the city of Jyväskylä, in Keltimäki suburban area 1,300 inhabitants are using this collection system. Wastes are separated at homes into four fractions: biowaste, paper & cardboard, glass & metal and the rest, called 'mixed waste. Wastes are collected into 1,000 – 4,000 litre plastic bags placed in holes dug in the ground, with only one metre of the bag above ground. In one block of flats, after three months of collection, the amounts of wastes were as follows: mixed waste 45.8% (57 kg/m^3 ; 65 m^3=3.7 t), paper & cardboard 25.7% (116 kg/m^3; 19 m^3=2.2 t); biowaste 25.4% (530 kg/m^3; 4m^3=2.1 t) and glass & metal 3.1% (180 kg/m^3; 1.4 m^3= 0.25 t). All wastes are utilized: biowaste is composted in a drum composter, 'mixed waste' is burned in a pulp and paper mill furnace, paper is used for recycling, and glass & metal are separated and used as raw material. This Molok deep collection system is already used in several communities in Finland, Norway and Sweden as well as in some Western European countries.

Now the new environmental legislation and more detailed regulations published by the Ministry of the Environment call for 50% of wastes to be reused or recycled and only 50% landfilled by the year 2000. In response to this the source separation, collection and composting of biowaste will be introduced in all Finnish cities and major towns. Already this has begun in Helsinki, Tampere, Lahti, Oulu and Mikkeli. The presently preferred treatment method is centralized composting in open air windrows, and the amount of biowastes to be collected in 1995 is about 15,000 tonnes. It is currently estimated that only 3% of the detached houses (20,000) within the organized waste management system will choose composting in their own small composters, but waste collection fees can be expected to influence this number.

The biowaste accumulation in the Helsinki Metropolitan Area (HMA) into 250–litre surface bins, is measured to be 41 kg of SSSB/inhabitant/a, with an efficiency of separation of 47% (Salo 1995). In the Keltinmäki project the accumulation is 49 kg of SSSB/inhabitant/a. The efficiency of the collection is not yet determined. If it can be assumed that the amount of wastes generated per inhabitant is the same all over the country than in Helsinki and Jyväskylä, we can estimate that the whole of Finland, with its 5 million inhabitants, would produce with 50–55% efficiency 200,000 – 250,000 tonnes of SSSB (with a dry matter content of 50–60%).

Composting of agricultural and forestry wastes

About 14 million tonnes of animal manure and 4 million tonnes of straw is generated annually in Finland (Mäkelä-Kurtto 1995). Traditionally these have been composted in early spring in windrows or heaps – although the process was not necessarily intended or thought to be composting. Now the manure is mostly processed as sludges, and the amount that is actually composted is difficult to estimate. Wood harvesting residues are generated annually 15 million tonnes, their utilization degree is poor, only 2% according to the estimates of Ministry of the Environment.

Reactor composting. The experiences with the two Dano reactor composters in the 1960's were failures. Too much was expected of the equipment, and one individual in Helsinki kept complaining about the foul smell, which evidently came from the drum (Lundström 1993). The upshot, however, was that a cylindrical reactor composter that refines animal waste into fertilizer was developed in Finland in 1970´s, and this now provides an economically profitable means of production. Various companies, Biofacta the most prominent, have built 100 drum composters for Finnish cattle farms for the treatment of manure, and quite recently also for municipalties for the treatment of sludges (Kangas 1995). Drum composting would seem to be especially appropriate for very wet materials, sludges with moisture content of 75% can be composted in a reactor with peat as the major bulking agent. The homogeneous mixture produced in drum will compost in windrows better than mixtures made only with a front end loader.

Compost fertilizers are currently based primarily on chicken faeces and residues. Biolan, which is the major manufacturer of such organic fertilizers in Finland, runs four production lines: drying and granulation of chicken faeces 25,000 m^3/a; composting of chicken faeces, loosely packed or press baled in plastic bags for fertilizers 10,000 m^3/a, production of mull by composting of ground bark, clay and sand and mixing this with composted chicken faeces 30,000 m^3/a; production of fertilized peat (peat + composted chicken faeces) 70,000 tonnes/a (Haukioja 1995).

Production of carbon dioxide for greenhouses. A commercial application of composting in horticulture is in the experiments of trying to utilize in warm seasons of carbon dioxide produced by the composting of straw bales in greenhouses. Carbon dioxide is of key importance in greenhouses and the cost of the gas is high (about 10 mk/ m^2). Difficulties have occurred in the collection and regulation of the carbon dioxide stream.

Composting of industrial wastes

From industrial wastes 5.4 million tonnes or 50% of this is organic fractions (see Table 1). Composting of these wastes in Finland is currently marginal, only bark and wood wastes are utilized to some extent as bulking agents. For many industrial waste fractions, composting would be a suitable alternative.

Slaughter wastes of reindeers. The foodstuff and beverage production wastes in Table 1 include the wastes (3,000 tonnes) of 120,000 reindeers slaughtered annually. Although the amount is rather small, this is a significant waste fraction considering that it is generated in Lapland – a very sparsely populated part of the country. Previously slaughtering was carried out on a small scale in many places, and the wastes were often simply buried in the snow and left for nature to take care of. When, as a lead up to joining the EU the slaughtering was concentrated into a few large slaughterhouses (10,000 – 20,000 animals/a), the waste had mainly to be transported to landfills. There are few fur farms near the slaughterhouses, which could take care of the wastes in a natural way. Composting of the waste is the logical alternative, the humus soil produced finds a welcome use in agriculture.

Table 1 Groups of industrial organic wastes and their amounts in Finland.

Waste group	Amount (1000 t)	Recycled (%)	Landfill (%)	Other (%)
Plant and animal waste	249	60,5	15.5	24.0
Foodstuff and beverage prod.	58	62.0	26.2	12.1
Vegetable and animal fat proc.	56	80.7	9.6	10.7
Carbohydrate refining wastes	6	2.8	6.5	90.7
Animal feed wastes	3	17.8	78,7	3.5
Textile and leather prod.	41	5.6	93.2	2.4
Wood wastes	2,493	92.8	6.8	0.4
Bark	1,362	92.2	7.4	0.5
Cellulose, paper & paperboard	788	50.4	45.0	4.7
Waste paper	192	56.8	29.0	14.1
Paper & paperboard	138	47.8	51.7	0.0
Total	5,386			

Pulp and paper mill biosludges (PPMB) from wastewater treatment are generated in the amount of 360,000 tonnes of dry matter annually. This is the most abundant material in Finland available for composting (more than twice the dry matter content of sewage sludges). Nearly half of the collected sludge is transported to landfills run by the mills, while the other half is burned in the mills as low grade fuel.

Only a few composting studies have been carried out on PPMB. In tests on the use of sludge in agriculture, the composted product was found to be more suitable than the untreated sludge (Campbell et al. 1995, Mäkelä–Kurtto et al. 1992, Ruhanen 1992). There are nevertheless major factors arguing against composting: the huge amounts of material to be processed, the large storage and handling areas required, and the need for large quantities of bulking agents. The marketing of the compost might also prove difficult. Spreading of composted material in forests to increase the humus content of forest soils is not seen as a viable solution. The good effects of such an operation are not appreciated and the costs would be high. Spreading of untreated biosludge, in turn in forests could be considered to violate the traditional Nordic right of public access to forests and the right freely to collect berries.

Composting of toxic wastes and toxicity of composts.

The composting of toxic wastes in Finland has been concentrated on destroying of polyaromatic hydrocarbons (PAH) contaminated soils (Lilja & Uotila 1995), oil residues of service stations (Puustinen et al. 1995) and the phenolics in contaminated sawmill wastes and soils. In conjunction with this, research has been going on in the universities of Helsinki and Turku to develop microbes able to metabolize phenolics effectively.

We found higher concentrations of pentachlorophenol (PCP), polychlorodibenzodioxins (PCDDs) and polychlorodibenzofurans (PCDFs) in SSSB compost than in PPMB compost, although all the concentrations were 'acceptable' (Hänninen et al. 1995). The reason for this might have been that the wood chips used as bulking agent in the biowaste compost contained waste wood impregnated with pentachlorophenol. According to the literature, lumber impregnated with PCP is one of the most probable sources of PCDDs and PCDFs in biowaste composts. It is calculated that 1.8 kg of lumber impregnated with a 10% PCP solution could contaminate 1,000 kg of compost with 0.02 mg/kg of octachlorodibenzodioxin (Harrad 1991). Evidently organochlorine phenolics, which are harmful to the environment when they got into watercourses, are not present in PPMB composts to such an extent as to limit the use of composted sludges.

Toxic heavy metals. The toxicity of composts is usually determined by its total heavy metal content. Based on the literature survey and on our own results it seems, however, that the availability to plants of harmful heavy metals, especially cadmium and mercury, decreases when the composted material is used as growth media (de Haan 1981, de Haan and Lubbers 1983). More studies are needed to determine whether, through the increased use of compost humus, we could decrease the amount of cadmium in our food (Stölzer et al. 1994, Hänninen and Mäkelä-Kurtto 1995).

According to the National Food Administration (NFA) Phosphorus fertilizers are considerd to be the major source of cadmium in Finnish agricultural soils

(1995). NFA further estimates the average accumulation of cadmium in Finnish food to be 9.5 (g/daily portion of food, which is 16% of the recommendated maximum limit. Varo (1984) estimates it be 13 (g, assuming an average meal having an energy content of 12 MJ. About 57% of the cadmium is obtained from cereal products. One third of the cadmium in food is thought to come from air and the rest from the soil.

The amount of cadmium in the PPMB of the Äänekoski mills near Jyväskylä is slightly more than 3 ppm, and in the ash it is about 15 ppm. The trees the mills process comes were growing in a relatively unpolluted area in Europe. Where does the cadmium come from? In general, the trapping of cadmium from flue gases is rather poor. Is perhaps cadmium being kicked into the atmosphere, when burning is done at high temperatures in order to avoid dioxin emissions? Are we cycling cadmium through high temperature burning?

Perspectives of composting

For the incineration of wastes there is only one plant in Finland, in Turku. In the late 1980's and early 1990's HMAC had plans to build a plant for the Helsinki area, but environmentalists and environmental organizations were afraid that incineration would effectively stop waste recycling (Pohjanpelto 1991). This fear is shared internationally (Young 1991). HMAC did not in the end build an incineration plant, but instead started composting.

One plant in Vaasa is treating municipal wastes anaerobically. The plant is advanced and incorporates modern technology, but the wastes it is processing are poorly fractionated. The quality of these wastes can be described as equal to the mixed municipal waste that was used in composting plants in Europe in the 1960's and 1970's.

The compostable biowaste potential of Finland (as well as all other paper producing countries) lay on the wastes of pulp and paper production. If the black liqueur of pulp and paper processing is taken into consideration the biowaste potential is increased to a global dimensions. However, at the present fiber production technology this waste group is burned in order the recycle the cooking chemicals, and its energy potential is utilized, too. The Ministry of Environment is planning to levy costs on the landfilling of PPMB (which fraction do not any more contain cooking chemicals), even in the case the landfills are owned by the plants. If such measures are realized the interest on the treatment of PPMB waste e.g. by composting will in the near future increase.

The role of composting in Finland in waste management is expected to increase, and it is not expected to be threatened by either incineration or anaerobic treatment. All are important waste treatment methods, for which symbiotic development is desirable. A danger may lurk, however, in the too eager effort to turn waste management from a cost incurring problem into a profitable business. Profits can accrue only if collection fees are levied on individual households. In Sweden high fees have already led to a reduction in amounts of waste collected, and communities are concerned about making their investments profitable (Erikson 1995). In Finland the right to dispose of one's waste without or with a minimal charge has long been taken for granted, and levying a too high fee might cause amounts of collected wastes to diminish. Right at the outset this may create an important obstacle to the enterpreneur running a business: can there be guarantees of a continuous supply of the raw material, and of achieving a profitabile return on investments within a narrow space of time?

Peoples' attitudes towards waste recycling are relatively favourable in Finland now. All measures should be directed to encouraging these attitudes, because the collection of SSSB fractions for recycling and reuse largely depends on attitude. An obvious danger that needs to be avoided is that householders diligently sorting their biowaste should come to see their efforts as futile if the biowaste (or other waste fractions) ends up on the dump anyhow.

The quality and image of the SSSB compost are important. It should be acceptable for high quality use in home gardens. Means and methods to identify and to standardize the 'good quality' needs to be developed. Chemical identification of the compounds in the organic matter of the compost, and following the changes in composition as the process proceeds, are of key importance in determining the quality of the compost. Chemical analysis of the humus has shown the importance of aliphatic, especially carbohydrate derived-structures, to be underestimated and the importance of aromatic structures to be overestimated. For this, methods for further fractionation and determination of the humus compounds need to be developed. Good quality would guarantee good demand of SSSB compost.

Cradle-to-grave concept. Now, it is no longer sufficient to produce materials with functionality, production efficiency, and economics as the only design criteria. The 'cradle-to-grave' concept of material design requires planning for ultimate disposal of the material in an ecologically sound manner. There is a need to design and engineer materials with consideration for their method of disposal. The attributes of recyclability and biodegradability should be built into the materials, but without loss of the performance characteristics. Separate infrastructures must then be developed for materials in order they then end up in the appropriate processing. Recyclable materials will need to be collected and transported to a recycling facility, to be processed into the same or new products. Similarly, biodegradable materials will need to be collected and sent to a facility where, along with other wastes they undergo biodegradation into compost. (Narayan and Snook, 1994).

Acknowledgement

Dr. Kay Ahonen is thanked for revising the English of the manuscript.

References

Campbell, A. G., Zhang, X. and Tripepi, R. (1995). Composting and evaluating a pulp and paper sludge for use as a soil amendment/mulch. Compost Science & Utilization, Winter 1995, 84–95.

Central Statistical Office (1994). Environment Statistics. Environment 1994:3.

CHSW, City of Helsinki Sewage Works (1994). Annual Report 1993.

Erikson, L. (1995). Communities fight about wastes. Teknisk Tidskrift 1995:6, p. 4. In Swedish.

FAMW, Finnish Association for Municipal Works (1967).Waste management of Finnish cities. Joensuu. 170 p. In Finnish with English abstract.

Haan de, S. (1981). Results of municipal waste compost research over more than fifty years at the Institute for Soil Fertility at Haren/Groningen, the Netherlands. Neth. J. Agric. Sci. 29, 49–61.

Haan de, S. and Lubbers J. (1983). Microelements in potatoes under 'normal' conditions, and as affected by microelements in municipal waste compost, sewage sludge, and dredged materials from harbours. Instituut voor Bodemvruchtbaarheid, Haren–Gr., the Netherlands. Rapport 3–83, 22 p.

Harrad, S. J., Malloy, T. A., Ali Khan, M. and Goldfarb, T. D. (1991). Levels and sources of PCDDs, PCDFs, chlorophenols (CPs) and chlorobenzenes (CBzs) in composts from a municipal yard waste composting facility. Chemosphere 23 (2). 181–191.

Haukioja, M., Hovi, A. and Rajala, J. (1983). Composting. Tammi, Helsinki, 116 p. In Finnish.

National Food Administration (1995). There are only little of cadmium in the food of the Finns. Official statement to news agencies. Helsingin Sanomat 20.3.1995. In Finnish.

Hukkinen, J. (1994). The institutional prerequisite for sustainable waste management in Finland. In: Kurki-Suonio and Heikkilä (eds). The requirements for sustainable development in Finland. Tammi, Helsinki 711–746. In Finnish.

Hänninen, K., Huvio, T., Veijanen, A., Wihersaari, M. and Lundström, Y. (1993). Occupationl hygiene of windrow composting. VTT Publications 776, Espoo, Finland, 102 p. In Finnish with English abstract.

Hänninen, K., Heimonen, R., Miikki, V., Lilja, R. and Malinen, H. (1995). Drum composting of biosludges and solid biowastes. VTT Publications, submitted. In Finnish, English abstract.

Hänninen, K. and Mäkelä-Kurtto, R. (1995). Windrow composting and use of source separated biowaste. The publication series C of the HMAC, Submitted.

Hänninen, K., Tolvanen, O. Veijanen, A. and Villberg, K. (1994). Bioaerosols in windrow composting of source separated biowastes. In Proc. 8th Europ. Conf. on Biomass for Energy, Environment, Agriculture and Industry, Vienna, Austria, 3–5 October 1994. 6 p.

Kurki-Suonio, I. and Heikkilä, M. (eds) (1994). The Requirements for Sustainable Development in Finland. Tammi, Helsinki 896 p. In Finnish.

Lettenmeier, M. (1994). Roskapuhetta. Ministry of Environment. Helsinki. In Finnish.

Lilja, R. and Uotila, J. (1995). Bioremediation of PAH contaminated soil. Proc. Int. Symp. 'The Science of Composting', Bologna, Italy, May 30–June 2, 1995.

Mäkelä-Kurtto, R., Sippola, J. & Jokinen. R. (1992). Industrial waste sludges and their utilization in agriculture. MTT, Finnish Agricultural Research Centre, Jokioinen 51 p. 39 app. In Finnish.

Mäkelä-Kurtto, R., Sippola, J., Hänninen, K. and Paavilainen, J. (1995). Suitability of composted household waste of Helsinki Metropolitan Area for agriculture. Proc. Int. Symp. 'The Science of Composting', Bologna, Italy, May 30–June 2, 1995.

Narayan, R. and Snook, J. (1994). The role of biodegradable materials in solid waste management. In: Johnson, M. A. and Samiullah Y. (Eds.) Waste Management and Recycling International 1994 , 40–43. Sterling Publications Limited, London.

National Food Association (1995). There are only little of cadmium in the food of Finns. Helsingin Sanomat 20.3.1995.

Paatero, J., Lehtokari, M. and Kemppainen, E. (1984). Composting. WSOY, Juva, 268 p. In Finnish.

Pohjanpalo, R. (1991). The knock-out of waste incineration plant. Helsingin Sanomat 20.12. p. B1.

Puustinen, J. Jorgensen, K.S., Strandberg, T. and Suortti, A–M. (1995). Bioremediation of oil contaminated soil from service stations. Publications of the Water and Environment Administration. Series A 208, 42 p.

Ruhanen, M. (1992). Cultivation experiments with the waste water sludge of a pulp mill. Yhtyneet paperitehtaat Oy Joutseno. 9 p. 8 app. In Finnish.

Salo, M. (1995). Source separation of biowastes in Northern Helsinki. Follow-up report 1994. HMAC Publication series C 1995:5, 15 p, 7 app. In Finnish.

Stölzer, S., Fleckenstein, J. and Grabbe, K. (1994). Die Immobilisierung der Schwermetalle Blei und

Cadmium durch Komposte. Mull und Abfall 9, 551–560.
Varo, P. (1984). Mineral Nutrient Statistics. Otava, Helsinki. In Finnish.
Young, J. E. (1991). Reduce waste and save material. In Lester B. (Ed.) State of the world '91.
Worldwatch Institute. Naturvårdsverket Förlag, Sweden, pp. 39–54.

Personal Communications

Ahmio, K. (1987). City of Joensuu.
Haukioja, M. (1995). Biolan Oy, Kauttua.
Jyränkö, P. (1986). City of Helsinki.
Lundström, Y. (1994), 1995. City of Helsinki.
Kangas, J. (1995). Biofacta Oy, Korpilahti.
Katainen, E. (1995). Vapo Oy, Jyväskylä.
Paavilainen, J. (1995). HMAC, Helsinki.
Mantsinen, R. 1995. Humuspehtoori Oy, Ruutana.
Mäkelä–Kurtto 1995. Agricultural Research Centre of Finland, Jokioinen.
Vasara, E.–H. 1995. Kekkilä Oy.

Perspectives and State of the Art of Composting in France

J. M. MERILLOT – Ademe, square La Fayette BP 406
49004 Angers Cedex France

Summary

The present French situation of waste management is the result of traditionnal treatment plants, sorting-composting MSW composting plants for instance, which are still running, and new perspectives coming from the achievement of the national policy (law n° 92–646 published in 1992). This transition period can be an opportunity for composting development if the three levels of analyze are correctly assessed as well as waste management alternative and as agricultural and ecological pertinence. For this assessment, better knowledge on impacts and new tools of management (including monitoring and control) are required. Up to now, we can consider that, in France all the conditions of composting development are present depending on a common will to do it. The Ademe will propose an adequate working scheme for the next 5 years, hoping that its partners will agree and correctly consider the effort that must be done.

Introduction

In France, composting has been traditionnally used for municipal solid wastes. In 1989, about 1.1 millions tons of MSW (7.5% of the French MSW production) were treated to produce 650,000 tons of compost. The 75 'sorting-composting' plants used to run either slow composting (open air windrow turning – 42 plants) or accelerated composting (33 plants). The evolution between 1985 and 1989 shows a slow decline of composting (Table 1), partly because of compost marketing problems. During the same period, sludge composting has remained marginal with only 5 to 7 sludge composting plants of the first generation. Since 1985, the ANRED (now Ademe) has tried to develop composting either by improving 'sorting-composting' plants and the quality of urban compost or with new applications, to yard waste or animal manure for instance, or new strategies for sludge composting. But, the competition with landfilling didn't play in favour of more investments, new technologies ...

Table 1 Evolution of Treatment of MSW from households in France (1985 vs 1989)

	number of plants	raccorded population total	%
Thermal Treatments	+5.8 %	+16.4%	+10.5%
Biological Treatments	−16.5%	−1.6%	−6.3%
Landfilling	+41.9%	+29.0%	+0.7%

In 1989, the ANRED has published a technical document on sorting-composting of domestic wastes which analyzed the main problems of these kind of plants: the efficiency of sorting devices applied to such an heterogeneous product: garbage. The development of Quality Certificate for Urban Compost has also been promoted but failed in improving the marketing situation of these products.

Finally, in 1992, the national policy for waste management has been precised with the publication of the law n° 92–646:

- landfilling is no longer a treatment process but a storage stage dedicated to the 'ultimate waste'
- 'ultimate waste' is a waste resulting or not from waste treatment and which cannot be any more treated for recycling or to decrease their pollutant potential, considering the actual technical and economical conditions
- wastes management plans must be realized (deadline 1996) on a department level for municipal (and similar) wastes (excluding toxic and hazardous waste – regional plans).

Considering these events, what can be the place of composting in the future in France? To answer this question, three levels must be analyzed:

i)   the management of wastes through composting
ii) the conception and running of composting plants
iii) the process of composting

Composting: from waste management to treatment plants

The French policy for waste management considers:

- reuse and recycling of wastes: waste derived products
- energy recovery through incineration: waste derived energy

as acceptable options. As far as compost is a waste derived organic fertilizer, composting is an acceptable option. But, there is also a specific regulation on waste from packaging, with a specific financial mechanism: Ecoemballage, which aims to help recovery and recycling of this part of the waste flow.

For domestic wastes, the main problem is on source separation and mainly on the number of bins in households. The last evaluation of departemental plans has shown that municipalities prefer the two bins following scheme:

- source separation of clean materials vs the rest,

− developpement of materials sorting centers (150 in 2002) ...
− combined with incineration and energy recovery (twicefold treatment capacity). Only few experiments consider source separation of organic matter. But, composting is choosen for yardwaste treatment from private and public gardens and green spaces.

Sludge management will depend on the access of sludge to agriculture, mainly through spreadings of stabilized products, i e after lime, or thermal, or biological treatments, including composting. In some cases, more sophisticated schemes are studied, mainly co-treatment (co-composting) of organic wastes from different origins (municipalities, agro-industries, agriculture ...).

For animal manure, composting is studied as a way to improve their management either on a farmscale level or for marketing. The aim is clearly to answer to the new constraints on nitrogen load limitations coming from nitrate regulations.

In 1993, a national survey of composting plants has given the following results:

− 73 MSW composting plants (no source separation)
− 13 sludge composting plants (+1 project), co-composted with different bulking agents (sawdust, bark, straw...)
− 30 yardwaste composting plants (+21 projects)
− 16 farmyard manure composting plants
− 10 other kind of wastes or mixed composting plants

Future development for composting will depend on two trends:

a) the emergence of departemental schemes for organic wastes management, whatever the waste origins and including the fate of composts
b) the attraction of domestic organics source separation, compared with other alternatives and depending on the number of bins in households

Thus, the Ademe is working on:

a) the concept of biological management of waste, including but not exclusively composting and compost impacts. The aim is to define an acceptable alternative strategy, based on sound scientific knowledge and easy to present and justify.
b) the role of farmers within this concept, as partners and the possibility of a specific mechanism to guarantee the final use of waste derived fertilizers. It is not only a problem of compost quality, but also a problem of agricultural practices.
c) Standards for different types of compost and Quality Insurance applied to this kind of waste management

Composting: from treatment plants to technologies

The fate of the existing 'sorting-composting' plants will depend on the strategic choices of each departmental waste management plan. All these plans must be achieved for the end of January 1996. But, it will also depend on their efficiency and of the state of their compost market. The other point is the flexibility of the composting technologies and their ability to compost either source-separated organic waste or other kind of waste. Most of these plants are owned by municipalities, which limits the possibility of new uses. Their treatment capacities range from less than 4,500 MSW tons/year to more than 60,000 MSW tons /year.

Table 2 presents the different technical diagrams of these plants. Different technologies are used for each stage. For instance, shredding can be done with 'rotating shears' or 'shredding trommels'. Sorting devices are numerous and reflects the difficulties of this stage. 7 types of technologies (magnetic, densimetric, air-classification, sieving ...) has been analyzed and evaluated. Slow composting differs mainly by the windrow-turning machineries and by the management of turnings. The technologies for accelerated composting include towers or cells or rotating drums with different kinds of forced-aeration control.

Table 2 Technical diagrams of 'sorting-composting' plants of MSW

	DIAGRAM 1 (15 plants)	DIAGRAM 2 (10 plants)	DIAGRAM 3 (9 plants)	DIAGRAM 4 (7 plants)
1°	GRINDING	GRINDING	GRINDING	SHREDDING
2°	SORTING	COMPOSTING	COMPOSTING	COMPOSTING
3°	COMPOSTING	SORTING	MATURATION	MATURATION
4°	MATURATION	MATURATION	SORTING	SORTING

The efficencies of these technical chains has been assessed with two types of parameters: the part of organics in compost and the part of non-organics in rejects. The results show that to have a relatively good quality of composts, a significant part of organic matter must stay in rejects. A final assesment of sorting-composting plants is carrying out on the efficiency of sortings on the different trace-elements flows.

Other composting plants are rather small plants. Treatment capacities range mainly from 5000 to 10,000 tons of waste per year, and investment costs from 2 to 7 millions Francs. The ratio between Investment and Treatment capacity goes from 500 to 800 F/ton. These plants are running at 30 to 60 % of their treatment capacity, with treatment costs between 150 to 300 F/ton. As far as the production of compost is 35 to 65 % of the treated waste, it is difficult to compare the different plants on their production characteristics.

The different technical chains are as follows:

a) for yardwaste composting

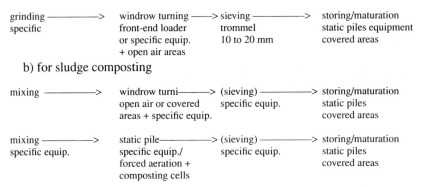

```
grinding ————————>   windrow turning ——> sieving ——————>   storing/maturation
specific              front-end loader     trommel           static piles equipment
                      or specific equip.   10 to 20 mm       covered areas
                      + open air areas
```

b) for sludge composting

```
mixing ——————————>   windrow turni———————> (sieving) —————> storing/maturation
                      open air or covered    specific equip.  static piles
                      areas + specific equip.                 covered areas

mixing ——————————>   static pile———————> (sieving) —————>   storing/maturation
specific equip.       specific equip./     specific equip.    static piles
                      forced aeration +                       covered areas
                      composting cells
```

For this new composting plants, the problem of composting management is crucial: odour control, leachates, mixing monitoring, retention time ... and is the present field of improvement.

Composting: from technologies to the process

The process of composting is too much considered as a natural process, which works by itself. The management of the microbial ecosystem, related to nutrients inputs, to energy control and to environmental conditions (moisture, oxygen, temperature ...) must be much more strict than it is up to now.

The Ademe tries to develop Research and Development on composting: simulation equipments, process modelling and composting trials.

A first work on composting simulation has been done by the INSA of Lyon. After a first period dedicated to the conception of the equipment, tests of organic mixtures have been carried out. Some improvements are required mainly on the thermodynamic aspects as well as new tests of organic wastes. The aim is to predict the behaviour of products during composting. Composting trials applied to pig slurries mixed with different kinds of carbon bulking agents are realized in Brittany by the C A T of Quatre Vaulx. The aim is to compare forced aeration vs windrow turning and the ability of the bulking agents to compost with a maximum load of pig slurries. This work must be achieved for next June.

The part of R and D on composting is too much low and the Ademe intends to develop it whith Research programmes during the next 2 or 3 years. The problems of nitrogen behaviour and fate, of odour controls, of pathogens control and of air emissions need more scientific works.

The other point is technologies development. The future development of composting will come from yard waste composting. But, the required technologies for YW composting are based on grinding /windrow turning/sieving. It works pretty well because of the 'ideal'composting kinetic of YW. But, mixed with sludge or with source-separated organic wastes, troubles arise because of a deficiency of process control due to windrow turning limitation. There is a technological breakpoint from windrow turning to forced aeration. So, the Ademe will develop R and D on forced aeration in the next two years: air production and distribution, monitoring equipments and strategies ...

At longer term, a microbial management may be needed for some specific uses of composting or for specific compost qualities (relation with soil ecosystem).

Composting and composts

The application of composting to different types of waste shows that the characteristics of compost are mainly inherited from the waste inputs (graph 1 and 2), even if composting influences numerous parameters as the content and quality of the organic compounds, the moisture, particles size, nitrogen flow, pH, the pathogenic potential, conductivity...

Refering to the French waste management policy, the problem of treatment processes comes after the problem of waste fate. Compost utilization is much more important than compost production. The need of quality parameters related to uses is obvious and supposes to have numerous references from field trials. A major point is to define a soil quality policy in order to control trace-elements flows.

One question for the future is how far must go the transformation process in order to fit agricultural uses, and thus, what will be the limits of composting, facing this production goals.

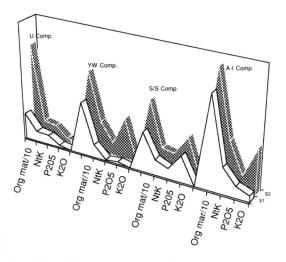

Figure 1 Profiles of different composts according to the Quality of the initial wastes

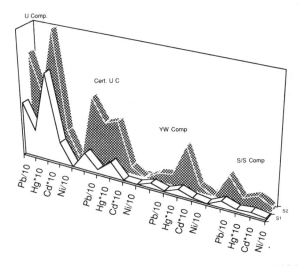

Figure 2 profiles of different composts according to the Quality of the initial wastes

References

ISWA, AGHTM, FNADE (1994). Symposium International sur le Traitement des Déchets Pollutec Lyon 18–20 Octobre 1994

ANRED (1990). Sorting/composting of Domestic Waste

Agence de Bassin Loire-Bretagne, Agence de Bassin Seine-Normandie, Ministère de l'Environnement (1984). Compostage des Boues Résiduaires Urbaines

ANRED (1987). Guide de l'Utilisateur des Amendements Organiques

DURANCEAU Nathalie (1994). Le Compostage: techniques générales et opérations réalisées en France.Rapport d'Etude

WIART J.(1994). Qualité et Commercialisation des Composts de Déchets Verts en France Document de Formation

Ademe (1994). Les Déchets en Chiffres – collection Données et Références

Composting of Agricultural Waste in Denmark – in Respect of Potential, Industrial Process Technology and Environmental Considerations.

MOGENS HEDEGAARD – Project manager. M.Sc.
I. KRÜGER AS – Sofiendalsvej 88 DK – 9200 Aalborg

Abstract

An amount of at least 1.000 mio.tons of slurry from animal husbandry in Europe should, based upon environmental considerations, be subject to centralised treatment in order to recycle nitrogen and utilise the energy potential.

The process technologies until now are lacking behind in order to fulfill ecological requirements to nitrogen recycling and utilisation for fertilising purposes, whereas the energetic utilisation by means of biological processes has been highly developed .

The existing paradox, where costs of BOD reduction in urban areas are accepted to be over 100 times higher than what is allowed in the animal husbandry sector, even under very negative ecological impacts by denitrification and high consumption of fossil energy, can only be overcomed through sector integrated waste treatment allowing energy neutral operations etc..

The NRS–SEABORNE technology should be a modern innovative mean to go for in urban/agricultural integration within waste treatment, ensuring ecological sustainability as well as clear advantages expressed in pure financial terms at the moment.

European potential for centralised treatment of agricultural waste.

The major potential emerges as residues from animal husbandry.

As long as milk, meat and eggs etc. are in demand hudge amounts of slurry will occur e.g. for 1 ltr. of milk app. 3 ltr. of slurry and for 1 kg of meat app. 13 ltr. of slurry. More industrialised systems in animal husbandry are provoked out of a common public desire to obtain steady relatively lower prices of animal products for food.

Thousands of productive animals gathered together in industrialised farming systems on the other hand causes troubles to an environmental cross compliance between production, fertilising and substitution of inevitable losses of plant nutrients.

Composting has in the history of farming been the traditional way of ensuring soil fertility by recycling, even that nitrogen has been a limiting factor to production just until the availability of fossil energy in abundance came up in this century. It is thus in an *extremely short period* within the history of mankind that recycling not has been a must, solely as a reason for survival of human beings.

In a work for the European Commission, DGXVII (Lit.ref. 1.) we have examined the european potential which could become a subject to centralised digestion of animal wastes.

The total amounts of animal wastes in the European Union, prior to the enlargement in 1995, were found to be – calculated as slurry with 6% VS and listed in order of magnitude:

France:	425 mio. tons / year
Germany:	300 mio. tons / year (except the new 'Länder')
England:	215 mio. tons / year
Italy:	200 mio. tons / year
Ireland:	125 mio. tons / year
Spain:	120 mio. tons / year
Holland:	110 mio. tons / year
Belgium:	65 mio. tons / year
Denmark:	50 mio. tons / year
Portugal:	30 mio. tons / year
Greece:	15 mio. tons / year
Luxembourg:	3 mio. tons / year
Total:	1.658 mio. tons slurry / year, (6% volatile solids)

The slurry from animal husbandry represents in many regions of the Union an energy potential excessing 250 kg oil equivalents/ha.

Our survey for DGXVII has proved that app. 60 % of the above amounts should be subject to centralised digestion and recycling measures for plant nutrients. This finding has been based upon combining environmental and energetic points of view with the allocation in regard of density per ha for the slurry potential.

Centralised treatment measures for around 1.000 mio. tons / year presents an enormous task of engineering, but seems inevitable if the environmental concerns about recycling and emissions shall be met, ensuring supply and price level for animal products to the consumers.

A great paradox exists among what we are willing to pay for cleaning up waste water of human origin and what is demanded of the farming system as allowable costs for its animal waste, although it basically is about the same stuff.

1.000 kg BOD emerges yearly from around 50 adult human beings equivalent to what comes from 30 pigs or 1 milking cow.

Sewage plants for human beings have costs of establishment around 10 ECU per kg BOD treated yearly and the yearly costs of operation can come to 15 ECU / kg BOD. It seems somewhat unrealistic to expect that the agriculture can solve its slurry handling problems in an environmental sustainable way for costs *less than 1 %* of what is accepted costs of cleaning up after human beings, but that is as a matter of fact what the european farming industry is up against recently, where recycling and the environment have been put into a political focus. If the previous mentioned costs of waste water treatment from human beings should apply to animal husbandry prices for 1 ltr. milk and 1 kg meat should be added respectively 2 ECU and 8 ECU !

The paradox related to the accepted high costs in cleaning up waste water from human beings also includes the 4 – 5 doubled consumption of fossil energy when removal of N & P is required. Can it be an environmental target to pollute the air, even without substantial recycling of N, in order to have more clean water ? It should hardly be the case and the conclusion remain that more sustainable methods of recycling and low emission for all wastes, including the agriculture, should be in high demand.

Applied industrial process technology and development of new innovative processes e.g. the NRS–SEABORNE technology.

The amounts in question calls for drastic reductions in volumen besides the concerns of preserving nitrogen on plant available forms. Both the aerobic and the anaerobic way can reduce BOD, but hardly cope with the desired reduction in volumen. In DK as in most other european countries a number of physical and biological methods have been put into operation e.g. destillation, reverse osmosis, nitrification and denitrification. They are all costly in terms of energy with e.g. reference to the high costs of purificating human residues in sewage plants mainly based upon the biological processes.

Under parallel session B 5 tomorrow at noon, some innovative processes under patent pending will be presented of the inventor dr. N.O.Vesterager from Germany. It is not up to me to make anticipations about this presentation on its technical aspects, but it shall be stated that the processes of dr. Vesterager´s are fulfilling major requirements to ecology and economy as regards treatment of agricultural waste and even can facilitate the gateway to co-digestion and energy-neutral cleaning up after human beings. The point is, in particular for the process to treatment of slurry from animal husbandry – the NRS process (N=nitrogen, R=recycling, S=system), that the innovations of dr.Vesterager´s are complementary to usual physical and biological processing, but solely is based upon the chemical premises of the nitrogen itself. We have followed the work of dr. Vesterager for 6 years on an advisory basis as regards the practical aspects to an agricultural implementation. At the moment it is close to a first demonstration plant and the financial competitiveness is seen in the following way:

NRS-40 economy in upgrading slurry from animal husbandry.

Background: 6 years of research and technical development, partly with EU subsidies. The NRS process can be applicated without any pretreatment or downstreams an usual aerobic or anaerobic processing. Posttreatment in the aerobic way can come in question.

Preassumptions: 30.000 m3 slurry/year NRS treated. 2/3 of that with $NH4MxPO_4$ precipitation and 1/3 with recirkulation of $MxHPO4$ and precipitation of $C_6H_{12}N_4$.

Costs of construction, 1´st demo.plant: 0,8 mio.ECU

Costs of construction, following plants: 0,7 mio.ECU

Depreciation: 15 years, rate of interest: 6 % p.a., maintenance: 2 % p.a..

The NRS-40 plant is with buildings and independent

installations, but connected to exstern heat- and electricity supply.

By integration with a sewage plant and co-digestion of sewage sludge, solid organic household waste and slurry from animal husbandry, including CPH generation, the costs of construction per m3 treated for the NRS part can be very low, cut down to less than relatively half of the above.

Costs ECU/m^3 slurry NRS upgraded

	NRS-40 1´st demo. 30.000 m3/år	NRS-40 Following 30.000 m3/år	NRS-40 Integrated 120-150.000 m3/år
Variable	2,0	2,0	1,6
Capacity	0,5	0,5	0,2
Capital	2,8	2,3	1,4
Total	5,3	4,8	3,2

Comparative european slurry costs under 'good' application.
(9 months storage, trailing hose system, 50 % N utilisation)

	ECU/m^3
Storage	3,5
Field application*	1,8
N loss **	0,5
Total	5,8

* Differential costs, equal amount of plant nutrients applicated in solid fertiliser incalculated.
** By NRS-40 the N utilisation increases from 50%——> 80 %.

NRS-40 ecological advantages, beyond the above pure financial terms, can in common be quantified to app. 3 ECU/m3 slurry and would be financial relevant, if the farming industry equal to other industries to full extend was underlaid the 'polluter pay' principle.

The NRS costs represents less than 1% of what remains social acceptable costs by cleaning up the very same stuff after human beings and in the case of normal sewage plant operations there will occur negative ecological impacts due to denitrification and hudge consumption of fossil energy.

The NRS-SEABORNE technology of dr. Vesterager´s represent a missing link to implement sector integrated waste treatment in a sustainable ecological way under financial competitiveness far beyond what to-day is accepted in urban waste water treatment.

Environmental impacts of the NRS-SEABORNE technology within treatment of agricultural wastes.

Some disadvantages incurred by storage of slurry in environmental and agricultural terms can be listed without any attempt to financial quantification:

- surface emission of NH4 during time of storage
- CH4 emission during time of storage
- problems of bad odours from mainly emissions of volatile S compounds
- increases in amount caused by rainfall
- potential danger of uncontrolled slurry outlets to the environment,(concentrated pollution, expensive to clean up)
- high peak in agricultural work pressure by the application within only 3 months.
- problems of social acceptance by the agricultural slurry application in a time of the year, where the farmland is at a peak of its natural beauty.

Environmental benefits as regards a better social acceptance of the agriculture by avoiding bad odours from slurry handling can hardly be quantified in financial terms. Reduced emissions and effluence of nitrogen can better be dealt with.

As indicative figures it should be stated that 1 kg nitrogen more recycled pr m3 slurry saves 17 kWh in fossil energy in reproducing the equivalent amount in commercial fertilisers. In the same way removal of 1 kg nitrogen in waste water, in order to lower the polluting impact by discharging excessive nitrogen to the water environment, demands app. 7 kWh fossil energy equivalents. Altogether it seems realistic to claim savings of more than 20 kWh in fossil energy equivalents pr kg nitrogen recycled from slurry, which also can be expressed in terms of reduced CO2 emission, possibly subject to a coming environmental levy within the European Union. Reproduction of 1 kg N causes emission of app. 4 kg carbondioxide under input of the required fossil energy equal to app. 60 MJ/kg N.

As regards environmental benefits emerging from reduced CH4 emissions it shall only be stated that methan as a greenhouse-gas has to be considered as several times worse than carbondioxide – a natural consequence of differences in chemical/physical properties.

Denitrificaton in the soil can, especially by overloaded slurry fertilising, be remarkable causing emission of N20, which as an greenhouse gas is considered to be over 20 times so bad as CH4, which again is considered to be 10–20 times worse than CO_2.

Based upon the above it does not seems unrealistic to assume the more common and quantifiable environmental benefits to amount to at least 3 ECU pr m3 slurry treated by the NRS technology.

A number of probable advantages by the new technology related to chemical/physical circumstances around citrate/water solubility and plant up-take of nutrients, all as connected to preservation of groundwater resources, have not been taken into account but these could even be the most striking impact of the NRS technology in the future as the polluting effects of water soluble commercial fertilisers appears more and more.

An attached appendix shows an attempted N mass balance for the agricultural system for emphasising the relations. The approximative figures applies to situations in the northern part of the European Union and is put up after a cross examination and evaluation of datas of environmental investigations from the last 10 years.

Appendix:

N-mass balance, agriculture app..
kg Nha, 220–220

| | N fixation, rainfall etc. | NH3 evaporation, denitrification etc. |
| 30 | 75 |

Fertilisers *)   130   10   Sales of plant products

Reserves of plant nutrients in soils

+/- zero ?

Fodder **)   60   25   Sales of animal product

85   Groundwater   25   Streams

*) Commercial fertilisers

References to literature:

1) Krüger (1992): Centralised digestion of animal manure and determination of the most favourable European regions for the energy use of animal waste. Technology. CEC report, 366 p.. Vesterager, N.O. & Hedegaard, M.

2) (1991): Organic fertiliser research and assessment study on organic fertilisers. CEC, JOULE reports, 124 p..

3) (1991): Gewinnung von Energie aus Abfällen am Beispiel einer Biogas-Anlage. 3. Langeooger Gespräch, p.148 – 160.

4) (1992) : Experimental organic fertiliser study, CEC, JOULE report, 115 p..

5) (1993): Soil fertility and cultivation of energy crops, CEC, AIR report, 170 p..

6) (1993): Coordinated R & D activity in the biomass sector. CEC (INRA), JOULE exten. report, 105 p..

7) (1993): Chemical separation process to slurry from animal husbandry and industrial organic waste. CEC, VALUE report, 225 p..

8) (1993): Verfahrenstechnik und Ökonomie bei industriellen Biogasanlagen. KTBL, Darmstadt, p. 60–74.

9) (1994): Compost and biodigestion within European Network on Sweet Sorghum. CEC, AIR reports, 120 p..

10) (1994): Die Umwandlung von Flüssigmist in einen organischen Dünger. KTBL, Darmstadt, BMFT, Bonn, p.140 – 150.

Composting in Italy: Current State and Future Outlook

GIANNI ZORZI, SILVIA SILVESTRI, ANDREA CRISTOFORETTI – Istituto Agrario di San Michele all'Adige (Trento, Italy)

Abstract

In this paper the existing composting plants in Italy, either in operation or that are about to come into operation in a short-to-medium time, are localized and the problems associated with two different approaches to composting are discussed: either from household waste that is not source separated or following the recovery and processing of sorted biomasses.

A general overview shows a large number of 'conventional' plants (n° 42), most of which were designed in the late '70–80s'. These must undergo deep, albeit gradual, changes in their productive process or, alternatevely, the organic component is to be biologically stabilized before dumping. The existing processes will be mantained only if there is evidence of actual opportunities to profitably use the compost.

Selected waste processing plants are rising fast (today they are at least 33) and some of them (at least the most efficient ones) have gained large market shares, due to their striking agronomic and environmental features.

These advances, however, may be in vain, if the existing legislation, not encouraging the production of high-quality compost, remains in force. Although, however, this seems to be about to be amended, unacceptable restrictions are likely to be introduced by the Ministry for the Environment in terms of environmental criteria related to the building of plants and compost quality standards (with special regard to the levels of some heavy metals, such as copper, zinc and lead).

Generalities

The processing of municipal solid waste for the purpose of compost production restricts the opportunities to put the resulting products to an efficient agricultural use, these often being of poor quality.

The large number of plants that are currently in operation nationwide (at least 42) must urgently come to grips with the problem of compost underuse in agriculture, either through the identification of other possible applications, such as the daily and final covering of controlled landfills, the reclamation of degraded areas and the sanitation of polluted sites, or through a process of biological stabilization of the waste organic matter, so that this can be then safely dumped, or finally retrofitting these plants to cater for the latest production requirements, as with other plants that already exist or are in the design stage (n° 6).

The possibility of using municipal waste compost as a farmland fertilizer has proven to be virtually non existent or not feasible, due to a number of factors, mainly resulting from the current complex and conflicting regulations, requiring, among other things, that technical and administrative prescriptions are strictly followed and, thus, disencouraging farmers from using compost (prior control of soil conditions, limitations in the usable compost amounts, etc.), as well as to the prevailing conditions in some national areas, where the naturally existing or induced content of heavy metals is higher than is laid down by law, and finally to the likelihood that further legislative constraints are introduced.

In this regard it should be emphasized that at present very few plants turn out municipal solid waste compost that comply with the existing regulations, and are therefore to be considered exceptions (Zorzi and Urbini, 1994).

It should also be noted that the actual composition of waste make it less suitable for the recovery of organic matter from unprocessed waste, due to the fact that this is increasingly mixed with larger and larger amounts of foreign fractions. Indeed, even though the organic matter proportion has remained nearly unchanged in absolute terms in recent years (a decade ago organic matter accounted for 35-40% of a daily waste production, estimated to be equal to 800 g/inhabitant; today the organic matter component is of the order of 25–30%, while the unit waste production has gone up to 1000–1200 g/per unit), the increase in unwanted fractions, such as glass and plastics, may considerably deteriorate the quality of compost from municipal waste (AA.VV., 1994).

It is beyond doubt that any attempt not to landfill the organic component of waste failed not only because of the motley nature of waste, but also of the inefficient sorting and final enhancement technologies, the poor understanding of plant processes and the poor management.

This led to the general consensus that only through the source separation of waste with a high organic component, compost can be obtained that is beneficial for plant crops and environmentally safe. This goal can be achieved through separate collection programmes, that should best be implemented first in large producers of waste, or at least not solely in households.

Any such programme is to be devised while keeping in mind the individuals to be involved in the different steps, finding out whether they are willing to reorganise their waste collection services and identifying in advance the compost center that can handle the selectively collected waste, in the amounts and at the times as are convenient for the collection service.

The separate collection of wet waste, as was provided for in the Ministerial Decree 29.5.91, was never put into practice by the Italian Regional Authorities and just very few of them have taken some measures, failing coordination and planning. If, on the one hand, several municipalities in Piedmont, Lumbardy and Veneto, and to a lesser extent even in Emilia-Romagna and Tuscany, have steered the pathway of organic component separation even in household waste, not always exist authorized facilities to accomodate these materials, which are often meant for landfills.

Based on its new concept, composting plays the role of a production technology, whereby products largely fulfilling the market requirements can be won from a large variety of selected waste. This could, therefore, at the same time cover the farmers' needs for organic conditioners and help solve waste disposal problems. The failure of many municipal waste programmes brought a negative feeling about composting to farmers. Hence the proposal that high-quality compost products should be renamed, and be referred to in the fertilizer legislation (Act L 748/84) as green or mixed composted conditoners.

In other European countries the conventional plants underwent dramatic retrofitting changes, so as to process only or nearly only source separated waste.

On account of the high investments made to build the over 40 conventional plants that will shortly become operational in this country (so far a total of between 900 and 1000 billion Lit was spent), a wide-ranging analysis and a global proposal are needed according to duidelines fixed at national level, thus avoiding to fix local situations only, as was actually the case in a few instances. The wealth of technologies deployed in the existing plants, as is required by the complex nature of untreated waste, should not be wasted, indeed retrieved and enhanced, through low-cost investments, in order to process selected waste.

Should the present operational approaches not be revised and things remain as they are, one should find out whether, among the possible alternative uses of compost (intermediate filling of landfills), it is best to perform short biostabilization and sanitation processes of the waste organic matter (25–30 days) rather than a long term composting. The former approach is known to afford advantages not only in relation to quantities and volumes, but also in terms of reduced production of leachate and biogas, materials sanitation, and low emissions of foul smells. In addition the plant capacity would be greately increased as a result of the shorter leadtime.

If other possible applications of municipal wast compost are to be considered, such as reclamation of degraded sites, a thorough knowledge of all critical cases is needed, as well as a survey of any local short or medium term plans developed or implemented in the fields of lanscape architecture and environmental protection. At the same time all those concerned with waste recovery should be bound to use the resulting compost.

Few (n° 6), although particularly significant, plants elected to compromise, and set up, either in a temporary or final way, two different waste processing lines: one intended for the production of compost from unsorted municipal waste or for the biostabilization of organic matter, the other for the biological transformation of selected waste into top-quality compost.

Separate collection of organic matter from catering operations or commercial businesses, and possible from households, as well as the recovery of waste from fruit and vegetable, flower, fish and neighbourhood markets, of prunings or grass mowings, of selected agroindustrial sludge and sewage sludge, of waste from food, textile, paper and wood manufacturing plants, and of crop and livestock residues are a must if we are to obtain raw materials to produce a high-quality compost that sells easily on the market.

Hence the need or opportunity to develop storage and treatment platforms, which can accomodate waste of different nature and origin, thereby providing a comprehensive approach, not limited to municipal waste, to the disposal requirements of a given catchment area. Such waste, particularly if from large urban areas and production units, are frequently source separated, or can be easily separated, and therefore can be recovered following just a few minor adjustements to the collection and delivery systems.

Of the utmost importance, prior to any biological process, is the identification of guiding principles, whereby the suitability of any waste for processing into quality compost can be evaluated.

For these programmes to be successful and profitable, enterprises should develop a deeper understanding of composting techniques through an ongoing management of plants, i.e. not limited to the design and implementation phases. In this way they could submit more advanced technical tenders than they do now.

After completion of the biological process, users must be assured as to the high quality of compost. To this end a voluntary compost producers' association has been established (Consorzio Italiano Compostatori), who process only strictly selected waste and have adopted a seal of quality, to be granted solely to productions that fulfill the agronomic and environmental standards set by its technical-scientific committee.

Compost plants processing source unseparated municipalwaste

A total of 42 plants for the production of compost from unsorted municipal waste, and in some cases added with sewage sludge from waste water treatment plants, were built or are about to come into operation throughout the country (table 1). Most of these were designed in the past decade (and some even before), but a number of them have not been completed or tested yet. What amazes, and even concerns us, is that several plants, for different reasons, have been for some years now in an advanced state of development, with no estimate whatsoever of when they are going to be completed or start operation (Sulmona and Vallo di Diano 95%, S. Maria Capua Vetere, Tempo Pausania, Col San Felice 90%, Reggio Calabria 75%, Pescara 30%, Ostellato 55% etc.). Consequently chances are that the current technologies become quickly obsolete and inefficient and get into a state of disrepair.

Table 1 Composting plants treating MSW or MSW and Sewage Sludge (still working or to be set up in the short-medium time)

Plants	Technology	N°	Capacity ton/year	ton/d	Products	Compost ton/year
Piemonte						
Cuneo	Peabody		45000	150	compost;RDF	9000
Novara	Emit		75000	255	compost;RDF	15000
SUBTOT.		2	120000	405		24000
Lombardia						
Ceresara (MN)	Daneco		48000	160	compost;RDF	10000
Pieve Coriano-MN	Daneco		63000	210	compost;RDF	12500
SUBTOT.			111000	370		22500
Friuli						
S.Giorgio di Nogaro (UD)	Ferrero		70000	250	compost;RDF	25000
Udine	Daneco		80000	280	compost	16000
Villa Santina (UD)	Daneco		25000	90	compost;RDF	8500
SUBTOT.		3	175000	620		49500
Alto – Adige						
Bolzano	Weiss-Kneer		135000	450	compost	35000
Bressanone	Bühler		22000	75	compost	5000
Pontives	Vöst Alpina		18000	60	compost	4000
SUBTOT.		3	175000	585		44000
Veneto						
Feltre (BL)	db		36000	120	compost	8500
Schio (VI)	Slia – Snam		40000	160	compost;energy	8000
SUBTOT.		2	76000	220		16500
Emilia Romagna						
Ozzano (BO)	Weiss–Kneer		33000	110	compost;RDF	8000
Ostellato (FE)	db		39000	130	compost	9500
SUBTOT.		2	72000	240		17000
Toscana						
Massa Carrara	db		70000	250	compost;RDF	15000
Pistoia	Slia		60000	210	compost	12000
SUBTOT.		2	130000	460		27000
Marche						
Ascoli Piceno	Secit		78000	260	compost	19000
Fermo (AP)	Secit		63000	210	compost	15000
Pollenza (MC)	Secit		55000	180	compost;energy	12500
SUBTOT.		3	196000	650		46500
Umbria						
Perugia	Sorain -		115000	380	compost;RDF	15000
Foligno (PG)	Cecchini		38000	110	compost;RDF	6000
SUBTOT.		2	153000	490		21000

db: De Bartolomeis

Table 1 Continued

Plants	Technology	N°	Capacity ton/year	ton/d	Products	Compost ton/year
Lazio						
Col S.Felice (FR)	Sorain C.		160000	600	compost;RDF	28000
Terracina (LT)	Slia		45000	150	compost;RDF	9000
SUBTOT.		2	205000	750		37000
Abruzzi						
Giulianova (TE)	Cons.coop		28000	95	compost	7000
Pescara	Emit – '		60000	200	compost;RDF	15000
Sulmona (AQ)	Emit		24000	80	compost;RDF	5000
Vasto – Cupello (CH)	'Daneco		42000	140	compost;RDF	9000
SUBTOT.		4	130000	515		36000
Campania						
Caserta	db		103500	345	compost;RDF	13500
S.Maria Capua Vetere (CE)	db		49000	160	compost;RDF	9500
Vallo di Diano (SA)	db		18000	60	compost;RDF	4000
SUBTOT.		3	163500	545		27000
Basilicata						
Matera	Secit		19000	62	compost	5500
Tursi (MT)	Secit		24000	80	compost	5500
SUBTOT.		2	43000	142		11000
Calabria						
Cosenza – Rende	db		60000	215	compost;energy	14000
Lamezia Terme (CZ)	Italimp.		45000	130	compost;RDF	11000
Reggio Calabria	db		75000	260	compost;RDF	18000
Rossano Calabro (CS)	Snam		45000	130	compost	11000
SUBTOT.		4	232000	755		54000
Sicilia						
Regalbuto (EN)	db		36000	120	compost	8500
Trapani	db		30000	120	compost;RDF	8000
SUBTOT.		2	66000	220		16500
Sardegna						
Cagliari	db		150000	600	compost;energy	30000
Macomer (NU)	Ferrero		45000	160	compost;energy	10000
Tempio Pausania–SS	db		30000	120	compost	6500
Villacidro (CA)	db		40000	140	compost from anaer. digestion	8500
SUBTOT.		4	265000	1020		55000
TOTAL		42	2336000	7907		505000

Note: some values are the average of the capacity in relation to the seasonal incomings; the compost production of several plants has been estimated.

The first comment is that there are no certainties as to the future development of these plants, and, as a result, to the strategies to increase their effectiveness and ability to fulfil the desired goals. It should also be emphasized that some of them have been abandoned for years and will require much effort to be brought into operation again (such as Novara and Ozzano).

The processing capacity of all these plants is 2,336,000 tons waste/year, equivalent to 7,907 tons/day; their average unit size is 56,975 tons/year, equal to 192 tons/day. The estimated compost output exceeds half a million tons a year. It therefore appears absolutely critical to find adequate applications for these materials, as was pointed out earlier on.

Composting plants from selected waste

In harmony with the current trends in composting, 33 composting plants have been or are about to be built in Italy. These are mainly located in the northern and central areas of the country (just in a few regions, however, in the centre), whereas there is virtually none in the south of Italy and in the islands (table 2). 22 of them process sludges from sewage and/or agrofood water treatment plants, which proves not only the want of alternative waste disposal methods, but also the possibility to win compost with remarkable agricultural properties from this waste, especially when mixed with other organic matters and, in particular, lignocellulosic waste (table 5).

The environmental quality of products is most often ensured through a strict sorting procedure of inputs (sludges and other biomasses). Mention is made here not only to the plants situated in Piedmont and Veneto, which are compelled to abide by Act 4558 of 23.6.86 and Act 4978 of 6.9.91, but also to the processing plants that have adopted an even more stringent self-discipline than the ones mentioned above, such as Trento and Eco-pol of Bagnolo Mella (table 4).

The year-long experience of the Trento plant in the production of compost from selected sewage sludge (Zorzi et al., 1992) was treasured up by many with the aim of optimizing the way this particular kind of waste is processed. Striking examples are provided by Agrinord in the Verona district, Eco-pol in the Brescia area, Ecopì of Alessandria and possibly the Consorzio del Bacino dello Scrivia (that has just been tested). The Trento plant receives sludges from 12 waste water treatment plants serving small and medium urban communities (out of a total of 65 in operation the provincial area). The average weights of heavy metal concentrations in the sludges are the following: Cd 0,23, Cr 22, Ni 15, Pb 40, Cu 157, Zn 533 mg/kg dry matter. The whole process is constantly fine-tuned and monitored by Istituto Agrario di San Michele all'Adige, that is responsible for the performance of repeated physical, chemical and biological analyses of products as weel as for their testing on different crops.

Table 2 Composting plants treating selected wastes in Northern and Central Italy (still working or to be set up in the short time)

PLANTS	CAPACITY ton/year	TREATED WASTES	COMPOST ton/year
Piemonte			
Casalcermelli (AL)	12500	sludge, pruning, barks	6000
Tortona (AL)	33000	sludge, wood chips, barks, pruning	6000
Alba – Brà (CN)	26975	sludge, industrial waste, pruning	7350
Roccavione (CN)	1500	sludge, pulp waste, poultry litter	700
Saluzzo (CN)*	7300	sludge, FPW and yard waste	4000
Verzuolo (CN)	10000	barks and manures	8000
Ghemme (NO)	10000	sludge, manures, agricultural waste	2000
Verbania (NO)	5000	yard waste, source separated MSW *	1220
Ghislarengo (VC)	28000	sludge and industrial waste, pruning	8000
Torino – Italconcimi	10000	yard waste from public green areas	8000
Torino – AMIAT*	54000	FVMW, pruning, source separ. MSW	20000
SUBTOT.	198275		71270
Lombardia			
Bagnolo Mella (BS)	25000	sludge, FPW and pruning waste, FVMW, source separated MSW	8000
Castiglione Stiv. (MN)	4000	sludge, FPW and pruning waste	1300
Milano – AMSA*	43000	FVMW, pruning, source separ. MSW	16500
SUBTOT.	72000		25800
Veneto			
Isola Scala (VR) – Agrinord	35000	sludge, pruning, FVMW,poultry litter source separ. MSW, mushroom litter	20000
Isola Scala (VR) – Agrofertil	35000	sludge, yard waste	15000
S. Bonifacio (VR)	16000	sludge, poultry litter and manures	6000
Sorgà (VR)	6000	sludge and yard waste	2500
Vigonza (PD)	9000	sludge, pruning, grape stalks,agric.w.	2500
Trecenta (RO)	12000	sludge, animal residues, agric.waste	3700
Castelfranco (TV)	9000	pruning and grass mowing	3100
Spresiano (TV)*	5000	pruning, grass mowing, FVMW	1700
Mira (VE)	21000	sludge, source sep.MSW, agricultural waste, manures, industrial residues	7800
SUBTOT.	148000		62300
Province of Trento	20000	sludge, barks, pruning, grape stalks, FPW, cotton, source separated MSW (experim.)	6000
SUBTOT.	20000		6000
Emilia-Romagna			
Rimini – AMIA	18000	source separated MSW, pruning, sludge, PVW, SHW	4500
Piacenza	29000	sugar beet residues and pruning	10000
Vignola (MO)	5500	pruning, source sep. MSW, sludge	1800
Soliera (MO)	experimental	sludge, pruning, source separated MSW, manures, grape stalks	–
Faenza (RA)			
Cesena (FO)*	5000	pruning, source sep. MSW, FVMW	1750
Imola (BO)*	3600	sludge, pruning, source sep. MSW	1000
SUBTOT.	61100		19050
Marche			
Senigallia (AN)	7500	sludge and sawdust	1100
SUBTOT.	7500		1100

Table 2 Continued

PLANTS	CAPACITY ton/year	TREATED WASTES	COMPOST ton/year
Toscana			
Campi Bisenzio (FI)	9300	pruning, grass mowing, source separ. MSW	3500
SUBTOT.	9300		3500
TOTAL	516175		189020

FVMW= Fruit and Vegetable Market Waste
FPW=Food Processing Waste
PVW=Processing Vegetables Waste
SHW=SlaughterHouse Waste
* plants to be realized
Note: some data related to the compost production have been estimated

Table 3 Italian composting plants having double trend

Plant	Technology	Capacity ton/year		Capacity ton/day		Compost ton/year	
		MSW	SW	MSW	SW	MSW	SW
Alessandria	Ecologia	30000	12000	100	40	6000	4000
Cedrasco (SO)	Ecologia	30000	15000	100	50	6000	4500
Spresiano (TV)	not realized	70000	20000	230	65	*	7000
Carpi (MO)	Snam	72000	6500	300	25	15000	2000
S.Agata Bolognese (BO)	Castalia – Unieco	45000	22000	170	70	10000	10000
Sesto Fiorentino (FI)	Degremont	110000	8000	358	27	30000	2500
TOTAL		357000	93500	1258	277	61000	26000

SW=Selected Wastes
* Biostabilization of the organic fraction for 30 days

Likewise the Rimini plant has become a reference point for the operators producing compost from organic waste separately collected from hundreds of catering businesses.

A brilliant example of good management is provided by the firm Maserati in Piacenza, which, using oversimplified facilities and technologies, composts waste from sugarbeets mixed with vegetable materials.

The development of two large plants designed to handle waste from marketplaces and other large producers is raising great expectations (AMIAT in Turin and AMSA in Milan).

It should, however, be noted that not all of the surveyed plants accurately screen their inputs and the biological process, in particular, is carried out rather carelessly.

A composting plant must be responsible for all recovery steps: from collection and previous checks of materials to the final destination of output, including testing the impact of compost on crops.

The regions where composting from selective waste is most widespread are Piedmont (11 plants, including the AMIAT plant in Turin that was recently subcontracted) and Veneto (9 plants for now, but more to come).

Table 4 Acceptability limits adopted for the incoming wastes

Parameters	Regione Piemonte	Regione Veneto	Province of Trento	Eco-pol
pH	5.5 ° 8			
Moisture (fresh matter)	_ 80%			
Organic substance (dm)	_ 40%		_ 40%	
Salinity (meq/100g)	_ 200	200		
S.A.R. (with salinity >50)	_ 20			
Volatile Phenols (mg/kg dm)	_ 10			
Surface actives (mg/kg dm)	_ 100			
Chlorides (with salinity>50) (mg/kg dm)	_ 5000			
SO_4–S (mg/kg dm)	_ 10000			
Cd tot. (mg/kg dm)	20	10	0.6	< 5
Cr III (mg/kg dm)	1000	750	50	< 250
Cr VI (mg/kg dm)		10		< 1
Hg (mg/kg dm)	5	10		
Ni (mg/kg dm)	250	300	25	< 100
Pb (mg/kg dm)	600	750	55	< 200
Cu (mg/kg dm)	1000	1000	200	< 300
Zn (mg/kg dm)	3000	3000	750	< 1200
As (mg/kg dm)	10	10		
B (mg/kg dm)	60	60		
Se tot. (mg/kg dm)	5	5		
fm=fresh matter				
dm=dry matter				

S.A.R.= Sodium Adsorption Ratio

The total operating capacity of all plants considered is over half a million tons/year, with a compost output approaching 200,000 tons/year, equal to about 400,000 m³/year (table 2). The market for the highest quality composts is doing well, and indeed is looming large in terms of increased capacity and market demand. The current sales price for top-quality products is about 20,000 Lit/m³ free plant, even though quotations of up to 25–28,000 Lit/m³ are quite common for bulk products. These are remarkable figures, if one considers that unenriched young peats are currently commercially available at 45–50,000 m³ and that no agreement yet exists between plants to harmonize supplies.

The most common applications for products with outstanding agronomic properties (high total and free air space, water volume, little shrinkage) and low or comparatively low salinity and pH values are in the horticultural-flower and forestry nursery sectors and mushroom cultivation. For the preparation of adequate media for cultivation in containers or benches compost is usually mixed with peats in a proportion of 25–25% in volume. In the case of recreational use of compost, mushroom growing and seedling nurseries the ratio of compost and peats goes up to 1:1. Compost is widely used as it is in grasslands and for open ground organic fertilization, especially in the case of fruit and vine-growing, where the product is not only spread throughout the area but is also filled into plant-holes and used for mulching.

Compost plants based on different approaches

Table 3 shows 6 plants where compost is produced according to two different approaches. The first one concerns the processing of the municipal solis waste organic fraction retrieved by mechanical separation on the spot; the resulting compost is of lower quality (standard compost). In the case of the Spresiano plant the production line is designed to biologically stabilize the organic component before dumping materials into a landfill. The second approach, which usually features smaller capacities than the first until separate collection becomes more widespread, has to do with composting of source separated waste and is meant to produce a high-quality compost.

Following the introduction of information and education programmes, a more accurate identification of waste sources with a high natural organic component, including non municipal ones, and the development of collection services, the first stream of waste will gradually shrink and the second will expand. This newest approach can be applied also to conventional plants and marks a big step forward in the production of higher quality compost.

The potential capacity of these plants is 357,000 tons/year of aggregate waste and 93,000 tons/year of selected biomasses. Compost output is 61,000 tons/year and 26,000 tons/year respectively, that add up to the figures shown in the previous chapter.

Legal considerations

In Italy, in spite of the commendable, albeit lengthy process of revision of the current regulations on compost production and use, the provisions of the Interministerial Commission responsible for the enforcement of Art. 4 of the Decree of the President of the Republic 915/82 are still in force (table 6).

At the same time Act 748/84, issued by the Ministry of Agricultural, Food and Forestal Resources, set new standards on fertilizers and outlines different agronomic properties of conditioners from municipal solid waste (or mature compost) than are set forth in the Decree of the President of the Republic 915/82, in that it requires that just a few agronomic prerequisites be met, and makes no mention of environmental requirements.

If on the one hand DPR 915/82 pursues the aim to prevent an easier use of soil as dosposal site for every kind of waste, on the other hand the foreseen procedures block whatever possibility to use the best quality compost (AA.VV., 1994).

Table 5 Analytical characteristics of compost from different plants (values are expressed on a dry matter basis with exception of moisture, pH and EC)

PLANTS PARAMETERS	TRENTO		AGRINORD		SCRIVIA	ECOPOL		AMIA RIMINI		AMIU (MO)–CRPA	
	A	B	A	B		mixed	green	1	2	S+barks	S+litter
Moisture %	38.7	34.5	–	–	49.5	46.6	42.0	29.2	37.8	38.4	56.0
pH	7.73	7.66	7.7	8.2	6.4	7.6	6.8	7.6	6.8	7.2	6.9
EC (25°C)µS/cm	1530	2242	2200.0	3600.0	1447	939.0	950.0	–	–	–	–
Ashes %	32.00	37.59	27.1	21.7	33.5	–	–	38.2	40.2	49.6	54.1
Org. C %	37.22	33.51	38.8	45.0	30.9	24.1	22.9	28.3	27.5	34.5	34.1
Org. Substance %	64.18	57.78	66.9	77.6	66.5	41.3	40.0	48.8	47.4	57.7	60.0
Humific. Rate %	18.03	20.15	*	*	24.1	37.4	34.7	23.7	29.8	15.9	13.2
Humific. Index %	12.11	12.72	**	**	–	–	–	–	–	–	–
C Humic/Fulvic Ac	2.26	2.17	–	–	2.5	–	–	–	–	–	–
C/N2.26	16.36	16.8	12.2	8.0	25.3	14.0	15.3	21.7	18.3	23.0	21.3
N %	2.29	2.29	1.6	2.9	1.2	1.7	1.5	1.3	1.5	1.5	1.6
P (P2O5) %	2.49	2.84	–	–	0.3	2.0	0.3	1.1	0.8	1.7	3.0
K (K2O) %	0.97	0.81	–	–	0.2	1.5	0.4	0.8	0.3	0.9	2.1
As mg/kg	–	–	1.0	1.5	1.4	1.8	1.7	nv	nv	–	–
Cd mg/kg	1.28	0.75	4.7	5.0	1.2	<3.0	<1.0	nv	nv	1.6	2.0
Cr mg/kg	36.1	51.8	53.7	63.8	59.0	131.0	64.0	62.0	112.0	133.0	127.0
Hg mg/kg	–	–	0.1	0.3	1.2	0.8	0.5	nv	2.0	–	–
Ni mg/kg	15.1	16.8	43.2	37.0	37.0	30.0	23.0	32.0	90.0	46.0	48.0
Pb mg/kg	44.5	52.9	66.0	79.5	108.0	130.0	95.0	114.0	38.0	136.0	160.0
Cu mg/kg	124	157	143.0	208.0	118.0	217.0	147.0	318.0	200.0	247.0	222.0
Zn mg/kg	380	480	587.0	993.0	610.0	556.0	260.0	200.0	462.0	1056.0	1279.0
Germinat. Index %	87.47	66.25				95.0				76.4	97.3
Growth Index %	–	–				180.0					
Bulk density g/l	284.00	290.57				366.0					
S.D. %	10.36	8.8				16.0					
T.P.S. %	84.27	84.14				84.0					
Air Volume %	24.29	31.11				29.4					
Water Volume %	59.98	53.02				70.8					

AGRINORD A= soil improver; B= soil improver with chicken manure

S= sludge

AMIA 1= food waste compost ; 2= vegetable compost with low % of sewage sludge

* Humic carbon A= 21% dm; B= 23.9% dm; **1#humic fraction A= 16% dm; B= 14.5% dm; nv= not valuable

B.D.= Bulk Density; S.D.= Shrinkage Degree; T.P.S.= Total Porosity Space

Thus new high-quality composting procedures were advocated. For quite some time now a national legislation has been demanded, which systematically brings together all tentative and temporary provisions adopted by some regional authorities, with a view to promoting the free trade and use of quality products, that are to be regarded to all intents and purposes farming aids capable of increasing the physical, chemical and biological fertility of soils (table 6).

Table 6 Quality limits of compost – National and regional regulations

	DPR 915/82	L 748/84	Regione Lombardia	Regione Piemonte	Regione Veneto
Moisture %	< 45	–	–	≤ 40	< 50
pH	6÷8.5	–	< 8.5	6÷8	5.5÷8.0
EC (25°C) µS/cm	–	–	–	–	–
Salinity meq/100g dm	–	–	–	≤ 50	< 50
Soluble Clorures mg/kg dm	–	–	–	≤ 2000	–
Soluble Sulphate mg/kg dm	–	–	–	≤ 5000	–
Volatile Phenols mg/kg dm	–	–	–	≤ 10	–
Ashes % dm	–	–	–	–	< 40
Organic Carbon % dm	–	–	–	–	> 25
Organic Substance % dm	> 40 (%dm)	> 20 (%fm)	–	≥ 40 (% dm)	–
Humic Carbon % on total C	> 20 *	–	–	> 20 *	> 20 **
1ªhumic fraction % on total Carbon	–	–	–	–	> 25 **
Humic ac./fulvic ac. Carbon	–	–	–	> 1.5 *	–
C/N	< 30	< 30	< 30	< 20	< 25
N % dm	> 1	< 2	–	> 1.7	> 1.5
NH_4–N % dm	–	–	–	< 0.06	–
NO_3–N % dm	–	–	–	> 0.04	–
P (P_2O_5) % dm	> 0.5	–	–	> 1	–
K (K_2O) % dm	> 0.4	–	–	> 0.7	–
Zn mg/kg dm	< 2500	–	< 400	≤ 1500	< 1250
Cu mg/kg dm	< 600	–	< 200	≤ 500	< 300
Ni mg/kg dm	< 200	–	< 50	≤ 150	< 150
Pb mg/kg dm	< 500	–	< 200	≤ 350	< 200
Cr tot. mg/kg dm	< 510	–	< 150	≤ 500	< 151
Cr VI mg/kg dm	< 500	–	–	–	< 150
Cr III mg/kg dm	< 10	–	–	–	< 1
Cd mg/kg dm	< 10	–	< 3	≤ 5	< 5
Hg mg/kg dm	< 10	–	< 2	≤ 2.5	< 3
As mg/kg dm	< 10	–	< 5	≤ 2.5	< 5
Se mg/kg dm	–	–	–	–	< 5
B mg/kg dm	–	–	–	_ 40	< 100

* extractable with $Na_4P_2O_7$ 0.1 M and NaOH 0.1 M
** exctractable with NaOH 0.1 M

Legislative inconsistencies curb the expansion of the industry, in that enterpreneurs, willing as they are to make large investments in composting plants, take a 'wait and see' attitude, pending adeguate legislative measures.

To set appropriate standards for the production of high-quality compost amounts to giving fair and full consideration to an universally renowned industrial biotechnology designed to process natural-occurring organic matters under aerobic conditions.

Composting is a fully fledged industrial procedure using different raw materials, in well-defined weight ratios, in order to ensure the best possible development of biological processes through suitable technologies and to attain the desired qualitative standards.

An extensive research conducted by Istituto Agrario di San Michele all'Adige on a wide number of compost samples from several kinds of accurately selected organic wastes, including sewage sludge (288 samples from 124 plants), from households as well as from industrial, commercial, business, agricultural and livestock breeding operations, revealed excellent agronomic and environmental qualities of the resulting products, that are comparable or even superior to conventional commercial organic fertilizers (peat, cattle manure, chicken manure). In these products, despite some differences due to the starting materials, considerably lower levels of heavy metals were found than in municipal solid waste compost (i.e. 1/3 lower for zinc and down to 1/9 for lead). If these concentrations are referred to the latest trends in the Eec legislation, a good accordance can be found between required standards and actual values, with the exception of the very stringent limits set, or should we rather say pursued, by some countries in relation to some potential pollutants. Such limits, moreover, can hardly be assessed, these pollutants being largely spread in the environment. Comforting is the fact that limits are exceeded only in the case of the least hazardous, or even beneficial metals, such as copper and zinc, for which a higher tolerance should be envisaged (Gasperi and Zorzi, 1994).

What strategic role could composting play, if used only in a minority of compost lots? It would not relieve the final disposal site of large masses of waste, it would not promote the dissemination of the method and would not acquire any market shares for compost products.

It should be remembered that the commonly used organic fertilizers are indeed obtained from other residues. This proves that an improper usage of the term 'waste' is usually made, unless its use is associated with a contamination hazard.

From this standpoint quality composts should be viewed as resources that not only restore soil fertility, but, even more importantly, allow to close the nutrient cycle, whereby whatever comes from the soil is finally returned to it (Sequi et al., 1992).

If the Ministry of Environment is to consider quality composting an important link in an integrated waste disposal system (and enough convincing evidence that it is so comes from the above-mentioned processing capacity of plants and from the high, never fully satisfied demand for composting products), it should issue new technical standards and, at the same time, refrain from criticism and remove the constraints provided for in its draft regulations, as well as adjusting the limits to such values as are actually achievable in Italy, with special regard to zinc, copper and lead (the proposed values are 400 and 100 mg/kg dry matter for the first and the following elements, respectively). This ajustment is certainly feasible, considering that copper and zinc play a very minor role in air pollution and lead is contained in many vegetable materials. It is believed that concentration of about

180–200 mg/kg dry matter for copper, 500–600 for zinc and 150 for lead are very far from the present legal standards (copper 600, zinc 2500, lead 500 mg/kg dry matter).

Conclusions

Composting is a fast growing industry. There is increasing awareness that the only chance for compost products to become firmly established on the market is associated with the controlled biological treatment of source separated waste and, consequently, the need arises to find outlets for the outputs of approx. 40 conventional plants processing unsorted waste.

Possible applications in the landscape architecture and environmental protection sectors are dependent on well thought out plans, in which particularly degraded areas are identified and the short and medium term compost requirements are calculated. In addition the contractors executing the land reclamation project must commit to the use of compost.

If that is not the case, plants should be completely revisioned, so that incoming streams are separated and discrete production lines can be operated (standard compost from source unseparated waste and quality compost from accurately selected biomasses), with a view to giving more and more priority to the most advanced line.

Another feasible and indeed noteworthy alternative to standard compost production is provided by the biostabilization of the organic fraction of the site-separated waste prior to landfilling.

Conversely, the present scenario in the quality composting sectors is extremely reassuring as to the latest approaches taken by over 30 plants.

Adeguate regulations and quality certification systems play a pivotal role at a time when increasing attention is given to composting, as, through an accurate and strict process management, they can ensure hightest user and environment full protection.

The use of low metal waste allows to minimize the environmental hazards of a procedure that was –and still is– perceived a convenient shortcut to get rid of huge amounts of materials.

Highly-professional and serious plants are not an occasional occurrence, but are becoming increasingly numerous. The existing installation, technological and management solutions are reliable and long-established. New impetus may stem from the removal of the contraddictions, uncertainties and inadequacies of the present legislation.

References

AA.VV. (1994). Ingegneria della trasformazione in compost. Rapporto finale delle Giornate europee di studio sull'ambiente. Bari, 26–27 ottobre 1994, Ed. CIPA, Milano.

Gasperi, F., Zorzi, G. (1994). Lo standard qualitativo di compost da rifiuti organici selezionati. Gea, Ed. Maggioli, n° 4, 36–48.

Sequi, P., Figliolia, A., and Benedetti, A. (1992). Utilizzo di acque reflue e fanghi in agricoltura: la chiusura del ciclo degli elementi nutritivi. Acque reflue e fanghi, Ed. Centro Scientifico Internazionale, Milano.

Zorzi , G., Gasperi, F., Cristoforetti, A., Silvestri, S., Pinamonti, F., Nardelli, P. and Piccinni, P. (1992). Composting of waste and refuse biomasses in pilot and real scale plant. Uses and marketing of produced composts. Proc. of the ISWA annual conference, Amsterdam, 127–146.

Zorzi, G., and Urbini, G. (1994). Stato dell'arte e prospettive del compostaggio in Italia e negli altri paeesi europei. Proc. Giornate europee di studio sull'ambiente. Bari, 26–27 ottobre 1994, Ed. C.I.P.A., Milano, 137–166.

State of the Art and Perspectives for Composting in the United States of America

JEROME GOLDSTEIN – Editor, BioCycle Publisher,
Compost Science & Utilization

Abstract

During the period from 1985 to 1995, the increased number of composting facilities in the United States of America was primarily motivated by high landfill tipping fees and bans that prevented the landfilling of vegetative organic residuals such as leaves and grass clippings. In the case of waste water biosolids, composting was part of a beneficial reuse strategy that kept sludge from being dumped in oceans and rivers as part of clean water programs. In the 1990s, increasing attention is being given to the utilization of a high quality compost in sustainable agriculture and horticulture. Specifically, compost products are gaining recognition as an economical and environmental substitute for chemical inputs including toxic pesticides. More researchers are proving the relevance of compost applications in sustainable agriculture programs. While much progress has been made in the production and application of quality compost, there still remains great challenges if the potential widespread application of compost is to be realized.

Introduction

Significant changes are taking place in the utilization of organic by-products. Twenty-five years ago, there were only small numbers of composting projects to divert such materials as leaves and grass clippings from landfills; waste water biosolids from being dumped in oceans and rivers; and livestock manures from creating nitrate and odour problems. Today in the United States, as the statistics below indicate, there are broad-based public and private initiatives in the composting of source separated organics.

According to the *BioCycle* 1995 State Of Garbage In America Survey, the number of yard trimmings composting facilities had reached 3,202. When the first survey was conducted in 1988, the figure was 700 facilities. Thus in the last seven years, the number of yard trimmings composting facilities has risen 391 percent. The 1994 *BioCycle* annual survey of biosolids composting facilities in the United

States reported that there were 318 projects. Of the approximate 200 facilities that are operational, 92 utilize aerated static pile methods; 44 have an in-vessel system; 43 use windrows; 11 use aerated windrow; and seven compost in static piles. The balance of the projects remaining are in construction, permitting, design or planning stages. A 1994 *BioCycle* survey of centralized facilities that compost mixed or source separated municipal solid waste specified 17 operational and 34 in various stages of development. A significant finding of the survey was the impact of the Supreme Court's flow control decision in May, 1994 (Carbone vs. Clarkstown). That decision essentially eliminated designated measures implemented by solid waste management facilities specifying where solid waste should be taken for processing.

As pointed out by Richard Kashmanian of the U.S.E.P.A., composting has steadily been increasing as a waste management practice in agriculture. On-farm residuals that are being composted in increasing quantities include manure and crop residues as well as poultry and other livestock carcasses.

Emphasis on end product use

For many farmers, making compost and applying it to soils is an integral part of the transition to more sustainable and profitable crop production. Farmers are changing methods because of steadily growing concerns over high cost of chemical inputs, declining soil fertility, increasing problems with ground water contamination and side effects on their personal health from pesticide applications.

Correspondingly, as solid waste managers in cities and states have come to recognize the role of composting as a way to divert organic residuals from landfills, the importance of producing a quality compost with high value in agriculture has become more widely recognized. Many research findings explain how the nutrient cycle that links urban residuals with agricultural needs can be fulfilled.

Research centers are spending more time and effort to help agriculture and society achieve the ambitious goals of environmental quality, social responsibility and economic success. Increasingly research is being directed towards biological control, soil improvement and organic waste management. When it comes to 'best management practices' for organic residuals, composting is cited as an effective method.

Recognition of the role of composting in the utilization of organic residuals

State agencies – specifically agricultural and environmental departments — are increasingly compiling data on the amount of organic wastes generated in food production and processing industries. Following are some examples of data being published to indicate the potential of compost that could be made available:

Florida: According to the University of Florida's Institute of Food and

Agriculture Sciences, animal manures generated annually total 533,500 tons; yard trimmings represent 3.1 millions tons of compostable material. To illustrate the potential uses for compost in Florida, agricultural specialists provide these estimates of various categories: pasture – two million acres; citrus – 791,290 acres; vegetables – 418,000 acres; field crops – 581,000 acres; annual forest plantings – 170,000 acres; sod farms – 47,500 acres; greenhouse crops – 7,200 acres; lawns – 1,000,000 acres; mine lands – 248,000 acres.

Louisiana: Solid waste specialists at the Agricultural Center of the Louisiana State University System supply these figures on agricultural residuals generated annually in Louisiana. Cotton gin trash – four million tons; sugar cane bagasse – 2.5 million tons; sugar cane filter press mud – 360,000 tons; rice hulls – 1.2 million tons; poultry carcasses – 5,500 tons and poultry litter – 1,000 tons.

Kansas: A survey of 171 Kansas agribusinesses to calculate the types and volumes of waste generated at their facilities turned up 96 different types of discards totalling over 46,000 tons per year. As defined by the Kansas Industrial Extension Service and the Agricultural Engineering Department at Kansas State University, Kansas agribusinesses are those enterprises whose primary process products are associated with the food, grain and/or livestock industry.

Arkansas: More than one billion broilers were produced in Arkansas in 1992, making the state number one in poultry production in the United States. It also meant that too much poultry litter was produced in the western two-thirds of the state. An option to reduce threats to surface and ground water quality is to transport some of the two billion pounds of litter generated annually to the Delta region of the state. A program creating a Poultry Litter Marketing and Utilization Project led to the transport of more than 100,000 tons of litter from western to eastern Arkansas in 1993.

Maryland: On the Eastern Shore of Maryland, as well as in other parts of the Delmarva Peninsula, the poultry industry is highly concentrated. About 517 million chickens are raised annually; broiler houses average 50,000 head. Approximately five tons of manure for 1,000 chickens are produced per year. According to Lew Carr and Herb Brodie of the University of Maryland, annual by-products include 646,250 tons of poultry litter; 24,800 tons of mortality; 12,480 tons of hatchery waste; and 15,510,000 gallons of dissolved air flotation skimmings.

Alabama: The average poultry farm in Alabama produces 180,000 birds per year in six to eight week cycles. As it leaves the broiler house, each bird leaves behind about five pounds of litter composed of its droppings and the bedding spread out to absorb it. In Alabama this is equivalent to about two million tons.

Coordinated approaches to improve knowledge of the composting process

A significant approach to documenting and communicating critical information about composting was taken by a consortium of United States government agencies and nonprofits on the subject of bioaerosols and their impact on public health. The organizers of a special workshop in January, 1993 included the U.S. Department of Agriculture, the U.S. Environmental Protection Agency, the Composting Council, and the National Institute for Occupational Safety and Health. The workshop was held at the Soil Microbial Systems Laboratory of the U.S. Department of Agriculture in Beltsville, Maryland. The specific question addressed by participants was: 'Do bioaerosols associated with the operation of biosolids or solid waste composting facilities endanger the health and welfare of the general public and the environment?'

Dr. Patricia Millner of the Laboratory edited the final report and prepared the following summary:

'The 25 scientists and engineers drawn largely from regulatory and research agencies attempted to examine the full spectrum of potential bioaerosol agents and impacts, including actinomycetes, bacteria, fungi, arthropods, protozoa, and organic constituents of microbial and plant origin and not just those that might arise from the fungus *Aspergillus fumigatus*.

"To the best of our knowledge, this was one of the first attempts at viewing the comparative health impacts of such a broad spectrum of bioaerosols from different sources of decomposing organic materials (e.g. grass clippings, wood chips, food and household wastes, agricultural wastes, and biosolids) in the environment. As such, the report on this effort helps establish a scientifically reasoned basis for evaluation of health impacts from bioaerosols associated with the processing and handling of biologically degraded materials at composting facilities compared with other sources, and helps set the stage for future advances in knowledge about this important subject.

'Several conclusions reached by the working group included:

1) The general population is not at risk to systemic (i.e., whole body, generalized, as in circulatory, lymph, etc.) or tissue infections from compost associated bioaerosol emissions.

2) Immunocompromised individuals are at increased risk to infections by various opportunistic pathogens, such as *A. fumigatus*, which occurs not only in compost but also in other self-heated, organic materials present in the natural environment.

3) Asthmatic and 'allergic' individuals are at increased risk to responses from bioaerosols from a variety of environmental and organic dust sources, including compost. *A. fumigatus* is not the only or even the most important bioaerosol of concern in assessment of risk for ODTS, MMI and HP (extrinsic allergic alveolitis) associated with exposure to dust from organic materials. The amounts of airborne allergens that sensitize and subsequently incite asthmatic or allergic episodes cannot be defined with current information available, especially given the wide variation in host sensitivity, the numerous sources of natural environmental exposure, and the diversity of constituents and bioaerosols. Prospects for such precise definition are limited in the short-term because of these factors.

4) In spite of the fact that some types of bioaerosols can cause occupational allergies and diseases, and that some of the same types of bioaerosols are present in the air at facilities that compost organic materials, available epidemiological evidence does not support the suggestions of allergic, asthmatic, or acute or chronic respiratory diseases in the general public at or around the several open air and one enclosed composting sites evaluated.

Hence, the answer that emerged to the question posed at the beginning of the workshop is: *"Composting facilities **do not** pose any unique endangerment to the health and welfare of the general public."*

This approach of establishing an objective consensus of well-founded research views that relate to composting will be used frequently in the future. It is predicted that the results of such efforts will increase the acceptance of composting as a major waste management practice throughout the world.

Bioremediation projects using compost

Bioremediation of soils contaminated with pesticides, hydrocarbons, explosives and other pollutants by application of the composting process is rapidly advancing in the United States. While much data being accumulated emanate from anecdotal reports of private companies, research at university centers is verifying the effectiveness of compost methods. The USDA Soil Microbial Systems Laboratory has reported, for example, its research for appropriate disposal/detoxification of pesticide containing waste at remote border sites. Reports the USDA research team:

'Biodegradation studies were conducted using soils from cattle dipping waste pits that were contaminated with high levels of coumaphos. Results showed that the coumaphos in all of the soils could be rapidly biodegraded in soil slurries using indigenous soil microorganisms.'

At two U.S. military installations in the Northwest, large-scale tests are continuing to use compost to remediate soils contaminated with explosives. Windrows are made with dairy manure, wood chips, potato waste, alfalfa and 30 percent contaminated soils. Water is added and each pile composts for about 30 days. The demonstrations are being conducted at the Umatilla Army Depot in Hermiston, Oregon and the U.S. Naval Submarine Base site in Bangor, Washington. At Umatilla, treatment reduced initial average contaminant concentrations of 1,574 ppm TNT to 4 ppm; 944 ppm RDX to 2 ppm; and 159 ppm HMX to 5 ppm in around 40 days. In Bangor, bench scale studies demonstrated that composting reduced concentration of TNT in one kilogram of soil from 822 ppm to 8 ppm after 60 days of treatment. According to Harry Craig, EPA Region 10 in Oregon, a study estimates treatment costs at 40 to 50 percent less than on-site incineration for quantities from 1,200 to 30,000 tons.

At the Seymour Johnson Air Force Base in North Carolina, a method was developed for on-site bioremediation of soils contaminated with petroleum products.

The method uses windrowed layers of 75 percent contaminated soil with 20 percent yard trimmings compost and five percent turkey manure. The base uses a compost turner for thorough mixing; windrows are covered with a vinyl-coated nylon tarp for approximately 30 days.

In 1985, scientists at Utah State University, led by Dr. Steven Aust, began successfully using white rot fungi (WRF) *Phanerochaete chrysosporium* to degrade hazardous materials such as pesticides and PCBs. White rot, a naturally occurring degrader of lignin, commonly is found in compost piles. Aust intensively studied the fungi, determining that the easily grown organism could mineralize complex organic chemical compounds into carbon dioxide. The research led to formation of Intech One-Eighty Corporation which has an exclusive licensing arrangement with Utah State to set up commercial applications of WRF. Currently, Intech is considering combining its white rot fungi technology with the type of tunnel composting originally used in the mushroom industry.

Other companies in the U.S. are involved with using the compost process to clean up soils. In Rouge River, Oregon, J.A. Pinckard – professor emeritus in microbiology from Louisiana State University – writes that his firm, Bioremediation Technology Services, has been working with such companies as Vicksburg Chemical (Vicksburg, Mississippi) and National Helium (Liberal, Kansas). According to Pinckard, his research in the 1940s at LSU demonstrated that rain spoiled legume hay buried under cotton rows suppressed soilborne diseases of cotton. His early work led to use of agricultural wastes for cleansing soils in Florida, California, and elsewhere.

'A number of agricultural wastes may be composted to build consortia of microorganisms which may then be applied to a contaminated soil for cleansing it,' writes Dr. Pinckard. 'We chose to use cotton gin trash because of its consistent suitability for our purposes.' For the past six years, his company has been marketing a product known as BTS Soil Amendment.

Current research projects focused on use of varied feedstocks

Reflecting the increased awareness of the need for reliable data on combining different organic feedstocks, researchers throughout the United States are focusing on specialized projects. Following is a brief description of some of this research:

University of Maryland – Leslie Cooperband at the Wye Research and Education Center directs a project that combines municipal solid waste compost and poultry litter to create a slow-release organic fertilizer. The project will analyze the compost characteristics of varying combinations of these materials and the nutrient release characteristics when applied to farm fields.

Tennessee Valley Authority – Improving fertilizer quality of broiler manure is the goal of research which is specifically designed to demonstrate ways to use litter for fertilizer, feed and potting mixes.

Agricultural use of municipal wastes

The increased emphasis on agricultural utilization of organic residuals from municipal, industrial and animal sources is accelerating the flow of composted products to the marketplace. These trends are clarifying specific research needs to obtain well-documented data on volumes and characteristics of feedstocks, and their impacts on crop production and soil conservation.

Recently, the United States Agricultural Research Service (ARS) drafted a list of research needs to be addressed in order to 'ensure efficient and environmentally safe utilization of readily available waste materials.' Following is a summary of the ARS list:

A national data base listing the amounts produced and the agronomic characteristics of major wastes generated; Analytical methods to estimate the levels of nutrients and toxic components in wastes and amended soils; and Assessment of the fate and effects of trace elements, synthetic organics and pathogens in wastes on soils, plants, animals and humans. (A risk assessment pathway approach similar to the one used to develop regulations for land application of biosolids will be needed, note the ARS scientists.) Approximately 75 percent of the nitrogen in animal wastes is lost before it is available for crop use. Appropriate research would improve the understanding of basic chemical and biological processes in wastes and waste mixtures, resulting in designs for storage and surface application to minimize losses of objectionable gases and bioaerosols. A clearer understanding is needed of such factors as aeration, temperature, water content, inoculation and mixing on levels of pathogens, beneficial organisms and viable weed seeds in composts.

Research is needed to blend, mix or cocompost different wastes to yield final products with desirable characteristics for agricultural or horticultural uses. According to ARS staff, information on the concentrations, chemical reactions and bioavailability of beneficial and potentially hazardous components of wastes will be needed to develop mixing and composting procedures which can eliminate pathogens and toxins, reduce availability of toxic trace elements and enhance nutrient availability in 'designer waste' end products.

Anecdotal evidence on the multifaceted beneficial results of compost applications on crops and soils has traditionally been a significant factor. The personal experiences of farmers, commercial plant growers and backyard gardeners have played a key role in sustaining interest in compost use for many decades.

An example 'case study' is illustrated by Jack Pandol, Jr. of Pandol & Sons, Inc., who last year became Undersecretary of California's Environmental Protection Agency. Pandol and Sons farms 6,000 acres of vegetables, tree fruits and grapes in California's San Joaquin Valley. They began applying a compost made from manure and cotton gin trash four years ago, seeking to reduce pesticide use by 50 percent over a five year period, and actually achieved an 80 percent reduction in three years.

Flynn Rainbow Nurseries of Fallbrook, California grows 250 acres of container plants – six to 10 million annually – which are marketed throughout the United States. According to nursery owner Mick Welti, compost has been the significant factor in decreased use of commercial fertilizer and prevention of nitrate runoff into nearby streams. Yard trimmings compost at up to 50 percent of the container mix has allowed the nursery to stop using a sterile growing medium 'that needs to be fed fertilizers and protected with fungicides,' notes Welti. 'Using compost reduces the fungicide bill quite a bit. And while I can't quantify it, our water usage has been reduced as well. There is less nitrogen runoff – which is a major benefit.'

As this paper indicates, there has been a steady increase in the number of research projects in the United States – as well as throughout the world – that are documenting the practical experiences of growers concerning the methods needed to produce a consistently high quality compost and the beneficial effects of using such materials.

The Present Situation and a New Trend on Composting in Japan

MAKOTO SHODA – Research Laboratory of Resources Utilization, Tokyo Institute of Technology,4259 Nagatsuta, Midori-ku, Yokohama, Japan

Abstract

Recycling of municipal wastes has been attracting attention and some practical recycling system has been established in Japan. The trend in disposal methods of municipal wastes in Japan will he explained. The garbage is an appropriate material for composting and thus, many composting plants have been constructed for the past ten years. The fever of the compost production by using garbage, however, is cooling down, mainly because of high construction and operation costs, quality change of municipal waste, contamination of plastics, ill-treatment of malodorous exhaust gas, etc. Characteristics and problems of a large-scale composting plant at Tokyo Metropolitan Government will be a typical example of the big city composting facility.

At present, a small-scale home composter is getting popular especially in local areas. Some local governments prepare a subsidy system to promote the usage of such a composter at each house. The aim of this promotion by local governments is to reduce the amount of garbage disposed from each house.

There are two types of composter. One is a simple plastic bin-type container which is placed in the outdoors. Another one is a new electric type equipped with mixing, heating, and amended with bulking agent to accelerate composting reaction rate. This electric one is also operated outdoors. Although there is a need of a lot of modification of the composters for further proliferation, this trend will be one possibility to loose burden of the waste treatment expenses in local governments.

Introduction

The treatment of municipal wastes is a serious problem in many big cities in Japan. The main purpose of controlling waste discharge in local governments is focused on discharge suppression and reutilization of renewable wastes. For recycling of the wastes like cans, glasses and papers, many practical achievements are materialized. On the other hand, the garbage or organic wastes produced is considered to be an appropriate material for composting and composting facilities have been built in several cities to reduce waste load to incineration facilities. However, the expectation for composting is fading away recently, because concrete step to solve the waste disposal problem is not established yet. Recently, an electric or bin type home composting equipments are attracting attention and some local governments are promoting such home-base composting as alternative to the large scale composting operated in local government.

The demand for compost

The total demand for organic matter in Japanese farmland is estimated to be about 2.5 million tons on dry basis. The organic matters produced from cattle farming in Japan reach about 2.8 million dry tons. This data indicate that if these organic wastes are properly treated to composts and recycled, enough amount of composts can be produced by using the cattle excreta alone. However, the number of cattle farmers are decreasing due to low price products imported from U.S.A. and Australia and their farming areas are isolated from ordinary farmland and housing areas due to environmental problems. This eventually cut off the recycling system of organic matters which was established previously between them. Thus, as an alternative of organic matters from cattle farming, municipal wastes and organic wastes from food industry are of attention for compost production.

On the other hand, intensive supply of inorganic fertilizers into land reduced the fertility of land and this caused the frequent occurrence of plant diseases in farmland. This introduced the excessive utilization of chemical pesticides or fungicides, which caused the breakdown of ecological system. In this sense, the introduction of organic fertilizer or composts to the land is being promoted.

Hence, the demand for composting is getting popular.

Municipal waste composting

Large-scale composting facilities

In 1991, the total amount of municipal waste produced was about 46 million tons and the waste output per person per day was about 1.1 kg. The municipal wastes are disposed of by incineration, direct reclamation, recycling as resources, composting and others. The chronological change of these treatment methods is shown in Fig. 1. About 73 of the municipal wastes are treated in the incineration, 23 are reclaimed and others including composting are only 4%. Iron, aluminium, glass etc. recovered as resources from collected municipal waste are recycled and the recycled amount is about 3.4 in I 990 of total amount of disposed waste. This means that the actual percentage of composting is only 0.2% of total amount of disposed waste. This trend is almost constant in these recent years.

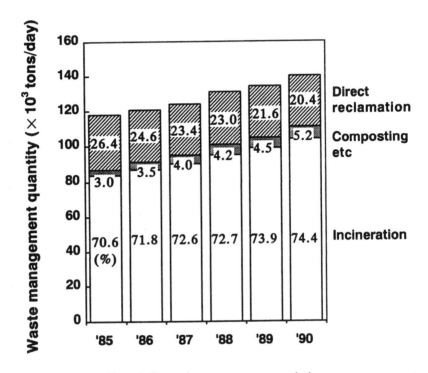

Figure 1 Changes in waste management methods

The percentage of municipal garbage in total amount of municipal wastes is about 24 % in average and the water content is 70%. Thus, 3 million tons of organic matter is produced on dry basis and can be used for composting. In 1976, a big boom occurred in construction of composting facilities of municipal wastes. However, this boom faded down due to ill-separation of waste materials and thus strong refusal of use of such contaminated compost in agricultural areas. In 1986, the second boom occurred and 39 facilities were constructed and at present only 29 facilities are in operation as shown in Fig. 2. There is a big difference between the actual numbers of composting facilities and the number in operation. Only 70% of the facilities are in operation.

The Tokyo Metropolitan Government constructed the composting plant in 1985 with construction cost of 0.73 billion U.S.$.

The initial specification of the composting plant was as follows:

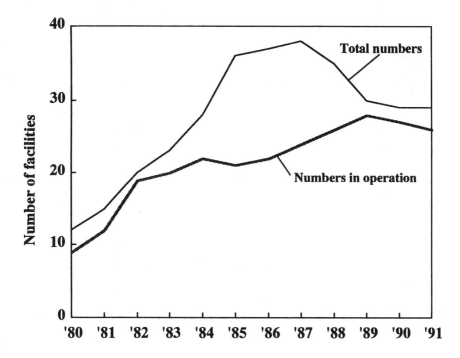

Figure 2 Number of municipal waste composting facilities in Japan (Kimura, 1991)

Capacity of treatment of municipal wastes: 50,000ton/day, Production capacity of compost: 5 ton/day, Composting time: 30 days, Supplementary equipments: mechanical sorting machine, compost pelletizer, chemical deodorizing plant etc. As shown in Table 1, the productivity of compost was fairly lower from the initial operation and maintained at lower level. The following problems were considered to be the main reasons for this low productivity.

Table 1 Composting Operation at Tokyo Metropolitan Composting Plant

	1985	1987	1989	1991
Municipal wastes delivered (ton)	14,620	11,953	13,801	14,282
Municipal waste composted (ton)	3,317	1,589	3,060	4,087
Compost produced (ton)	560	340	288	442
Compost productivity (%)	3.8	2.8	2.1	3.1

(1) Change of quality of wastes.
The municipal waste consists of 88 % of combustible refuse and 12 % of incombustible refuse. The former includes garbage 5.7%, papers 33.4%, plastics 21 .4%,

wood and bamboo 8.4%, fibers 5.0%, and others 14%. The latter contains stones 4.6%, shell and eggshell 3.6%, soil and sand 2.7%, and metals 1.2%.

The wastes delivered into the composting plant are the mixture of all municipal wastes mentioned above and the relative ratio of plastic bags, cardboard, plastics, woods etc., became higher compared with garbage and the separation or sorting of these are incomplete. This insufficient separation caused frequent clogging in the machines and thus mechanical trouble and breakdown. This is one of the main reasons in low operation efficiency.

(2) Quality of compost.
The increased content of papers, and plastics like foaming polystyrene in municipal wastes led to the deterioration of quality of the compost.

(3) Corrosion of the plant.
The evolution of high concentration of malodorous gases like H_2S, ammonia and other gases and moisture accelerated corrosion of machines and conveyers. Thus repair expenses increased and operation was suspended frequently.

(4)Inefficient removal of malodorous gases evolved during composting.
The chemical method(acid-alkaline treatment of gases) which was constructed for the treatment of exhaust gas from the plant was inefficiently operated mainly due to higher gas concentrations evolved. Charcoal adsorption treatment column was installed, but the moisture in the exhaust gas reduced the removal efficiency of gases, and another supplementary charcoal column was installed. The operation cost of the treatment hindered the efficient operation.

(5) Cost for operation.
The production cost of compost which excludes depreciation cost of machines was as high as 460 $/ton.

(6) Demand for compost.
The compost was planned to be used for agricultural areas, but the high cost of the land and urbanization in Tokyo area made farming in Tokyo area impossible. Thus, the main users of the compost were citizens in Tokyo and the demand for the compost was unstable. The final decision to stop the operation due to the problems mentioned above was made in 1993.

However, the quality of the compost was fairly good. The content of ingredients like glasses, plastics, etc. was less than 1 % of unit dry weight of the product. The growth of the vegetables in the farmland where the compost was introduced was 25 % higher than those treated with chemical fertilizer or cow-dung compost. The contamination levels of heavy metals like mercury, arsenate, and cadmium were below the regulated values. The C/N ratio of the compost product was less than 12, which was the recommended value for the municipal compost.

The most effective point in this composting plant is that this plant played an important role as an educational campaign to demonstrate the importance of recycling in this affluent society to Tokyoites.

Small-scale home composting equipment

As mentioned above, a large-scale composting plant especially in big cities have a low efficiency of production of compost. As an alternative to the large scale composting, small home composters are being popularized recently. There are two types of small composting equipments. One is a bin-type plastic composter and the other is an electric composter equipped with mixing and heating. In some local governments, the subsidy system is introduced for the local residents to promote the use of these home composters.

More than 2 million units of simple bin type composters(100-300 liters in volume) which cost in the range of 50 to 150 U.S.$ have been reported to be sold in Japan. This number is almost equivalent to 20 % of houses where this composter is available. However, the statistical data which showed the reduction of disposed garbage in proportion to the sold number of composters are not fully collected. However, local governments hope that the proliferation of these composters will eventually loose burden of the waste treatment expenses in the local governmental budget.

Commercial appearance in market of electrical composters is a new trend in Japan. More than 30 kinds of such composters are in markets or in planning to make a debut on market. As the variety of these instruments are so large and the new trend has just Jstarted, it will take time to evaluate the effect of the machines. One standard characteristics of the composter is as follows:

The machine generally is equipped with mixing,aeration and heating and this will accelerate the composting reaction and reduce the time of composting.

The size is 30cm in width × 44 cm in depth x 80cm in height and composting rate is about 1 k~ay.

Separation of shells, eggshells or plastics is essential and supposed to be conducted before introducing the garbage into the reactor. The base bulking materials consisting of sawdust, or woodchips or other solid materials with a large surface area are prepared and about 30 kg of them are placed in the composter beforehand. Some machines supply a kind of seed containing mixture of microorganisms prepared by each manufacturer. When the garbage is put into the composter, mixing and aeration automatically promote the degradation of garbage. The exhaust gas is expelled into outdoors through a deodorizing treatment unit. When the outdoor temperature is below 10 C, the heater installed inside the reactor operates to keep the optimal temperature for microbial degradation of organic matter. As most of the inside materials are the bulking agent prepared beforehand, almost complete degradation of organic matter is proceeded. The bulking agent is replaced with a new one every 4–6 months.

These composters cost in the range of 500 $ to 1 800$.

By the survey conducted by a local government on the usage of these machines by local residents reported the following problems.

1) The treatment capacity for one family is about 500 g to 1 kg per day. When the amount is too small or too large, the composting is of low efficiency.
2) The evolution of malodorous compounds is inevitable in composting. The electric composter has a deodorizing unit, but still the odor problems cannot be solved completely.
3) In summer, the operation is fairly well but in winter, the degradation rate is lowered due to low atmospheric temperature.
4) The newly designed composter is not like a rice cooker or microwave ovens because care or personalized attention is essential in order to maintain efficient microbial reactions. This is a burden for city people who are busy and have a job.
5) The buyers for these composter are at present restricted to the people who own an individual house with yards or gardens where the composter can be set and the composted product can be treated in their own yards. This means that the machine is still insufficient for indooor usage which is essential for people who live in apartment houses.

Future's aspect

There will be several future's aspects to be solved if composting is properly conducted.

i)   In a large-scale composting in big cities, the rational design and economical operation are needed based on source separation,and the materials used for composting should be restricted in quality.
ii) A medium-scale composting reactor should be developed for a complex of apartment areas.
iii) Compact and inexpensive home-composter, especially indoor-types should be manufactured. (iv) Development of efficient deodorizing methods are urgent not only for a large-scale but also a small-scale composting.
v)   For recycling of organic matter, as well as technical development, system analysis and system synthesis are essential by the cooperation among local government, private companies, residents who produce municipal wastes and the users of composts and optimal treatment system should be established.

Composting: Experiences and Perspectives in Brazil

JOAO TINOCO PEREIRA NETO[1] – Universidade Federal de Viçosa, Brazil

Abstract

The present reports on many aspects of composting in Brazil with particular emphasis on the main problems which contributed to the discredit of this important process in many areas of the country.

Consideration, is made of the constant increase in the use of selective collection which can persuade facilities to increase the use of composting in waste management.

Finally, aspects of solid waste composting in Brazil, related to its perspectives, the low cost techniques and recycling are considered.

Introduction – The Main Constraints and Contradictions

Brazil produces an average of 90,000 tonnes of re~se per day, of which approximately, 80% is disposed of in open dumps. Generally only the major cities have a regular system of collection followed by uncontrolled landtilling or badly managed composting plants.

Open dump disposal, as well as the use of landfills, has led to several problems for the environment and public health with considerable impact on the poor communities surrounding the major cities.

The interest in composting arises from the fact that the refuse produced in the country typically contains an average of 60°/° putrescible organics (deteriorated food, vegetables and other organics). This amount of putrescible material when disposed of in open dumps or uncontrolled landfill, forms a natural 'habitat' for disease vectors such as, flies, cockroaches, rats, 'vultures, dogs, etc. The effect of these is enhanced with a population which suffers from malnutrition (25% of the population in Brazil). Another problem associated with the organic putrefaction on these sites is the great amount of lechate generated, often flowing out to streams used for domestic services, water for animals, irrigation, etc.

[1]Professor at the Universidade Federal de Vicosci-MG, Brazil of the University Public health Section, Consultant of WHO/UN (joint FAO/UNEP/UNHCS)

An additional facet to this complex problem is the decrease in agricultural performance as a result of the decrease in soil fertility due to lack of organic content. To compensate for this loss of fundamental fertility there is an increasing demand for expensive inorganic fertilizers. The high rate of loss of organic matter from soils in tropical areas is a particular feature of the hot climate. The effect of reducing soil productivity is further magnified due to the growth of population which means more people to feed and more pressure on the exhausted soil.

These many factors related to the uncontrolled disposal of waste and the need to improve soil productivity gives rise to a situation where low cost composting come into its own.

The Problem (A Composting Rush)

As soon as it was seen that composting could be an answer to the domestic solid waste problems in Brazil a big rush occurred and composting was seen as the solution for all sanitary and environmental problems related to it. However, unfortunately, it was also seen as a technique worthy of exploitation by greedy speculators and suppliers. This fact brought to Brazil several technologies, mostly emphasising only the equipment side (enormous industrial electromechanical plants) and neglecting the biological side of the process.

Some systems claim to be able to solve the whole problem associated with the solid waste using only composting plants. Others promised to compost the organic fraction of the MSW in only (4) days. Both of these claims were of course untrue. Other plants, locally built (using adapted technology), also neglected the biological side of the process. As a result there spread in the country, a strong 'business' based on composting using inappropriate technologies which formed the basis of large scale projects, and several systems were built based on the wrong concepts about composting. This fact lead to several poor quality products on the market, reducing the credibility of the process in the country. A fact that aggravated this problem was the lack of regulations and legislation for composting, which meant that any inert and dark organic material could be called 'compost'.

The Main Composting Systems in Use in Brazil

At present Brazil has 74 composting plants spread throughout the country, 19% of these are large fully mechanised systems (150 to 1,200 tonnes/day), of which 8 are Dano, 3 Triga, I Bartolomeis, I Danbraz and 1 Viçosa. The others (81%) are technologies adapted locally called 'simplified systems' and basically consist of ditches (for waste reception), conveyor belts (for hand sorting), a crusher (for particle size reduction) and a screen for the final product (to take off the rejects).
A new system which seems to have been well accepted in the country, for mainly small communities (5–20 thousand inhabitants) has been developed by Viçosa

University, which has already built two plants. It is a low-cost composting plant which can be built in three months, using a range of available construction materials such as: bricks, concrete, blocks, soil cement etc. These plants only use man power (typically 6–10 operators) and consist of a refuse reception area (small sloping concreted area), a concrete 'table' for sorting and a composting pad. A small shed is used to store tools and recyclables, and each site has toilets and an office (1).

The Plant Operation

In the case of the 74 composting plants previously mentioned, a current problem is the plant operation. In general all these plants are associated with strong odours, vector (flies) attraction and lechate production. On the compost pad the material is found under anaerobic putrefaction and the composting time is no longer than 30 days. Maturation has never been considered as part of the process.

In the plants using the D'Anio drum systems at the front end with a few exceptions, it is believed by the operators that after an average time of 3 days in the drum the organic is composted and, in many plants, the material is sold named as 'compost' or 'raw compost' (2). The Triga systems has been really the worst in terms of sales promotion and process claims one. In the advertising of it's Brazilian associate they affirm that the 'Triga process makes compost from MSW in only four (4) days.'

This aspect shows how developing countries are suffocated with imported and inappropriate technologies, some of which are not used in their countries of origin. The plants built under the simplified system using an adapted technology, which are the most used plant in the country, although presenting an interesting concept also lack an understanding of the biological needs of the process, and the great majority have not been properly operated. As a result of this, approximately, 40% of the compost plants in Brazil, are closed.

A Good System to be Used in Brazil

A considerable amount of research work carried out by the Universidade de Viçosa-UFV, on several composting plants running in Brazil using different systems has shown that these plants do not produce a stabilised and pathogen free end product.

A comprehensive study was made by UFV on a Dano plant located in Belo Horizonte city (2). The findings were that the material (out of the drum), sold as compost, even though it was well homogenised by the drum action, was far from a stabilized and sanitized material. The pH remained acid (5.8), the final volatile solids was 71% (the same as the initial value) and the bacteriological contamination was considered high (E. coli 10^7 c.f.u/g).

Another study was made (with the support of the Universidade de Brasilia and its

bacteriological laboratory) of the Brasilia composting plant which uses Triga technology. The results showed that the material obtained, out of the 'hygienizer silo',
supposed to provide a pathogen free product, also had high levels of contamination
(e.g. E coli levels between i0~ and 10- cf u./g). In both cases the material presented clear evidence of anaerobiosis (3).

This work carried out by Universidade de Viçosa has as its sole objective to
contribute to the science of composting, and the work was not funded by any rival
composting company. This work represented part of a research programme on
recycling and composting, which began in I 987, under an agreement with Leeds
University in the UK. The topics studied, included the following:

1. methodologies for refuse and organic waste characterization;
2. sampling and monitoring process methodology;
3. low cost composting technologies for domestic waste;
4. low cost composting technologies for agricultural organic wastes;
5. forced aeration composting using blowing, sucking and a hybrid mode of
   aeration
6. vector and lechate control;
7. pathogen survival studies;
8. maturation and humification of organic wastes
9. use and application of organic compost;
10. operational procedures in small and large scale operations;
11. extension courses for technicians and plant operators;
12. planning, design, construction and building of recycling and composting
    plants, and
13. technical assistance to farms and city councils.

From the LFV experience of composting it was evident that the best system for
composting the organic fraction of MSW in Brazil is the windrow. It is a very suitable system for Brazil because of' the average city's population (60,000 inhabitants) and the fact that land and labour requirements are not a problem. The
treatment of the organic matter can be done by a simple system mentioned earlier, which is able to provide a good material to be composted with particle size
around 20-50 mm. Once on the composting pad, the pile can be constructed and
turned, mechanically by an ordinary front-end loader.

In order to avoid the usual problems associated with this process (such as odour,
flies and operational control) a complete investigation was made, where the turning was planned such that the piles remained aerobic and did not reach excessively
high temperatures. Working from this data the most appropriate cycle (compared
with other cycles turning every 5 and every 8 days) was found to consist of turning
every 3 days in the early stages, after which the turning was extended to every 6
days until the maximum core temperature achieved was less than 40°r (a condition
achieved on average at around 75 days). The material was then taken and stored
for the period of maturation or curing (30 to 50 days). The geometrical configuration of the pile constitutes one of the most important factors in relation to the cross

sectional temperature distribution. It was found that this configuration had to be modified. depending on the stage of degradation, to achieve a thermal balance between the heat produced in the mass and the heat lost to environment. This made it possible to achieve a balance point with a maximum temperature of around 60°C. This temperature was found to be good both to enhance biodegradation and maximise pathogen inactivation.

From the extensive research work on this particular system, a methodology was developed (2), which was shown to be more effective than the conventional windrow process. However, the effectiveness will only be achieved by the following procedures:

(i)   the pile should be covered with 'nature compost at each turning until flies and odour are no longer a problem (generally after the first week);
(ii) turning should be carried out every 3 days for the first 30 days;
(iii) the water content should be in the range 45-55% and adjusted if necessary during turning;
(iv) the operating temperature should be kept in the range of 55-65°C in order to speed up breakdown by making appropriate changes in the pile cross section during the turning process; and
(v)   the composting must be carried out in two phases, the end of the first phase being achieved when the pile core temperature does not rise about 400C, following this the maturation phase which should take place in static piles (unturned for 30-50 days or more).

Using these procedures the following observations were typical:

(i)   the average reduction in volatile solids during composting (90-130 days) was around 45%, considered to be high for this type of turning process. The C/N ratio starting between 38 and 45 showed a final value around 12 reflecting a high degree of degradation and humification; and
(ii) with respect to pathogen inactivation the previous results were confirmed about the difficulty of elimination organisms during the first phase of the process (4,5,6) however, in the subsequent phase of maturation a satisfactory sanitisation was achieved (E. coli <102 c.fu./g from a starting value of more than 10^6 cfu/g).

Conclusions

Composting is recognised as one of the oldest method for biological waste treatment. It is also one of the most effective process for recycling organic waste intended for use in agriculture. Brazil seems to be an ideal country for the use of this system due to its waste disposal problems and the soils need for organic material. However it must be recognised that more effort must be made in order to achieve better dissemination of experience in research and indicate information on successful projects.

Table1 Characteristics of the MSW from Belo Horizonte and Vicosa Brazil.

Component	Units	Belo Illorizonte*	Vicosa**
Organic Material	% total wt	54.06	81.10
Paper	%totalwt	12.50	5.28
Cardboard	% total wt	6.00	3.02
Plastic	% total wt	5.90	4.28
Metal Cans	% total wt	2.70	2.13
Glass	%totalwt	3.15	1.91
Fabric	% total wt	6.20	0.85
Rubber	% total wt	1.07	0.12
Metal	% total wt	0.60	0.03
Wood	% total wt	1.30	0.58
Leather	% total wt	1.09	0.08
Ceramic and Stones	% total wt	I .60	0.20
Soil	% total wt	2.50	
Other	% total wt	I .33	0.42
pH	–	5.30	5.10
Volatile Solids	% d.w.	72.00	79.20
Carbon	% d.w.	38.00	43.50
Nitrogen	% d.w.	1.03	1.31
P2 0-	% d.w.	1.30	1.50
C~ ratio	–	37.00	33.00
Faecal Streptococci	c.fu/g	7 x 10-	3×100
Total Coliforms	c.fu/g	2 x 108	5×10-
E.Coli	c.fu/g	–	$3 \times I0$~
Density	kg.ml	254	310

* Industrial city ($1,5 \times 100$ inhabitants)
** Rural city ($5 \times ION$ inhabitants)

The M SW characteristics, shown in Table 1, indicate the great potential for recycling and composting. The agricultural potential for compost, the climatic favourable conditions, the availability of land and labour makes composting an important process for use in developing countries. In these countries the use of composting will not only recycle organic wastes, but provide treatment for a great amount of contaminated materials which brings enormous benefits for poor communities such as:

1. a product which is stabilised and bacteriologically safe and can be used as organic ferti1iser;
2. it breaks the disease cycle related to the MSW, improving public health;
3. it eliminates pollution associated with MSW and rural wastes which enhances the environmental quality of the community;
4. it helps soil fertility and hence agriculture productivity which helps to improve nutrition, and
5. it improves different social aspects in terms of opportunities to obtain a better standard of living.

The potential for the use of composting in Brazil can be promising if a great effort is made, mainly by the government, with policy and legislation to cover the control of the system, its operation and its final products.

In Brazil, more than 30 Municipalities are using a source separation system (Table 2) and valuable components have been reused through industrial processing. This fact will certainly contribute to the improvement of the use of composting in the country.

References

I. Pereira Neto, J.T. (1994) 'The Role of Recycling and Controlled Composting in Waste Management for Low Income Areas in Developing Countries' Paper presented at the European Conference on Sludge and Organic Waste - Wakefield, –. In the Proceedings of the Conference.
2. Resende, A.A.P and Pereira Neto, J.T. (1993) 'Estudo e Avalia~ao da Eficiencia de Uma Usina Dano de Compostagm: Processo de Producao do Composto' Presented at the 17th ABES Conference. In the Proceedings of the Conference.
3. Internal Report of LFV (1989) 'Laboratorio de Engenharia Sanitaria e Ambiental - LESA– Vicosa –MG.
4. Pereira Neto, J.T., and Stentiford, E.l. (1989). 'A Low-Cost Controlled Windrow System'. In the Proceedings of the International Symposium on 'Compost Recycling of Wastes', Athens, October 1989.
5. Stentiford, E.l. Taylor, P.L., Leton, T.G. and Mara, D.D. (1985). 'Forced AerationComposting of Domestic Refuse and Sewage' Journal of the Institute of Water Pollution Control, Vol.84, No 1, p,23-32.
6. Pereira Neto, J.T., and Stentiford, E.l (1986). 'Comparative Survival of Pathogen Indicators in Windrow and Static Pile Compost Systems'. In Compost-Production Quality and Use, Elsevier Appl. Science, pp.276-195, Udine, Italy.

Structural Changes to a MSW Composting Plant in Accordance with Modern Waste Management Concepts in Italy

Dr. Ing. DIETER SCHÖNAFINGER– Environmental Engineering Consultants Innsbruck – Bolzano – Milano Defreggerstr. 18, A–6020 Innsbruck

Summary

Since mid-1993 an already existing conventional MSW (municipal solid waste) composting plant (serving 80.000 population) is being redesigned as a center for recycling and treatment. This comprises not only the construction of a biowaste composting plant according to the latest state the art in engineering for the recovery of reusable, separately collected organic components of domestic waste, but also the stabilization of residual waste according to the concept of 'cold pretreatment'. The technical conception of the plant allows capacities for a texture-adapted treatment of sewage sludge, too.

The domestic waste composting facility Sciaves/Alto Adige has been in use for more than 10 years and is processing approx. 20,000 tpa of MSW. The essential components of the plant are the mechanical treatment (comminution, homogenization, sieving), conditioning and landfilling of screen rejects (facultative production of RDF), and the controlled decomposition of the organic-rich fine fraction which will be deposited or used for remediation measures.

The objective of the new conception and the technical adaption being carried out at present is to treat residual wastes to enable environmentally sound disposal and to treat and subsequently utilize source separated waste streams (compost, 'dry recyclables'). The result will be a modern, integrated waste management center with several lines of processing.

Most of the already existing equipment can be further used for the processing line 'cold pretreatment' and only few modifications of the process have to be carried out. Since the technical optimization of the process is mainly based on a simplification of the procedure it can also be regarded as a 'low-cost-treatment'. It is envisaged to reduce degradable components of non-utilizable wastes (no incineration plant is projected medium term) by biologic treatment before landfilling and thus establish an inactive state of the material.

On the basis of a preliminary study on the large-scale separation of organic wastes from industrial and domestic sources, an integrated composting facility was planned within the existing plant, which is destined for the sole production of well marketable quality compost. During the determination of the optimal treatment procedure two crucial settings were considered by the use of semimobile and modular enlargeable plant components: firstly the slowly raising amount of biowastes in the next years and secondly the need for a flexible adaptation on varying input materials. Thus it is also possible to selectively treat less contaminated sewage sludge as well as the organic fraction of domestic waste. The central part of the facility is an intensive rotting phase in an enclosed environment which guarantees an optimal process control and a minimization of emissions (especially odour).

The presented project can be regarded as a model for the adaptation of existing, but technically not up-to-date biological waste treatment systems to facilities which are suitable to modern waste management strategies. It shows that realistic approaches for development are applicable to many disposal regions in Italy which are comparable to the described situation.

In Alto Adige esiste l'ordine politico di riutilizzare i rifiuti urbani biogeni dopo averli adeguatamente preselezionati presso il produttore iniziale. L'idea fondamentale di questo modo di procedere viene seguito già da tempo dalle Comunità di Val d'Isarco e Alta Val d' Isarco che gestiscono dal 1985 presso Sciaves un impianto di compostaggio. Le autorità competenti hanno dato ora l'autorizzazione per adattare questo impianto alle esigenze attuali. Sono ogetto della progettazione sia alcune modifiche nello schema d'impianto per facilitare la gestione (tra l'altro la sostituzione necessaria di alcuni blocchi logori) che la realizzazione di un centro di raccolta differenziata primario. Quest' ultimo comprende una zona di trasbordo per materiale riciclabilie ed una semplice unità per la selezione dei rifiuti ingombranti ed industriali.

Ambito d' azione relativo alla gestione dei rifiuti

Come future destinazioni per un impianto come quello di Sciaves esistono le seguenti alternative:

1. Riorganizzazione per trattare poi i rifiuti biogeni e stoccare i rifiuti residui non pretrattati in discarica
2. *Riorganizzazione per pretrattare ed inertizzare i rifiuti residui e i fanghi di depurazione* prima dello stoccaggio in discarica. Trattamento dei rifiuti biogeni in un impianto a parte.

La seconda variante assume maggior importanza, in quanto l'obiettivo delle richieste legislative è la riduzione delle emissioni come percolato, biogas, ecc. che sono fenomeni tipici legati allo smaltimento dei rifiuti non pretrattati in discarica. Là dove l'incenerimento dei rifiuti (il quale soddisfa ampiamente queste esigenze) non è impiegabile, i trattamenti biologici come quello di Sciaves sono di grande importanza nell'ambito della gestione dei rifiuti se vengono utilizzati per il pre-trattamento dei rifiuti residui o, con una parola di moda, per il 'pretrattamento freddo'. Il vantaggio di un tale 'pretrattamento freddo' consiste nella riduzione per via biologica della parte biodegradabile dei diversi rifiuti residui prima del loro smaltimento in discarica, in questo modo si ottiene un materiale da scaricare con un comportamento non attivo, come nel caso dell' ossidazione termica dei rifiuti. Rispetto lo stoccaggio di rifiuti residui non pretrattati é possibile ottenere con tale tecnica un aumento del peso volume e del tempo utile della discarica ca. del 40%.

Trattamento dei rifiuti residui

Coll' adattamento tecnico di un impianto di compostaggio per rifiuti urbani come quello di Sciaves è possibile soddisfare in larga misura le esigenze suddette. Si tratta di apportare leggere modifiche nelle procedure tecniche, preferibilmente da realizzare tramite appalto concorso. Le modifiche significative, che sono riportate nelle due seguenti figure, riguardano la sostituzione del mulino a martelli logoro con uno a bassa velocitá di rotazione (con questo è possibile triturare anche tra l'altro anche frazioni di rifiuti ingombranti) e la vagliatura preliminare dei rifiuti residui per tener lontano dal mulino le componenti fini responsabili del logora-mento. Per ottenere la riduzione delle attuali emissioni di odori sgradevoli sono previsti diversi interventi tecnici (p.es. incapsulamento della zona di consegna dei fanghi di depurazione, mescolatore rifiuti residui – fanghi). Per l'adattamento della tecnica di fermentazione alle nuove esigenze si offre l'appalto concorso; da prevedere è in ogni caso un processo ampiamente automatizzato. La tabella sot-tostante riporta le quantità del materiale d'ìngresso relativo al trattamento dei rifiuti residui. Tali quantità, considerando di ca. 12 settimane il tempo utile per la matu-razione in cumulo, necessitano di un'area pari a più di 3000 m2; un valore che entra nel valor massimo della superficie utile (3.600 m2) dell' attuale capannone.

Table 1 Pretrattamento freddo Sciaves – sommario quantità rifiuti e volumi

Frazioni di rifiuti	tonn/a	m³/a
Rifiuti residui (pretrattati)	ca. 10.000	ca. 22.000
Fanghi di depurazione (1.500 tonn di materia secca/a, min. 80 % contenuto d'acqua)	≈	
Spazzatura stradale, Materiale sgrigliato e residui da dissabbiatore da impianti di depurazione, terreni contaminati, ecc.	ca. 4.000	ca. 6.000
Totale		
*) volume sommato reale	ca. 23.000	ca. 33.000 *)

Il ciclo operativo riportato di seguito lo si può inserire nel caso dell' utilizzazione di una frazione come RDF; componenti residui non biodegradabili ed ad alto potere calorifico sono RDF ai sensi del modello provinciale di gestione dei rifiuti.

Compostaggio dei rifiuti biogeni

Riguardo questa parte é necessario apporre due condizioni marginali:
1. Le quantità di rifiuti biogeni senza impurità non aumenteranno nei prossimi anni bruscamente ma successivamente.
2. Specialmente per gli impianti di trattamento dei rifiuti biogeni si può notare un rapido sviluppo tecnico.

Ambe due le condizioni marginali suggeriscono di progettare questa parte nel modo più flessibile possibile per ottenere così un adattamento ottimale (riguardo le spese) alle quantità d'input continuamente crescenti e non fissarsi ora su un sistema che forse tra due anni potrebbe risultare superato.

È necessario dunque adattare dal punto di vista edile l'esistente capannone destinato attualmente al raffinamento del compost e al centro di raccolta differenziata ed acquistare inoltre i necessari macchinari (tritturatore, vaglio) indipendentemente dal futuro sistema di compostaggio. Nel progetto é previsto inoltre l'aggiunta di fanghi di depurazione a basso carico inquinante con tutte le dovute misure di prevenzione riguardo l'emissione di odori sgradevoli (unità di fermentazione chiuse con depurazione aria in filtro biologico).

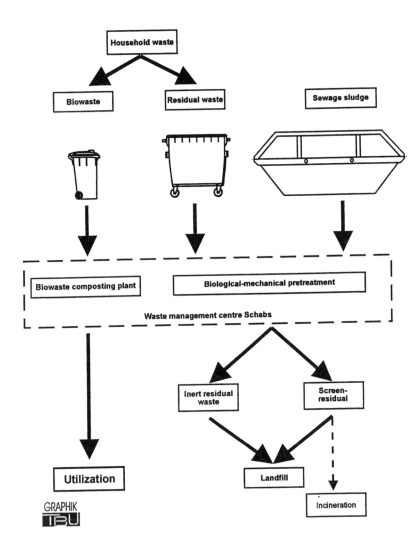

figure 1 Illustrazione schematica del riordinamento del trattamento dei rifiuti nel bacino d'utenza dell'impianto di compostaggio

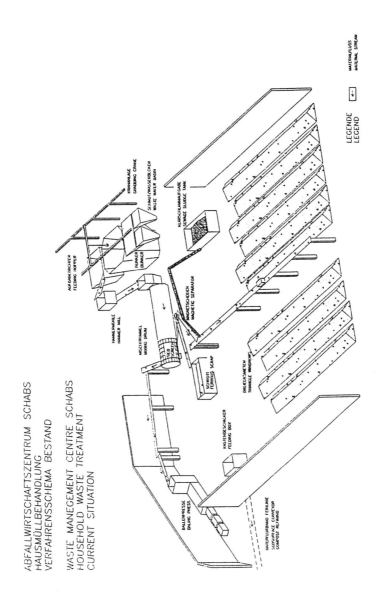

figure 2 Impianto di compostaggio per rifiuti urbani – schema esistente

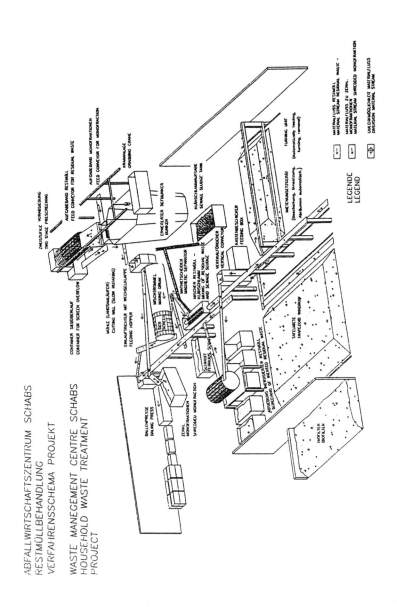

figure 3 Impianto di trattamento meccanico–biologico di rifiuti residui – schema progetto

The Power of Composting; The Power of Partnership.

JAN BEYEA, Ph.D. – Chief Scientist National Audubon Society*, 700 Broadway NY NY 10003, USA 212-979-3073 Fax: 212-353-0508

Partnerships with environmental groups, industry, and the public sector have proven to be a very effective way to advance source-separated composting of residential and commercial organic waste. The partnership concept has provided an opportunity to gather meaningful data and understanding needed to address key environmental, economic, and institutional questions that hinder broader acceptance and implementation of composting. In the United States, two innovative partnerships have shown that collaboration can achieve these goals. One partnership, 'Compost for Earth's Sake,' involves the Grocery Industry and the environmental group, the National Audubon Society, working with local governments on pilot projects investigating source-separated composting. In addition to gathering data on collection performance, economics, and public acceptance, the team is investigating compost contamination from a multi-stakeholder perspective, both by measuring compost quality directly and by projecting long-term soil concentration levels that might result from sustained use. The second partnership, 'Food for the Earth,' is a partnership with representatives from the food service industry, the US Composting Council, and the National Audubon Society. In addition to testing the feasibility of behind-the-counter separation of organics, the food-service team is investigating the options for composting grease, which is an important disposal issue for the industry.

Yard and leaf composting programs are very popular in the United States. The composting of home and commercial organics, however, is in its infancy. The advancement of composting in the United States, beyond yard and leaf composting, requires broad public acceptance. In order to gain that acceptance, it will be helpful if both business and environmentalists support it. This will insure that the same positive message is heard wherever the public turns, helping to forge a social consensus on the value of organics recovery.

In the United States, partnerships with environmental groups, industry, and the public sector have proven to be a very effective way to produce consensus and advance source-separated composting of residential and commercial organic

---

*The National Audubon Society is a U.S., non-governmental environmental organization with a membership in excess of 600,000. Our interest in composting arises out of a desire to get clean, humic-like materials, which would otherwise be buried or burned, back to the land. This, we believe, will help maintain and restore soils, for the benefit of birds and other wildlife.)

waste. The partnership concept has provided an opportunity to gather meaningful data and understanding needed to address key environmental, economic, and institutional questions that hinder broader acceptance and implementation of composting.

No matter how charismatic and persuasive, no individual from one sector can produce broad-based consensus alone. That means individual groups supporting composting should be willing to go to great effort to get consensus from the key actors in composting. Development of consensus requires listening carefully to the needs of the key constituencies and adopting a position that is as inclusive as possible. This can be a difficult path to follow for those leaders who have gotten to where they are by being stubborn and going their own way, swimming against the tide that keeps down new ideas.

Fortunately, there are techniques to build consensus, even in a litigious society like the United States. One of the best conflict resolution techniques, in fact, is the formation of partnerships which include traditional opponents.

How do you get disparate constituencies to form partnerships? Obviously, there have to be some common interests. And there must also be a safety valve, which protects the core beliefs of the constituencies. This safety valve can be an 'agreement to disagree' on issues that unalterably divide the partnership.

I have been involved in forming two such partnerships in composting, and they are working.

The first is Compost...for Earth's Sake (CFES), a partnership of the National Audubon Society with the U.S. grocery industry (which is made up of groups like the Grocery Manufacturers of America and the Food Marketing Institute and individual companies like Procter & Gamble, Quaker Oats, Nestle's, Kraft, Scott Paper, Hannaford Brothers, Krogers, Pricechoppers, Lucky Stores, Shoprite and others). The goal of CFES is to promote compost collection in homes and grocery stores. CFES has undertaken a number of collection and composting pilots as well as joint publications, including a 'how-to' guide for establishing pilot programs. We have prepared a 'thought' piece on the importance of composting and a paper on the soil contaminant issue, both of which are now circulating for review.

The second partnership is Food for the Earth (FFE), a partnership of the Composting Council, the National Audubon Society and the food service industry (with groups like McDonald's, Cargill, Sweetheart cups, James River, and International Paper). All of this is in association with the National Restaurant Association. The goal of this partnership is to promote composting in the food service industry. This group is also working on pilot programs.

We believe that pilot programs are needed in each part of the country, despite the fact that successful pilots have been done elsewhere and that successful programs are already operating in other countries. In addition to having completed successful pilots in Connecticut and California, the partnerships are in the midst of a commercial pilot in Minnesota, as well as a residential drop-off program in New York State. We have been working with a Florida Solid Waste Authority to assist them in exploring composting operations in their community. We are also working with yard waste composters in Ohio and Vermont who are in the process of adding commercial organics to their mix.

What common interests led to the formation of the CFES and FFE partnerships? A major commonality was an eagerness to find a waste diversion technique that would extend traditional recycling. A second was to get compost used for land restoration. A third commonality was support for source-separated composting even though some of the involved parties might support mixed waste composting in other forums.

(To Audubon and other environmental groups, 'doing composting right' means source separating at home or store, with actual composting taking place at home or at a central facility. U.S. environmental groups cannot at this time support mixed waste composting because of concern about contamination with household hazardous waste.)

A fourth commonality was that composting of some sorts of paper was appropriate and desirable (for example, soiled paper and paper that would not be recyclable for many years). There was agreement on this point, although the parties have differing views about exactly which types of paper are acceptable for composting. Everything has worked because of a willingness to work on common elements, while agreeing to disagree on divisive issues.

The goal of both of these programs is to make source-separated composting a reality in the United States without interfering with recycling and even helping the economics of recycling while at the same time keeping environmentalists , industry and government happy. Sound like an impossible dream? Well there's a vision out there that can make it work in the United States; a vision of people separating their compostable organics at home, restaurants and supermarkets into smoothly integrated collection systems; a vision where traditional recycling of paper would be given the top priority; a vision of composting providing a very high diversion rate when coupled with traditional recycling. Compost would be produced in filtered facilities when the operation is situated anywhere near homes.

With this joint vision, we have found in our partnerships that it is pretty easy to gain ideological acceptance in the communities where we have worked. This was true in our Connecticut pilot, where editorial writers and participants all wanted the efforts to continue beyond our experiments. When we had to get an emergency permit from the state of Connecticut, we got tremendous and quick support. In our California pilot, we also saw great community acceptance.

We are also finding that state solid waste and environmental agencies are quite receptive to this vision, especially in the crowded Northeast.

Examples of the vision can be found all over the country: at supermarkets in Washington and New Jersey, where separation takes place in the store and composting goes on in large wind rows. Cracker Barrel restaurants has shown that behind-the-counter separation of compostables is easy at its family-style restaurants. McDonald's has shown the same to be true for its quick-service restaurants.

These activities around the country are the kinds of examples that show everyone moving in the same direction. These activities show that there's something in composting for everyone. When you add pictures of backyard composting and community yard and leaf composting as well as the idea of compost going to restore degraded lands and help community gardens, you have a powerful and exciting vision of getting organics back to the land to serve a useful social purpose. Now we need to be able to show a lot more examples from around the country so we can extend ideological acceptance nationwide. The question becomes: How do we do that?.

A major strategy for reaching the goals of both CFES and FFE is to find champions in the community, government agencies and businesses. A champion in someone who believes in the potential of organics recovery, who can see beyond the obstacles and who is willing to speak out in favor of composting and take supportive actions. We need to find them and provide them with both psychological and technical support.

There are also potential champions who with a little education and acquaintance with the vision will become champions. There are hundreds and thousands of such people out there in government agencies, environmental groups and the recycling community. These are the people who are going to be there at the crucial moments, which occur all too often in the U.S., when public acceptance hangs in the balance in a community.. These are the people who will speak in defense of composting at a critical public meeting or in a critical planning meeting in a public agency or at a crucial business meeting. The recycling community has lots of potential champions, although they may not know it at this point. The recycling community is a natural ally for economic and preservation reasons.

As we have learned in our California collection pilot, wet/dry bag combinations can reduce the system costs in a community doing curbside recycling. When solid waste is collected in an organics container that has all of the messy stuff like food, soiled paper, cat litter and diapers (all of which can be composted in closed vessel facilities), it is very easy to handle the rest of the waste stream in one or more dry containers and easy to pull out the recyclables. The only residual is household hazardous wastes and large material that need not be collected very often. This means you can reduce the number of truck passes, dramatically reducing the system costs of a typical trash/recycling system, which today tends to have lots of redundancy. As pressure builds to cut costs in materials recycling programs (which seems to be happening more and more around the U.S.), the recycling community will be more and more receptive to composting beyond leaf and yard waste. If the traditional recycling community becomes supportive, it will really help with acceptance of centralized composting.

To maintain long-term ideological acceptance of composting, there are going to have to be markets for the final product. Thus, markets are more than an economic issue – markets are also an acceptance issue. An example of how to address markets in a consensus fashion is offered by the Coalition of Northeastern Governors (CONEG), which has established a Compost Committee. The committee has decided to focus its first efforts on developing model product specifications for compost. I serve as the group's co-chairman.

Right now, state and local governments have a difficult time in purchasing compost because of the absence of consensus language on product specifications. By bringing together the major actors in the composting field, CONEG hopes to fill the gap, as well as insure that specifications are uniform across the Northeast so finished compost can freely travel across borders. Adoption of model language by CONEG would not only help local government in the Northeast, but also provide guidance for commercial customers as well as government bodies outside the region.

As I said in the beginning, the first step in gaining public acceptance is getting ideological acceptance. This has been proven easy to do when one go into a community with industry and environmental partners to help with source-separated composting, and when we all agree to put paper recycling above composting in the hierarchy.

Gaining local acceptance of communities for the siting of compost facilities is more difficult, but we have learned that it, too, is helped by the partnership approach.

I am optimistic for the long run because we have a great vision. However, it is a hard task that we've all taken on, namely to promote composting as a major handler of waste. It cannot be done alone. Environmentalist and industry must get down in the trenches together.

Composting Plant in the City of Forli'
The Public Administration Experience

T. GIUNCHI – Administrative Board of the Province of Forli', Italy
G. VERONESI – University of Bologna, Italy
G. ZECCHI – Local Health Units ('USL'), Forli', Italy

Summary

After illustrating the reasons behind the decision to support the development of composting plants and after enumerating the facilities already present in the Province of Forli', the authors of this paper proceed to describe the main problems encountered in the operational phase of the plants and the solutions and legislature proposed, with a view to guaranteeing, in the future, an activity that conforms to the environment requirements of the population that resides close to the plants.

Introduction

The Province of Forli' has a high density of animal-farms, in particular, battery and ground poultry- farms. The fact that little land was available and suitable for spreading, plus the fact that the population is scattered, together with the fact that greater attention is being paid to environment issues and, in particular, to those resulting from the zootechny industry, has determined a general awareness of the problems arising from environment disharmony due to the presence of refuse deriving from animal-farms and, consequently, the need to identify possible solutions to these problems.

In particular, while on the one hand the Municipal Authorities sought to limit the creation of new plants by increasing restrictions and controls, on the other, the technicians of the Administrative Board of the Province of Forli' and the Local Health Units, by detecting the more serious problems on hand, worked towards identifying, amongst a series of technological proposals available, those that were the most qualified to restore balance to the production activity already present in the area.

After many attempts, some of which were not satisfactory (for example, the case of plants for the drying and combustion of chicken dung), the conclusion was reached that *composting* was the direction to take to solve various problems; thus, on the one hand, various proposals for plant organization were favoured, whilst on the other, a collaboration project with the research institutes was activated for the monitoring and testing of the concrete possibilities available to solve the problem.

Composting

The first approaches at studying this technology took place towards the middle of the '80s. The technology certainly presented itself as a system having different positive characteristics such as low consumption of electricity and the possibility of transforming the bio-mass into a mature conditioner that could be utilized in agriculture without causing all the health-hygiene problems arising from the utilization of animal dung.

As transportation even at great distances was possible, both from an economic and sanitary point of view, the problem of local eccess of fermentable elements and bearers of fertility, with consequent negative effects on the environment, was solved.

However, a certain number of negative circumstances took place which were basically, but not solely, ascribable to the strong bad-smelling emissions which had a determining importance in numerically limiting the diffusion of this solution. At that time, in fact, it was necessary to risk utilization of technologies that were not the result of national research programs, but which derived from tests and uses carried out in areas that had characteristics quite different from the territory under examination, and, therefore, which were not always suitable for the operational conditions of the site to be utilized.

In fact, the first plants that were constructed also created problems of bad-smelling emissions deriving both from the technologies adopted as well as from management shortcomings; notwithstanding this, the actions that were undertaken led to the belief that the real solution to the environment problems already present was contained in the utilization and diffusion of such technology - after appropriate re-examination and updating in order to respect the health and environment requirements of a territory that was highly populated, also bearing in mind the economic-management compatibility of the technology with the existing breeding facilities.

Difficulties encountered in identifying suitable sites

Once it became apparent, also with the help of scientific literature, that the production of compost to be used for agronomic purposes furthers, at the same time, both the safeguard of the environment and the improvement of soil fertility, the Administrative Board of the Province of Forli', together with the Public Health Departments and the world of the manufacturers, carried out research, tests, and promoted the creation of plants for the production of compost.

The public offices were responsible for defining the criteria to be adopted in identifying suitable sites for the plants, whilst the private bodies were responsible for the construction and the management of the same plants.

The problems that arised from the running of the first plants caused alarm in the public offices that in the meantime were ready to authorize the construction of new plants, and in the inhabitants of the areas that had been selected to locate the plant, so much so that in certain cases, anti-project committees were created. Even in the presence of extremely degraded situations, in which the creation of a plant would have in any event brought an advantage with regard to the existing situation, and in which the incertitude of the Public Board was more accentuated and this interacted with the strong opposition of the inhabitants, even in this case it became impossible to construct new plants, even in the presence of new improvement proposals with respect to the exisitng ones which had proved to be defective and difficult to implement.

To this we must add the fact that not always did the builders of the plants behave correctly (the plants at times proved to be unreliable and had the wrong dimensions), and likewise behaved the persons managing the plant, who did so both for the financial problems encountered in utilizing mixing material having a high cost, as well as for their real technological incapacity in correctly managing the plant. In the meantime, however, technologies evolved, knowledge of the most correct management techniques increased, and this led to the construction of other plants; at the present time, different composting plants are present on the territory, and this circumstance has also increased the experience of the authors regarding the identification of the most important factors for the management of processes devoid of negative effects.

If, on a theoretical basis, it has been possible for quite some time to define the conditions for an optimal project and functioning mode, operationally-speaking, a large number of possible conditioning factors capable of varying the emission situation and therefore the environment acceptability of the plants, must be taken into account.

Initially, these essentially provided a solution to poultry breeding problems, and only subsequently were other types of organic refuse, both mixed or not, taken into consideration.

Existing plant typologies, even if they vary from each other, take into account the original interest for poultry residues, because plants equipped with troughs and turning-over machine are the more frequent, even if other types of technologies exist, with a wide variety of organic products that can be treated, compost that can be obtained, and finally disposed of.

Current experiences

The plant for the composting of poultry dung and, in lower quantities, of special assimilable refuse, of the AGROFERTIL company, in S. Sofia, was created in the '80s and originally provided answers to the problems linked to the treatment of poultry dung; subsequently its license was extended to encompass the treatment of other types of assimilable refuse such as: urban biological sludge, food industry refuse, refuse from abattoirs, wastes from paper mills, wood shavings, vegetable refuse and food residues.

The technology that was adopted, which derived initially from Japanese proposals and which was subsequently modified on account of mechanical insufficiencies, proposes mainly to treat animal dung through an exothermic process that allows dehydration through evaporation and stabilization of organic material, carried out inside greenhouses. The plant comprises 4 troughs measuring 105m x 12m in which aeration is obtained by turning-over and advancing of the material through the use of special machinery.

Frequent moving of the material (twice daily) would favour dispersion of thermal energy if the process were not contained within greenhouses having controlled replacement of air and humidity. For the start-up of the process, there is a requirement for a maximum humidity rate of 70%, therefore the dung and other refuse are mixed with each other before starting the fermentation phase.

The plant is able to treat 25,000 metric tons/year of refuse that is composed of poultry litter (14,000 metric tons/year) and poultry (egg-laying chickens) dung (9,000 metric tons/year); the remaining quota originates from diverse organic refuse. The compost produced is available for use in agriculture, both loosely or in sacks, or in pelletized form.

Other companies have acquired their own composting plant for the disposal of poultry dung; the first of these companies to do this was ARRIGONI whose facility dates back to 1987, and which can treat 40 metric tons/day of poultry dung mixed with a substratum to reduce its humidity, thereby allowing fermentation.

Composting takes place in troughs and the material under fermentation undergoes turning-over in order to aerate it, and the troughs are enclosed within a structure similar to a greenhouse. The end of the process is obtained once a level of 15%–18% residual humidiy is reached, to allow for pelletization of the compost.

The plant is comprised of 3 troughs having a length of 45m each, surmounted by tracks on which the turning-over machine runs; fermentation time is 35 days. Other similar plants can be found at the AVIZOO company, in Savignano, at BERARDI, in Borghi, at CAICONTI, in Bagno di Romagna, and at GUIDI, in Selvapiana.

A different type of treatment is that applied to the litter piggeries having an external composting reactor at the AGRICOLA CANALI farm in Piano di Spino di Meldola. In this plant, pig litters are treated. Before the entry of the pigs in the piggery, the litters, composed of a stratum 80cm deep of wood shavings and sawdust, are mixed with liquids plus special enzymes so as to allow the enzymes to act and destroy the manure when the pigs settle on them.

In this case, it is the litter that acts as a composting plant, the top stratum blocks thermal dispersion, whilst in the lower stratum, metabolic processes are started that increase the temperature. Oxygenation of the material is supplied both by the animals and by the personnel who periodically overturns the material, thereby burying the fresh dung (once weekly) and inoculating new enzymes and bacteria. After approximately 15–18 months, the top stratum (25cm–35cm) is removed and is renewed with other fresh wood-cellulosic material; the removed litter represents the residue of the breeding activity.

The part of this farm that has not been able up until today to house this type of deep litter, produces at the current time diluted sewage that undergoes composting in an external litter contained in a waterproof trough surmounted by a machine that distributes the sewage and the enzymes. Everything is enclosed to avoid direct contact with atmospheric agents.

The AMIA plant, in Rimini, for composting of pre-selected and special assimilable organic refuse, which was created in 1989, and is located near the dump of Ca' Baldacci, gathers within differentiated storage troughs, urban wastes, sludge, cellulosic material (originating from certain urban areas), from which are singularly extractable the quantities required in order to obtain diversified dosage of the components within the proportioning devices that control the C/N ratio and humidity level. The resulting mixture which is distributed in heaps in the trough so as to allow for a double aeration action leads to a composting cycle of 35–45 days.

To date, the plant has operated from June until September, a period which coincides with the tourist season during which time there is the greatest production of organic refuse. The quantities of waste being treated has increased from year to year and has now reached the level of 905 metric tons (in 1991); when the plant reaches full planning configuration, a volume of at least 4,750 metric tons of urban organic waste will be treated. Compost produced in 1991 was 181 metric tons with a residual humidity rate of 32%–33%, and a production/refuse treated ratio of 20%; the loss of water is estimated at 60%–65% of the initial quantity treated.

All this was made possible through daily differentiated collection of organic refuse with special containers having a capacity of 240 l. Always with regard to the urban area where this collection takes place, 50 l containers for the collection of glass are distributed and the remaining quota of urban refuse is made through 360 l containers. The compost produced is utilized for public parks and large-scale farming.

The processing plant of food industry refuse of the FRUTTADORO company of Cesena will be able to treat a total of 240 metric tons by end–May 1995; refuse utilized results from the processing of food, vegetables and sludge deriving from the biological treatment of industrial waters. Its technology is the classic technology in which heaps are aerated through the overturning technique; the experimental part of this plant accelerates and controls the different phases of the composting process by utilizing enzymatic and bacterial mixtures in the heaps.

In this part of the plant, the dimension of the composting trough is 22,5 sq.m and can house tests carried out on a heap having a dimension of 6m x 3m and a height of 3m–3,5m, which is subjected to fermentation cycles and to 3–4 turnings-over, one every 15 days, thus the composting period is of 45–60 days.

Operational considerations

As stated beforehand, operation of the above-described plants, especially those associated with poultry breeding and which are generally based on the technique of trough-composting with turning-over machine, has brought out numerous problems resulting from plant deficiencies and bad management. Without entering into an individual analysis of the functioning of each plant, the negative aspects that were encountered can be grouped in the following categories:

1. Breakdowns in the machinery.
2. Production capacity which is inferior to the one stated.
3. Fermentation problems.
4. Ammonia emissions and other bad-smelling products.

The analysis of these events has allowed the authors, in the majority of cases, to identify the effects and causes, and therefore to foresee a series of cautionary measures to be applied when realizing new plants.

Breakdowns in the machinery.

Breakdowns in the machinery have proved to be more frequent in the latest and more mechanically– sophisticated machinery compared with the simpler machinery of the past.

The nature of the breakdowns (breaking of rotational parts, breaking of movement arms, problems with electric cables, electrical problems in the command panels, etc.) points out how often the mechanical systems are underdeveloped for the required exertions or else are scarsely protected with respect to highly corrosive environment conditions.

Furthermore, many breakdowns are ascribable to lack of maintenance and, above all, to lack of lubrication, both due to oversight of the person running the plant, as well as to a real lack of evident lubrication points on the machinery itself.

Also, if the initial breakdown is of a minor entity, the same can trigger a series of effects and, above all, can cause halting of the process and could require human intervention in conditions which are often hygienically unacceptable.

Halting of the process in fact quickly sets off anaerobic phenomena in the biomaterial, with consequent bad-smelling emissions, while often it is necessary to see to repairs directly in the fermentation troughs and, at times, having to manually remove the fermenting material to allow access to the turning-over machine.

These conditions must therefore be avoided, or at least must be drastically reduced, both for hygiene reasons in the workplace, as well as for the negative effects that often arise in the surrounding area following blockage of the fermentation process.

It is therefore necessary to create systems that are capable of guaranteeing, for the majority of the breakdowns possible, and in any circumstance, transportation of the turning-over machine to the extremities of the trough, in an area which permits greater access.

Furthermore, a maintenance plan must be prepared and rigidly observed, whilst, in the plant testing phase, conformity to safety norms, which are not always observed, must be verified.

Production capacity inferior to the one stated.

This has proved to be a nearly constant characteristic in the various plants examined, mainly due to the fact that, at time of planning, the volume of mixing products to be added for the correct balance of the reactions was often not taken into consideration.

As will be explained further on, it appears evident that, for a correct management of the plant, it is necessary to guarantee certain conditions on the initial material, which can be summarized in an apparent density not above 700 g/dmc, humidity not above 70%, and C/N ratio not below 15–20.

These conditions are not generally found in the raw material being treated, therefore it is necessary to add other products (straw, sawdust, or other wood-cellulosic residues) to allow for correct initial balance.

This entails an obvious increase in the volume of the total composted mass which, depending on varying conditions, can reach an increase of over 50% with respect to the basic product, and from this derives, in general, an incapacity to treat the total product available, with consequent management problems.

Another deficiency aspect has often resided in the movement capacity of the turning-over machines. If we exclude the case, which was encountered, of machinery having completely insufficient dimensions, we often find that there is an inadequate moving capacity of the material to be treated which is due to an insufficient couple in the turning-over machine.

Though these situations usually occur on account of excessive compactness and density of the material employed (therefore, material that does not meet the above-stated standard conditions – a requirement which cannot be excluded on an operational level), these situations lead to blockage of the movement and require direct human intervention. The same, furthermore, are stimulated by mechanical coupling between the electric engine and rotor, thus, in the event of a sudden increase in the torque, a slowing of the rotation takes place with exit from peak r.p.m.

Utilization of hydraulic transmissions and automatic control systems of the movements according to the torque would allow reduction, if not elimination, of this problem. Even automatic controls on the loading and on the mixing of the incoming materials, which would guarantee a mixture which is closer to standard values, would contribute to the solution of such problems which, we repeat, often require manual 'demolition' of the mass that has been formed.

Fermentation problems.

The fermentation process takes place spontaneously, but requires the observance of certain conditions if it is to be correctly carried out.

We have already mentioned composition of the material, above all for what concerns permeability to the air (correlated with the apparent density) and humidity. Experimental situations have occurred with densities above 1,2 kg/dmc and humidity above 80%.

In these conditions, in addition to the problem of compacting of the material, which results in the operational incapacity of the turning-over machines, anaerobic phenomena are favoured which, due to the hydrolysis processes that accompany them, further accentuate the negative situation. Fermentation is blocked, the material is 'disactivated' and the only possibility of activating it anew is to drastically mix it with dry material, an operation that often requires human intervention.

Similar conditions are favoured by prolonged breakdowns which lead to halts of over 48 hours, which give rise to processes of excessive cooling of the mass caused by insufficient protection from adverse atmospheric conditions and from recondensation of the evaporated water onto the material itself.

These phenomena have been observed in the case of light coverings, as in a greenhouse, which were once utilized because it was erroneously thought that light panels would be effective insofar as they would capture and utilize solar energy.

Obviously, fermentation problems can occur because of disinfectants or sterilizing agents present in the material that is subjected to fermentation; in this case, however, it has been observed that it is useful to recycle part of the fermented product and use it as 'triggering' material, even if this at times requires manual intervention.

Ammonia emissions and other bad-smelling products.

Maintaining a correct C/N ratio (between 20 and 30) is a guarantee from possible ammonia emissions. In operational terms, at least in the presence of residues that are rich in nitrogen, such as fecal or animal-farm residues in general, this ratio is not easily reached because in the originating material the ratio is usually around 8–10.

Consequently, the mass of wood-cellulosic additive to be used should exceed 100% of the main mass, thus bearing negative consequences on plant productivity and on costs (insofar as the additive in general can be found only upon payment). Furthermore, even the quality of the composted product, having levels of nitrogen below 3%, would qualify it as a conditioner and not as a fertilizer, thereby considerably lowering its market value.

Consequently, in practical terms, the operational C/N ratio does not exceed 15, as this would cause emission of ammonia into the atmosphere. Inside the reaction sheds, the value of concentration of ammonia is around 40–50 ppm, which is not very high but superior to those allowed for the presence of human operators, as during the mixing phase peaks of over 250 ppm can be reached.

These emissions, which do not permit the presence of humans inside the sheds and which therefore lead to the requirement for global automation of the processes (an event that has nearly never occurred), do not generally create problems for the surrounding environment. In fact, due to the high diffusivity of the gas, in general, an immediate remixing with the surrounding air takes place, with consequent lowering of the levels within the norms in the immediate surroundings of the facilities.

On the contrary, many problems have been caused by bad-smelling emissions which were due to long-chain composts deriving from decomposition of nitrogenous and sulphurized organic products. These are materials having a high molecular weight, with low diffusing capacity, which can suffer from atmospheric phenomena such as thermal changes or draughts, and which can cause, even at a distance, concentrations of bad-smelling gases which are even superior to those found in the vicinity of the facilities.

To this category of products can be ascribed the series of phenomena, currently existing, which have led to problems of acceptance on the part of the inhabitants who are seriously subjected to smells which are often extremely nauseating. Unfortunately, existing legislature is not easily applicable, while the same monitoring of smells is costly and extremely criticizable.

Experience has shown that these situations are the result of bad plant management (aeration blockage, anaerobic pockets, insufficient C/N ratio), or derive from the moving of originating material which is already in an advanced anaerobic state when it reaches the composting plant. Several attempts were made to find a solution by using enzymatic products, pre-mixings, etc., but they all proved useless, and at times even adverse.

On the basis of the tests carried out, we have come to the conclusion that possible solutions consist in placing particular attention on the initial fermentation phase, by opportunely mixing the incoming material with important quotas of recycled material and facilitating air-material contact, thereby ensuring, in any circumstance, the possibility of treating the expelled air, through suitable biofilters.

This obviously implies a covering that is designed expressly, and therefore it is not always possible to adapt existing facilities. Because the cost of utilizing biofilters has proved to be high, it is opportune that a methodology limiting the use of such structures be created, to be used only when it is indeed necessary.

Conclusions

From this brief summary, all the tests that have been carried out by the various public boards are not entirely evincible to allow for a correct indication as to the most suitable methodology for the recycling of solid organic wastes. The economic importance of animal farms and related transformation and food industry activity is evident as it guarantees a high standard of living and is labour-intensive for many municipalities of the Province of Forli'.

On the other hand, the need to guarantee a high quality of life and respect for an enjoyable environment has convinced us that it is worthwhile to verify the possibilities offered by technology, and to intervene on them in order to suggest, if not impose, a whole series of adjustments capable of guaranteeing correct plant management, convinced as we are that industrial and operational projects can be created from such an activity. These would no longer be borrowed from other countries, such as Japan, as was the case in the majority of the plants described above, but would be the result of tests and applications of conditions found both on a national as well as on a European level.

Digestion by WAASA Process of Optically Separated Waste

RUNE WESTERGÅRD – Avecon International Ltd Oy
Vaasa, Finland

Introduction

Household waste has been treated by digestion in city of Vaasa, Finland since 1990. The plant capacity is 25.000 t/a, which is corresponding to biowaste from 200.000 inhabitants.

Part of the waste treated at the plant is source separated by a new method. The system is based on using plastic bags of different colours, starting in the kitchen. The bags are detected by optical sensors and automatically separated from each other. The biowaste from households is digested in thermophilic condition. Special components are necessary for successful digestion of household waste and such components have been developed and patented by AVECON Ltd in cooperation with the waste treatment company in Vaasa.

General

Ab Ekorosk is a waste management company owned by five communes in Finland. Source separation was introduced in the area with 50.000 inhabitants at June 1st 1993. The waste treatment system practised at Ab Ekorosk is based on an unique separation concept of household waste. The waste is separated into two different coloured bags in the kitchen. Wet waste, consisting of food, wet paper and other organic waste, is placed in black bags, and dry waste, like packages, plastic and paper is placed in non-black bags. In the optical sorting plant the two different waste types can be automatically separated. The wet fraction is suitable for digestion and the dry fraction as a fuel in energy plants.

In the digestion plant in Vasa this source separated waste is treated by the Waasa Process. The RDF, consisting mainly of non-recyclable paper, plastics and textiles is used as fuel in a plant using coal and peat as main fuel. Up to 30% RDF mixed with other fuels is allowed in Finland without classifying the process as waste incineration.

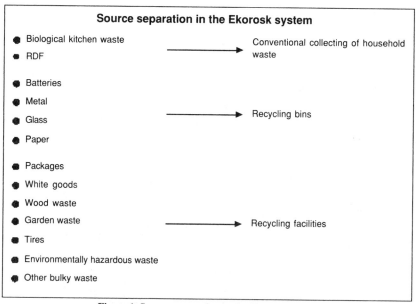

Figure 1. Source separation in the Ekorosk system.

Choise of source separation model

The consept named OptiBag is based on source separation of household waste, with households placing the various types of waste in different coloured bags. Ab Ekorosk Oy has built up a model (see Figure 1) in which the waste is divided into

a digestable fraction and other waste, which is collected on the premises. In addition, there is separate collecton of paper, glass and batteries at local recovery stations. Bulky waste and environmentally hazardous waste are collected at larger
recovery stations.

Before, during and after the introduction of the system, the inhabitants of
Jakobstad were given systematic information in source separation of household
waste. Effective information is necessary for achieving a satisfactory result.

The optical sorting plant (OptiBag) permits a continuied utilization of conventional, compressed single compartment trucks for collection, as well as one bin
per household. Thanks to these advantages the collection costs can be kept low
and the hygienic level high. Studies show a significant cost reduction using the
OptiBag method in comparison to alternative methods.

Experience of the OptiBag system since 1990 in Sweden

The separation results are excellent, with over 90% being sorted correctly by
households. The automatic separating system has also produced very good results,
with 99% of the bags being separated correctly.

The advantages of the OptiBag system compared with other solutions are:

– conventional collection trucks and bins can be used
– lower investment costs
– high flexibility to meet future requirements
– superior hygiene and working environment.

The system's flexibility allows source separation to be easily extended with several
fractions which can be placed in bags of different colours.

The Vaasa-plant for anaerobic digestion of MSW

The waste treatment company ASJ Ltd, owned by the commune of Korsholm and
the city of Vaasa has operated an anaerobic digestionplant for treatment of
mechanically sepa-rated Municipal Solid Waste and sewage sludge since the
spring of 1990. After a few years experience the ASJ-company decided to extend
the biological treatment capacity and a second line, using the further developed
Waasa-process was built during 1993.

The yearly energy-balance at the Vaasa plant is the following (design figures):

– The plant has a total design capacity of treating the waste from 200.000 inhabitants. The waste is devided into three streams in a conventional pre-treatment
plant; RDF (55%), organic materia (40%) and ferrometals including bulky
waste (5%).

- RDF (Refuse Derived Fuel), heat value of 14,5 MJ/kg, a total energy content of 133.000 MWh/a
- Bio waste, 150 m^3 biogas/ton$_{input\ of\ waste}$, heat content of 6,5 kWh/m^3, gives a total energy output of 25.000 MWh/a
- The internal energy consumption of the digestionplant is about 7.000 MWh/a including the heat needed during wintertime with an ambient temperature of −25 (C for several months. Calculations show that if operated in a Central European climate the plant would consume less than 10% of the energy provided by the biogas.

The total energy surplus for the ASJ-plant when including the energycontent of the RDF is 150.000 MWh/a.

The digestion plant is processing about 40% of the total household waste and also sludge from the sludge treatment plant in Vaasa, some industrial organic waste and some manure from the agriculture.

The end-products from the biological treatment are digestate (compost), biogas and water. The compost is used as a fertilizer, cover material on old disposal places and for green areas. The biogas is utilized in a gas-engine for electricity and heat production. A part of the electricity is used in the plant and the remaining part is sold to the local electric network. The excess water from the process is pumped into the municipal network and is treated at the sewage treatment plant in Vaasa.

Experience

The availability of the mechanical and electrical systems in the ASJ digestion plant has been over 95 % since mid 1990 . The biological process has worked without disturbances.

The biogas quality has been stabile and the analyzes made during 1992–1993 gives a mean gas quality as follows:

CH_4	65%
CO_2	34%
N_2	<0,5%
H_2S	<, 5–12mg/m^3
Cl^-	<0,5mg/m^3
F	<0,5mg/m^3
NH_3	<0,05mg/m^3
Humidity	1%

The heavy metal content in the compost has been measured since the start and the results have improved by time. A serie of analyzis (n=6) made in 1993 gave the following content of heavy metals.

Zn	1250 mg/kg	TS	Stdev	258
Cu	200 mg/kg	TS	Stdev	22
Pb	190 mg/kg	TS	Stdev	83

Cr	67 mg/kg	TS	Stdev	4
Ni	32 mg/kg	TS	Stdev	5
Cd	3,1 mg/kg	TS	Stdev	1,0
Hg	1.4 mg/kg	TS	Stdev	0,2

The heavy metal content in the compost depends direcly on the quality of the incoming waste. The excess water from the prosess is pumped directly, without cleaning, to the municipal waste water treatment plant.

Digestion plant in finland extended by the waasa process

Choise of process

The treatment company ASJ decided on the Waasa Process developed by Avecon International Ltd Oy. In 1989 Avecon started to change and develop the process in Vaasa together with the plant owner ASJ.

More than doubled treatment capacity per reactor volume is expected compared to the first line. With the experience from the first line a strong technical development has taken place in Vaasa. Avecon has developed technical solutions for increased efficiency. One example is the Twin reactor solution, which means that the reactor consists of a pre–chamber and a bacteria injection system. In a conventional reactor there is always a risk of emptying untreated organic material, but the Twin reactor eliminates the risk of short circuit. The new solutions shorten the retention time and create a digestate of better quality. The second line is a thermophilic Waasa Process, which is the most influential factor in achieving a shorter retention time.

Avecon has developed a special component named a Mixseparator for handling the temperature and water content variation problems; it also has many other functions. The household waste consists of many different materials and, even when cource-separated, the waste must be thoroughly separated further before digestion. The fine separation is most effective in the wet stage and this is carried out in the Mixseparator.

European interest

The Waasa Process has aroused interest in many European countries. Avecon has signed licence and co-operation agreements with Ingg. De Bartolomeis for Italian market and Thyssen Still Otto Anlagentechnik for the German market and with Holland Construction Group (HCG) for the Benelux countries. Co-operation partners for other markets are being sought in 1994 and 1995. Interest in the Waasa Process is based on the reference plant and the level of development (its theoretical biochemical research is a cooperation venture between Avecon and the University of Tampere). As in most technical fields product development in anaerobic digestion technology is a continuous process necessary for future success.

Rune Westergård is Director of Process Technology at Avecon International Ltd Oy, Vaasa, Finland

Sources:

Optical sorting of source separated household waste: The OptiBag concept by Anders Wahlquist.

B3 Composting as an Integrated System of Waste Management

The Co-Treatment of Municipal and Industrial Wastes

Dr R P BARDOS – WASTE Division, CRBE, Nottingham Trent University, Burton Street, Nottingham, NG1 4BU, Dr S FORSYTHE – Dept Life Sciences, Nottingham Trent University, Clifton Campus, Nottingham, NG11 8NS and Dr K WESTLAKE – Centre for Hazard and Risk Management, University of Loughborough, Loughborough, LE11 3TU.

Abstract

Regulatory pressures are forcing major changes on the waste disposal industry. The proposed EU Directive on Landfilling of Waste, if implemented in its current form, will ban the co-disposal of industrial and municipal wastes in landfill. These developments indicate a likelihood that co-disposal will no longer be available as a disposal option for industrial wastes in the long term.

Above-ground co-treatment of industrial wastes with sewage sludge and/or municipal solid waste *prior* to landfill is an approach to enhancing the efficiency of waste disposal to landfill and reducing reliance on incineration. It also provides an alternative to disposal to land where the wastes are considered too toxic. Degrading municipal waste and sewage sludge have been shown to promote the degradation of many toxic organic compounds and fix other organic and inorganic substances, for example as sulphides or into humic matter. These processes are used to justify the practice of co-disposal to landfill. Unfortunately, these processes are not optimised and cannot be controlled in landfill and hence co-disposal has come to be viewed unfavourably in many countries. Optimisation of these processes in an above-ground treatment system would have many advantages, including:

1 Stabilisation of environmentally damaging substances,
2 Reduction of waste volumes prior to landfill,
3 Generation of recoverable biogas,
4 Detoxification of hazardous substances in a low temperature biological process,

without the input of additional fossil fuel or the likelihood of generating dioxins and furans,

5   Generation of a benign stabilised residue,

6   Biological sanitisation of the treated waste stream,

7   Disposal of a waste with a reduced capacity to generate landfill gas and possibly improved geotechnical performance.

Introduction

The co-treatment of waste is the use of one waste in the treatment of others, or the combination of a variety of waste types within one process to achieve some treatment advantage. Historically, incineration and landfill processes have commonly had co-treatment applications. Co-treatment applications for biological processes, anaerobic digestion and composting, have been less well investigated, although aerobic processes are applied to the treatment of contaminated soil. However, they may offer significant advantages in some cases over co-incineration and codisposal.

It is not the purpose of this paper to describe a proprietary system or specific research approach, but rather to demonstrate the potential use of biotic and abiotic processes that occur in composting and anaerobic digestion in the combined treatment of municipal and industrial wastes and so stimulate research in the field. As well as being a valid treatment approach in its own right, we suggest that co-treatment offers an opportunity to alleviate the process limitations of codisposal landfill in an above ground treatment system.

Background

For all waste management practises, the ideal is the achievement of sustainable wastes management, within the framework of sustainable development. The acceptance of sustainable development for waste treatment is a pre-requisite to the acceptance of co-treatment as a waste management practice, and is therefore discussed in some detail in Text Box 1. Parts of the co-treatment approach are already enshrined in the waste management policy of several countries which still find themselves disagreeing on waste management practice at the most fundamental levels. Co-treatment may be a viable compromise position for these various waste management philosophies, and should therefore be of direct interest to the European Commission and European proponents of biological waste treatment processes, such as ORCA.

Industrial and municipal waste management strategies appear to fall into four broad areas:

1   Pre-treatment of all wastes before landfill to render them as inert as possible

and limit as far as possible biological activity in the landfilled waste [5,8]; (For example, in Germany the maximum allowable organic content in any landfilled waste will be 5% by the year 2005, which, with the present state of the art, leaves thermal treatments as the only practical pre-treatment [39]);

2 Operating landfills to achieve complete and permanent containment (the so called 'dry tomb') reducing ingress of water and eliminating as far as possible emissions from the site [39], and where possible segregating wastes into different types, in particular using *monofill* sites for hazardous industrial wastes, or wastes which might conceivably be recyclable at some time in the future [5,8];

3 Using landfill cells as active bioreactors to accelerate decomposition rates and landfill gas production, and minimise the time the landfilled wastes are a potential pollution source, and hence reduce the lifetime of landfill containment [5,8,24,35,58,59];

4 Codisposal of municipal and compatible industrial wastes is widely practised in the UK and USA, and to a lesser extent in other European countries [24,35,58,79], some of which used the approach more widely in the past [60]. It is salient to note municipal solid wastes (msw) also contains hazardous substances. For example, msw in the USA is thought to contain 0.2 to 5% by mass of hazardous substances [76]. Codisposal of msw and sewage sludge is practised in many countries, and appears to accelerate landfill gas generation [18.93].

Codisposal sites accept municipal wastes (eg refuse and/or sewage sludge) and permitted liquid and solid industrial wastes [32]. Such codisposal has been common practice in Europe and North America for many years. Recent data [35] indicate that leachate from efficiently managed codisposal landfills is similar to that from those accepting only municipal waste, lending further support for the argument that the biological attenuation processes of codisposal can effectively treat a wide range of industrial wastes. Indeed the view of the UK Department of the Environment [35] is that for certain waste streams, codisposal may represent the Best Practicable Environmental Option (BPEO) [74]. Effective codisposal landfills are regarded as stationary fixed film reactors [24,59], in which mixing of substrate and bacteria is the rate limiting step in the degradation of the industrial wastes.

The effectiveness of codisposal landfill sites as a treatment is limited by waste heterogeneity, moisture availability, temperature and pH [21,40,45,58,75,92], which can result in wastes remaining undegraded for significant periods [73], or otherwise degradable contaminants such as phenol being persistent problems in landfill leachate [92].

Within Europe there is significant political pressure to phase out codisposal. Although moves to achieve this in the proposed European Directive on Landfill [26] are being strongly resisted by the UK, the long term future for codisposal in Europe is uncertain [5–8]. The proposed Directive on Landfill will, if introduced, ban the codisposal of wastes to landfill, except at those sites with existing codisposal licenses, who will be allowed to continue until those sites are complete (subject to meeting set criteria for landfill control). Within the UK, approximately 2,070 landfill sites receive some combination of domestic, commercial or industrial waste, and of these, approximately 360 landfill sites are licensed for codisposal [35]. In Germany, most hazardous industrial wastes that are disposed to land are monodisposed in various containment facilities, including disposal to underground salt mines, which are exempted from the provisions of the Directive [5,8].

Concerns about codisposal, and indeed biologically active landfills in general, include the following:

- Generation of landfill gas (which may be a major source of atmospheric methane and also contains trace levels of carcinogens), recovery of landfill gas from sites is incomplete and landfill gas may be generated for many decades [4,5,8,39];
- Poor control of degradation processes in landfill sites, some operators may be unable to achieve methanogenic conditions in their landfill sites [79];
- The impact of landfill emissions, such as landfill leachate [38,58,60] but also arising from the volatilisation of industrial chemicals such as cyanides [59]
- The possibility of corrosion of clay liners by leachates from active sites [22].

Whilst critical attention has been focused on biologically active and codisposal landfill sites, the disadvantages of other approaches have been ignored.

- The extreme pretreatment requirement in Germany presupposes the general acceptance of thermal treatment processes by the public [39];
- Landfilling of 'inert' wastes, may also generate methane [3];
- The leachates from monofill sites, or sites largely composed of incinerator residues are likely to have a greater pollution potential than those from bioreactor landfills [35];
- The sorption capacity of wastes landfilled after thermal pretreatment for inorganic and undegradable pollutants will be significantly less than that of landfilled waste containing organic matter[1] [59,76], and the hazardous substances themselves will be more concentrated. Not only does this increase the pollution potential of the site, it may take many hundreds of years for such a site to become reusable or for containment to be relaxed [38]. It also makes stringent demands on containment design because of the enhanced risks of damage if containment fails.

These considerations do not include wider issues of environmental cost and benefit. For example waste management based exclusively on intensive pre-treatment and monofills, there is likely result in greater use of waste transportation to centralised high cost (thermal) pre-treatment facilities or monofills. Furthermore, such landfills commit our descendants to their care for many hundreds of years, which is not compatible with any notion of 'sustainability'.

In broad terms, at present countries appear to be choosing one of two options: pretreatment and long term containment of landfills, or using landfills as bioreactors and taking advantage of this for codisposal. The two approaches are seen as mutually antagonistic [5,8]. Unless extreme positions are taken by both sides this antagonism is unnecessary. Co-treatment offers a means of combining the advantages of both approaches, whilst also reducing their disadvantages.

Co-treatment Processes

A co-treatment system could include a variety of process steps. For example, co-treatment could employ an anaerobic first stage, including biogas recovery, with an aerobic second stage to maximise nitrification. Co-treatment facilities could be part of an overall waste management strategy where recyclable materials were recovered as far as feasible from msw, trade and industrial wastes, and the residues were treated together to maximise the flexibility and capacity of scarce landfill resource. Residual activities from co-treatment processes would be likely to persist through the early stages of treated waste disposal, and this activity could be regarded as the closing phase of the co-treatment. It is conceivable that with good landfill design and careful optimisation of the end point of the above ground process, the residual activity of the infilled wastes would create a stable long term environment within the landfill site maintaining pH and redox conditions at optimal levels for the retention of immobilised substances and the continuation of long term degradation processes. The aim of this would be to achieve as rapid a stabilisation of the pre-treated and infilled waste as possible.

Technologies which might be applied to co-treatment are already well established. The application of composting to treating municipal wastes is well established [43]. The use of anaerobic digestion is well established for wastewater and various industrial wastes [80] but has yet to be widely used for the treatment of msw, although there have been pilot and full scale initiatives in Europe and North America [27,68,71,80]. Combinations of anaerobic digestion with composting [51,57] or combinations of biologically active landfilling with anaerobic digestion (ie using a landfill cell as an acetogenic bio-reactor followed by an invessel methanogenic reactor [44,96]) have been less well studied.

A variety of biotic and abiotic processes taking place in these biological treatment systems have the potential to destroy, immobilise or transform hazardous organic and inorganic substances, for example:

Biological processes: Biodegradation of organic substances to simple mineral compounds and biological transformation of organic substances into less toxic or less mobile forms;

Abiotic processes: Immobilisation of organic and inorganic substances in humic materials; Immobilisation of inorganic substances by biologically generated or mobilised ions such as phosphate ions, sulphide ions under anaerobic conditions, or bicarbonate under aerobic conditions; and neutralisation of pH and consequent reductions in heavy metal mobility.

Box 1 Sustainable Development in Waste Management

'Sustainable development' has been defined in 1987, in the report of the World Commission on Environment and Development (The Brundtland Report) as 'development which meets the needs of the present without compromising the ability of future generations to meet their own needs'. This report identified the characteristics of sustainable development as:

- The maintenance of the overall quality of life
- The maintenance of continuing access to natural resources and
- the avoidance of lasting environmental damage

These concepts are also iterated in the 5th EC Action Programme [26] which states that "sustainable development implies putting in place a policy and strategy for continued economic and social development without detriment to the environment and the natural resources on which human activity depends'. The concept of sustainable development recognises that both economic and environmental factors affect }quality of life', and that any new development must account for the environmental cost as well as the *economic* cost.

Among other things, sustainable development implies that:

- the environment and its natural resources must be recognised as the basic foundation of all human activity, and their satisfactory guardianship is a precondition for sustainable development [26].

The concept of sustainable development was addressed at the United Nations' Conference on Environment and Development in Rio de Janeiro in 1992. Chapter 21 of the conference's Agenda 21 specifically addresses 'Environmentally sound management of solid waste and sewage-related issues' and was written in response to the conference assembly's decision that environmentally sound management of wastes was among the environmental issues of major concern in maintaining the quality of the Earth's environment, and especially in achieving environmentally sound and sustainable development in all countries. Agenda 21 states that environmentally sound waste management must go beyond the mere safe disposal or recovery of wastes that are generated and seek to address the root cause of the problem by attempting to change unsustainable patterns of production and consumption. Four major waste-related programme areas were identified. These were:

- Minimising wastes
- Maximising environmentally sound waste re-use and recycling
- Promoting environmentally sonnd waste disposal and treatment and
- Extending waste coverage service

In keeping with the theme of sustainable wastes management, the European Commission, in its 5th action Plan [26] has identified a waste management hierarchy as follows:

- Prevention of waste
- Recycling and re-use
- Safe disposal of remaining waste in the following ranking order
  - combustion as fuel
  - incineration
  - Landfill

The overall aim of the above is to achieve more appropriate waste management techniques that minimise the impact upon the environment, and account for the environmental cost (short-term AND long-term), and not just the economic cost.

Biological Processes

Biodegradation describes the decomposition of an organic compound into smaller chemical subunits through the action of organisms, and both aerobic (oxygen requiring) and anaerobic degradation pathways exist, although there are some differences in which types of compound will degrade under aerobic compared with anaerobic conditions [2,63]. Principally micro-organisms (bacteria, fungi and actinomycetes) are responsible for practically useful bio-remediation processes. Completely degraded compounds are said to be mineralised, and the end products might be carbon dioxide, water and chloride ions for a chlorinated hydrocarbon degraded under aerobic conditions. Guthrie [47] describes this as 'ultimate biodegradation. He defines 'acceptable biodegradation' as implying breakdown to below toxic levels, and 'primary biodegradation' as a structural change in the parent molecule, which is more commonly referred to as 'biotransformation'.

Biodegradation may proceed via enzymic activity on compounds adsorbed into cells or through the activity of extracelluar enzymes active outside the confines of the cell. Cells may also use enzymes to generate free radicals or peroxide ions that attack organic compounds, particularly insoluble compounds. In many cases organic compounds do not enter microbial cells since the compounds are either sorbed to soil surfaces, are too large or are physically incapable of being sorbed into cells. More complex compounds may not be completely degradable by single organisms, and are degraded by *consortia* of organisms, and in some cases may not be completely degradable in any circumstance. Some organic compounds may be coincidentally degraded as a result of microbial activity against other substrates, *cometabolism*. For example lignin degradation processes may also degrade complex organic compounds such as polynuclear aromatic hydrocarbons (PAHs) [1,10,12,33,82].

The use of biological processes for treating toxic substances is perhaps best advanced in the bio-remediation of contaminated soils. At present practical soil bio-remediation processes are limited to the degradation of fairly readily degradable contaminants: mononuclear aromatic (eg benzene toluene, ethylbenzene, xylenes); simple aliphatic hydrocarbons (eg mineral oils, diesel fuel) and lower PAHs (2,3 and 4-rings) [14]. However, full scale applications of bio-remediation to treat more complex contaminants (such as pentachlorophenol) are taking place, and some success is reported by the process operators [53]. Biodegradation of a variety of recalcitrant compounds, including polychlorinated compounds has been demonstrated under laboratory conditions [17,29,70]. Examples of composting or similar processes applied to the treatment of contaminated soils are reviewed in Text Box 2.

Aerobic decomposition processes appear to be more thoroughly investigated than anaerobic processes [19]. There is some dispute about both the general effectiveness of anaerobic decay processes, which are perceived by some as being slower and/or less complete than aerobic processes in soil or landfill [11,32,58]. Other researchers argue that the potential destruction of organic substances under optimal anaerobic conditions is or must be the same as for aerobic processes [19], and the anaerobic decomposition of a variety of organic pollutants such as chlorinated solvents and aromatics PAHs and PCBs has been demonstrated [46,48,90].

Concerns about the use of biological processes in waste treatment include:

- Their susceptibility to inhibition by toxic contaminants, for example heavy metals. However, aerobic biodegradation processes appear quite robust to inhibition by toxic substances [53,54], and methanogenic activity in landfills appears to resist heavy metal toxicity [59].
- The mobilisation and release of potentially toxic partially degraded contaminants and the mobilisation of inorganic pollutants, for example by methylation under anaerobic conditions, the production of ligands, or indeed the microbial oxidation of sulphides under aerobic conditions [12,15]. Immobilised contaminants, may only remain sorbed if suitable stable conditions of redox and pH can be maintained in the landfill site where residues are disposed of. Notwithstanding this limitation, the possibility of immobilisation in landfilled co-treatment residues seems to be a clear advantage over the potential mobility of contaminants in thermally pretreated wastes.

Abiotic Processes

These processes include physico-chemical processes such as adsorption to surfaces and also immobilisation by biologically produced materials. Organic matter sorbs heavy metals and organic substances [37,59]. There is experimental evidence to suggest that the extractability of heavy metals from msw decreases over the course of composting, even although absolute concentrations increases as organic matter is degraded [88].

There is some evidence that PAHs may be irreversibly adsorbed into humic matter [65,66,85,87], through a variety of biological and non-biological processes. Indeed, the possibility exists that humus naturally contains subunits similar to PAHs although this observation may be an artifact of the digestion processes used in the analyses. [65,66]. Soil organic matter also appears to be important in promoting abiotic processes of contaminant degradation for some organic compounds and in co-metabolic degradations of organic compounds [37,94].

Some industrial waste materials may themselves sorb dissolved species such as heavy metals [9]. In addition some inorganic contaminants, such as lead, may gradually become immobilised as insoluble phosphates by precipitation with phosphate ions and this has been proposed as a means of dealing with contaminated sites [78].

Box 2 Composting and similar processes used in land remediation

Full scale treatments where excavated soil is placed in windrows or piles and subjected to forward (blowing) or reverse (suction) aeration akin to aerated static pile composting are in use in North America and European countries [31,67,69] and simple windrow turning processes have been applied in Germany. Alternatively, contaminated soil may be contained in some form of lined receptacle [91]. The aerated contaminated soil windrows may also contain amendments, either added through mixing before the pile is constructed or through irrigation, which is also used to adjust moisture content.

Amendments that may be mixed into the soil include organic matter (such as bark, straw, sewage sludge) [13,50,53]; microbial inocula; proprietary agents to enhance contaminant availability and protect micro-organisms from contaminant toxicity such as 'Daran,end' or 'Biocrac' [28,30,81] and structural materials such as pulverised debris [13,20,50].

Organic matter may be added for one or more of the following reasons: to supply an active microbial population; to add structural material (see below); to add microbial nutrients; and to add water [41,50,52,95]. Organic matter may also have the benefit of sorbing and rendering contaminants less toxic [85]. This sorption may also inhibit biodegradation according to some studies [49]. However, sorption of contaminants to surfaces may also promote their degradation by both biological and non-biological processes [37]. There is uncertainty over whether the addition of readily biodegradable organic matter assists contaminant degradation through co-metabolism or interferes with contaminant degradation through the diversion of microbial activity to easier substrates [10].

A good structure in the windrowed soil is important to process performance by ensuring adequate aeration and irrigation of the pile. Structural materials tend to be non-biodegradable (pulverised debris) or slowly biodegradable (eg bark, straw or wood chips]. The incorporation of structural materials may also have the benefit of allowing a greater recycling rate of on site materials (eg debris), or the inclusion of an additional waste stream in the treatment process (eg straw or sewage sludge).

Mixing is important for all et situ approaches to allow them to exploit their major advantage over in zilu approaches which is improved accessibility of the contamination to the treatment. In some et situ venting approaches mixing oilly takes place when the pile is established [62]. In some cases biodegradation in windrows is not assisted at all by venting; regular mixing is used to ensure aeration and the exposure of fresh surfaces to microbial decay [50,6 I].

In some cases process temperatures may reach the thermophilic range [95] although full scale designed use of thermophilic composting in contaminated soil treatment is rare. Most experience is related to the treatment of TNT contamination [55] which, incidentally also appears to be degradable under anaerobic conditions [72]. Pilot scale studies of the co-composting of sewage sludge and gasworks soils have also taken place, along with observations of the degradation of recalcitrant organics such as PCBs through composting [41,52,56]. For municipal waste composting a major thermophilic process control difficulty is the control of moisture content, since the amount of heat generated is usually sufficient to cause excessive drying of the windrowed materials [86] particularly at the core of the pile where aeration is employed. One interesting possibility is the use of mushroom composting reactors, where air is recirculated through the compost pile, after controlled cooling via a heat exchanger. Fresh air is then drawn in oniy to support microbial processes and not for cooling. This not may not oniy ease the management of water balance (since cooling of the exhaust air need not be to fresh air ambient levels), but also helps ensure a more even process temperature since edges can be kept warm [42].

Determining mass balances for contaminants is important in process evaluation [83]. In some cases it may be that dilution of the soil mass with pulverised debris and/or losses through volatilisation/leaching may be responsible for a significant part of observed contaminant losses.

Codisposal Studies

Decomposition and attenuation of organic compounds has been demonstrated under simulated landfill conditions in mixtures of msw. Tests fall into three general types [59], column tests, lysimeter and microcosm studies, and observations and measurements made on landfill sites.

A variety of column and lysimeter tests, in which the migration of substances through columns packed with msw, have been carried out internationally [35,59,76]. Degradation of a proportion of added organic compounds and sorption of both inorganic and organic substances has been demonstrated. Sorption processes can be reversible, although the extent of any such reversal under landfill conditions is not clear.

Microcosm studies, to simulate microbial activity in landfills, have demonstrated the possibility of degradation of a variety of organic compounds such as aliphatic compounds, some plastics and PAHs [22,58,59.75]. Measurements using landfilled wastes have demonstrated evidence of microbial assays using assays of enzyme activity [45], although there is some evidence that such activity, for example degradation of cellulose in newsprint, is suboptimal [16,34].

Although proponents of codisposal would argue that these observations support the view that effective codisposal is possible [53,59,79], there is little doubt that codisposal at some landfill sites has caused significant contamination of leachates by organics arising from industrial wastes, and/or adversely affected methanogenic processes, for example by changing the pH of the saturated zone, in some landfill sites [22,92,93]. Whilst it is possible that poor performance of codisposal may be due to poor site management [79], it is perhaps more realistic to recognise that control of such a vast and heterogeneous, static *in situ* system as an entire landfill site or cell is difficult to achieve and its performance hard to predict. Use of an above ground system would allow true process control and the mixing of substrates. It would also allow monitoring of both the treatment and the composition of treatment residues prior to their disposal.

Processes on the Landfill Periphery

Landfill disposal is the final stage of a co-treatment system, in which biologically stabilised wastes are landfilled within a containment system designed to permit the continuation of residual biological and abiotic activity. Emissions (ie landfill gas and leachate) might be expected to be reduced compared with untreated msw. For instance, leachates from aged composts have lower chemical oxygen demand and total dissolved solids than landfill leachates [36]. Ideally such emissions could be dealt with by passive means, in principal enhancement of naturally occurring processes of decay and attenuation that occur in the vicinity of landfill sites. These processes are poorly understood [32] but include biological processes (ie anaerobic processes in areas where biological oxygen demand exceeds supply, and aerobic processes in groundwater where dissolved oxygen is present) and physical and chemical processes: these include precipitation of inorganic hydroxides along with co-precipitation and sorption of other substances; adsorption and ion exchange; mechanical filtering and buffering.

Enhancement of naturally occurring decay and attenuation processes are widely used in other environmental applications such as: the use of wetlands for the treatment of acid mine drainage [89] and the use of reed beds for the treatment of industrial and municipal waste-waters [23] and the use of low level fertiliser additions to stimulate biodegradation of oil spills on beaches [64]. Attempts have been made to passively enhance these processes beneath landfills by careful design of drainage and bottom layers [77]. Enhancement of natural or intrinsic decontamination processes is likely to be a longer term and lower input approach, but requires much greater knowledge of fundamental soil processes, how these are affected by site specific conditions and what factors are likely to be rate limiting for a particular site.

Environmental Costs and Benefits of Co-Treatment

Table 1 is an illustration of the environmental benefits (advantages) and costs (disadvantages) in comparison with the four major waste management approaches discussed earlier in the paper. These are: codisposal as currently practised, landfill after *stringent* pretreatment aimed at eliminating biological activity (ie thermal pretreatment), long term 'dry' landfills including monofills, and bioreactor landfills (excluding codisposal sites). Each of the five approaches has been ranked in terms of its likely environmental impact under a variety of categories, with a ranking of 1 representing 'good', and a score of 5 'bad'. For example, with regard to waste transportation: codisposal and bioreactor landfills score 1 as their sites are the most flexible, co-treatment 2 as co-treatment plants are less flexible so conceivably more waste will have to be transported between plants, and stringent pretreatment and long term landfill / monofills 3.

The following attributes are considered: the likely generation of methane rich landfill gas and the likely generation of high COD leachate over time; the potential for hazardous leachate generation if containment is breached; the duration that isolation monitoring and control of the site will be required until stabilisation; the energy and resource requirements for facility construction and operation and maintenance; the likely impact of disposal strategy on the prospects for re-use of the site once the waste is stabilised or if it is stabilised; and the likely environmental impact of waste transportation and preliminary treatment for example at transfer stations that each approach would require.

Table 1 Illustrative Environmental Cost Benefit Analysis for Co-treatment

Attribute	Co-treatment and Landfill	Stringent Pretreatment	Bioreactor Landfill and Landfill	Dry Landfill	Codisposal
Landfill gas generation	2[1]	1	3	1	3
Leachate generation	2	5	3	1	4
Impact of containment breach (ie potential generation and impact of leachate)	1	3	2	3	3
Duration of containment requirement (rate of site stabilisation)	1	3	2	3	3
Use of resources and energy in facility construction	3	4	2	3	1
Use of resources and energy in facility maintenance	4	5	2	3	1
Possibilities for site re-use	1	2	1	2	2
Requirement for waste transportation and transfer stations	2	3	1	3	1
Total	**16**	**26**	**16**	**19**	**18**
Ease of environmental verification and process validation	1	1	3	2	4
Total including verification score	**17**	**27**	**19**	**21**	**22**

Note Landfill gas emissions may be sufficiently slight to allow oxidation of most methane to carbon dioxide using passive control measures, which would make this score 1, and hence the total 15.

Although this table is only illustrative, the apparent cost (ie high score) of using stringent (ie thermal) pretreatment as a matter of course before landfill is striking. At first glance the other four approaches are broadly similar in their environmental cost. The lowest scoring approaches are bioreactor landfills and co-treatment. However, bioreactor landfills have yet to be demonstrated for an operating lifetime in practice, whereas co-treatment offers the real advantage of verifiable process performance in short timescales.

Conclusions and Recommendations

The potential of established biological waste treatments for municipal and industrial wastes to the combined treatment of these wastes has yet to be fully explored. Studies of the behaviour of wastes in codisposal sites and observations of their use in other environmental applications, such as contaminated soil treatment, indicate their potential to not only reduce waste volumes and reduce potential leachate and landfill gas volumes per unit of input waste, but also their capacity to ameliorate industrial wastes.

The potential of the co-treatment approach should be more fully investigated, both in terms of its likely environmental benefits and costs, but also to determine the requirements for optimal process performance.

References

[1]   Alexander, M. (1977) Introduction to Soil Microbiology. 2nd Edn. John Wiley and Sons.
[2]   Alexander, M. (1981) Biodegradation of chemicals of environmental concern. *Science* 211 132–138.
[3]   Anon (1991) Gas hazards identified at 'inert' waste landfills. *ENDS Report* (203) 9.
[4]   Anon (1993) Landfills under pressure as methane emissions soar. *ENDS Report* (217) 7.
[5]   Anon (1994) DoE's shift on co-disposal. *ENDS Report* (238) 29–31.
[6]   Anon (1994) End of the road for co-disposal.30 *ENDS Report* (234) 40–41.
[7]   Anon (1994) Landfill Directive's surprises on co-disposal financial security. *ENDS Report* (233) 34–36
[8]   Anon (1994) Staking out the battleground over the future of landfill. *ENDS Report* (236) 17–20.
[9]   Apak, R. (1992) Proposal: Sorption and solidification of selected heavy metals and radionuclides from water onto unconventional sorbents. Presented at the NATO/CCMS Pilot Study First International Meeting: Evaluation of Demonstrated and Emerging Technologies for the treatment and Clean-up of Contaminated Land and Groundwater (Phase II), Budapest, Hungary, October 18–22, 1992.
[10]   Armishaw, R., Bardos, P., Dunn, R., Hill, J., Pearl, M., Rampling, T. and Wood, P. (1991) Review of Innovative Contaminated Soil Clean-Up Processes. Report LR 819 (MR) Warren Spring Laboratory, Gunnels Wood Road, Stevenage, SG1 2BX, UK, ISBN 0856246778.
[11]   Atlas, R.M. (1981) Microbial degradation of petroleum hydrocarbons: an environmental perspective. *Microbiol. Rev.* 45 180–209.
[12]   Atlas, R.M. and Bartha R. (1987) Microbial Ecology Fundementals and Applications. 2nd Edition. Benjamin/Cummings Pub Co Inc. ISBN 0201 003 007.
[13]   Bachhausen, P. (1990) Experiences with microbiological on-site decontamination of soil and building debris contaminated with solvents. pp 983–988 In Contaminated Soil '90. (Arendt, F., Hinsenveld, M. and van den Brink, W.J. Eds). Kluwer Acad. Pub., Dordrecht, the Netherlands. ISBN 07923 10586.
[14]   Barber, S.P., Bardos, R.P., van Ommen, H.C., Stapps, J.J.M., Wood, P.A. and Martin, I.D. (1995) WASTE 92 Area XI: Contaminated Land Treatment: Technology Catalogue. Prepared under *Waste 92, Area IX*. Report for Waste Management, DGXI, Commission of the European Communities, Rue de la Loi 200, B-1049 Brussels, Belgium.
[15]   Barkay, T., Tripp, S.C. and Olson, B.H. (1985) Effects of metal-rich sewage sludge application on the bacterial communities of grasslands. *Appl. Environ. Microbiol.* 49 333–337.
[16]   Barlaz, M.A., Schaefer, D.M. and Ham, R.K. (1989) Bacterial population development and chemical characteristics of refuse decomposition in a simulated sanitary landfill. *Appl. Environ. Microbiol.* 55 (1) 55–65.
[17]   Barriault, D. & Sylvestre, M. (1993) Factors affecting PCB degradation by an implanted bacterial strain in soil microcosms. *Can. J. Microbiol.* 39 594–602.
[18]   Beker, D. (1989) Environmental aspects of landfilling sludge. pp 325–336 In Sewage Sludge Treatment and Use, New Developments, Technological Aspects and Environmental Effects. (Dirkzwager, A.H. and L'Hermite, P. Eds). Elsevier Appl. Sci. London and New York.
[19]   Berry, D.F., Francis, A.J. and Bollag, J.M. (1987) Microbial metabolism of homocyclic and heterocyclic aromatic compounds under anaerobic conditions. *Microbiol. Rev.* 51 (1) 43–59.
[20]   Bewley, R., Ellis, B., Theile, P., Viney, I. and Rees, J. (1989) Microbial clean-up of contaminated soil. *Chem. Ind.* 4 Dec. 1989 (23) 778–783.
[21]   Bognor, J.E. (1990) Controlled study of landfill biodegradation rates using modified BMP assays. *Waste Manage. Res.* 8 329–352.
[22]   Bracke, R. and Puettmann, W. (1993) Bio-geochemical investigations for the assessment of long

term stability of mineral sealings in waste disposal sites. pp 267–276 In Contaminated Soil '93. (Arendt, F., Annokkee, G.J., Bosman, R. and van den Brink, W.J. Eds). Kluwer Acad. Pub., Dordrecht, the Netherlands. ISBN 07923 23289.

[23] Brix, H. and Schierup, H-H (1989) Sewage treatment in constructed reed beds – Danish experiences. *Wat. Sci. Technol.* 21 1665–1668

[24] Carson, D.A. (1994) Full-scale leachate-recirculating msw landfill bioreactor assessments. pp 75–79 In 20th Annu. RREL Research Symp. Abstract Proc. EPA Report: EPA/600/R–94/011.

[25] CEC (Commission of the European Communities) (1991) Proposal for a Council Directive on the Landfill of Waste (91/C190/01) *Official Journal of the European Communities*, 22 July 1991, C190/1–18. Amended proposals COM(93)275. CEC, Brussels.

[26] CEC (Commission of the European Communities) (1992) Towards Sustainability–A European Community Programme of Policy and Action in Relation to the Environment and Sustainable Development, COM(92)23 final – vol. II. CEC, Brussels.

[27] Cecchi, F., Traverso, P.G., Mata-Alvarez, J., Clancy, J. and Zaror, C. (1988) State of the art of R&D in the anaerobic digestion process of municipal solid waste in Europe. *Biomass* 16 257–284.

[28] COGNIS (1995) Biocrac information. COGNIS Soil Cure GmbH, Bldg Y 20, D–40191, Dusseldorf, Germany.

[29] Collard, J.M., Corbisier, P., Diels, L., Dong, Q., Jeathon, C., Mergeay, M., Taghavi, S., Vanderlelie, D., Wilmotte, A. & Wuertz, S. (1994) Plasmids for heavy-metal resistance in Alcaligenes eutrophus CH34. *FEMS Microbiol. Rev.* 14 405–414.

[30] Conroy, D. (1993) Grace Dearborn's Daramend™ technology is being monitored by EPA in Ontario. The Bioremediation Report, Vol 2(11), November 1993, Published by King Communications Group inc, 627 National Press Building, Washington, D.C. ISSN 1064-2455.

[31] Craig, H. and Sisk, W. (1994) The composting alternative to incineration of explosives contaminated soils. *EPA Tech Trends* EPA Pub. EPA 542–N–94–008.

[32] Crawford, J.F. and Smith, P.G. (1985) Landfill Technology. Butterworth, UK.

[33] Criddle, C. (1993) The kinetics of cometabolism. *Biotech. Bioengineering* 41 1048–1056.

[34] Cummings, S.P. and Stewart, C.S. (1994) Newspaper as a substrate for cellulolytic landfill bacteria. *J. Appl. Bacteriol.* 76 196–202

[35] Department of the Environment (1994) Waste Management Paper No.26F: Landfill Co-disposal – A Draft for Consultation. HMSO, London.

[36] Diaz, L.F. and Trezek, G.J. (1979) Chemical characteristics of leachate from refuse-sludge transport. *Compost Sci. Utilisation* 20 (3) 27–30.

[37] Dragun, J. (1988) The fate of hazardous materials in soils. Part 3. *Haz. Mat. Control* 1 (5) 24–43.

[38] Ehrig, H-J. (1988) Water and element transfer from landfill sites. pp 37–48 In Proc. Intern. Workshop on Impact of Waste Disposal on Groundwater and Surface Water. Aug. 15–19, Copenhagen. NAEP, Copenhagen, Denmark. ISBN 87503 77078.

[39] Federal Ministry of the Environment (1993) Environmental Policy in Germany. Technical Instructions on Waste from Human Settlements (TA Siedlungsabfall) and Supplementary Recommendations and Information. Federal Ministry of the Environment, Nature Conservation and Nuclear Safety, PO Box 120629, 53048 Bonn, Germany/

[40] Ferguson, C.C. (1993) A hydraulic model for estimating specific surface area in landfill. *Waste Manage. Res.* 11 227–248.

[41] Finstein, M.S. (1989) Activities on composting as a waste treatment technology at the Department of Environmental Science, Rutgers University. *Waste Man. Res.* 7 291–294.

[42] Finstein, M.S. (1992) Composting in the context of municipal solid waste management. *Environ. Microbiol.* 1992 pp 355–374. Pub. Wiley–Liss Inc.

[43] Finstein, M.S., Miller, F.C. and Strom, P.F. (1986) Waste treatment composting as a controlled system. pp 363–398 In Biotechnology 86, Volume 8: Microbial Degradations. (Rehm, H.J. and Reed, G. Eds), VCH, Weinheim, Germany. ISBN 089573 0480.

[44] Ghosh, S. (1985) Solid-phase methane fermentation of solid wastes. *Trans. ASME* 107 402–405.

[45] Grainger, J.M., Jones, K.L., Hotten, P.M. and Rees, J.F. (1984) Estimation and control of microbial activity in landfill. pp 259–273. In Microbial Methods for Environmental Biotechnology. Soc. Appl. Bacteriol. ISBN 0122950402.

[46] Grbic-Galic, D. (1989) Microbial degradations of homocyclic and heterocyclic aromatic hydrocarbons under anaerobic conditions. *Dev. Ind. Microbiol.* 30 237–253.

[47] Guthrie, R.K. and Davis, E.M. (1985) Biodegradation in Effluents. *Adv. Biotechnol. Processes* 5 149–192.

[48] Hanson, R.S., Tsien, H.C., Tsuji, K., Brussaeu, G.A. and Wackett, L.P. (1990) Biodegradation of low molecular weight halogenated hydrocarbons by methanogenic bacteria. *FEMS Microbiol. Rev.* 87 273–278.

[49] Hatzinger, P.B. and Alexander, M. (1995) Effect of aging chemicals in soil on their biodegradability and extractability. *Environ Sci Technol* 29 (2) 537–545.

[50] Henke, G.A. (1989) Optimised biological soil reclamation on contaminated sites (In German). *WLB Wasser Luft Boden* 3 54–55

[51] Hofenk, G., Lips, S.J.J., Rijkens, B.A. and Voetberg, J.W. (1983) Two phase process for the anaerobic digestion of organic wastes yielding methane and compost. *Solar Energy R&D EC Series E* pp 315–322.

[52] Hogan, J.A., Toffoli, G.R., Miller, F.C., Hunter, J.V. and Finstein, M.S. (1988) Composting physical model demonstration: mass balance of hydrocarbons and PCBs. Presented at Intern. Conf. Physico-chemical and Biological Detoxification of Hazardous Wastes. 3–5 May 1988, Atlantic City, USA.

[53] Holroyd, M.L. and Caunt, P. (1994) Fungal processing: a second generation biological treatment for the degradation of recalcitrant organics in soil. *Land Contamination Reclamation* 2 (4) 183–188

[54] Jensen, V. (1977) Effects of lead on biodegradation of hydrocarbons in soil. *Oikos* 28 220–224.

[55] Johnson, J.H. and Wan, L.W. (1994) Use of composting techniques to remediate contaminated soils and sludges. pp 131–134 In 20th Annu. RREL Research Symp. Abstract Proc. EPA Report: EPA/600/R–94/011.

[56] Johnson, J.H. and Wan, L.W. (1994) Use of composting techniques to remediate contaminated soils and sludges. pp 131–134 In 20th Annu. RREL Research Symp. Abstract Proc. EPA Report: EPA/600/R–94/011.

[57] Kayhanian, M., Lindenauer, K., Hardy, S. and Tchobanoglous, G. (1991) Two-stage process combines anaerobic and aerobic methods. *BioCycle* 32 (3) 48–53.

[58] Kjeldsen, P. (1988) Degradation within waste disposal sites. pp 25–36 In Proc. Intern. Workshop on Impact of Waste Disposal on Groundwater and Surface Water. Aug. 15–19, Copenhagen. NAEP, Copenhagen, Denmark. ISBN 87503 77078.

[59] Knox, K. and Gronow, J. (1990) A reactor based assessment of co-disposal. *Waste Mange. Res.* 8 255–276

[60] Krebs, H., Rubio, M.A., Debus, O. and Wilderer, P.A. (1990) Development and operation of a pilot plant for the biological treatment of the Hamburg-Georgserder landfill leachates. pp 1113–1120 In Contaminated Soil '90. (Arendt, F., Hinsenveld, M. and van den Brink, W.J. Eds) Kluwer Acad. Pub., Dordrecht, the Netherlands.

[61] Lapinskas, J. (1989) Bacterial degradation of hydrocarbon contamination in soil and groundwater. *Chem. Ind.* 4 Dec. 1989 (23) 784–789.

[62] Lei, J. and Pouliot, Y. (1994) Final Report: Combined Scrubber bed/biofilter for the removal of volatile organic compounds from contaminated soil in venting process. Presented at the NATO/CCMS Pilot Study Third International Meeting: Evaluation of Demonstrated and Emerging Technologies for the treatment and Clean-up of Contaminated Land and Groundwater (Phase II), Trinity College, Oxford, United Kingdom, September 12–16, 1994.

[63] Leisenger, T. and Brunner, W. (1986) Poorly degradable substances. pp 475–513 IN Biotechnology 86, Volume 8 Volume 8: Microbial Degradations. (Rehm, H.J. and Reed, G. Eds), VCH, Weinheim, Germany. ISBN 089573 0480.

[64] Lindstrom, J.E., Prince, R.C., Clark, J.C., Grossman, M.J., Yeager, T.R., Braddock, J.F. and Brown, E.J. (1991) Microbial populations and hydrocarbon biodegradation potentials in fertilised shoreline sediments affected by the T/V *Exxon Valdez* Oil Spill. *Appl. Environ. Microbiol.* 57 (9) 2514–2522.

[65] Lotter, S., Brumm, A., Bundt, J., Herrenklage, J., Paschke, A., Steinhart, H. and Stegmann, R. (1993) Carbon balance of a PAH-contaminated soil during biodegradation as a result of the addition of compost. pp 1235–1246 In Contaminated Soil '93. (Arendt, F., Annokkee, G.J., Bosman, R. and van den Brink, W.J. Eds). Kluwer Acad. Pub., Dordrecht, the Netherlands. ISBN 07923 23289.

[66] Mahro, B. and Kaestner, M. (1993) Mechanisms of microbial degradation of polycyclic aromatic hydrocarbons (PAH) in soil-compost mixtures. pp 1249–1256 In Contaminated Soil '93. (Arendt, F., Annokkee, G.J., Bosman, R. and van den Brink, W.J. Eds). Kluwer Acad. Pub., Dordrecht, the Netherlands. ISBN 07923 23289.

[67] NATO Committee on the Challenges of Modern Society (1993) Demonstration of Remedial Action Technologies for Contaminated Land and Groundwater. Final Report. 2 volumes. Published United States Environmental Protection Agency Report EPA/600/R–93/012 (a,b&c).

[68] Ng, A.S., Wong, D.Y., Stenstrom, M.K., Larson. L. and Mah. R.A. (1983) Bioconversion of classified municipal solid wastes: sate of the art review and recent advances. pp 73–106 In Fuel Gas Development (Wise, D.L. Ed) Boca Raton CRC. **

[69] Oberbremer, A. and Petersen, R. (1993) Recycling biologically reclaimed soils. pp 1191–1193 In Contaminated Soil '93. (Arendt, F., Annokkee, G.J., Bosman, R. and van den Brink, W.J. Eds). Kluwer Acad. Pub., Dordrecht, the Netherlands. ISBN 07923 23289.

[70] Odesto, P., Amerlynck, P., Nyns, E.J. & Naveau, H.P. (1992) Acclimatization on a methanogenic consortium to polychlorinated compounds in a fixed film stationary bed reactor. *Water Sci. Technol.* 25 265–273.

[71] Pauss, A., Nyns, E.J. and Naveau, H. (1984) Production of methane by anaerobic digestion of refuse. pp 209–222. In EEC Report EUR 9347.

[72] Preuss, A., Fimpel, J. & Diekert, G. (1993) Anaerobic transformation of 2,4,6–trinitrotoluene (TNT). *Arch. Microbiol.* 159 345–353.

[73] Rathje, W.L., Hughes, W.W., Archer, G.H., Wilson, D.C. and Cassells, E.S. (1989) Digging in landfills. Presented at 5th Annu. Conf. Solid Waste Manage. Mat. Policy. New York City, January 1989.

[74] RCEP (Royal Commission on Environmental Pollution) (1988) *Best Practicable Environmental Option.* 12th Report. HMSO, London.

[75] Rees, J.T. (1980) The fate of carbon compounds in the land disposal of organic matter. *J. Chem. Tech. Biotechnol.* 30 161–175.

[76] Reinhart, D.R., Pohland, F.G., Gould, J.P. & Cross, W.H. (1991) The fate of selected organic pollutants codisposed with municipal refuse. *Res. J. Wat. Pollut. Control Fed.* 63 (5) 780–788.

[77] Robinson, H.D. and Lucas, J.L. (1984) Leachate attenuation in the unsaturated zone beneath landfills: instrumentation and monitoring of a site in southern England. *Water Sci. Technol.* 17 477–492.

[78] Ruby, M.V., Davis, A. and Nicholson, A. (1994) *In situ* formation of lead phosphates in soils as a method to immobilise lead. *Environ. Sci. Technol.* 28 646–654

[79] Rushbrook, P.E. (1990) Co-disposal of industrial wastes with municipal solid wastes. *Res. Conserv. Recycling* 4 33–49.

[80] Saw, C.B. (1988) Anaerobic Digestion of Industrial Wastewaters and Municipal Solid Wastes – a Position Study. LR 672 (MR) Warren Spring Laboratory, Gunnels Wood Road, Stevenage, SG1 2BX, UK, ISBN 0856245321.

[81] Seech A.G. and Marvan, I.J. (1992) *In situ*, on site bioremediation of wood treatment soils containing phenols and PAHs. Paper presented at the GASRep/DESRT 2nd Annual Symposium on Groundwater and Soil Remediation. March 25–26, 1992, Hotel Vancouver, Vancouver, British Columbia.

[82] Slater, J.H. (1981) Mixed cultures and microbial communities. pp 1–24 IN Mixed Culture Fermentations (Bushell, M.E. and Slater, J.H. Eds) Academic Press.

[83] Smith, J.R., Nakles, D.V., Sherman, D.F., Neuhauser, E.F., Loehr, R.C. and Erickson, D. (1989) Environmental fate mechanisms influencing biological degradation of coal-tar derived polynuclear aromatic hydrocarbons in soil systems. pp 397–405 In Third Intern Conf. on New Frontiers for Hazardous Waste Management. Proc. Sept. 10–13, 1989, Pittsburgh, USA. US EPA Report: EPA/600/9–89/072.

[84] Sposito, G. (1989) The Chemistry of Soils. Oxford Univ. Press, UK. ISBN 019 504 6153.

[85] Stegmann, R., Lotter, S. and Heerenklage, J. (1991) Biological treatment of oil contaminated soils in a bioreactor. pp 188–208 IN On-Site Bioreclamation Processes for Xenobiotic and Hydrocarbon Treatment (Hinchee, R.E and Olfenbuttel, R. Eds) Butterworth-Heinemann.

[86] Stentiford, E.I. and de Bertoldi, M. (1988) Composting – process technical aspects. In Proc CEC Workshop on 'Compost Processes in Waste Management.' Neresheim, Germany, 13–15 Sept. 1988. (Bidlingmaier, W. and L'Hermite, P. Eds) E Guyot SA, Belgium, CEC, ISBN 2872630198.

[87] Stottmeister, U. (1993) Capabilities and limitations in the performance of microbial remediation processes. pp 1219–1227 In Contaminated Soil '93. (Arendt, F., Annokkee, G.J., Bosman, R. and van den Brink, W.J. Eds). Kluwer Acad. Pub., Dordrecht, the Netherlands. ISBN 07923 23289.

[88] Traina, S.J., Logan, T.J. and He, X–T. (1992) Fate and Transport of trace metals in the subsurface

and surface environments following land application of msw compost. pp 43–47 In The Composting Council's 3rd National Conference Tech. Symp. Vista Hotel, Washington DC, November 11–13, 1992.

[89] United States Environmental Protection Agency (1993) Handbook for Constructed Wetlands Receiving Acid Mine Drainage. Emerging Technology Summary EPA Pub.: EPA/540/SR–93/523. Full publication available from NTIS.

[90] van Dort, H.M. and Bedard, D.L. (1991) Reductive *ortho* and *meta* dechlorination of a poly-chlorinated biphenyl congener by anaerobic micro-organisms. *Appl. Environ. Microbiol.* 57 (5) 1576–1578.

[91] von Wedel, R.J., Hater, G.R., Farrell, R. and Goldsmith, C.D. (1990) Excavated soil bioremedi-ation for hydrocarbon contaminations using recirculating leachbed and vacuum heap technolo-gies. pp 240–252 In Proc. 1990 EPA/A&WMA Intern. Symp. Hazardous Waste Treatment: Treatment of Contaminated Soils. VIP-14, Air & Waste Manage. Assoc. Pittsburgh, USA.

[92] Watson-Craik, I.A. and Senior, E. (1991) Low-cost microbial detoxification – science or serendip-ity? *J. Chem. Tech. Biotechnol.* 50 127–130.

[93] Watson-Craik,I.A., Sinclair,K.J. & Senior, E. (1992) Landfill co-disposal of wastewaters and sludges. pp 129-169. In `Microbial control of pollution', SGM Symposium 48, Ed. Fry, Gadd, Herbert, Jones & Watson-Craik. Society for General Microbiology, Reading, UK.

[94] West, C.C. (1994) Natural organic matter supports reductive dechlorination of PCE. *EPA Ground Water Currents* Issue 10. EPA Pub.: EPA-542-N-94-009

[95] Williams, R.T. and Ziegenfuss, P.S. (1989) Composting of explosives abd propellant contami-nated sediments. pp 204–216 In Third Intern Conf. on New Frontiers for Hazardous Waste Management. Proc. Sept. 10–13, 1989, Pittsburgh, USA. US EPA Report: EPA/600/9–89/072.

[96] Wise, D.L., Leuschner, A.P., Levy, P.F., Sharaf, M.A. and Wentworth, R.L. (1987) Low capital cost fuel gas production from combined organic residues – the global potential. *Res. Conserv.* 15 163–190.

NOTES:

1) To obtain Warren Spring Laboratory reports, contact Caroline Wood, AEA Technology, National Environmental Technology Centre, Glasgow Office, Kelvin Road, East Kilbride, Glasgow, G75 0RZ. Tel 03552 42626, Fax 03552 33355.

2) NATO/CCMS conference papers are abstracted in a forthcoming publication of the United States Environmental Protection Agency: Interim Status Report: NATO/CCMS Pilot Study on Research, Development and Evaluation of Remedial Action Technologies for Contaminated Soil and Groundwater (Phase II). February 1995, Draft. Available from: Technology Innovation Office, OSWER, US EPA (OS11OW), 401 M Street, SW, Washington DC 20460, USA.

1Whilst such sorption is reversible, eg by EDTA #33#, it does enhance the contaminant buffering capacity of the waste – containment system.

The Importance of Waste Characteristics and Processing in the Production of Quality Compost

G.M. SAVAGE – CalRecovery, Inc. Hercules, California USA

Abstract

The quality of compost produced from municipal solid waste or selected waste components therefrom is determined primarily by the characteristics of the waste components and by the processing methods. Consequently, a knowledge of the characteristics of the waste components and of the fundamentals of preprocessing and composting technologies is required in order to design a system to produce a compost to a desired set of specifications. The specifications can be a result of regulations, or market conditions, or both. The characteristics of the waste to be composted can be manipulated at the point of collection (e.g., by separate collection of the predominantly biodegradable waste components) and during subsequent processing. This paper focuses on the characteristics of various waste components and of mixtures thereof, the key fundamentals of processing relevant to composting, and the influence of waste characteristics on compost quality. Besides the discussion of applicable fundamentals, the characteristics of compost produced from mixed solid waste organics and from mixtures of source separated components are compared both on a theoretical basis and related to some actual data. The discussion provides information relevant to the determination of the advantages and disadvantages of using mixed solid waste organics or source separated wastes as compost feedstocks.

Introduction

The production of compost of high quality from mixed municipal solid and fractions derived therefrom can be an important factor in the acceptability of this form of processing as a means of recycling post-consumer wastes. Compost of high quality means a material that satisfies regulatory conditions, market specifications, or both.

Two predominant factors govern the quality of compost produced from solid wastes, namely the characteristics of the parent material and the methods of processing. The manipulation of the characteristics of solid wastes to produce a high-quality product can start as early as the point of generation, e.g., separation at the source (i.e., source separation), of only the biodegradable organic materials from other wastes set out for collection, and separate collection and processing of them. Alternatively, processing of mixed solid wastes can be used to produce a compost product, wherein all of the manipulation of characteristics occurs subsequent to collection. Manipulation of the biodegradability ('compostability') of materials can occur as early as the design of the products that contain the materials. In fact, product formulations are currently being designed to enhance their ability to be composted or recycled. The focus of this discussion, however, is the manipulation of material properties after the useful life of the materials.

Waste Characteristics

Many of the characteristics of the compost product have their genesis in the characteristics of the parent waste. For example, the concentration of glass and of heavy metals in a compost produced from waste materials is related to the concentration of these materials in the solid waste feedstock. Processing can be and is used to change the properties of the solid waste feedstock to those that are needed for a high-quality compost product. Processing is used here in the broad sense, i.e., any form of manual or mechanical methods of processing, including manual segregation of waste materials by the generator prior to collection.

The characteristics of solid waste also impact the performance of manual and mechanical processing systems, the design of composting operations, and act as a determinant of the operating conditions of the systems. Characteristics that impact system performance, operating conditions, and design include the particle size distribution and the ratio of carbon to nitrogen of the solid waste feedstock. The ratio of carbon to nitrogen in the feedstock, to a large extent, determines the magnitude of the need for a nitrogen source. If sufficient nitrogen is not inherently available in the solid wastes, the length of time required for the composting process will be unduly long.

The particle size distribution influences the type of equipment selected for handling (e.g., the infeed conveyors must be of adequate dimension to contain and transport the materials), the degree of segregation of oversize (i.e., non-processible) waste prior to processing, and the selection of the appropriate size reduction equipment.

A typical composition of mixed solid waste is given in Table 1. The carbon and nitrogen content and metal concentrations found in some waste components are shown in Tables 2 and 3, respectively. The average metals concentrations shown in Table 3 are the results of component analyses conducted on waste components collected from one eastern and from one western U.S. location. The maximum and minimum concentration of one metal (lead) are also shown in the table to illustrate the range of concentrations. As opposed to the other components listed in

Table 1, the Other Organic category requires some additional explanation since it will be shown later to significantly impact compost quality. This category includes textiles, wood, and fine organic materials that cannot be visually classified because of their small particle size.

Table 1 Typical Composition of MSW

Component	Wet Wt. %
Corrugated	7
Newspaper	9
Mixed Paper	22
Yard Waste	15
Food Waste	8
Other Organic	9
Plastic	9
Ferrous	4
Aluminum	1
Glass	8
Other Inorganic	8
	100

Table 2 Carbon and Nitrogen Contents of Some Solid Waste Components (% Dry Wt.)

Element	Corrugated	Newspaper	Mixed Paper	Yard Waste	Food Waste	Other Organic
Carbon, C	45.5	48.8	44.0	49.3	41.7	48.8
Nitrogen, N	0.16	0.10	0.43	3.00	2.80	1.75
C:N	284	488	102	16	15	28

Table 3 Metal Concentrations in Some Solid Waste Components (Mg/kg, Dry Wt.)

	Metal	Number of Samples	Corrugated	Newspaper	Mixed Paper	Yard Waste	Food Waste	Other Organic
Avg.	Cadmium	18	0.88	0.91	0.76	0.69	0.30	6.19
Avg.	Chromium	18	39.2	36.9	42.2	124.6	27.0	821.5
Avg.	Copper	18	30.9	44.1	62.4	73.4	18.2	54.3
Avg.	Mercury	18	0.042	0.021	0.024	0.027	0.007	1.835
Avg.	Nickel	18	16.6	25.9	20.8	38.5	14.2	41.6
Avg.	Zinc	18	51.5	46.8	175.7	205.6	25.9	2167.6
Avg.	Lead	18	21.9	28.9	31.4	36.1	201.5	91.7
Max.	Lead	18	101.0	106.0	131.0	106.0	3050.0	436.0
Min.	Lead	18	0.1	1.2	4.1	0.1	0.2	3.1

Relevant Fundamentals

Several factors, such as particle size distribution, heavy metal content, and carbon/nitrogen ratio of the feedstock have been mentioned previously to illustrate the relevancy of waste characteristics to process design and to the quality of the compost end product. Practically speaking, two general areas within the waste management system envelope are conducive to manipulation of the characteristics of the end product – the generator and the processing operation. Of these two, the generator can usually exercise the greater degree of quality control if properly educated and motivated.

Some mention must also be made of the role of collection as it pertains to composting of wastes. The role is primarily one of transport in the case of collection of source separated biodegradable materials. In the case of collection of mixed solid waste, while the role again is primarily one of transport of waste, an important consequence is the fostering of contaminating events, such as the breaking of glass containers and scattering of glass throughout the load, and the rupturing of metal containers holding fluids during compaction and unloading activities. The broken glass is difficult to remove by processing. Similarly, the contents of the ruptured metal containers can include toxic organics and heavy metals which have been dispersed among the materials in the load of waste and which, once dispersed, are difficult or impossible to remove from the other materials or to render innocuous through microbial activity.

The initial carbon/nitrogen ratio (C:N) plays an important role in determining the length of the composting process. A stable compost has a C:N of about 20: to 25:1 for a solid waste-derived compostable mixture; the achievement of a value in this range requires different lengths of time, depending on the initial C:N – a higher initial C:N requires more time for the organic material to stabilize than a mixture having a lower ratio, all other conditions being equal.

The composting process for the organic fraction of solid waste requires proper conditions. The proper conditions include a mixture with a moisture content of about 55%, with sufficient structural integrity to support porosity and maintenance of aerobic conditions, and containing an adequate population of microorganisms and nutrients. Also, a high percentage of biodegradable materials is advantageous in order to minimize contamination of the compost product and production of process residues. These aforementioned conditions can be met through judicious selection of wastes to be processed, through supplements (e.g., addition of water or nutrients), or a combination of these means.

The composting process converts biodegradable materials into predominantly intermediate solid products, CO_2, and H_2O. The process of conversion also applies to some toxic organics, but excludes a number of chlorinated organics (e.g., PCBs) that are recalcitrant to aerobic processes of conversion. The compost process does not alter the molecules of heavy metals that are in the waste. In fact, a concentrating effect occurs since the mass of the solids is reduced (through the conversion of a portion of the biodegradable material to gaseous CO_2 and water), and the mass of metals remains constant.

Important Process Related Factors

Three of the more important considerations for determining feasible solid waste-derived organic feedstocks are moisture content, carbon/nitrogen, and concentrations of metals. Of the three, moisture content is the least significant, as explained subsequently. Except in the case of the food waste component which usually has a moisture content above 60%, most organic components typically have moisture contents below 40%. To conduct aerobic composting, a moisture content of about 55% is optimum. Consequently, moisture typically must be added if non-food waste organic components or mixtures of them are to be composted in an efficient manner. Thus, consideration of moisture content normally is not a driving consideration in defining which components should be selected for collection, processing, or both.

On the other hand, the carbon/nitrogen and metal concentrations have substantial significance in the selection of feasible components and mixtures of them. In the simplest of terms, carbon/nitrogen is important because, to a large degree, it governs the duration of the composting process to stabilize the organic materials. An initial C:N of about 40 is optimum for solid waste-derived organics (i.e., containing cellulosic materials) from the standpoint of efficient use of time and equipment.

Compost Characteristics

The component characteristics given in Tables 2 and 3 can be used to estimate the characteristics of various mixtures of components as compost feedstocks and of the resultant compost product. It should be borne in mind that the estimates are only as accurate and reliable as are the data concerning the component characteristics.

Seven alternative feedstock mixtures have been constructed as examples and analyzed in terms of appropriateness for composting and of the quality of the composted product. The mass loss of organics is assumed to be 40%, as a result of microbial activity. The quality parameter studied is metals concentration. The composition of the seven feedstocks are shown in Table 4. The mixtures range from all of the organic components (Alternative Feedstock No. 1) to individual components (i.e., Alternative Feedstocks 5, 6, and 7). The component compositions of the feedstocks listed in Table 4 are derived from those in Table 1.

Table 4 Composition (%) of Feedstock Alternatives

Component	1	2	3	4	5	6	7
Corrugated	7	7					7
Newspaper	9	9					
Mixed Paper	22	22	22				
Yard Waste	15	15	15	15	15		
Food Waste	8	8	8	8		8	
Other Organics	9	—	—	—	—	—	—
TOTAL	70	61	45	23	15	8	7

The carbon/nitrogen ratios of the feedstock alternatives are shown in Table 5. The impracticality of composting corrugated in the absence of a source of nitrogen (e.g., food, materials, yard waste, or sludge) is reflected in its carbon/nitrogen ratio of 284. The relatively low carbon/nitrogen ratios of Alternative Feedstocks 4, 5, and 6 identify them as potential sources of nitrogen in those instances where nitrogen is required to reduce a high initial carbon/nitrogen ratio of a feedstock to an acceptable one.

Table 5 Estimated Carbon and Nitrogen Content (% Dry Wt.) of Feedstock Alternatives

Chemical Element	1	2	3	4	5	6	7
Carbon	46.3	45.9	45.4	46.7	49.3	41.7	45.5
Nitrogen	1.35	1.29	1.71	2.93	3.00	2.80	0.16
C:N	34.2	35.5	26.6	15.9	16.4	14.9	284.4

Table 6 Estimated Concentration of Chemical Elements in Composts for Various Feedstock Alternatives (Mg/kg Dry Wt.)

Chemical Element	1	2	3	4	5	6	7
Cadmium	2.37	1.20	1.09	0.92	1.15	0.50	1.47
Chromium	262.2	98.9	111.6	151.1	207.7	45.0	65.3
Copper	88.6	88.3	97.0	90.3	122.3	30.3	51.5
Mercury	0.428	0.040	0.037	0.033	0.045	0.012	0.070
Nickel	44.6	40.9	42.5	50.1	64.2	23.7	27.7
Lead	97.2	89.0	105.3	156.1	60.2	335.8	36.5
Zinc	653.5	216.9	265.1	238.5	342.7	43.2	85.8

The estimated metals concentrations of the feedstock alternatives are given in Table 6. The results in the table indicate that Feedstock Alternative No. 1 in general would have the higher concentration of metals, particularly chromium, mercury, and zinc. These metals are disproportionately contributed to Feedstock Alternative No. 1 by the Other Organics category. The composition of this category was described earlier. Materials included within the Other Organics category as defined herein could be expected to be set out with the more conventional organic materials (i.e., paper, food, yard waste, etc.) as one component of a wet/dry collection system. This analysis shows the potential magnitude of the quality difference that might be expected between a collection and/or processing system that targets all organic categories of the waste stream versus the targeting of 2, 3, or 4 component categories.

The estimated metal concentrations of compost, as determined by this analysis, are generally similar to and generally support the trends of metal concentrations found in composts produced in Germany from source separated MSW and from yard waste and agricultural wastes [Golueke and Diaz 1991]. As one example, defining Alternative Feedstock No. 1 is a source separated MSW compostable feedstock, this analysis reports nickel and zinc concentrations of 653.5 mg/kg and 44.6 mg/kg, respectively, for the compost product versus 408 mg/kg and 29 mg/kg, respectively, reported for a comparable feedstock in Golueke and Diaz 1991. Complete comparisons with the compost metal concentrations and feedstocks reported are not possible since the feedstocks (both the theoretical feedstocks defined herein and the German feedstocks) obviously are not identical. In fact, in the case of chromium, this analysis predicts concentrations in source separated MSW organics and yard waste that are 8 to 16 times those reported in Golueke and Diaz 1991. However, for the same metal (chromium), this analysis and the German data indicate that the concentration in the compost produced from source separated MSW organics is about 1.3 times that in compost produced from yard wastes.

Bearing in mind the range of lead concentration given in Table 3 for various organic components, the concentration of lead among samples of composts produced from the feedstock alternatives could be expected to differ by a factor of 3 to 15. This large potential for variation can have substantial repercussions in terms of meeting regulatory standards on a sample-by-sample basis.

Comparing the strictest metal standards (i.e., the Minimum Value) shown in Table 7 with the estimates given in Table 6 indicate that compost produced from any of the seven alternative feedstocks will have trouble meeting on average one or more of the limits, in particular cadmium, copper, and zinc. In some locations, even the purest of individual organic categories may not be able to meet the strictest metals limits, especially if the maximum concentration measured among samples is the governing metal criteria.

Table 7 Compost Standards (Mg/kg Dry Wt.) [1]

Chemical Element	U.S. EPA	Holland	France	Austria	Germany	Blue Angel	Minimum Value
Cadmium	39	1.5	8	4	2	1	1
Chromium	1200	100		150	400	1000	100
Copper	1500	50	400	100	75	50	
Mercury	17	1.5	8	4	1.5	1	1
Nickel	41	50	200	100	50	50	41
Lead	300	150	800	500	150	100	100
Zinc	2800	250	1000	400	300	250	

Conclusions

Composts produced from materials derived from the solid waste stream will have different characteristics depending on the characteristics of the parent materials, market specifications, regulatory limitations, and type of processing. Manipulation of compost characteristics can commence at a number of points after the wastes are generated. Thus, source separation is a possible initial manipulation. Manual and mechanical processing also can be used after collection of materials for the purpose of producing compost with the desired or required characteristics. The characteristics of several solid waste-derived composts have been shown to have substantial and important differences based on the types of feedstocks and the methods of processing.

The results presented herein indicate the magnitude of metals concentrations of compost produced from a variety of mixtures of organic components of the waste, including single component feedstocks (e.g., yard waste). The analysis indicates that the inclusion of the Other Organics material category in a compost feedstock (such as might occur in a wet/dry collection system) could result in substantially greater metal concentrations in some cases than for other combinations of components.

However, depending upon the regulatory criteria and methods of sampling and analysis, even the purest of material categories and combinations of them may not meet some of the allowable limits for metal concentrations.

References

Golueke, C.G. and L.F. Diaz (May 1991). "Source Separation and MSW Compost Quality." BioCycle, 32(5), 70-71.

[1]   Holland through Blue Angel standards—World Wastes, Aug./Sept., 1993.

The Role of Biological Treatment in Integrated Solid Waste Management

Dr PETER R. WHITE – Procter & Gamble Ltd., Newcastle Technical Centre, Whitley Road, Longbenton, Newcastle-upon-Tyne, NE12 9TS. UK.

Abstract

Biological treatment plays an important role in solid waste management since it can effectively treat a significant proportion of municipal solid waste, as well as certain industrial wastes. However, no one treatment method can manage all waste materials in an environmentally effective and affordable way. Biological treatment therefore needs to be part of an overall integrated waste management (IWM) system. Making such IWM systems both environmentally and economically sustainable requires that both their overall environmental burdens and economic costs can be predicted, and then optimised. This paper show how a Lifecycle Inventory (LCI) tool can be used to assess overall sustainability and optimise the role of biological treatment within integrated solid waste management. It suggests that further integration of both solid and water-borne CTwaste treatment can lead to additional environmental and economic improvements.

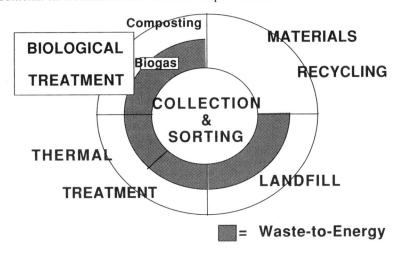

Introduction

Biological treatment, whether the aerobic process of composting, or the anaerobic process of biogasification, represents a valuable opportunity for solid waste management. Organic materials suitable for biological treatment arise from many industrial processes, such as food production and processing, brewing, and paper, leather, wool and textile production (BfE, 1991). A large proportion of municipal solid waste (MSW) also consists of biodegradable material which can also be treated biologically. In Europe, for example, food and garden waste can contribute up to 52% of MSW in some countries, whilst food, garden and paper waste together can account for up to 80% (see data given in White et al., 1995). Biological methods present an opportunity, therefore, to treat a significant proportion of both municipal and other solid wastes.

The benefits of biological treatment

The benefits of biological treatment are two-fold:– where markets for energy and products exist, biological treatment represents a method for valorising part of the waste stream; it also acts as a pre-treatment of solid waste prior to final disposal.

Valorisation

Both composting and biogasification produce a stabilised organic material that may be used as a compost, soil improver, fertiliser, filler, filter material or for decontaminating polluted soils (Ernst, 1990). The key point that distinguishes the compost-like product, from a residue that needs to be disposed of, is the presence of a market for the material.

In addition to a compost-like product, biogasification also recovers value from waste as biogas, which can be sold as gas, or burned on-site in gas engine generators to produce electricity. Around 90-150 Nm3 of gas, with a calorific value of 21–29 MJ per Nm3, can be produced per tonne of organic waste treated (White et al., 1995). There is normally a market for this product, at least for the electricity. Since the gas can be stored between production and use for electricity generation, export of power into the national grids can be timed to take advantage of high energy prices during peak consumption periods.

Pre-treatment for disposal

Volume reduction
Breakdown into methane and/or carbon dioxide and water can result in the decomposition of up to 75% of the organic material on a dry weight basis (BfE, 1991). On a wet-weight basis, the weight loss is of the order of 50%. As paper, food and garden waste together can contribute up to 80% of MSW, the potential for reducing the volume of landfill required for final disposal is considerable.

Stabilisation
Since much of the decomposition has occurred during the biological treatment, the resulting material is considerably more stable than the original waste. If subsequently landfilled, this material will produce less landfill gas and leachate than the original waste, so alleviating potential problems elsewhere in the waste management system.

Sanitisation
Both composting and biogasification are effective in destroying the majority of pathogens that are typically present in the feedstock. Aerobic composting is a strongly exothermic process, generating elevated temperatures of 60–65°C over an extended period of time, which is sufficient to ensure the destruction of most pathogens and seeds. Biogasification processes are only mildly exothermic, but may be run at an elevated temperature of 55°C (thermophilic process) by the addition of heat. This temperature, plus the anaerobic conditions, is usually sufficient to ensure that the residue is pathogen-free.

Optimising the biological treatment process

There are many routes to optimise the biological treatment of solid waste, whether by composting or biogasification. Selecting the appropriate process type and feedstock material are two basic choices. Taking the feedstock material, for example, it has been shown that using a broader feedstock definition for source-separated material that includes paper (non-recyclable paper and paper products) can offer processing advantages over treatment of a more narrowly defined vegetable, food and garden waste (VFG) feedstock.
The advantages can include:

- Reduced production of leachate during the process (Verstraete et al., 1993)
- Reduced requirement for bulking agents (Haskoning, 1991)
- Improved carbon:nitrogen ratio for the composting process (Jespersen, 1991)
- Increased organic content of the final compost (ORCA, 1991)
- Reduced salt level of the final compost produced (Fricke, 1990)

To consider the effect of any proposed improvement, however, it is necessary to look at the overall biological treatment process (Figure 1) to consider the inputs and outputs of the process. Optimisation consists of increasing the production of useful products (gas/energy and marketable compost), whilst reducing emissions (to water, air and land) and the consumption of energy. There are always two sides to the optimisation process: minimising the environmental burdens of biological treatment, and keeping the overall cost to an acceptable level relative to other available waste treatment options.

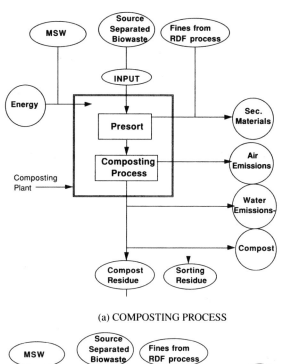

(a) COMPOSTING PROCESS

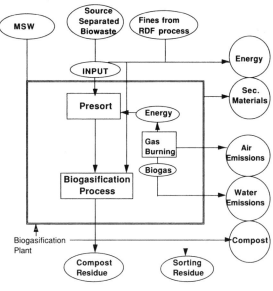

(b) BIOGASIFICATION PROCESS

Figure 1 Flow Diagrams for Typical Composting and Biogasification Plants

Biological treatment as part of an integrated solid waste management system

Whilst the composting or biogasification process can itself be optimised, biological treatment does not operate in isolation. It relies on the existence of an appropriate collection and, in some cases, sorting system to provide the input materials in a suitable form. Collection and treatment are inter-dependent. For example, if the biological treatment process is designed to accept a narrowly-defined feedstock, the collection system and public education programmes must ensure that only the desired materials are collected. Conversely, if only commingled materials are collected, then the biological treatment plant must be designed to deal with this feedstock. Biological treatment must therefore be considered within the overall waste management system.

Despite its value, it is clear that biological treatment alone cannot effectively manage all materials in the waste stream, since a significant proportion of materials will not be decomposed by such processes. Similarly, there will be residues from biological treatment that need to be treated further by other technologies, such as incineration or landfilling. Overall, therefore, it is becoming accepted that no one single treatment method can deal with all materials in the solid waste stream in an environmentally sensitive way. What is needed is an integrated waste management (IWM) system that uses a range of treatment methods, including biological treatment, materials recycling, thermal treatment (burning of refuse-derived fuel (RDF), packaging-derived fuel (PDF) and/or mass burn incineration) and landfilling. The elements of an integrated waste management system are shown in Figure 3. Following an appropriate collection and sorting system, a combination of such treatment methods should be able to deal effectively with all materials in the solid waste stream.

Biological treatment can, therefore, play an important role within integrated solid waste management.

Optimising the overall iwm system

As it is part of the larger solid waste management system, any changes in the biological treatment process will have effects elsewhere. Taking the earlier example of including paper in the feedstock for composting, this will have effects on the collection and sorting operation, and also on the overall performance of the waste management system.

Reported advantages of collecting a broader 'biowaste' fraction that contains paper include:

Reduction of malodours during collection (White et al., 1995)
Reduction of seepage water during transport and storage
Reduction in seasonal variability in the amounts of biowaste collected
Increased acceptance by participating households (Boelens, 1995).

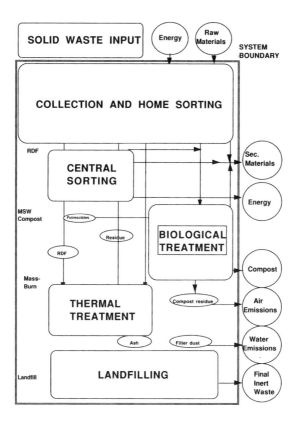

Figure 2 The role of biological treatment in an integrated solid waste management system.

Including paper in the biowaste fraction is also likely to have the overall effect of increasing the total amount of municipal solid waste that is valorised (ORCA, 1991).

Because of the interactions between the different elements of a solid waste management system, the best approach is to optimise the overall system. The objective should be sustainable waste management, which can be broken down into both environmental and economic aspects.

Environmental sustainability involves reducing the overall environmental burdens associated with managing society's solid waste, both in terms of resource usage (energy and materials) and the release of emissions to air, water and land.

Economic sustainability requires that the overall waste management cost is acceptable to all sectors of the community which is served.

If we want to optimise the overall system against these objectives, we need to be able to predict both the overall environmental burdens and economic costs of waste management systems. The developing tool of Lifecycle Inventory (LCI) can be used for this purpose. We have applied the lifecycle concept to solid waste management to produce a Lifecycle Inventory for the management of municipal solid waste (White et al., 1995). The details of the solid waste system modelled in this LCI are shown in Figure 3.

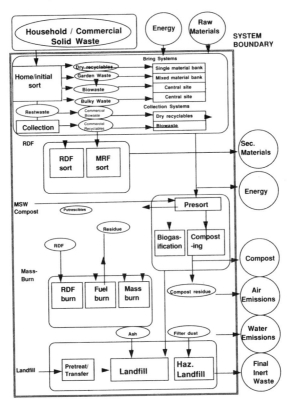

Figure 3 Detailed Materials and Energy Flows in the Lifecycle of Solid Waste.

Whilst individual processes, such as composting or biogasification, need to be optimised, optimising the overall IWM system involves choosing which waste management options should be included, and how they best fit together. Focusing on biological treatment, for example, how do the overall environmental burdens and economic costs for a solid waste system that relies on landfill only compare with those for a system using mixed waste composting? How does this compare with a system that uses incineration rather than composting? How would the overall performance alter if biogasification were used instead of composting, or if a source-separated feedstock were used instead of mixed waste? What would be the overall effect if there was a market for the compost so produced, versus if there were no market and the compost residues needed to be landfilled?

To be able to carry out such 'What if ...?' calculations a flexible, yet easy to use, LCI tool is required. We have created such an LCI tool as a spreadsheet. The user defines the amount and composition of the waste to be managed (this can be on the scale of a town, city, region or country) and selects the waste management options desired. The options available are shown in Figure 3. The spreadsheet will then calculate the overall environmental burdens in terms of energy consumption and production of emissions to air, water and land, and the overall economic cost.

A hypothetical case of how this can be applied is shown in Figures 4–6. This example takes an area containing one million households, with the waste generation and waste composition characteristics of France. It then considers five different ways of managing this waste, using a combination of biological treatment, incineration with energy recovery, materials recycling and landfilling. Biological treatment options include composting of commingled MSW (both with and without the presence of a market for the composted material) and composting of a source-separated biowaste fraction (with both narrow and broad definitions). For each scenario, results for just three parameters are given here: overall energy consumption, final landfill volume and global warming potential of resulting air emissions. The LCI tool itself will give the full range of emissions as well as energy consumption. Full details of these analyses are given in the book Integrated Solid Waste Management: A Lifecycle Inventory (White et al., 1995), though not included here due to lack of space.

Using an LCI tool of this kind, it is therefore possible to optimise an overall Integrated Waste Management system, in terms of both environmental and economic sustainability, Thus it is possible to consider the effects of including any form of biological treatment on the overall performance of a solid waste management system.

Details of Waste Management Systems compared in Figures 4–6.

1. Basic System: commingled collection of household waste followed by landfilling.
2. Commingled collection and mass-burn incineration of household waste.
3. Commingled collection and composting of household waste, with market for the compost.
   a. As in case 3, but with no market for the compost.
4. Separate collection and composting of biowaste, landfilling of restwaste.
   a. As in case 4, with an integrated collection system.
   b. As in case 4, including washing of the biobin.
   c. As in case 4, but including paper in the biowaste collection.
5. Separate kerbside collection of dry recyclables and incineration of the restwaste.
   a. As in case 5, with an integrated collection system.
   b. As in case 5, with a close-to-home bring system for dry recyclables.
   c. As in case 5, with a central bring system for dry recyclables.

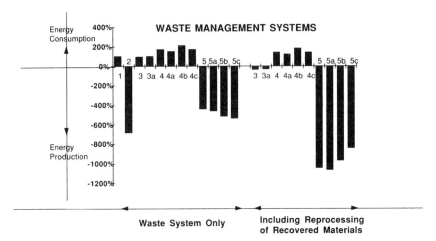

Figure 4 Overall energy consumption of different solid waste management systems.
Note: Energy consumption (thermal energy equivalent) indexed relative to basic waste management system (Case 1).

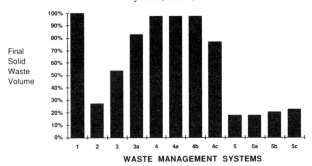

Figure 5 Overall final solid waste produced by different solid waste management systems.
Note: Final Solid Waste Volume indexed relative to basic system (Case 1)

Figure 6 Relative global warming potential (GWP) due to air emissions from different solid waste management systems.
Notes: 1. Global Warming Potential (GWP) is shown relative to the basic waste management system (Case 1)
2. GWP values used are CO2 = 1, CH4 = 35, N2O = 260. (Values for 20yr timescale, IPCC, 1992)

Further integration of waste management

This paper has dealt with the role of biological treatment within integrated waste management systems that manage solid waste, and in particular, municipal solid waste. It has shown that by taking an holistic approach, it is possible to optimise the overall system both environmentally and economically. Such integrated systems, however, are also capable of managing other types of wastes. In particular, IWM systems employing biological treatment are capable of handling sewage sludges that arise from the treatment of water-borne wastes. Sewage sludges can also be treated thermally, by incineration with municipal solid waste. There are other connections between the two waste systems:– solid waste and water-borne waste can represent alternative disposal routes for some waste items such as paper products (tissues/hygiene products). Since there is this overlap between the solid waste and water-borne waste systems, it makes sense to plan their treatment together in an overall integrated strategy for all wastes. Benefits of this further level of integration include the prevention of problem-shifting (i.e. so that improvements in water-borne waste management do not give rise to greater problems in the management of solid waste or *vice-versa*), plus the greater efficiencies, economies and flexibilities of scale.

Conclusions

Composting and biogasification both present good opportunities to manage significant amounts of municipal and other forms of solid waste.

Biological treatment cannot effectively treat all materials in a municipal solid waste stream, so needs to be part of an integrated solid waste management system.

Whilst it remains important to optimise individual biological treatment processes, real, overall environmental improvements will be achieved by optimising the overall IWM system against the goals of environmental and economic sustainability.

Lifecycle Inventory, and in particular a user-friendly LCI tool, can be used to optimise IWM systems and predict the effects of wider use of biological treatment within integrated waste management systems.

Further integration of waste management systems, to include both solid and water-borne wastes, offers potential for further overall environmental and economic improvements.

References

BfE (1991) Bundesamtes für Energiewirtschaft. *Vergärung biogener Abfälle aus Haushalt, Industrie und Landschaftspflege*. Schriftreihe des Bundesamtes für Energiewirtschaft Studie Nr. 47.

Boelens, J. (1995) *Paper in Biowaste: Effects on household biowaste collection, composting process and compost quality.* Proceedings of ORCA Congress 1995. Organic Reclamation and Composting Association, Brussels.

Ernst, A–A, (1990). *A review of solid waste management by composting in Europe.* Resources, Conservation and Recycling, 4: 135–149.

Fricke, K. (1990). *Grundlagen der Kompostierung,* EF–Verlag, Berlin.

Haskoning (1991) *Conversietechnieken voor GFT-afval,* NOH, 53430/0110.

IPCC, (1992). Intergovernmental Panel on Climate Change: 1992 IPCC Supplement. IPCC Secretariat, World Meteorological Organisation, Geneva, Switzerland.

Jespersen, L. (1991). *Source separation and treatment of biowaste in Denmark.* Verwerkingsmogelijkheden en scheidingsregels van groente-, fruit- en tuinafval, Koninklijke Vlaamse Ingenieursvereniging (K.V.I.V.).Syllabus studiedag 7 maart, 1991.

ORCA (1991). *Composting of Biowaste - The Important Role of the Waste Paper Fraction.* Solid Waste Management, An Integrated Approach part 8. The Organic Reclamation and Composting Association, Brussels, Belgium.

Verstraete, W., De Baere, L. and Seeboth, R-G. (1993) *Getrenntsammlung und Kompostierung von Bioabfall in Europa.* Entsorgungs Praxis, March 1993.

White, P.R., Franke, M. and Hindle, P. (1995) *Integrated Solid Waste Management: A Lifecycle Inventory.* Blackie Academic and Professional. 362pp.

Acknowledgement.

The figures in this paper are taken from *Integrated Solid Waste Management: A Lifecycle Inventory.* by White, P.R, Franke, M. and P. Hindle. Published by Blackie Academic and Professional. 1995.

Results of Laboratory and Field Studies on Wastepaper Inclusion in Biowaste in View of Composting

DE WILDE B., BOELENS J. and DE BAERE L. –
Organic Waste Systems NV Dok Noord, 4 B-9000 Gent
Belgium

Abstract

A comparative study was run during 13 months on two biowaste definitions involving both lab tests and field surveys. A narrow biowaste definition, allowing only biogenic wastes was compared to a broad biowaste definition, including compostable man-made products, such as non-recyclable wastepaper and diapers. Two similar real-life test areas with each about 425 inhabitants were defined in a semi-urban area North of Antwerp. During the whole test period the amount of curbside waste, this is biowaste and restwaste (the 'non-biowaste'), was continuously and precisely measured and also analysed regularly (twice per season) for composition. At the start, middle and end of the test, surveys were held with a questionnaire for the population of each test area. In each season of the year, bench-scale aerobic composting experiments were run to evaluate the influence of both biowaste definitions on the composting process and the compost end product.

The introduction of source-separated waste collection resulted in an overall landfill diversion of 43% for the narrow biowaste definition and 46% for the broad biowaste definition. The contamination of biowaste (about 3%) was low for both definitions, the restwaste (or so-called grey waste) still contained a lot of organics (34 to 66%). Supposing that the collection and the appropiate disposal of organics could be improved to 95% efficiency (compared to about 60% currently), the landfill diversion could be increased to 59% for the narrow and 74% for the broad definition. Whereas the average efficiency of separate collection of organics is about 61%, it is 49% for non-recyclable paper and even about 20% only for certain categories of compostable, non-recyclable paper. Apparently some more education or a better system of recognition and identification is needed to improve the collection efficiency of man-made compostables.

The acceptance and goodwill of the population was significantly higher for the broad biowaste definition, especially in the summer months. The yearly, overall composition of the total curbside waste (biowaste and restwaste combined) is 17% kitchen organics, 47% yard waste, 4% recyclable paper, 13% non-recyclable paper and 19% non-compostables. It must be mentioned that glass, paper and large yard waste are collected separately by a voluntary bring-system. The broad biowaste typically contained 16.0% paper (of which 2.9% was recyclable) versus 2% for the narrow (0.3% recyclable).

The aerobic composting process was improved by expanding the biowaste definition through easier moisture control, better aeration and a more tempered pH evolution. A significant difference was seen for NH_3 and corresponding odour emission, being much lower in the broad definition. The quality of the compost produced was similar and acceptable for both biowaste definitions.

Introduction

In recent years a strong trend has developed in several European countries towards source separated household waste collection and subsequent composting or biogasification of the organic biowaste fraction. In several cases the biowaste fraction is limited to natural, biogenic organic material only although several other man-made or manufactured products could be considered to be compostable and given access to the biobin (or biowaste).

At the same time, a parallel move arose to recuperate and recycle as many waste products as possible. Well-known examples are the Töpfer-law resulting in the Green Dot and DSD system in Germany, the similar Eco-emballages in France and Fost+ in Belgium, and the packaging directive on a European level (Club de Bruxelles, 1994). The Flemish Masterplan on Solid Waste 1991–1995 stipulates that waste products which can be 'recuperated' may not be landfilled or incinerated (OVAM, 1994). In all these regulations composting of waste products is considered a form of material recycling.

Materials and methods

In order to evaluate the effect of source separated household waste collection and more specifically of biowaste definition on the diversion of household waste from landfill, a comparative field study on biowaste definition was carried out in the area of the 'intercommunale' IGEAN, northeast of Antwerp, Belgium. The study lasted 13 months, from the beginning of September 1993 to the end of September 1994.

The main objectives were to investigate the effect of biowaste definition on:

– Appreciation, participation rate and sorting behaviour of households
– Quantity and composition of Biowaste/Biowaste Plus and restwaste (impact of seasonal variations)
– Landfill diversion of household waste
– Sorting behaviour and performance

- Composting processing
- Compost quality

To achieve the objectives, 2 test areas in a semi-urban region were selected, each with a population of + 140 households, identical demographics and maximum similarity. In one area, Wuustwezel, a narrow biowaste definition (kitchen + garden waste) was applied and in the other area, Essen, an expanded biowaste definition (kitchen, garden and non-recyclable paper waste). Further on the narrow biowaste definition will be referred to as 'Biowaste' and the expanded biowaste definition as 'Biowaste Plus'.

During the first month of the study mixed household waste was collected and analysed to establish baseline data. After that the source separated household waste system was applied in which every week one waste collection was run, alternating between Biowaste/Biowaste Plus (in 120 L rigid biobins) and restwaste (in plastic bags). The amounts of waste collected were recorded as well as the participation rate of the households. Per season 2 detailed sorting tests on both Biowaste or Biowaste Plus and restwaste were done. Once per season a representative sample of the respective biowaste fraction was composted at pilot plant-scale (200 L) and the quality of the produced compost was determined according to the German RAL GZ 251 analytical methods (BGK, 1994).

At start of the study the households were well informed through teasers and newsletters. To evaluate the reaction of the population throughout the project 3 polls were carried out at start, middle and end of the project.

Results

Participation rate and reaction of the population

The introduction of the source separated household waste collection system in substitution for a mixed household waste collection system was very well accepted and supported by the population, irrespective of the biowaste definition. At mid term, the population connected to the Biowaste scheme, requested change over to Biowaste Plus.

Expanding the biowaste definition with the non-recyclable paper fraction resulted in a higher and more consistent participation rate (= number of households using biobin divided by total number of households), + 22% on the average, and fuller biobins in comparison to Biowaste collection (Figure 1). As a result, the households with the Biowaste Plus definition asked for larger biobins.

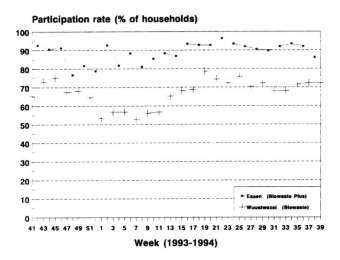

Figure 1 Measured Participation Rate of Households in Wuustwezel (Biowaste) and Essen (Biowaste Plus).

The level of comfort was much higher for the Biowaste Plus definition than for the Biowaste definition. Most complaints of households with the Biowaste collection were about odor, vermin and cleanliness. As a result about 6% of the household stopped separated biowaste collection during summer (versus < 1% for Biowaste Plus collection). Finally, households in both areas urged for a more frequent collection.

Household waste production

The amount of household waste to be collected at curbside was 60% of the total household waste stream. About 40% was already collected through container parks (paper, glass, recyclable paper, construction debris, oversize garden waste, metals, plastics,...).

The introduction of the source separated collection system resulted in both test areas in more household waste being collected (+ 15–30% versus baseline). The results of the weekly weighings showed that a major part of the Biowaste or Biowaste Plus could be diverted from landfill by composting/anaerobic digestion. The measured yearly landfill diversion percentage was 44% in case of Biowaste collection and 51% in case of Biowaste Plus collection (increase of 16%) (Figure 2). These results are comparable to those obtained in other studies (Spencer, 1993). Finally, biowaste production was subject to strong seasonal effects. In winter 20 to 40% less biowaste was produced. The amounts of restwaste produced remained fairly constant throughout the year.

Composition of waste, sorting behaviour and performance

Expanding the biowaste definition with the non-recyclable paper fraction did not lead to a higher amount of non-compostable products (contaminants, nuisance) in Biowaste. For both Biowaste definitions the sorting behaviour in the biobin was very good (3% impurities in Biowaste and 3.6% in Biowaste Plus). However, the purity of the restwaste was poor in both areas. In the 'Biowaste area' 34% of the restwaste consisted of Biowaste. In the 'Biowaste Plus area' 47% and 66% of the restwaste consisted of Biowaste and Biowaste Plus respectively. These results were striking, moreover because of the fact that even in winter about 40% of the rest-waste consisted of Biowaste, although less Biowaste was produced then and the biobins were on the average not completely full (impact of bi-weekly collection) (Table 1).

Table 1 Sorting Behaviour of Households (% of Incorrect Disposal).

Season	BIOWASTE			BIOWASTE PLUS	
	Biobin	Restbag	Biobin	Restbag (Biowaste Plus)	Restbag (Biowaste)
AUTUMN	1.7%	36.5%	2.5%	69.3%	49.0%
WINTER	4.1%	39.9%	4.3%	61.9%	40.4%
SPRING	3.4%	28.9%	3.5%	70.4%	55.0%
SUMMER	3.1%	29.5%	3.9%	60.2%	40.2%
TOTAL	3.0%	33.6%	3.6%	65.6%	46.5%

Assuming that the incorrect disposal of Biowaste/Biowaste Plus in the restwaste fraction could be reduced to 5%, the potential landfill diversion in Wuustwezel (Biowaste) would be 61.1% and in Essen (Biowaste Plus) 82.3% (increase of 35%).

The compostable fraction amounted to 84.2% of the total curbside waste stream in Wuustwezel (Biowaste) and 81.5% in Essen (Biowaste Plus). Of this the paper/board fraction amounted to 15.8% in Wuustwezel (Biowaste) and 16.9% in Essen (Biowaste Plus). This paper fraction consisted of 22.2% of so called recyclable paper in Wuustwezel (Biowaste) and 23.1% in Essen (Biowaste Plus). These figures show that expanding the biowaste definition is not disadvantageous to recyclable paper collection efforts (Table 2).

Table 2 Composition of Biowaste, Restwaste and Total Curbside Household Waste (Yearly averages in %).

Component	BIOWASTE			BIOWASTE PLUS		
	Biowaste	Restwaste	Total	Biowaste Plus	Restwaste	Total
Compostables	**99.1**	**66.7**	**84.2**	**96.4**	**65.6**	**81.5**
Kitchen waste	38.1	31.8	35.2	16.0	18.3	17.1
Yard waste	58.9	1.8	32.7	64.2	28.1	46.8
Compostable paper/board	2.0	32.1	15.8	16.0	17.8	16.9
Recyclable	0.3	7.3	3.5	2.9	4.9	3.9
Non-recyclable	1.7	24.8	12.3	13.1	12.9	13.0
Other compostables	0.1	1.0	0.5	0.2	1.3	0.7
Non-Compostables	**0.9**	**33.3**	**15.8**	**3.6**	**34.4**	**18.5**

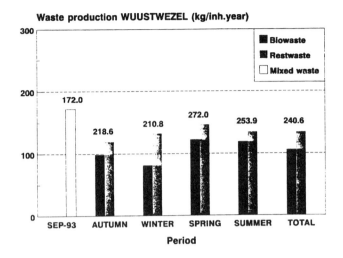

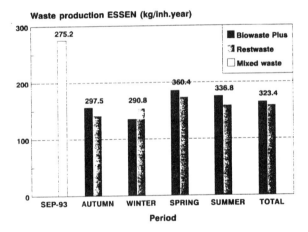

Figure 2 Amounts of Household Waste Produced per Inhabitant and Extrapolated to a Period of a Full Year. (The Figures on Top of the Bars are the Sum of Restwaste and Biowaste in the Respective Season).

The sorting performance was good to very good for garden waste and non-compostables with more than 70 to 90% of correct disposal. For kitchen waste the sorting performance was only 50 to 60%. This is a strong indication that kitchen waste was disposed of into the waste fraction to be collected first, be it biobin or restbag without any conscious sorting.

For the biodegradable/compostable products, other than garden or kitchen waste, the sorting performance varied a lot. For easily recognizable products (e.g. diapers) it was rather good (+70%). However, for products that are more difficult to recognize as 'compostable' e.g. cardboard packages, the sorting performance was less good (22 to 52%). These results show that an even better information system is needed so that compostable products are better defined and sorting rules easier to apply for the households.

Composting process

In general it was easier to compost Biowaste Plus than Biowaste. Temperature could be better controlled and pH remained more stable. Especially in winter it was very difficult to compost Biowaste, with a large amount of leachate, heavily loaded with organic material, coming out of vessels. For Biowaste Plus, on the contrary, water had to be added to ensure a good composting.

During composting the NH_3 emission was 2 to 3 times higher for Biowaste than for Biowaste Plus. This is important because NH_3 attributes to the phenomenon of acid rain. At the same time a better nitrification occurred during Biowaste Plus composting which is an indication of a better composting process (Figures 3 and 4).

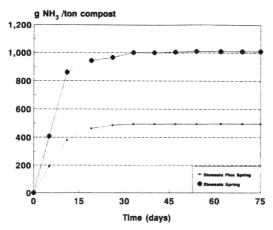

Figure 3 Cumulative NH3 Emission During Composting of Biowaste Collected in Spring.

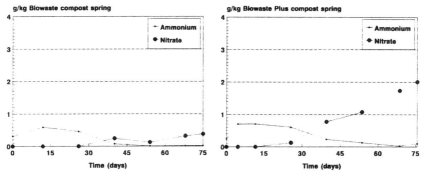

Figure 4 Evolution of NO and NH During Composting of Biowaste Collected in Spring.

Expanding the biowaste definition did not prolong the duration of the composting process. For both Biowaste and Biowaste Plus composting CO_2 production and organic matter degradation slowed down after 30 days. At the end of the test (75 days) there was no difference in CO_2 production and the % degradation of organic matter.

Refining of the compost resulted in a much bigger > 25 mm fraction of Biowaste Plus compost. However, this fraction consisted for a great deal of top and back sheets from disposable diapers. Based on the screening results the final landfill diversion could be calculated. In Wuustwezel (Biowaste) a landfill diversion of 42.6% was obtained, in Essen (Biowaste Plus) 45.6%. Assuming a 95% purity in the restwaste fraction a theoretical landfill diversion of 59.0% and 73.5% respectively could be obtained (Figure 5). Figure 6 shows the distribution of the household waste stream including the recycling efforts through container parks.

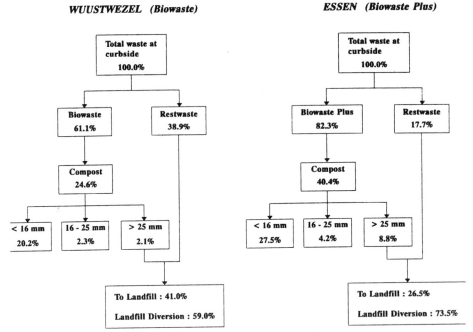

Figure 5 Total Potential Landfill Diversion.

Compost quality

Expanding the biowaste definition has no effect on the level of impurities in the marketable (< 16 mm) compost fraction. Large impurities such as the remainder of plastics or disposable diapers are almost entirely screened off over 25 mm and therefore do not end up in the marketable compost fraction.

Expanding the biowaste definition has no effect on the chemical quality of the marketable compost. On the contrary, Biowaste Plus compost had a higher organic matter content, a better nitrification and a slightly lower salt content. The heavy metal content of both Biowaste and Biowaste Plus compost was well below the limits. The Biowaste Plus compost had a lower zinc and copper content than Biowaste compost (Table 3).

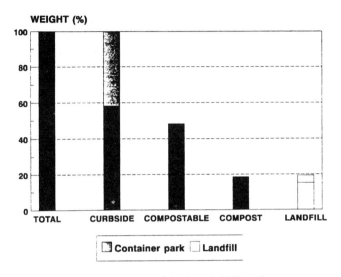

WEIGHT (%)

☐ **Container park** ☐ **Landfill**

Figure 6 Distribution of the Household Waste Stream.

At the end of 75 days of composting the Biowaste Plus compost was in general more stable than Biowaste compost, in spite of the addition of a non-recyclable paper fraction. This shows that expanding the biowaste definition does not lead to longer composting. With regard to phytotoxicity there was no difference between Biowaste and Biowaste Plus compost.

Table 3 Chemical Analyses of the Marketable Compost Fraction.

Parameter		YEARLY AVERAGE	
		Biowaste	Biowaste Plus
TS (%)		64.6	61.2
VS (% on TS)		23.6	28.1
Kj-N (g/kg TS)		12.4	12.7
NO-N (g/kg)		0.25	0.41
NH-N (g/kg)		0.00	0.03
C/N		9.5	11.1
pH		7.4	7.4
Ec (mS/cm)		3.04	2.95
Heavy metals	Zn	172	143
(ppm on TS)	Cu	24	19
	Pb	47	48
	Ni	7	8
	Cd	1.0	1.1
	Cr	16	19
	Hg	0.08	0.09

Conclusions

From the comparative lab- and field study, it could be seen that the inclusion of wastepaper in biowaste resulted in a higher acceptance and participation of the population while the purity of the biowaste fraction was not significantly affected. In both cases with or without wastepaper, large amounts of biowaste were still disposed of through the grey, restwaste fraction. The introduction of separate biowaste collection resulted in a landfill diversion of 43% in case no wastepaper was included and of 46% with wastepaper inclusion. The potential landfill diversion after better performance could however be estimated at respectively 59 and 74%.

With regard to composting, the inclusion of wastepaper in biowaste leads to an easier and more stable process with much smaller NH_3 emission whereas the quality of the compost produced is comparable or even slightly better (higher content of organics, lower level of heavy metals).

Acknowledgements

The 'intercommunale' IGEAN selected the test areas, organized the source separated household waste collection and the information/motivation campaign and provided useful information about the total household waste stream.
The study was sponsored by Procter & Gamble ETC.

References

BGK (1994). Methodenbuch zur Analyse von Kompost. Bundesgütegemeinschaft Kompost e.V., Köln, Germany, 86 pp.
Club de Bruxelles (1994). Packaging and Packaging Waste in Europe. Study written for the Conference organized by Club de Bruxelles 10 & 11, 1994. 103 pp.
OVAM (1994). Masterplan GFT- en Groenteafval. Evaluatierapport 1991–1994 (Masterplan Biowaste and Garden Waste. Evaluation report 1991–1994), 137 pp.
SPENCER, R. (1993). European Collection Programs for Source Separated Organics. Biocycle, June 1993, pp. 56–59.

Fuel Recovery: Valorization of RDF and PDF

MARTIN FRANKENHAEUSER[1] &
HELENA MANNINEN[2]

Abstract

Energy recovery of used materials can be performed as mixed municipal solid waste (MSW) incineration or as fuel recovery for co-combustion with conventional fuels. Recovered fuels are refuse derived fuel (RDF) which is mechanically separated and processed from MSW, as well as packaging derived fuel (PDF) which is the source separated, processed, dry combustible part of MSW.

A one year long co-combustion test of RDF with peat and coal has been carried out in a 65 MW CFB power plant in Kauttua, Finland. The efficiency of the combustion process and corrosion behaviour of the boiler were of particular interest in this study. Five different PDFs were also tested. A wide analytical programme was carried out including solid and gaseous emission measurements.

The results are encouraging, showing that RDF and PDFs are technically and economically feasible and environmentally friendly fuels for co-combustion. Low CO emissions showed clean and efficient combustion. SO_2 emissions decreased, because part of the coal was replaced by RDF and PDFs. HCl emissions increased when the chlorine content of the fuel mixture increased. Heavy metals were concentrated to the fly ash in unleachable form. PCDD/F (dioxin) emissions were at the normal power plant level and far below the strictest incineration limit.

Long-term co-combustion of 10 % RDF did not cause any high temperature chlorine corrosion of the superheater (500°C) of the boiler. Soot blowing sequences did not change and no fouling was detected.

The results show that it is useful, technically possible and environmentally friendly to combine resource and waste management in the form of fuel recovery for energy production in solid fuel fired power plants.

Introduction

A packaging system is designed for efficient distribution, protection of the packed product and for saving energy and other resources in the distribution chain. Light

[1] **Borealis Polymers Oy**, P.O. Box 330, FIN-06101 Porvoo, Finland
[2] **Neste Oy Corporate Technology**, P.O.Box 310, FIN-06101 Porvoo, Finland

weight one-way packaging is often the best solution for long distance deliveries of consumer goods. 'Source reduction' of packaging waste is a continuous process, because packaging materials are being constantly developed for strength and weight reduction. Short range distribution can utilize re-usable packaging, and industrial raw materials are increasingly transported in bulk without packaging.

Modern lightweight packaging materials are often difficult to recycle. The packaging itself has a high surface-to-weight ratio and used packaging is often contaminated with product residues. This makes separate collection and recycling costly and environmentally burdening. Energy recovery is therefore often the preferable recovery option for post consumer packaging waste.

In the combustion process, organic carbon (C) is oxidized to carbon dioxide (CO_2). The release of CO_2 from combustion for energy conversion is a much discussed issue. It is important to realise that a high energy conversion efficiency in any combustion process gives a low fuel consumption and consequently low CO_2 release per unit energy produced. One should also remember that organic carbon under anaerobic landfill conditions is converted to landfill gas, mainly methane (CH_4), which is considered a much more severe greenhouse gas compared to CO_2. The nature of the released CO_2 is also debated. Carbon in wood based combustible materials is of recent origin and the use of these materials as fuel is therefore generally considered CO_2 neutral. This means that the released CO_2 will be assimilated by growing biomass within a reasonable time. The use of fossil fuels increases the content of CO_2 in the atmosphere.

Plastics are today mostly made of oil fractions which are less suitable for liquid fuel products. The production of plastics requires roughly the same amount of energy as contained in the oil fraction transformed to polymer. Plastics in packaging, automotive, construction, insulation, etc., however, during it's lifespan saves a many times greater amount of energy in the form of oil-based fuels, gasoline, diesel fuel and heating oil. The avoided fuel consumption due to use of plastics leads to a decreased extraction of crude oil and consequently to a decreased overall CO_2 release.

The use of modern lightweight packaging materials, paper, board, plastics and composites of these, is sustainable. Energy recovery of used combustible packaging is part of the larger system and it could be shown that net CO_2 release may even be negative when total system fuel consumption is taken into calculation, compared to a system using non-combustible packaging. This is, in most cases, also true when compared to open loop recycling of small post consumer packaging.

Source Separation for Fuel Recovery

Energy recovery from used products and packaging can be performed in two principally different ways. It is, however, common to consider only incineration of mixed municipal solid waste, when addressing the subject. This is often called Waste-To-Energy.

Energy recovery can also be realized as fuel recovery for co-combustion with conventional fuels. In this case, the recovered fuel is refuse derived fuel, or packaging derived fuel, (ref. CEN-CR 1460). RDF is mechanically separated from MSW. PDF is the source-separated, processed, dry combustible fraction, otherwise ending up in MSW. This solution is natural when the wet organic part of MSW is diverted to composting or anaerobic digestion for biogas production. The system is cost efficient compared to separate collection aimed at material recycling, because the yield per bin and collection trip is high. Fuel recovery has a low environmental impact because of the high collection efficiency and because of the high energy efficiency of conventional industrial boilers and utility plants.

Fuel recovery will contribute to resource conservation by substituting fossil fuels. Acceptable environmental standards for co-combustion can be based on proper sorting instructions, permitted fuel processing and maintaining or even improving the clean front end combustion of the boiler plant, meeting emission regulations set for the primary fuel of the plant. The scheme can be fulfilled under the concept of shared responsibility as suggested in the 'Packaging Directive'.

Full Scale Co-Combustion Test Power Plant and Fuels

Combustion tests were carried out in a 65 MW Pyroflow CFB boiler in the Kauttua power plant in Finland in order to verify the technical and environmental aspects of fuel recycling. The unit consisted of a furnace, two hot cyclones, a 500°C, 84 bar steam boiler and electrostatic precipitators (ESP) for dust control. Fuels normally used in the plant include peat, wood waste, coal and mixed scrap from a paper mill and a packaging production plant. The boiler load consisted of steam provided to the two plants, heat supplied to the district heating network of the municipality of Kauttua and electricity for the grid.

Polish coal and milled peat are the primary fuels of the power plant. The following recovered fuels were tested.

PDF/LPB   (liquid packaging board) – Separately collected post-consumer milk and juice cartons from Helsinki. The cartons were shredded to about half of their original size.

PDF/PE   Separately collected post-consumer polyethylene bottles. The material contained mainly bottles and canisters from gasoline stations and households, but also some film material was included. The material was shredded to a 40 mm mean particle size.

PDF/PET   One-way polyethylene terephthalate bottles from the soft drink industry. The material also contained polypropylene caps and it was shredded to a 10 mm mean particle size.

PDF/YTV   The Helsinki Metropolitan Area Council operates a separate collection scheme for wet organic waste. Metals and glass were separated from the remaining dry fraction of MSW by hand. The fuel contained a large amount of organic components in spite of the separate collection scheme. The material was shredded to a 50 mm mean particle size.

PDF/ Collected by the Ekorosk-company in Pietarsaari. The wet and dry
Ekorosk fractions of MSW were source separated into black and white bags. Both bags were collected in a normal one-bin system by a compacting lorry. The bags were mechanically separated by colour in the Ekorosk plant. The dry fraction was baled, transported to Stormossen (see below) and shredded to a 50 mm mean particle size.

RDF/ASJ Mixed MSW was collected and processed in the Stormossen, Vaasa mechanical waste sorting plant, where metals, glass and organic components were separated out. RDF was shredded to a 50 mm mean particle size.

Test programme

The research project was divided into two parts: a one-year long-term study and two short-term emission measurement periods. The long-term study was carried out from September 1993 to July 1994. Primary fuels were peat, coal and wood and the secondary recovered fuel was RDF. Availability of the boiler, efficiency of the combustion process, and corrosion susceptibility of the super heater were of particular interest in this long-term study.

Table 1 Test Matrix.

RUN	TIME	Secondary fuel	Amount %–thermal	Duration hours	Remarks
1	Nov/93	–	–	26	Peat+coal/ reference
2	Nov/93	PDF/LPB	13	24	
3	Nov/93	PDF/YTV	12	21	
4	Nov/93	RDF/ASJ	13	24	
5	April/94	PDF/PE	19	24	
6	April/94	PDF/PET	19	16	
7	April/94	PDF/Ekorosk	26	28	
8	April/94	RDF/ASJ	26	17	

PDF	= Packaging Derived Fuel
RDF	= Refuse Derived Fuel
LPB	= Liquid Packaging Board
YTV	= Helsinki Metropolitan Area Council
ASJ	= Ab Avfallsservice Stormossen Jätehuolto Oy
PE	= Polyethylene
PET	= Polyethylene terephthalate

One-day combustion test runs for RDF and PDFs from different sources were carried out during the emission measurement campaigns. Gaseous and solid emissions were extensively analyzed. Continuous flue gas analyses and other samplings were performed by the Combustion and Thermal Engineering Laboratory of the Technical Research Centre of Finland (VTT Energy). The samples were analyzed by the Department of Environmental Sciences of the University of Kuopio (organic components), University of Jyväskylä, Outokumpu Oy, Hans Ahlström Laboratory and Neste Scientific Services. The first measurement period took place in November 1993 and the second one in April 1994. The test matrix is presented in Table 1.

Peat and coal only were combusted for two days before the emission measurement campaigns. Test runs for each recycled fuel lasted 16–28 hours. The changeover to the new fuel mixture was made in the evening, the process was stabilized over night and the emissions sampling and on-line measurements started the next morning. The first run was a reference test employing peat, coal and wood waste as fuels.

Results

Long Term

Feeding and combustion of coarse, fluff RDF did not cause any drawbacks to the effective operation of the plant during the whole year of co-combustion. Visual inspections and wall thickness measurements of the boiler before and after the one-year co-combustion of 10 % RDF did not show any signs of abnormal corrosion of superheater (500°C) or boiler tubes. Corrosion probe tests (600 h) at 500°C and 550°C did not show corrosion or chlorine containing deposits.

Emissions

Heat value, moisture and ash contents of the tested PDFs and RDF were between the values of peat and coal. Sulphur and nitrogen contents of RDF and PDFs were lower and chlorine and certain heavy metal contents, especially chromium, copper, lead and zinc, were higher than the respective values of peat and coal. The basic analysis of the fuel mixtures are given in Table 2.

The feeding line limited the amount of secondary fuel to 25 % because of an increased number of CO peaks at higher rates, so the original goal of 30 % could not be reached. Operation at the upper limit caused uneven feeding and consequently somewhat unstable combustion conditions.

The basic level of carbon monoxide (CO) in flue gas (Table 3) was low (15-40 $mg/m3n$) in all tests proving clean and efficient combustion. Operation near to the maximum capacity of PDF and RDF feeding line caused CO peaks increasing the mean CO level near to 200 $mg/m3n$. Sulphur dioxide (SO_2) emissions were lower in co-combustions than in the reference test (560 $mg/m3n$). Nitrogen oxide (NO_x) emissions were at the same level in all tests (below 200 $mg/m3n$).

Hydrogen chloride (HCl) emissions increased from 20 $mg/m3n$ to about 150 mg/m3n when the chlorine content of the fuel mixture increased. HCl emission results are mean values of three samples. The scatter was rather high, indicating the heterogeneous chlorine content of RDF/PDFs and somewhat unstable feeding. According to Cl-balance calculations, the HCl content in the flue gas should have been higher in tests with high contents of recycled fuel which, on the other hand, contained larger amounts of Ca which binds chlorine. HCl should not be a problem in a plant using desulphurization technology. HBr was not detected and HF values were very low.

Table 2 Basic analyses of the fuel mixtures.

Run	1	2	5	6	3	7	4	8
	Ref	PDF	PDF	PDF	PDF	PDF	RDF	RDF
Secondary fuel		LPB	PE	PET	YTV	Ekor	ASJ	ASJ
%-thermal		13	19	19	12	26	13	26
Moisture, wt-%	43.5	40.3	39.0	36.7	41.5	37.2	41.9	38.9
Volatile, wt-%	67.7	61.1	64.2	68.4	57.7	61.4	59.6	61.6
Ash, wt-%	5.8	6.1	5.8	4.7	8.2	8.5	7.8	9.1
HHV, MJ/kg	23.0	22.7	24.0	22.0	22.8	21.6	22.5	21.9
LHV, MJ/kg	21.8	21.5	22.6	20.7	21.6	20.3	21.3	20.6
C, wt-%	60.1	59.3	57.7	55.1	59.5	53.4	58.9	53.7
H, wt-%	5.5	5.6	6.7	6.1	5.4	6.0	5.5	6.0
N, wt-%	1.0	1.1	1.6	1.6	1.3	1.6	1.2	1.6
S, wt-%	0.37	0.36	0.35	0.27	0.40	0.36	0.37	0.37
O, wt-% (as difference)	27.2	27.5	27.7	32.1	25.1	29.9	26.1	29.0
Cl, wt-%	0.06	0.07	0.10	0.11	0.20	0.33	0.18	0.29
Cr, ppm	24	26	22	20	43	58	36	55
Cu, ppm	<10	<10	<10	<10	10	38	12	61
Ni, ppm	11	12	11	10	15	16	15	18
Pb, ppm	31	33	36	32	66	236	50	157
Zn, ppm	17	18	35	38	110	124	121	105
S/Cl_2,mol/mol	14	12	8	6	4	2	4	3

All results in dry solids

Table 3 Flue gas analyses of the test runs.

Run	1	2	5	6	3	7	4	8
	Ref	PDF	PDF	PDF	PDF	PDF	RDF	RDF
Secondary fuel		LPB	PE	PET	YTV	Ekor	ASJ	ASJ
%-thermal		13	18	21	12	26	13	26
H_2O, %	15.8	15.4	15.4	16.1	15.5	16.4	15.5	16.3
CO_2, %	12.2	12.8	12.2	13.7	12.6	12.8	12.2	12.3
CO, mg/m³n, mean value[1]	49	73	58	155	44	185	44	160
CO, mg/m³n, basic level[2]	40	35	15	30	25	35	30	30
CH_4, mg/m³n	2	4	6	16	2	20	2	19
SO_2, mg/m³n	560	520	450	460	550	480	520	480
NO_2, mg/m³n	170	180	180	180	200	150	190	160
N_2O, mg/m³n	ND	ND	ND	ND	ND	ND	ND	ND
Particles, mg/m³n	2	3	5	2	2	5	3	4
HCl, mg/m³n	17	40	34	30	154	140	60	120
HBr, mg/m³n	<3.5	<3.5	<3.5	<3.5	<3.5	<3.5	<3.5	<3.5
HF, mg/m³n	0.5	0.8	0.3	0.4	1.2	0.4	0.5	0.3

Results corrected to 11 % O_2 dry gas 1) including peaks 2) basic level without peaks

Heavy metals did not volatilize into the gas phase, and due to the efficient dust separation by the ESP and consequently low outgoing dust load (5 mg/m³n), the total concentrations in the flue gas were well below present EC incineration limits.

Table 4 Heavy metals in flue gas

Run	1	2	5	6	3	7	4	8
Secondary fuel	Ref	PDF	PDF	PDF	PDF	PDF	RDF	RDF
		LPB	PE	PET	YTV'	Ekor	ASJ	ASJ
%-thermal		13	18	21	12	26	13	26
Hg, μg/m³n	<1	<1	<1	<1	<1	<1	<1	<1
Cd, μg/m³n	1	<1	<1	<1	<1	<1	<1	<1
Ni+As, mg/m³n	<0.02	<0.02	<0.02	<0.02	<0.02	<0.02	<0.02	<0.02
Cr+Cu+Mn+Pb mg/m³n	1.61	0.74	0.20	0.08	0.06	0.12	0.17	0.13

Fly ash (Table 5) showed enriched amounts of heavy metals. The EPA-TCLP tests showed no adverse leaching of these elements, and the ashes can be disposed of in normal landfills.

Table 5 Fly ash analyses.

Run	1	2	5	6	3	7	4	8
Secondary fuel	Ref	PDF	PDF	PDF	PDF	PDF	RDF	RDF
		LPB	PE	PET	YTV	Ekor	ASJ	ASJ
%-thermal		13	19	19	12	26	13	26
Main components	%	%	%	%	%	%	%	%
C[1]	15.0	12.2	7.2	7.7	12.2	5.9	6.1	4.8
H[1]	0.2	0.2	0.1	0.1	0.1	0.1	0.1	0.1
N[1]	0.4	0.3	0.3	0.3	0.3	0.3	0.5	0.2
S[1]	1.0	0.8	0.7	0.7	1.0	1.1	1.0	1.2
Cl[2]	0.02	0.01	0.07	0.05	0.03	0.38	0.12	0.22
Al	9.75	11.1	9.47	9.48	10.6	11.2	11.3	11.0
Ca[2]	4.42	4.52	6.49	6.19	4.85	8.53	6.40	7.91
Fe	6.50	6.30	10.2	10.9	6.15	8.22	6.20	8.57
K	2.05	2.06	1.49	1.42	1.97	1.80	2.07	1.81
Mg	1.63	1.68	1.89	1.76	1.62	1.94	1.76	1.79
Na	0.64	0.70	0.51	0.47	0.76	1.12	1.12	1.21
P	0.29	0.30	1.01	1.07	0.34	0.94	0.58	0.93
Si	17.8	18.5	16.6	16.4	18.5	14.9	18.8	16.4
Ti	0.45	0.48	0.61	0.57	0.50	0.85	0.64	0.70
S/Cl$_2$, mol/mol	141	151	22	31	85	7	18	12

[1] by Leco analyzer   [2] by chemical methods   All other by XRF

The emissions of PCDD/Fs, 'dioxins' (Table 6), were far below the strictest incineration regulation limit 0.1 mg/m3n I-TEQ in all tests. The dioxin values were close to the detection limit in most cases, but increased slightly at higher contents (26 %) of PDF and RDF. Increased frequency of CO peaks because of mechanical problems, increased chlorine content of the fuel mixture, decreased sulphur to chlorine ratio and most significantly increased content of heavy metals like copper and lead in the fly ash are the probable reasons for the increase.

Table 6 Polychlorinated dibenzo-p-dioxins and dibenzofurans (PCDD/Fs)

Run	1	2	5	6	3	7	4	8
	Ref	PDF	PDF	PDF	PDF	PDF	RDF	RDF
Secondary fuel		LPB	PE	PET	YTV	Ekor	ASJ	ASJ
%-thermal		13	19	19	12	26	13	26
PCDD/Fs								
Actual concentrations								
Flue gas, ng/m^3n	<0,01	<0,01	0,19	0,06	0,09	31,7	0,16	1,87
Fly ash, ng/g	0,85	0 31	1,94	1,11	1,17	38,3	0,96	18,4
I-TE								
Flue gas, ng/m^3n	<0,01	<0,01	0,02	<0,01	<0,01	0,02	<0,01	0,04
Fly ash, ng/g	0,01	0,01	0,06	0,06	0,01	0,77	0,01	0,37

Conclusions

The test programme was the third full scale co-combustion test co-financed by the Finnish Ministry of Trade and Industry through its LIEKKI combustion research programme and by the industry. All tests with scrap and real waste, RDF and PDF, at a co-combustion rate of around 20 % together with coal containing primary fuel have shown good combustion efficiency, sufficient heavy metals capture in fly ash and PCDD/F emissions clearly below the strictest suggested incineration regulation. This test has also shown the long term technical feasibility of co-combustion in a CFB system. It is important that the recovered fuel is processed to a physical form compatible with the primary fuel and fuel feeding system of the plant in order to accomplish a steady combustion without CO-peaking. The results indicate that fuel recovery is a valuable alternative for resource management complementing biological treatment of the wet organic fraction of MSW and separate collection for recycling materials to products with a real market value.

The results show that used packaging contain higher amounts of certain heavy metals than conventional fuels. These metals mostly do not play a functional role in the packaging material itself but come form colouring pigments and printing inks. Although it is shown that these metals are efficiently concentrated to ash and filter residues it can be concluded that a consumer driven reduction of excessive 'cosmetics' could even further decrease the environmental impact of all resource and waste management routes.

References

CEN-CR 1460 Packaging – Energy Recovery from used packaging

Frankenhaeuser, M., Manninen, H., Kojo, I., Ruuskanen, J., Vartiainen, T., Vesterinen, R., Virkki, J., (1993). Organic emissions from co-combustion of mixed plastics with coal in a bubbling fluidized bed boiler. Chemosphere, Vol. 27, pp. 309–316.

Frankenhaeuser, M., Hiltunen, M., Manninen, H., Palonen, J., Ruuskanen, J., Vartiainen, T. (1994). Emissions from co-combustion of used packaging with peat and coal. Chemosphere, Vol. 29, Nos. 9–11, pp. 2057–2066.

Griffin, R.D. (1986). A new theory of dioxin formation in municipal solid waste combustion. Chemosphere, Vol. 15, pp. 1987–1990.

Mattila, H., Virtanen, T., Vartiainen, T., Ruuskanen, J. (1992). Emissions from combustion of waste plastic material in fixed bed boiler. Chemosphere, Vol. 25, pp. 1599–1609.

Ruuskanen, J., Vartiainen, T., Kojo, I., Manninen, H., Oksanen, J., Frankenhaeuser, M., Formation of polychlorinated dibenzo-p-dioxins and dibenzofurans in co-combustion of mixed plastics with coal: exploratory principal component analysis. Chemosphere vol. 28, pp. 1989–1999, 1994.

A Database for I.W.M. Covering Recycling and Composting

JULIA HUMMEL – Reference Programme Coordinator, The European Recovery and Recycling Association

ERRA mission

ERRA was established in 1989 by industries involved in the manufacture and sale of packaging. ERRAs mission is to 'help organise economically efficient and environmentally effective packaging waste recovery and recycling'.

ERRA Network of Reference Programmes

ERRA initially helped to establish I 0 pilot projects in Europe. It has recently begun expanding its data analysis and currently it has 17 programmes in its network of reference programmes.

The programmes reflect diversity in location, housing type and technology, but are representative both at national and European level. They can be put into three broad categories:

Kerbside box e.g. Adur (UK), Dublin (Eire), Sheffield (UK), N.district of Messine (F) Kerbside bin/bag e.g. Dunkirk (F), Lemsterland (NL), Louviers (F), Saarland (D) Bring *close-to-home* e.g. Prato (I), Queijas (P), Barcelona (E), Pamplona (E), Athens (GR)

The range of materials collected in each scheme is dependant on local conditions, but they all collect dry recyclables including packaging (plastic, glass, paper, metal) and news print. A number also include organic (kitchen and garden) material as well.

ERRA collects data I:0m these programmes and analyses it in order to understand the factors that influence the success of a scheme in the recovery of materials for recycling or other forms of recovery.

The Critical Factors ERRA has identified are:

1. The composition and quantity of the waste being handled.
2. The amount of material that can and is be diverted for recycling.
3. The operational factors (convenience, efficiency, etc.)

4. The overall cost of the recycling operation within the total waste management scheme.

Steps to prepare useful analysis – standardising diversity

Collecting data in a database alone is not enough to understand the effectiveness of different approaches. It is a key step that provides the raw information. The real challenge facing ERRA was how to analyse data from such a wide range of systems, each reflecting local conditions, in a way that could give meaningful answers and lead to a better understanding of household waste recycling.

The steps that must be systematically taken to achieve the understanding are:

1. *Define the objective of the analysis.* The objective clearly must be to be able to understand how and why a programme is operating the way it does, thereby identifying strengths and weaknesses allowing for the introduction of changes to achieve better results.
2. *Clearly identify the boundaries for the analysis.* Locally, the starting and the end point might be obvious. However, if data is to be collected From a wide range of different programmes ERRA had to define clear boundaries.
3. *Identify' and define what measurements or results are needed.* The next step is to identify what the key impacts are that will influence the operation of programme. Once these have been identified, data relevant to them must be identified, defined and then collected in a standard way. Obviously, local conditions will have a role in what will influence the good operation of a programme, but it is possible to identify broad categories, within which the local influences can be placed.
4. *Standardise the data to eliminate local variation.* Having collected data even according to standard definition it is necessary to standardise it by calculating relationships that eliminate the different baselines that exist locally. Calculating percentages gives the best type of result for comparative purposes as it removes all the variations. The system of Key Ratio Analysis that ERRA has developed is based on this principle.
5. *Recognise the limitations of the quantitative analysis.* Finally, it is important to recognise the limitations that a purely quantitative analysis provides. Qualitative assessment is always needed to assess the validity of the results and to explain them.

Key Ratios

ERRA believes that the only way to compare programmes across Europe in a meaningful way is to use ratios that cancel out local factors that influence the results. For example, a major factor influencing the quantity of recyclable material collected is the quantity generated by the householders i.e. thrown into the waste.

Through the experience of analysing programmes, ERRA has defined a limited number Key Ratios that measure the factors that influence the results. The Key ratios can be considered. individually, but for a proper understanding of each programme they should be analysed together. These ratios are described in detail in the ERRA reference publication 'Programme ratios'.

Ratio analysis provides a method of comparing not only schemes that collect dry recyclables From households. The boundaries can be broadened to include, for example, commercial and industrial waste, and/or kitchen and garden waste. The boundaries must be clearly defined before attempting to compare schemes, but once the boundaries are set, ratios will put the different data sets on a common baseline removing the impact of local factors. The local factors will be used to explain differences in results between programmes, and must therefore be known to be used in the qualitative analysis of programmes.

For this reason ERRA collects description and background data on the demographics of the programme area, and descriptions of all the waste management activities that impact on the waste streams that ERRA is analysing (Primarily household). The database allows this data to be used directly in the analysis of the results.

The other data that ERRA collects is ongoing mass flow data, for example, the quantities collected, sorted and sold. This data can be collected and included in the database for any streams that are collected, for example dry recyclables, refuse or a separate organic or compostable fraction. Indeed, in a number of the programmes compostables are collected as a separate stream and the data collected and entered in the database in exactly the same way as for the other streams.

Table 1 Waste stream Quantity and Composition

	Total waste stream (kg/inhab/yr.))	Composition (in % of total)					
		paper	glass	metal	plastic	organic	rest
Dublin (Eire)	226	25	5	3	9	46	12
Adur (UK)	244	42	9	8	9	19	13
Sheffeld (UK)	300	33	7	11	8	26	15
Dunkirk (F)	344	24	17	3	6	26	22
Barcelona (E)	370	32	6	3	13	37	9
Lemstertand (NL)	380	29	6	3	5	46	7
Prato (I)	386	19	4	3	6	33	33
Saartand (D)	366	21	11	4	7	29	26
Pamplona (E)	387	25	7	2	6	46	14

Why Integrated Waste Management

The baseline used in the Key Ratios is always the total waste that is being managed . It is therefore clear to see why it is important to consider not only the dry packaging components. Indeed it is difficult to segregate these items both in the collection and the analysis. ERRA believes that the waste stream needs to be considered as a whole, and that the most effective and efficient solutions to handling the packaging fiaction will be found when an integrated waste management approach is taken.

Though the data can all be collected together it is important to understand that the boundaries are different for programmes that collect different materials, and cannot be directly compared except to illustrate the impact including or excluding particular materials has on the overall results.

The table above details the quantity and composition of the waste streams in a number of the ERRA programmes across Europe. It illustrates the need to consider more than just the dry recyclables for maximum diversion.

Before putting together an integrated waste management strategy it is essential to understand both the quantity and the composition of the waste that is being generated in the area concerned. The quantity of waste generated can be established From the quantities of waste collected regularly. The composition can be established by waste stream analysis. These results are only sample results, though, so great care and attention must be taken to ensure that representative samples are analysed. Verification of the representatives of the samples should be carried out by comparing the sample results with actual collection quantities.

Once the composition and quantity of waste are known these data can be recorded in the database. They provide the baseline for the Key ratio analysis and are therefore critical information. As the baseline is the whole waste stream and not just the specific materials the ERRA programmes collect the whole composition analysis must be entered.

Table 2 The Effectiveness of Different Types of Programme

		Potential diversion rate	Actual diversion rate	Recovery rate
Dry recyclables				
bring	Barcelona (E)	36%	8%	21%
bring	Prato (I)	25%	9%	36%
teraalde	Sheffield (U~	38%	15%	39%
kerbside	Dublin (F)	39%	16%	40%
keri'aId~ bring	Saarland (D)	40%	17%	43%
kerbside	Dublin (Eire)	30%	17%	50%
keri'aide	Adur (UK)	34%	20%	50%
kerbside	Lemstertand (NL)	38%	30%	83%
Dry recyclable and Organics				
bring	Bnsda high-rise (NL)	60%	37%	62%
kerbslde	Breda low-rise (NL)	76%	55%	72%
kerbside	Lemstertand (NL)	86%	69%	80%

The Effectiveness of different types of programme

Applying the Key Ratios to the raw data From the programmes, trends start to become visible between the different programmes. There are groupings between specific types of operation, bringlkerbsida, and in the materials covered by the programme, dry recyclables only or organics as well. The data above illustrate the need both for mass flow and context data, as well as a clear definition of boundaries, with the schemes that collect organics as well as dry recyclables clearly forming a distinct group.

The results suggest that bring systems, where most of the materials are collected mixed together and are sorted at a MRF, generally can be expected to achieve lower diversion rates, and be slightly less effective (recovery rate). However, the results From Prato suggest that bring systems can be almost as effective as kerbside systems. as none of the schemes are optimised it is too early to draw definitive conclusions.

The results from Sheffield seem to be out of the general trend if the programmes are considered with respect to recovery rate. The probable explanation for this is that the scheme is small and has only had very low investment. Problems have occurred in the past where the collections have had to be suspended due to the collection vehicle braking down. This has affected the morale of the householders, some of who stopped participating.

The most effective programmes appear to be the kerbside 'box' schemes. This is probably due to the high visibility of the box, and householder pride in putting out a The box of clean recyclables. In the other kerbside schemes, and indeed in the bring schemes, much larger, closed containers are used so the contents are not clearly visible. The critical neighbourly eye, nor the householder pride play roles in these schemes. The costs are reviewed later.

The data illustrate trends only, and more programmes are now needed to establish baseline reference points for the different types of operation.

It must be stressed that the purpose is NOT to find the 'winner' and then apply this type of programme every where. There is no single solution. Baseline references should be used to identity where improvements in existing schemes are possible, and to forecast results from programmes that are being planned.

Table 3 Cost of the ERA Pilot Programmes 1993 (Locs/currency)

	Total Local currency	Collection Total	Processing Total
Adur	380787	278801	100886
Prato	1057 mll.	354 mil.	673 mll.
Barcelona'	426 mil.	320 mll.	106 mil.
Sheffield	63883	63993	
Dublin	746954	386147	380707
Lemsterland	702495	804687	87808
Dunkirk	25.8 mil.	16.1 mil.	8.7 mll.

*contractor cost onft   + calculated from incremental cost

Programme cost

The cost data illustrates very clearly the problems in trying to compare different programmes, but again ratios can be used to reduce the impact of local influences.

As well as all the issues with respect to the size of the programme, quantity of the waste stream etc. there is the problem of currency. The simplest solution, to convert all the individual currencies into a single currency, for example the ECU., while standardising the units does not actually help understanding the real costs as even within Europe there are sometimes major fluctuations in exchange rates. Recent fluctuations, for example, resulted in the Barcelona project becoming suddenly 20% cheaper on conversion into ECU.

In addition, the total cost of a programme includes capital items, labour costs and other operating expenses. Just comparing the cost does not take account of differences in local or national taxes nor in the cost of living in different countries. It is common knowledge that gross salaries cannot be compared across national borders, and you would be considered foolish, or at best naive, to expect the costs of any operation in the Netherlands to be the same as a similar operation in, for example, Greece.

Finally, published costs generally reflect an accounted cost rather than an actual cost. The cost therefore is more a reflection of accounting practice, for example asset depreciation and profit and loss statement, than cash required for payment. Hence, comparisons of cost without standardisation are at best meaningless and at worst very misleading.

The first step towards a meaningful comparison is the collection of 'actual cost' data as opposed to 'accounting cost' data. To achieve this ERRA developed, with Coopers Lybrand, a system of standard data recording. ERRA has implemented this in most of its pilot projects, and has used the results to develop a cost analysis that standardises the data from different programmes and allows direct comparisons to be made.

This standardised cost data is collected and entered into the database for the total waste operation in the ERRA schemes, in other words, for the dry recyclables and the refuse streams..

Ratio Analysis of Costs

On closer evaluation of the question 'how much does it cost to recycle?' it becomes clear that this is not the real issue. The immediate evaluation that the enquirer does is to compare the answer with the cost of recycling elsewhere and the existing refuse disposal costs. Only alter this comparison will the conclusion be drawn that 'that is an expensive scheme' or 'this is a good value scheme'.

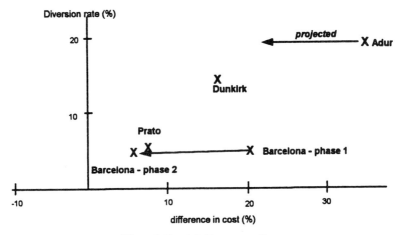

Figure 1 Trends in Programme Cost

The real question is therefore not the actual cost, but information demonstrating if the scheme is *affordable* How affordable anything is will depend on the difference between doing it or doing something different instead. So in waste management terms can be defined as the difference in cost between having a scheme that collects recyclable separately, and just collecting and disposing of the waste. Hence the ratio '% *difference the cost*'.

The calculation of % difference in cost also achieves the elimination of local variables through the calculation of a percentage. It also has the same boundaries as the Key Ratios that measure the mass flow results (diversion rate, recovery rate, etc.).

Initial Trends in Programme Cost

It is possible to plot the results of this ratio with the diversion rate to evaluate the results that are being achieved against the cost. In this way the different programmes and their results can eventually be compared. Unfortunately, today it is difficult to make definitive comparisons as programmes are generally not optimised, but eventually it will be possible to establish benchmark values for the different types of programmes.

The data above illustrates the improvements that programmes are currently making and the general trend in cost reduction. The cost reduction can be due to both internal and external factors, for example higher sales revenues. The explanation is essential if valid conclusions are going to be drawn.

From the analysis of the data ERRA has been able to develop a theoretical relationship between the cost and the results.

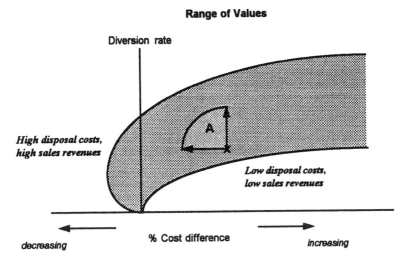

Figure 2 % Cost Difference vs Diversion Rate

How much Diversion is Cost Effective depends on the Cost of Disposal

For each programme area it should be possible to plot the range in which the programme should be with respect to cost and diversion. The range will depend on the cost and availability of different recycling, recovery and disposal options.

The programme results should fall into the shaded area on the graph. The exact position will depend on the actual conditions and waste management options selected as well as the efficiency of the specific programme.

Optimisation

Having established the position of a programme within this range it is then possible to track the programme to monitor it, over time, it is becoming more efficient. It is also possible to project the impact of changes that could make the system more effective.

In general terms, as a programme becomes more effective i.e. optimised, the effect should be indicated by a movement towards the left i.e. cost reduction, and/or a move up, i.e. an increase in diversion without additional cost. This field is indicated by the shaded area A.

This model can be used in the planning of the whole waste management system towards an optimum diversion to cost impact.

It is also possible to convert targets set in waste legislation into diversion targets, and thereby work towards cost effective methods of meeting the targets.

Data Management for I.W.M.

- Data Collected needs to cover background description, mass flow and cost
- Precise definition of the data is required
- Clear definition of the boundaries for the analysis is essential
- Ratios should be used to standardise data for valid comparisons to be possible
- The whole waste management system should be managed and hence analysed complete
- Data from all the streams should be stored in a single database to facilitate the analysis of the total system

Conclusions

ERRA has helped to establish a number of recycling programmes in Europe from which it has gathered data. Data describing the programmes, as well as mass flow and cost data relating to the waste management system defined by the boundaries of the analysis have been, and continue to be entered into a specifically defined database. ERRA analyses the data using ratios in order to minimise the impact of local influences on the results.

Experience has allowed ERRA to identify a limited number of Key Ratios that can be used to provide an understanding of the efficiency of a programme, as well as allow comparison with other programmes across Europe. The Key Ratios only required a small number of carefully defined data to be collected.

ERRA uses the Key Ratios to analyse primarily the collection and sorting of dry recyclable packaging materials, but where other materials, for example kitchen compostable waste are collected in the programme areas, ERRA has shown that the ratios can be used to analyse these programmes in the same way.

Having the data from all the waste streams in a single database permits the analysis of integrated waste management systems in a simple and coherent way.

Alternative Utilization of MSW Compost in Landfills

RAFFAELLO COSSU and ALDO MUNTONI –
Department of Geoengineering and Environmental
Technology, University of Cagliari

Abstract

In this paper the Authors discuss the possible utilization of compost in landfills as cover material and biofilter. Compost can act as control layer of water inflow and perform a precious buffering function; contact with compost can remove some heavy metals from leachate under basic conditions. Well cured, unrefined composts can be suitable due to favourable size distribution, low compactability and high permeability.

Introduction

To date, many of the composts produced in Italy from MSW rather than from organic matter deriving from separate collection do not meet legislative limits with respect to inerts, glass, paper, plastics and heavy metal content. Thus, the final destination of these composts is not use in agriculture, but disposal in landfills. Heavy refining of these composts in order to meet legislative limits would require very high costs and extensive modification of existing plants.

In the near future the new technical directives will probably foresee two types of compost:

- type A deriving from composting of organic matter from separate collection;
- type B deriving from composting of unsorted MSW.

Type A will be used in agriculture, while for type B it will be necessary to find alternative uses.

Some characteristics of compost should allow its utilization in MSW landfill construction and management (Figure 1) as:

- daily covering material;
- temporary cover material;
- final capping material;

– biofilter for leachate drainage system.

Several tests on utilization of MSW compost in sanitary landfills were performed at the Department of Environmental Technologies of the University of Cagliari. Six types of MSW composts are under experimentation in order to verify their possible utilization as daily, temporary and final covering material. Tests have been made and are in progress in order to define physical parameters of compost (such as granulometry, permeability, void index, compaction rate, composition), bio–chemical parameters and characteristics (stabilization level, heavy metal content, possible interactions with MSW landfill leachate and biogas).

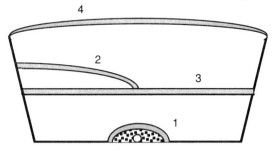

1	Biofilter for leachate drainage system
2	Daily covering material
3	Temporary covering material
4	Final capping material

Figure 1 Possible utilizations of compost in MSW landfills

Use of compost as covering material

Daily covering of waste is recommended by the current management procedures of sanitary landfilling in order to avoid dispersion of light fractions (paper, plastics) and feeding of animals (rats, birds). Daily covering leads to several disadvantages such as loss of volume for waste disposal and high costs if the material is not available near the facility. Moreover, field experience has shown that the main problems derive from the utilization of low permeability soil as covering material: in this case the landfill is divided into many hydraulically insulated 'pockets' where leachate and biogas accumulate. Consequences are: difficulties in draining leachate and biogas, side and surface escape of leachate and biogas, stability problems, limitation of water circulation inside the landfill and therefore inhibition of waste stabilization. For these reasons finding new materials for daily covering is becoming a pressing need.

The new materials should have the following main characteristics:

– to be easy to manage;
– high permeability;

– low costs;
– their use in landfill must not cause an increase of pollution, but, possibly, the opposite effect.

Well stabilized compost seems to have these characteristics, in fact:

– if the compost is well stabilized there is not a considerable increase of waste organic load; moreover compost does not attract animals and the mass losses for biodegradation are negligible;
– well stabilized composts have usually a low moisture content, thus they can not be compacted very well and mantain high void index and permeability; moreover the low permeability allows compost to absorb infiltration water and control water flow inside the landfill;
– compost could act as biofilter improving leachate and biogas quality by means of hazardous compounds removal.

The above mentioned characteristics could also allow compost utilization as:

– material for temporary covering before the final capping.
– material for final capping;

Final capping usually is not carried out until the waste mass is not yet settled in order to avoid damages to the components of the barrier system (membranes, geotextiles, geocomposites etc.). During this period it is necessary to cover waste with a temporary layer. Temporary covering could be necessary also when an intermediate sector of the landfill is completed. The material used for build up the temporary covering layer has to perform the following main functions:

– to insulate waste from outside environment (and it itself has not to be feeding material for animals);
– to control water inflow, avoiding high rate of infiltration, but allowing the permeation in   order to provide the water amounts necessary for biostabilization of waste;
– to limit biogas dispersion.

The outside component of the final cover system usually is a vegetative layer which has to perform the following functions:

– to protect the underlying layers;
– to allow the growth of plants and grass;
– to limit erosion phenomena.

Use of compost as biofilter for leachate drainage system

Keeping the efficiency of drainage systems is the main problem in managing sanitary landfills; failures of leachate collection systems are mainly caused by biofouling (formation of insoluble, consolidated deposits and incrustations).

Incrustations are made of calcium carbonate and iron sulphide; the metabolic activity of anaerobic bacteria proved to be the direct cause of these phenomena and formation of incrustations is prevalent during the early stages when the landfill is 'young' and the leachate is acid and characterized by high organic load. The problem can not be avoided completely, but it can be limited by means of internal reduction of organic load of leachate (Brune et al., 1991; McBean et al., 1993). This result can be achieved setting over drainages layers of unrefined compost which can perform a partial degradation of the organic compounds in leachate; moreover compost can act as buffering material limiting the acid phase and enhancing waste biostabilization. Another possible advantage of using compost as biofilter could be the removal of heavy metals from leachate due to adsorption or complexation phenomena. Laboratory experiments were conducted by Brune (Brune et al., 1991) by means of plastic columns filled with 30 cm of drainage material on top of which a further layer of 15 cm of compost was placed; highly and lightly loaded leachate were supplied during a period of 18 months; samples of leachate inflow and outflow were taken and analyzed for COD, Ca, Fe and Mg content. A general positive effect due to the presence of compost was observed: columns where compost was not placed showed quick bio-fouling of gravel. Compost layers were able to remove organic load, Ca and Fe ions from leachate, as reported in Figure 2.

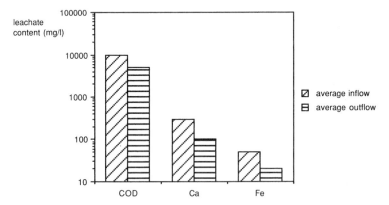

Figure 2 Average effects of compost on leachate quality (from Brune et al., 1991).

Brune reports of some problems occured with high loaded leachate: removal efficiency was lower due to the acid pH which inhibited bacteria activity and compost layers showed phenomena of consolidation and fouling with decrease of permeability. For this reasons, use of unrefined composts should be recommended, especially when contact with young leacahte is expected.

Table 1 summarizes the characteristics of compost and consequent possible utilizations in sanitary landfills.

Table 1 Compost properties related to possible utilizations in MSW landfills

		Compost properties					
		A	B	C	D	E	F
Compost	1		■		■	■	
utilizations	2	■	■	■	■	■	
(see Figure 1)	3	■	■	■	■	■	
	4	■		■			■

A Removal of hazardous compounds from biogas
B Removal of organic load and heavy metals from leachate
C Control of water inflow
D Buffering
E High permeability ·
F Enhancement of plant and grass growth

Materials and methods

Six types of compost produced in two different plants were tested:

- 6 months cured, unrefined compost (NR6);
- 2 months cured, unrefined compost (NR2);
- 6 months cured, refined compost (R6);
- 2 months cured, refined compost (R2);
- cured, unrefined MSW compost (NR);
- 'green' compost, produced by means of stabilization of waste deriving from cleaning of parks and gardens and from agricultural and zootechnical activities (CV).

Composts NR6, NR2, R6 and R2 were produced at the same facility. Compost NR6 and NR2 were preatreated by means of shredding, screening (rotary screen, 70 mm) and aerogravitational classification; the curing phase was carried out at an uncovered area provided with insufflation-aspiration device. Compost R6 and R2 were refined by means of aeraulic zig-zag classification and vibrating screening (15 mm).

Compost NR was preatreated by means of shredding (rotor mill), screening (rotary screen), glass and iron separation; the accelerated bioxidation phase was carried out by means of aired windrows (25 days) and followed by a curing phase of 50 days.

The following parameters were determined for each type of compost:

- pH;
- moisture content;
- volatile substances;
- total humic substances, humic and fumic acids;
- total organic carbon and organic substances;
- total N, total P (P_2O_5) and total K (K_2O);
- glass, inerts, plastics and iron content;
- heavy metal content.

Particle size distribution was determined taking into account the following size classes (mm):

- +25
- −25 +10
- −10 +5
- −5 +3

- −3 +1
- −1 +0,5
- −0,5

In order to define the stabilization rate, respirometric tests were carried out using a common BOD device; 40 g of compost were diluted in 400 ml of standard solution. Oxygen consumption due to aerobic degradation was measured for a period of 5 days.

Anaerobic digestion tests were performed to estimate the biogas production deriving from compost utilization in a sanitary landfill; 300 g of compost diluted according to a liquid–solid ratio of 2 were set inside artight glass bottles connected to a hydraulic gager in order to measure biogas production. The devices were placed in a thermostatic room regulated at a temperature of 30 °C. Gas flow was recorded daily.

Several leaching tests were carried out in order to define heavy metal mobilization and interactions between compost and MSW leachate; four types of leaching tests were performed using as extraction liquid: water acidified by means of CO_2 or acetic acid (recommended by Italian legislation), a acid phase and a basic phase MSW leachate. Leaching tests with MSW leachates were carried out following the same methods recommended for the test with acetic acid.

Eluates of the leaching tests were analysed and the following parameters were defined:

- heavy metal content;
- pH;
- COD and BOD_5;
- chloride and volatile acids (from C_2 to C_7).

In order to measure the permeability, compost samples were compacted into steel cells using a Proctor device. Compaction was performed according to ASTM prescriptions.

Compacted composts were saturated with water and permeability was measured by means of a variable head permeameter.

Results and discussion

Characterization

The values of VS, organic carbon and organic substances showed that unrefined composts have, as expected, an organic content lower than refined compost. This characteristic is positive for use of compost in landfill: infact the organic load of waste mass does not increase much. CV compost showed a high organic content.

N and P content showed that composts can introduce nutrients into the landfill, thus improving waste stabilization in view of the fact that these elements are scarce in waste.

NR6, R6 and CV composts have low moisture contents (10 – 20%), rendering them suitable for utilization in a landfill: in fact they are not an additional source of leachate and, on the contrary, can act as a control factor of water flow into the waste mass: considering a compost field capacity equal to 40% by weight and a moisture content equal to 14%, a layer of 1 m^2, 0,3 m thick and specific weight equal to 0,8 t/m^3, could keep 70 l of water.

Composts have an alkaline pH (7,5–8,6); for this reason they could be considered as buffering material capable of avoiding excessive decrease of pH due to the high production of volatile acids typical of the first phase of waste anaerobic degradation (low pH values inhibit waste stabilization) and limiting heavy metal mobilization and drainage incrustation.

Unrefined composts are characterized by high content of plastics, inerts and glass. The presence of inerts and glass and the absence of refining treatment in NR6, NR2 and NR allows a fairly well-balanced size distribution (Figure 3) which should allow good void index and permeability. For other reasons also CV compost is characterized by the prevalence of large particles.

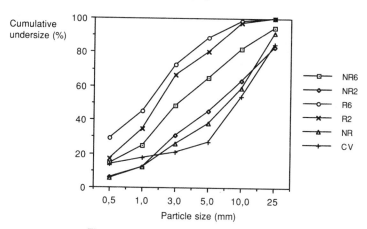

Figure 3 Compost size distributions.

Respirometric tests

In Figures 4 O_2 consumption measured during the respirometric tests is reported; results are referred to unit of compost mass. These data evidentiate how organic substance of composts from MSW is readily biodegradable, while slowly biodegradable matter (cellulose) is prevalent in CV compost. For composts from MSW, O_2 consumption is high during the first days but subsequently decreases, while the opposite occurs for CV compost. NR6 compost would appear to have a low organic content and therefore to be the more suitable for utilization in landfill.

mg O2/kg compost

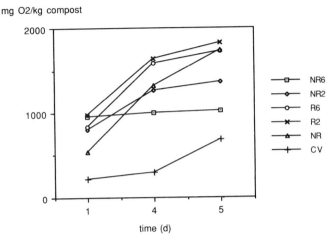

time (d)

Figure 4 Oxygen consumption for unit of compost mass.

Biogas production tests

In Figure 5 daily biogas production curves referred to unit of compost mass are reported.

ml/100 g compost d

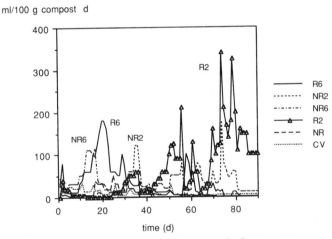

time (d)

Figure 5 Daily biogas production referred to unit of compost mass.

R2 compost was characterized by the highest daily production (350 ml/100 g compost d ≈ 3,5 Nm^3/t compost d) reached during the last days of experimentation; this delay in reaching the peak of production could be attributed to the high content of biodegradable matter which probably led to a consistent production of fatty acids and thus to inhibition of methanogenic bacteria due to acid pH values. The lowest daily production was given by CV, NR and NR6; the utilization of unrefined, well cured compost in a landfill will not lead to a consistent increase of biogas production from the waste mass.

Figure 6 shows cumulative biogas production curves referred to unit of compost mass.

Curves obtained confirm findings reported by daily production curves. NR2, R2 and R6 composts were characterized by negligible production during the first part of experimentation; subsequently the curves presented a flex point and production increased rapidly; gas production had not been exhausted following the 90 days of experimentation. The highest cumulative production was by compost R2 (5.500 ml/100 g compost). Gas production from compost NR6 started quickly but finished after 20 days; the gas production from compost CV (rich of slowly biodegradable organic matter) was low but not end up after the 90 days of experimentation.

The low gas production renders NR6 compost the more suitable for utilization in a landfill.

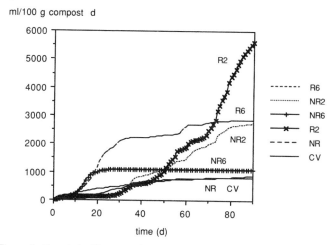

Figure 6 Cumulative biogas production referred to unit of mass of compost.

Heavy metal content in composts

Zn, Pb and Cu are metals found in the highest quantities, as verified also for MSW (Cannas et al., 1990; Muntoni A., 1994), due probably to the presence of dust from road and domestic cleaning, but legal limits for waste classification are respected. Compost CV has a considerably lower heavy metal content. Refining has not a remarkable effect on heavy metal content.

Leaching tests

Table 2 reports concentrations of metals found in eluates of leaching tests carried out using water acidified by means of CO_2 and acetic acid. A general low mobi-

Table 2 Metal concentrations in eluates of leachnig tests.

	Concentrations in eluates (mg/l)												Average contents in MSW leach. (mg/l)
	Test with CO_2								Test with acetic acid				
	NR	CV	NR6	NR2	R6	R2	NR	CV	NR6	NR2	R6	R2	
As	<0,03	<0,03	<0,03	<0,03	<0,03	<0,03	<0,03	<0,03	<0,03	<0,03	<0,03	<0,03	0,5 – 1,6
Cd	<0,01	<0,01	<0,01	<0,01	<0,01	<0,01	<0,01	<0,01	0,03	0,02	0,02	0,04	0,02-1,4
Cr III	<0,01	<0,01	0,01	0,02	0,03	0,04	<0,03	<0,01	0,02	0,02	0,03	0,07	2 – 1,6
Cr VI	<0,01	<0,01	<0,01	<0,01	<0,01	<0,01	<0,03	<0,01	<0,01	<0,01	<0,01	<0,01	0,2 – 1,6
Hg	0,001	0,001	0,001	0,004	0,001	0,004	0,001	0,001	0,002	0,002	0,001	0,035	0,005 –0,05
Pb	0,07	0,07	0,15	0,20	0,06	0,31	0,08	0,13	0,37	0,24	0,26	0,55	0,2 – 1,0
Cu	0,13	0,03	0,16	0,27	0,16	0,81	0,58	0,01	2,7	0,62	0,25	1,30	0,1 – 1,4
Ni	0,09	0,02	0,21	0,23	0,20	0,34	0,10	0,03	0,33	0,24	0,19	0,43	2,0 – 2,0
Zn	0,15	0,75	1,20	0,87	0,32	1,30	4,2	0,16	5,30	3,50	2,20	7,00	0,5-170

lization of metals can be stressed. The negligible mobilization of metals from compost to eluates is probably due to the buffer characteristics of composts: eluates presented a final pH of 7–8. Cu, Pb and Zn, the main metals present in composts, were leached in small quantities, probably due to their anphoterism which allows release only at very low or very high pH values, conditions which are not common in a MSW landfill. Concentrations of heavy metals are well below the common concentrations for MSW leachate.

Table 3 shows the values of several parameters measured for eluates of leaching tests compared to average values of MSW leachates. The buffering effect of composts is evident; this is surely a positive aspects in order to evaluate the utilization in MSW landfills; buffering capacity should limit the inhibition phenomena of methanogenic bacteria due to excessive decrease of pH, thus enhancing biostabilization of waste and limiting clogging of draining systems; moreover under basic conditions mobilization of heavy metals from solid matrix is reduced and removal of metals from leachate enhanced. The values of organic parameters (BOD, COD etc.) are well below values commonly found in acid MSW leachates and near to values typical of basic leachates produced from stabilized waste.

Eluates from tests using acid leachate and MSW compost show higher metal concentrations or similar than those of acid leachate itself: this means that the contact between compost and leachate facilitates mobilization of heavy metals from solid matrix to eluate; only compost CV seems to be able to slightly remove some metals (Cu, Ni and Zn) from acid leachate.

Tests with basic leachate show a remarkable removal of Cr III, Cu, Ni and Zn, in particular by compost CV: this means that contact between compost and leachates under basic conditions could improve inorganic load of basic leachates (see also Figures 7 and 8). This effect could be explained by adsorption or complexation phenomena which may prevail over mobilization effects probably because pH is not low enough.

Table 3 Main parameters measured for the leaching test eluates and average ranges in MSW leachates.

Parameters and tests	NR	CV	NR6	NR2	R6	R2	Average range in MSW leachate
pH							
– CO2	7,43	7,40	6,89	7,31	7,43	7,45	5,3 – 8,5
– acetic acid	5	5	5	5	5	5	
COD (mg O2/l)							
– CO2	1.389	1.155	1.394	1.524	1.848	1.889	1.000 – 70.000
– acetic acid	–	–	–	–	–	–	
BOD5 (mg O2/l)							
– CO2	189	155	230	274	428	419	500 – 30.000
– acetic acid	–	–	–	–	–	–	
volatile acids (mg C/l)							
– CO2	10	7	13	15	29	31	50 – 5.000
– acetic acid	–	–	–	–	–	–	

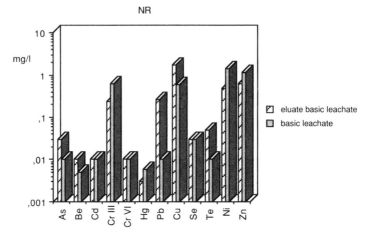

Figure 7 Concentrations of metals in basic leachate and in the eluate of leaching test carried out on compost NR using basic leachate as lixiviant.

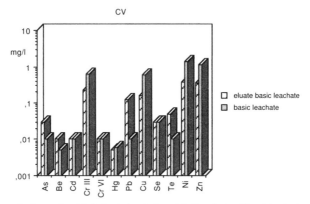

Figure 8 Concentrations of metals in basic leachate and in the eluate of leaching test carried out on compost CV using basic leachate as lixiviant.

Permeability tests

Figures 9 and 10 show compaction rate and permeability measured for the composts tested .

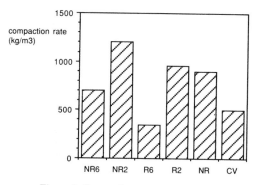

Figure 9 Compaction rates of composts tested.

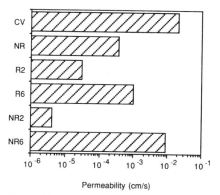

Figure 10 Permeability of composts tested.

It is evident that compost CV and R6 are the less compactable; this is probably due to the lower moisture content and to the particular physical structure of CV (similar to a fluff). The low compaction rate and the favourable size distribution allow compost CV to keep a high permeability and render it suitable to be used as cover material in MSW landfills. Also compost NR6 and R6 posses a sufficient permeability for use in landfills: the low compaction rate and, for compost NR6, the favourable size distribution, allow cured composts to keep high permeability.

Tests in progress

Currently, other tests are in progress in order to verify whether dangerous trace components of biogas (mercaptans, chlorinated components etc.) can be removed by means of flowing through compost layers. Several experiences (Figueroa and Stegmann, 1991; Figueroa, 1993 (a), Figueroa, 1993 (b)) have shown that methane can be oxidized to CO_2 and H_2O by a bacteria population growing in compost layers used as final capping of MSW landfills. The research programme aims to assess also whether trace components can be chemically or physically removed. An experimental device has been set up at a sanitary MSW landfill: biogas flows in columns filled with compost; samples are taken before and after the passage through the compost and analysis of gas composition are made by means of gas-mass-cromatography.

Conclusions

The experiment allows the Authors to assess that:

a) use of MSW composts in landfills brings the following advantages:
   - compost can act as buffering material enhancing waste stabilization and limiting clogging of leachate draining system;
   - compost can decrease the content of some heavy metals in leachate under basic conditions, while under acid conditions there is a negligible mobilization of metals from the solid matrix;

b) tests have to be carried out in order to evaluate the compost influence on the organic load of leachate;

c) green (CV) compost keeps a high permeability due to its favourable size distribution and particular physical structure; concerning MSW composts, in order to keep sufficient permeability values it is necessary to avoid refining treatments and improve curing phase;

d) well eured composts are more suitable to be used as cover material because:
   - they do not increase too much the organic load of the waste amount;
   - they have a lower moisture content which allow them to work as a control layer of water inflows;
   - they produce less biogas in anaerobic conditions;
   - they are less compactable and thus keep a high permeability.

Acknowledgement

The Authors wish to thank ASPICA and NUOVA GEOVIS srl for the support and Mirella Erriu for her precious co-operation.

References

Brune M., Ramke H.G., Collins H.J. and Hanert H.H., 1991. Incrustation processes in drainage systems of sanitary landfills, *Sardinia 91, Proceedings of the Third International Landfill Symposium*, 14–18 ottobre, S. Margherita di Pula, Italy.

Cannas P., Casu G., Cossu R., Granara F., Leuzzi F., 1990. MSW composition: influence on recycling. *Proceedings of the First International Symposium on Resource Recovery from Waste*, Imola 15–18 maggio.

Figueroa R.A. and Stegmann R., 1991. Gas migration through natural liners; *Sardinia 91, Proceedings of the Third International Landfill Symposium*, 14–18 ottobre, S. Margherita di Pula, Italy.

Figueroa R.A., 1993 (a). Methane oxidation in landfill top soils, *Sardinia 93, Proceedings of the Fourth International Landfill Symposium*, 11–15 ottobre, S. Margherita di Pula, Italy.

Figueroa R.A., 1993 (b). Landfill gas treatment in biofilters, *Sanitary Landfilling: Gas*, Christensen T.H., Cossu R. and Stegmann R. editors, Elsevier, in press.

McBean E.A., Mosher F.R. and Rovers F.A., 1993. Reliability-based design for leachate collection systems, 1993, *Sardinia 93, Proceedings of the Fourth International Landfill Symposium*, 11–15 ottobre, S. Margherita di Pula, Italy.

Muntoni A., 1994. Disposal and utilization of residues deriving from MSW pretreatments and coal combustion. *Ph.D Thesis*, Cossu R. and Ghiani M. tutors, Department of Geoengineering and Environmental Technologies, University of Cagliari.

B4 Bioremediation

Stabilization of Hazardous Wastes Through Biotreatment

L.F. DIAZ, G.M. SAVAGE, and C.G. GOLUEKE –
CalRecovery, Inc. Hercules, California USA

Abstract

Biological treatment of hazardous wastes is the application of microorganisms to break down the toxic compounds in the wastes into innocuous intermediates or end products. This paper discusses the basic requirements in order to use biological treatment to stabilize toxic compounds. In addition, the paper presents an overview of available types of treatment, including: *in-situ* remediation, treatment of liquid wastes, treatment of solid wastes, and the application of compost technology. Finally, a discussion is made of a case study in which oil refinery sludges were stabilized using a composting technology. In the case study, laboratory-scale experiments were first carried out to determine the viability of the composting process. The laboratory-scale experiments were followed by a series of pilot-scale tests, as well as extensive analyses of the gaseous and solid byproducts.

Introduction

Contamination due to industrial activity now is widely recognized as a potential threat to public health and the environment. Both industrialized and developing countries are facing serious problems associated with contamination due to toxic and hazardous wastes. Several countries have developed and implemented regulations to either limit or completely eliminate the production and release toxic chemicals to the environment. Unfortunately, significant environmental contamination already has taken place which will require various technologies and substantial resources to successfully remediate.

The need to clean up the contaminated sites, combined with the financial requirements associated with the cleanup, have encouraged the development of new technologies which emphasize detoxification and destruction of the contaminants. One of these technologies is bioremediation. Bioremediation is the application of microorganisms to detoxify and degrade toxic or hazardous compounds. Microbiological processes have been used for the stabilization of organic residues for a number of years. For instance, conventional municipal wastewater treatment relies heavily on the application of microbiological processes. These processes are engineered and controlled to accomplish specific goals and specifications.

In the context of this presentation, biotreatment is applied to the stabilization of toxic and hazardous compounds found in primarily soils, as well as in surface and groundwater. Thus, in general, the new application of microbiological processes differs from previous applications in terms of the types and concentrations of chemicals undergoing treatment and the medium in which the treatment may take place. Typically, a site is contaminated with complex mixtures of organic compounds. The concentrations of contaminants may vary substantially within the site. Frequently, inorganic residues such as metals also are present. The contaminated media adds another degree of complexity to the mixture in the waste. Consequently, bioremediation generally involves the handling of multiphasic, heterogeneous materials such as soils in which the contaminant may be present in association with the soil particles or dissolved in liquids.

Bioremediation technologies can be classified in two general categories: *in-situ* and *ex-situ*. *In-situ* techniques involve the treatment and stabilization of the contaminated material in place. *Ex-situ* are those processes which involve the collection, removal, and transportation of the contaminated material to another location for treatment.

The success of a bioremediation technique is dependent upon the following conditions: the presence of large enough populations of the correct type of microorganisms, and the proper conditions for the organisms to grow. The correct type of microorganisms are those which have the physiological and metabolic capabilities to break down the contaminants into harmless products. In many situations, the required organisms already are present at the site. In other cases, it may be necessary to introduce a particular culture to the material. The microorganisms must be in close proximity to the contaminants for degradation to take place. It is fairly common that the presence of a biodegradable contaminant leads to the establishment and enrichment of a culture capable of degrading it in the contaminated area. In situations in which the required populations are not present, a mechanism must be developed to bring the microorganisms into contact with the contaminants. Once the proper type of microorganisms are present, the environmental conditions must be adjusted to promote the growth and metabolic activity of the microorganisms. Environmental factors such as temperature, inorganic nutrients (primarily nitrogen and phosphorous), electron acceptors (oxygen, nitrate, and sulfate), moisture, and pH can be adjusted to optimize the environment for bioremediation.

Bioremediation offers several advantages over conventional treatment techniques. In most cases, bioremediation can be carried out on site, thereby eliminating liabilities and transportation costs. Similarly, industrial use of the site can continue while bioremediation is taking place. Biotreatment can result in the breakdown of the waste materials into carbon dioxide and water, thus permanently eliminating the waste and the liability associated with non-destructive treatment methods. Bioremediation can also be combined with other technologies allowing for the treatment of mixed, complex wastes.

Just like any treatment technology, bioremediation also has its disadvantages. For instance, some chemicals, such as highly chlorinated compounds and metals, are not readily amenable to biological degradation. In some situations, biodegradation may lead to the production of substances that are more toxic than the parent compound. Bioremediation is a scientific procedure which must be specifically designed to conditions prevalent at the site. Therefore, the capital investments required for site characterization and feasibility analysis for bioremediation may be higher than the capital costs associated with conventional technologies.

Degradation of Xenobiotics

Organic compounds are broken down through a series of reactions known as catabolism. In the process, a complex carbon compound is converted into carbon dioxide and water. Microbial breakdown of a complex organic compound does not always result in complete breakdown. Microbial degradation can transform a compound into one that is more toxic to the environment than the original compound. Thus, biodegradation studies which only measure concentrations of the compound of interest may not be totally accurate.

Petroleum Hydrocarbons (PHCs)

Petroleum and its products, such as diesel fuels and gasoline, are complex mixtures of organic compounds. The majority of the compounds found in petroleum products are hydrocarbons.

Aliphatic hydrocarbons are straight or branched chains of carbon atoms which have enough hydrogen to satisfy the valence requirements of the carbons. Depending upon the number of carbon-carbon bonds, aliphatic hydrocarbons can be classified into: alkanes, alkenes, and alkynes. Other important types of hydrocarbons are those known as aromatic compounds. In general, aliphatic hydrocarbons are easier to break down than aromatic compounds.

Straight-chain alkanes are primarily degraded through oxidation of the terminal methyl group. This is followed by cleavage of the molecule between the second and third carbon in the chain. The first reaction in the process consists of the addition of oxygen to the terminal carbon. Oxygen addition to the terminal carbon results in the formation of primary alcohol. Eventually, the primary alcohol is oxidized to a fatty acid. Anaerobic degradation of aliphatic hydrocarbons does not play a significant role in the degradation of petroleum hydrocarbons in the natural environment [Atlas 1988; Atlas 1991].

Aromatic hydrocarbons such as benzene, toluene, ethylbenzene, and xylene (BTEX) are found predominantly in light petroleum fractions such as gasoline, although they may be present in trace amounts in any type of petroleum product. These compounds also are widely used as industrial solvents and intermediates in chemical production. Benzene, toluene, ethylbenzene, and xylene are relatively soluble in water. The solubility of BTEX allows them to be transported away from spills, resulting in extensive soil and groundwater contamination. Because of their small molecular size and low boiling point, BTEX volatilize at ambient temperatures and as such may be substantial sources of air pollution. Aromatic hydrocarbons are highly toxic and are classified as carcinogens.

Biological degradation of aromatic hydrocarbons under aerobic conditions was first demonstrated in the early 1900s [Gibson 1984]. Since then, several studies have been carried out to investigate the biochemistry and genetics of the degradation of aromatic hydrocarbons. The results of these studies have demonstrated that bacteria and fungi are capable of degrading aromatic compounds under aerobic conditions. Biological degradation of these compounds can follow either one of several pathways.

Until recently, microbial breakdown of aromatic compounds was considered to be possible only under aerobic conditions. However, during the last ten years, a number of investigations have demonstrated that these compounds can also be degraded under anaerobic conditions. Many of these investigations have used enrichment cultures of mixed microbial populations.

Anaerobic degradation of aromatic compounds has been demonstrated to take place under methanogenic, denitrifying, and sulfate-reducing conditions [Suflita 1991]. Ferric iron and manganese oxide have also been shown to serve as alternate electron acceptors for the anaerobic degradation of aromatic compounds. The initial steps of aromatic degradation under anaerobic conditions are considerably different from those found under aerobic degradation.

Preliminary steps for system design

One of the most important steps in the design of a biotreatment system is the performance of biotreatability laboratory studies. These studies are important in determining whether or not the contaminants are biodegradable and if a large-scale biological process can be applied to the project.

One of the most difficult tasks in a site remediation project is the selection of the most appropriate treatment technology. This is due to the lack of reliable data on the treatability of different compounds. In addition, another complicating factor is the variability that may take place from site to site. Laboratory- and pilot-scale studies can play a critical role in determining whether or not biotreatment technologies can be applied in full-scale projects.

Treatment of specific compounds

Since pesticides and oily sludges have been demonstrated to be biologically degradable, the greater part of the sections that follow deal with those two groups of wastes.

Laboratory Experience

An extensive literature survey covering about sixty studies of microorganisms able to assimilate recalcitrant organic materials is presented in a report edited by De Renzo [Noyes Data Corporation 1980]. In the survey, it is pointed out that: 1) relatively little information is available in terms of descriptions of microorganisms capable of metabolizing recalcitrant materials; and 2) most of the studies dealt with monocultures and idealized carbonaceous substrates applied under carefully controlled laboratory conditions. Therefore, although information gained in those studies may perhaps not be directly applicable to the complex conditions characteristic of the compost process, it can serve as a foundation for studies of complex interactions among microbial populations, and for selecting a suitable microbial 'seed' either for starting or accelerating the composting of degradable organic wastes.

Petroleum hydrocarbons can be assimilated by more than 200 species of microbes. Indeed, fungi have been observed growing in jet fuels [Prince 1961] and in toluene [Nyns 1968]. Examples selected from this assortment are listed in Table 1.

Table 1 Microbial Groups Capable of Metabolizing a Hydrocarbon Substrate

Bacteria		Actinomycetes	Fungi
Achromobacter	Flavobacterium	Actinomyces	Aspergillus
Aerobacillus	Gaffkya	Debaryomyces	Aureobasidium
Alcaligenes	Hansenia	Endomyces	
Arthrobacter	Methanomonas	Nocardia	Cephalosporium
Bacillus	Methanobacterium	Proactinomyces	Cunninghamella
Bacterium	Micrococcus	Saccharomyces	Mycelia (Fungi imperfecti)
Beyerinckia	Micromonospora		Monila (yeast)
Botrytis	Mycobacterium		Torula (yeast)
Candida	Mycoplana		Torulopsis (yeast)
Citrobacter	Pseudomonas		Trichoderma (yeast)
Cellustomonas	Sarcina		
Colostridium	Serratia		
Corynebacterium	Apicaria		
Desulfvibrio	Spirillium		
Enterobacter	Thiobacillus		
Escherichia	Vibrio		

Compost Experience

The literature on the composting of hazardous wastes is relatively small. Probably the first such report is that by Rose and Mercer [Rose 1968], who investigated the composting of insecticides in agricultural wastes. They found that the concentrations of diazinon and parathion rapidly decreased by continuous thermophilic composting. Thus, the concentration of diazinon was reduced from about 3.3 ppm to less than 0.002 ppm within 42 days. The degradation of parathion amounted to 50% of its original concentration in 12 days. In an investigation of the compostability of a 5% to 10% TNT mixture, it was found that the TNT could be reduced to zero, or an acceptable level, within a relatively short time. Since then, other studies have been conducted to evaluate the feasibility of biodegrading soils contaminated with explosives [Argonne National Laboratory]. In a project involving the composting of sewage sludge [Epstein 1980], it was found that many of the aromatic compounds in the sludge were reduced to negligible levels. Results of research done by CalRecovery, Inc. indicate that concentrations of the oil in oil sludge were reduced to less than 75% of their original concentrations in less than a month. The breakdown of the oil occurred only when the environment was aerobic [Diaz 1994].

'Seeding,' Substrates, and Mechanisms of Breakdown

Composting hazardous wastes differs from routine composting in that 'seeding' of the raw wastes is an important feature. The seeding is done by recycling a portion of the compost product into the raw waste to be composted. A suitable ratio would be on the order of one part seed to nine parts raw waste. The seeding ensures the presence of a population of organisms capable of attacking the hazardous contaminant without incurring the need for a lag period during which the necessary population could develop.

Given the right conditions, hydrocarbons are broken down relatively readily. Aromatics are degraded more slowly than aliphatics and alicyclic nonaromatics. Microbes have enzymes (oxygenases) that convert polynuclear aromatics into unstable cyclic peroxide, which in turn is reduced to cisglycol. The cisglycol is further oxidized. The researchers who determined this pathway contend that in landtreatment, a period of 1 to 2 years might resemble the potential of soil to decompose these compounds [Overcash 1979].

Not all of the breakdown in a compost pile is biological in nature. Chemical and physical conditions established by the process can exert an important influence on the breakdown of certain organic compounds. Ultraviolet light, temperature, and pH level are three such conditions. Many biodegradable pesticides are so vulnerable to these factors that isolating and determining the role of microorganisms in the breakdown of the pesticide is very difficult [Paris 1975; Mount 1981]. Temperature and pH levels reached during the composting process are those at which such pesticides decompose most rapidly [Mount 1981]. Hence in composting, physical, chemical, and biological destructive agents combine to degrade the pesticides, thus magnifying the effectiveness of composting.

Research studies

A series of studies were carried to ascertain the degradability of API separator oily waste. The studies were conducted at the laboratory- and at the pilot-scales. The primary objectives of the studies were to: 1) determine the effectiveness of composting in the destruction of hazardous sludges generated in the process of refining petroleum; 2) isolate and classify the predominant microbial populations in the compost and test the microorganisms' effectiveness in treating the wastes; and 3) determine the potential of composting the material at the pilot-scale.

Laboratory-Scale Degradation Experiments

In the research program, four 12-liter glass jars were used as the reactors. The medium used was a mixture of composted sewage sludge, dewatered digested sludge, and woodchips. The sludges were obtained from a wastewater treatment plant, and were intended to serve both as a mass inoculum and as sources of nutrients. The woodchips served as a bulking agent.

The medium was thoroughly mixed with the oily wastes. The mixture of refinery sludge and medium was divided into two portions. A 10-liter aliquot of each of the two portions was placed in glass reactors. The reactors had been previously sterilized. One of the portions of the mixture of oily waste and medium served as 'control.' Medium used in the control was sterilized in an autoclave prior to being mixed with the respective test materials.

The reactors were placed in a water bath. The water bath was operated such that the temperatures of the reactor contents could be maintained within plus or minus 1°C of a desired temperature. The reactors were loosely covered to prevent airborne dust from contaminating their contents.

The parameters used for monitoring the process included: 1) extent of destruction of the test substances, and 2) odor and appearance of the reactor contents. The temperature of the reactors' contents was varied by changing the temperature of the water bath. The temperature was changed to simulate the temperature fluctuations found in conventional composting. The experiment was conducted for a 33-day period. The reactors were aerated manually by stirring the composting masses. The pH level was monitored and recorded periodically. No attempt was made to control the pH.

Sampling and Analyses

Samples for use in the second phase of the experiment were removed from all of the reactors. The appearance and odor of the individual cultures, as well as their loss in mass and volume, were observed and recorded on the days on which they were stirred. The concentration of oil and grease was determined according to Method 503D, 'Extraction Method for Sludge Samples' [American Public Health Association 1971].

Results of degradation experiments

During the first 13 days, the rate of destruction was more rapid in the active reactor. However, the downward trend for destruction paralleled that for the rise in temperature.

Odor and pH Level

Initially, the odor in the control reactor remained strongly oily in nature. After Day 5, the material began to take on the odor associated with anaerobiosis. Beginning about Day 11, the odor of ammonia became noticeable, and thereafter increased in intensity. However, by Day 24, it began to lose its intensity, such that by the 32nd day, the odor of ammonia was barely noticeable.

In the active reactor, during the first 6 days of the run, the predominant odor was the oily one. Thereafter, the odor became that of a faintly earthy odor that slowly increased in intensity until around Day 14. At Day 14, ammonia began to be noticeable and the odors of anaerobiosis became detectable on the 19th day. The 19th day coincides with the highest temperatures in the reactors. As the temperature was lowered beginning on Day 24, the anaerobic odors decreased. Finally, by the 31st day, the predominant odor was earthy.

During the course of the experiments, the pH level was on the order of 7 to 7.5 in all reactors. The only exception was the active oily waste reactor in which the pH level reached about 9 for only a short time, and then decreased to about 5.5.

Discussion – degradation experiments

The results demonstrated the importance of biological activities in the destruction of oily wastes. This was shown by the greater reduction by Day 13 in the concentration of oily waste in the active culture. However, the major difference was limited to the first 13 days in that after that time, the concentrations in both reactors were very similar.

The degradation that began on the first day and continued until the end of the experiment in the control was a function of the microbial population in the oily wastes. Since the oily wastes could not be sterilized, the contents of the control reactor were inoculated with the microbes in the oily waste. The rather high initial breakdown in the active reactor was due to the bacterial populations introduced with the composted and digested sewage sludges combined with those in the oily waste.

Laboratory-Scale Microbiological Experiments

The microbial program was conducted in two phases. The first phase involved the isolation of predominant microorganisms in samples taken from the active oily waste reactor. The second phase involved the challenge of the isolated microorganisms under aseptic conditions with the refinery wastes in order to determine the ability of the microorganisms to decompose the wastes.

In the isolation phase, microbial counts and isolation were conducted by using agar slants. The slants were incubated at 22°C to 37°C for 2 to 5 days.

The medium used in the challenge run was made up by adding sterile oily waste to sterile distilled water such that the concentration of oily waste in the mixture approximated that of the waste in the reactors. The oily waste was sterilized through gamma irradiation.

The oily waste medium was divided into seven 500-ml aliquots. Each of the aliquots was placed in a liter flask. Two of the flasks were inoculated with a single isolate, and three were inoculated with two isolates per flask. The seventh flask served as a control. In order to be able to account for the loss of oily wastes due to volatization, two additional controls were established; one of the controls contained distilled water, and the second, oily waste. All flasks were aerated by bubbling air through the cultures. The cultures were incubated at 35°C over a 13-day period. Initial and final concentrations of oil and grease were determined by chemical analysis.

Results and observations – microbiological experiments

The level of destruction of oil and grease is presented in Table 2. As the data in the table show, the 'net' percentage destruction over the 13-day challenge period was the highest in the flask containing isolate number 5, i.e., the degree of degradation attributable solely to microbial activity was 25.6%. Net destruction in the other flasks ranged from 0% to 5.5%.

Table 2 Results of Challenge Rounds – Oily Waste

Isolate	Net Oil and *Grease Degradation* [1] Cumulative Percent
5	25.6
1,2	0
3,4	0
6,7	2.4
8	5.5
E	0.5

[1] Net degradation = (gross)–(control over the 13–day period).

Discussion – microbiological experiments

The destruction of oily waste obtained with isolate 5 confirms the results obtained in the compost experiments. In this particular case, the degree of destruction of oily waste obtained in this run was similar to that obtained during the composting experiments.

The results obtained in this study can be compared with those experienced in conventional landfarming. Monthly rates of degradation of oily waste were calculated using data available in the literature on landfarming [Environmental Research and Technology 1983; Meyers 1980]. The results of the analysis indicated that landfarming achieved degradation rates between 4% and 11.9% per month. On the other hand, the compost reactors achieved rates between 30% and 66% per month. The data show the substantially higher rates of degradation that are obtained with composting [Environmental Research and Technology 1983; Meyers 1980].

In this study, the loading rate applied in the compost reactors was on the order of 1 to 3 lb oil and grease/ft^3 per month. Landfarming typically uses loading rates in the range of 0.2 to 1.1 lb oil and grease/ft^3 per month.

Pilot-Scale Experiments

The medium used for the experiment was obtained from an active composting pile used for treating non-hazardous residues. The toxic waste used in the experiment was API separator oily waste collected from a refinery. In the experiment, organic waste in the process of stabilization was loaded into a reactor. The material was thoroughly mixed three times per week. Water was added as needed in order to maintain appropriate moisture content. No other nutrients were introduced into the material. Oily waste was slowly introduced to the material in order to allow for acclimation of the culture. The total amount of oily waste added to the composting mass was 36 liters.

Samples of the gases released during the biotreatment process were collected for analysis. Representative samples of the mixture in the process of stabilization were taken before and during the addition of oil.

Results

The results presented in this paper are limited to the analysis of the oil and grease fraction before and after treatment. The analyses were conducted by means of gas chromatography. The main reason for the analyses was to characterize the oily waste in terms of the chain length of the hydrocarbons comprising the oil and grease fraction.

The results of the analyses are presented in Figure 1. The curves in the figure show that the hydrocarbons present in the oily waste had carbon lengths within the range of 8 to 29. The non-aromatic hydrocarbon species present in the waste had a high concentration of chain lengths in the neighborhood of 14 to 20 carbon atoms. The predominant aromatic species had chain lengths greater than 12.

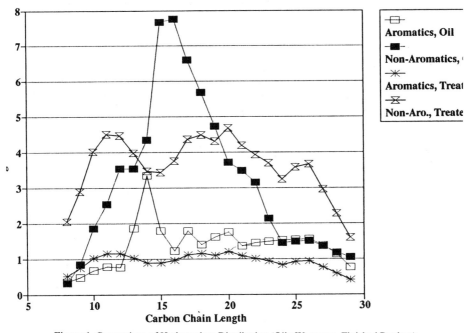

Figure 1 Comparison of Hydrocarbon Distribution (Oily Wastes vs. Finished Product)

The chromatographic scans of the final oily waste compost are also shown in Figure 1. The curves in the figure show relative uniformity between C10 and C26. Thus, the process resulted in the breakdown of the larger hydrocarbons into shorter chains.

Conclusions

The research demonstrated the technical feasibility of composting refinery sludges. Rates of degradation achieved and the size of the loadings applied in the research are substantially greater than those typical of landfarm treatment of oily waste. These factors contribute to a reduction of the land requirement for an oily waste compost operation in comparison with that for landfarming. Containment and biological treatment of oily wastes in totally enclosed reactors also substantially reduce the environmental monitoring required of land application sites and the problem of containment and treatment of the volatile constituents of the oily wastes.

Existing technology can be applied to the composting of hazardous biodegradable wastes through suitable modifications. The need to control the gaseous emissions from composting hazardous wastes demands the use of a system especially designed to meet it. Therefore, one of the modifications would be to install a mechanism for treating the gaseous emissions. The type of mechanism for the task would depend upon the nature of the gaseous substances.

An important constraint is the necessity of confining objectionable gaseous emissions within the reactor or enclosure until they are processed. This constraint rules out arrangements in which the only provision for gaseous emission control is by way of a ventilation system. Most of the current enclosed compost systems probably could be suitably to fit these requirements. Practically all available compost systems are readily adaptable to managing liquid and solid emissions.

Unless halogenated organic compounds are to be destroyed, an aerobic system would be the one of choice. If the halogenated compounds are to be composted, then an anaerobic phase would have to be included. If an oily sludge is to be composted, a bulking agent would be required.

The future of composting refinery sludges

In general, some of the alternatives available for the disposal of refinery sludges include: landfilling, landspreading, landfarming, and thermal destruction. Both landfilling and landspreading require that a specific amount of land be dedicated for the purpose. A certain amount of breakdown may take place during storage of the waste material due to natural decomposition. Oily wastes may also be tilled into the soil to promote additional decomposition. In most cases, the breakdown of toxic organic compounds may take on the order of months or even years [Ann Arbor Science Publishers 1981; Butterworth Publishers 1983; Norris 1980]. Since landfarming generally is not controlled, the process typically requires relatively large amounts of land.

Some organic residues can be used as substrates for the microbial activities that take place in composting [Epstein 1980; Pitter 1976]. Consequently, composting can be used for treating biodegradable hazardous wastes [Rose 1968; Goldstein 1983]. In addition, composting has become increasingly popular for the treatment of sludges from municipal wastewater treatment facilities [Golueke 1980].
Composting offers a major advantage over other available alternatives. The process, in essence, reduces the length of time necessary to detoxify the toxic compounds and stabilize the biodegradable components. Typically, simple storage or landspreading requires several months [Ann Arbor Science Publishers 1981; Butterworth Publishers 1983; SCS Engineers 1979]. Furthermore, if the containment site is not covered, the time required for stabilization becomes longer. On the other hand, a well-designed and managed composting operation does not have to be negatively impacted by the climate.

Aerobic composting can be a relatively simple operation in terms of equipment and manpower requirements, especially when compared with incineration.

Incineration of hazardous wastes usually requires the use of sophisticated air pollution control equipment and involves appropriate treatment and disposal of the ash. On the other hand, a composting operation normally affects a well-defined area. Consequently, the potential for adverse environmental impact by a composting facility is less than that by an incinerator.

References

American Public Health Association (1971). *Standard Methods for the Examination of Water and Wastewater*, 13th Edition.

Ann Arbor Science Publishers, Inc./The Butterworth Group (1981). *Decomposition of Toxic and Nontoxic Organic Compounds in Soils*. M.R. Overcash (editor), Ann Arbor, Michigan.

Argonne National Laboratory. Evaluation of the Feasibility of Biodegrading Explosives-Contaminated Soils and Groundwater at the Newport Army Ammunition Plant (NAAP), Report No. CETHA-TS-CR-92000, U.S. Army Toxic and Hazardous Materials Agency.

Atlas, R.M. (1988). Biodegradation of Hydrocarbons in the Environment, in Environmental Biotechnology Reducing Risks from Environmental Chemicals through Biotechnology. G.S. Omen (editor), Plenum Press, New York, New York, 211–222.

Atlas, R.M. (1991). Bioremediation of Fossil Fuel Contaminated Soils, in In Situ Bioreclamation. R.E. Hinchee and R.F. Olfenbuttel (editors), Butterworth-Heinemann, Boston, Massachusetts, 14–32.

Butterworth Publishers (1983). *Hazardous Waste Land Treatment*. K.W. Brown, G.B. Evans, Jr., and B.D. Frentrup (editors), Boston, Massachusetts, 430.

Diaz, L.F., G.M. Savage, and C.G. Golueke (1994). '*Biological Treatment of Refinery Sludges,*' in *Proceedings of the Eleventh Annual HAZMACON 94' Hazardous Materials Management Conference and Exhibition*. San Jose, California.

Environmental Research and Technology, Inc. (1983). *Land Treatment Practices in the Petroleum Industry*. Prepared for The American Petroleum Institute.

Epstein, E. and J.E. Alpert (1980). 'Composting of Industrial Wastes,' Chapter 15. *Toxic and Hazardous Waste Disposal, Vol. 4, New and Promising Ultimate Disposal Options*. Ann Arbor Science Publishers, Ann Arbor, Michigan.

Gibson, D.T., and V. Subramanian (1984). Microbial Degradation of Aromatic Hydrocarbons, in Microbial Degradation of Organic Compounds. D.T. Gibson (editor), Marcel Dekker, Inc., New York, New York, 182–252.

Goldstein, N. (1983). 'Hazardous Waste Landfill Benefits from Compost.' *BioCycle*, September/October, 30–31.

Golueke, C.G. and L.F. Diaz (1980). *Benefits and Problems of Refuse-sludge Composting*, Final Report to National Science Foundation, under Contract No. PFR–7917407, by Cal Recovery Systems, Inc.

Meyers, J.D. and R.L. Huddleston (1980). 'Treatment of Oily Refinery Wastes by Landfarming,' in *Proceedings of the 34th Industrial Waste Conference*. Purdue University.

Mount, M.E. and F.W. Oehme (1981). 'Carbaryl: A Literature Review.' Residue Reviews, 80, 1–64.

Norris, D.J. (1980). 'Landspreading of Oily and Biological Sludges in Canada,' *In Proceedings of the 35th Industrial Waste Conference*. Purdue University, 14.

Noyes Data Corporation (1980). Biodegradation Techniques for Industrial Organic Wastes. D.J. De Renzo (editor). Park Ridge, New Jersey, 10–103.

Nyns, E.J., J.P. Auguiere, and A.L. Wiaux (1968). 'Taxonomic Value of the Property of Fungi to Assimilate Hydrocarbons.' Antonie van Leeuwenhoek, J. Microbiol. Serol., 34, 441–457.

Overcash, M.R. and D. Pal (1979). Design of Land Treatment Systems for Industrial Wastes – Theory and Practice. Ann Arbor Science Publishers, Inc., Ann Arbor, Michigan, 246–255.

Paris, D.F., D.L. Lewis, and N. Lee Wolfe (1975). 'Rates of Degradation of Malathion by Bacteria Isolated from Aquatic Systems.' *Environmental Science and Technology*, 9(2), 135–138.

Pitter, P. (1976). 'Determination of Biological Degradability of Organic Substances.' *Water Research*, 10, 231–235.

Prince, A.E. (1961). 'Microbiological Sludge in Jet Aircraft Fuel.' Dev. Ind. Microbiology, 2, 197.

Rose, W.W. and W.A. Mercer (1968). *Fate of Insecticides in Composted Agricultural Wastes*, Progress

Report, Part I. National Canners' Association, Washington DC.

SCS Engineers (1979). *Selected Biodegradation Techniques for Treatment and/or Ultimate Disposal of Industrial Organic Materials.*

Suflita, J.M., and G.W. Sewell (1991). Anaerobic Biotransformation of Contaminants in the Subsurface. U.S. Environmental Protection Agency, Environmental Research Brief, EPA/600/M–90/024, Washington DC.

Starch Based Biodegradable Materials in the Separate Collection and Composting of Organic Waste

C. BASTIOLI, F.DEGLI INNOCENTI – Novamont, via Fauser 8, 28100 Novara, Italy

Abstract

Biodegradable plastics are a new generation of materials, still at an early stage of development. Their development is tied to the growth of composting infrastructures, to the definition of severe standards for biodegradability and compostability, and to the marketing of effective and truly compostable products. The data reported in this paper, show that Mater-Bi Z grades comply with the compostability requirements. The biodegradation studies at laboratory scale, the full scale trials, and the toxicity and compost quality tests prove the compostability of Mater-Bi Z grades according to the scheme followed by the different International Committees (ISR/ASTM, CEN, DIN, ORCA). Moreover, experiments of separate collection of organic waste organized in several European Countries with Mater-Bi bags were fully satisfactory. These data indicate that biodegradable materials are mature for a real industrial development starting from specific applications such as composting bags.

Introduction

Biodegradable plastics are a new generation of materials, still at an early stage of development, which retain their shape and properties while in use, but completely biodegrade when properly discharged.

While the plastics marketed ten years ago as 'biodegradable', constituted by synthetic polyolefin resins and small amounts of corn starch, were just disintegrable, today's degradable plastics meet stringent requirements for biodegradability. Many of the new plastics are actually made from natural sources. Exposed to the microbes and moisture in a compost pile, they break down, like paper, in carbon dioxide, water and humus; no toxic residues are left behind.

Degradable polymers alone are not going to solve the world's waste problems, but they can offer to waste producers additional options in situations where recycling is difficult or too expensive.

Expanded foams in place of polystyrene, composting bags, paper coatings, cutlery, diapers and other hygiene products are some of the possible applications of biodegradable polymers; many others can be foreseen if price targets, new commercial potentialities, and development of new or improved degradable resins are met.

At the moment, the main producers of biodegradable materials are: Novamont with starch based materials, Cargill with poly-lactides, Rohm and Haas with poly-aspartic acid, Union

Carbide with poly-epsilon-caprolactone, Zeneca with poly-hydroxy butyrate-valerate, Showa Denko with poly-butylen succinate, Eastman Chemical with cellulose acetate.

The growth of these materials in Europe is strictly linked to the development of source separation and composting of the organic fraction of municipal solid waste. In Germany, for example, the capacity for biowaste treatment has increased from 183750 tons/year at the end of 1992 to 425300 tons/year at the end of 1993. This evolution offers great possibilities to biopolymers on condition that they are truly compostable and do not jeopardize the quality of the compost.

International organizations such as the American Society for Testing and Materials in connection with the Institute for Standards Research (ASTM/ISR), the European Standardisation Committee (CEN), the International Standardisation Organisation (ISO), the German Institute for Standardisation (DIN), the Organic Reclamation and Composting Association (ORCA), are all actively involved in developing definitions and tests for biodegradability and compostability (ORCA, 1994; ASTM, 1993). They have already defined, at draft level, the basic requirements for a product to be declared compostable based on:

- Complete biodegradability of the product, measured through respirometric tests like ASTM D5338–92 and the CEN proposal (Degli Innocenti et al.) or Sturm test (OECD, 1981), in a time period compatible with the selected disposal technology (some months) (ASTM, 1993);
- No negative effects on compost quality and in particular no toxic effects of the compost and leachates to the terrestrial and aquatic organisms;
- Control of laboratory scale results on full scale composting plants (ORCA, 1994).

An important driving force for the development of biodegradable materials is the recently adopted European Directive on packaging waste (European Union, 1994). Composting of packaging waste is considered a form of material recycling whereas incineration with energy recovery is considered a form of recovery with lower priority. CEN (European Standardisation Committee) has been appointed to define the compostability criteria for packaging waste.

Moreover, in Germany, Entwurf LAGA–Merkblatt M10, has recently defined the list of products suitable for composting. This list includes fully biodegradable and compostable materials, according to the criteria of DIN FNK 103.2.3 on 'Biodegradable Plastics', based on test methods simulating watery and composting environments.

Aim of this work is to present an example of industrial development of biodegradable plastics, i.e. the starch based Mater-Bi Z grades, produced by Novamont, and an example of a specific application, the composting bags.

Composition and main features of Mater-BI Z grades

The main components of Mater-Bi grades belonging to the class 'Z' are starch and poly-epsilon-caprolactone. Minor components are natural plasticizers and compatibilizers.

Grade	Molding Technology	Use
ZI01U	Extrusion/casting/ film blowing/ injection molding	General purposes
ZF02U	Film blowing paper lamination	Diapers back-sheet,
ZF03U	Film blowing lamination	Bags, nets, paper

Under the trade mark 'Mater-Bi', Novamont produces other three classes of biodegradable materials (A, Y and V), all containing thermoplastic starch, characterized by different synthetic components (Bastioli in press; Bastioli et al. 1993a; Bastioli et al. 1993b; Bastioli et al.1992; Mater-Bi Technical Bulletins 1991, 1992a, 1992b, 1993).

The properties of the products, in comparison with LDPE are reported below, together with the standard methods.

Compostability of Mater-bi z grades

The compostability of these materials has been demonstrated according to the common approach followed by the different International Committees.

1. Intrinsic biodegradability of the material. This is shown by standard respirometric tests simulating composting conditions (ASTM D5338–92) and watery environments such as the urban depurator (OECD,1981; Molinari 1993; Molinari and Freschi, 1994; Freschi et al., 1994).

   These tests are intended to measure the degree of mineralization, namely the transformation of the test material organic carbon into carbon dioxide. The controlled composting test simulates a composting environment and compares the biodegradation behaviour of the test material with that of microcristalline cellulose, both at concentrations of 10% w/w. After 45 days the biodegradation of ZF02U, ZF03U, ZI01U, and cellulose was respectively of 74, 78, 100, and 85% (J.Boelens 1992a; J.Boelens, 1992b).

The OECD modified Sturm procedure is similar to ASTM D5209–91 and to the test of the Italian Decree DM 7/12/90. All Mater-Bi grades 'Z' showed after 56 days a biodegradation level quite similar to that of paper for food contact, according to what required by the Italian Decree to define an insoluble material as biodegradable. The biodegradation index was of 93% for ZI01U, 77% for ZF03U, and of about 73% for the reference paper.

2. Absence of negative effects of the test material on compost quality. This is demonstrated with terrestrial toxicity tests (Seed Germination as described by Zucconi et al. (1981), Terrestrial Plant Growth tests by OECD 207, Worm Acute Toxicity tests by OECD 208) and physical and chemical characterization of the compost obtained by controlled composting test and prepared according to the ASTM procedure 'for preparing residual solids obtained after biodegradability standard methods for plastics in solid waste for toxicity and compost quality testing'. All these tests, performed by the Belgian Company Organic Waste Systems (De Wilde and Boelens 1992) on the compost obtained by ASTM test D5338-92, demonstrated that Mater-Bi Z grades did not affect the compost quality and were not toxic. The compost obtained in presence of 10% of Mater-Bi ZFO3U, when added to agricultural soil, gave, in fact, results comparable with those for the reference compost, coming from the degradation of cellulose, and for the control compost.

3. Final cross-verification through full scale composting trials. The compostability of Mater-Bi Z grades was evaluated in composting plants located in different countries (Germany, Italy, Japan) using different technologies (static windrows, turned windrows, rotary fermenting reactors).

At the Scuola Agraria-Parco di Monza in Italy the degradation of ZI01U was tested in a yard static windrows (Favoino and Centemero, 1993). Pen caps in ZI01U were threaded with a plastic line and entangled in a plastic tubular net. Before every turning, the samples, located by sticks, were carefully removed by digging the surrounding composting mass, examined, and re-introduced in the compost. Examinations of the ZI01U pen caps were made at 20 days, 54 days and at the end of the standard cycle after 8 months. After 20 days the caps presented a strong degradation. At 54 days the phenomenon was more pronounced and the residue was constituted by fragments of 1 mm, soft and dank. At the end of the cycle, the nets were empty and only residual lumps were recoverable, stuck on the net with the typical appearance of compost. Other solid natural products in the heap, such as small pieces of wood, were yet not degraded, only the surface being slightly attacked. At the plant located in Limidi di Soliera, Modena Italy, belonging to the AMIU (Municipal Waste Treatment Department) of Modena, in a trench reactor a complete degradation of ZI01U was observed without any effect of the final quality of compost (Piccinini et al.).

At the Istituto Agrario San Michele all'Adige, Italy an experiment of source separated collection and composting of organic fraction using 100 liters bags of Mater-Bi ZF03U, was performed (Silvestri et al.). The compost quality parameters were within the specifications of the 'green' composts.

At the Nogi-Cho Recycling Center/Japan (Yoshida and Tomori, 1993) the composting behaviour of Mater-Bi bags containing organic wastes was followed during the process. No film-like material was found in the primary compost coming out of the rotary fermenting reactor. Only some knots of Mater-Bi bags were observed at 40 days. After 84 days there was no residue of Mater-Bi bags on the 5 mm opening sieves. The results of the experiment proved the compostability of Mater-Bi films under the standard 90 days treatment of the facility.

Production and use of composting bags made with mater-bi zf03u

An example of a successful application of Mater-Bi Z grades is represented by the compostable bags. ZF03U tubular film is produced by means of a traditional film blowing equipment for LDPE. The extruder is a 60 mm one, with a spiral head of 130 mm of diameter. The blow up ratio used is of 5 and the draw ratio of 3.8. The film thickness is between 25 μm and 40 μm. The plant throughput is of 80 kg/hour (Mater-Bi Technical Bulletin, 1992b). ZF03U bags are produced with traditional bag-making machines, increasing the hot knife temperature for the longitudinal sealing and adopting semicircular sealing profiles for the impulse sealing on the bottom.

Bags of different dimensions were used by thousands of citizens in different municipalities for tests of separate collection of organic wastes to be composted. In these experimental projects paper bags were compared with Mater-Bi bags to estimate, in addition to their compostability, the in use behaviour.

Here below the results obtained in the two municipalities of Furstenfeldbruck Bavaria and Korneuburgh/Austria are shortly summarized.

Furstenfeldbruck/Bavaria (Landratsamt Fürstenfeldbruck, 1993): an experiment of separate collection of organic waste performed by 2000 inhabitants with Mater-Bi and paper bags was followed for one year. In terms of in use performances (mechanical properties in wet environment, manageability, smell barrier, transparency etc.), 55.6% of the citizens preferred Mater-Bi bags, and 28.6% preferred paper-bags. The compost quality was found in line with the German specifications for compost.

Korneuburgh/Austria (Hauer W., 1993): This experiment was similar to the trial of Furstenfeldbruck. The citizens preferred Mater-Bi bags (87%) rather than the paper bags. The compost quality was estimated good.

Today more than 20 municipalities in Europe are using Mater-Bi Z bags with positive results.

Conclusions

The future of biodegradable plastics, a new generation of materials, is closely linked to the growth of composting infrastructures which can be considered, in agreement with the European Directive of packaging waste, as effective systems of recycling. The availability of a significant amount of biodegradable packaging waste to be composted can even strengthen the value of this technology. The acceptance of biodegradable plastics in composting, however, requires specific points to be met. First of all, the definition of standards for biodegradability and compostability, to provide clear decisional tools to the Authorities and to the other parties involved in production, disposal, and composting of biodegradable goods. Then, the industrial research & development must satisfy the market requirements by offering effective and truly compostable products. The data reported in this paper, show that Mater-Bi bags for composting comply with the requirements outlined by the International Organisms working on compostability criteria definition. Furthermore, experiments of separate collection of organic wastes organized in several European Countries with Mater-Bi bags, were fully satisfactory. In conclusion these data prove that biodegradable materials are mature for a real industrial development starting from specific applications such as composting bags.

Bibliography

ASTM (1993) Standards on Environmentally Degradable Plastics. ASTM subcommittee D2096. Publication Code Number (PCN) : 03–420093–19.

Bastioli C. Starch-Polymer Composites. In Scott G. (Ed. 'Degradable Polymers: principles and applications'. Chapman & Hall, London. In press.

Bastioli C., Bellotti V., Del Giudice L., Gilli G. (1992) Microstructure and Biodegradability of Mater-Bi Products. In: Vert M., Feijen, Albertsson A., Scott G. and Chiellini E. Biodegradable Polymers and Plastics. The Royal Society of Chemistry, pp. 101–110

Bastioli C., Bellotti V., Del Giudice L., Gilli G. (1993a) Mater-Bi: Properties and Biodegradability J. Environ. Polym. Degr., 1(3),181–191

Bastioli C., Bellotti V., Camia M., Del Giudice L., Rallis A. (1993b) Starch-Vinyl-Alcohol Copolymer Interactions. 3rd International Scientific Workshop on Biodegradable Plastics and Polymers, Osaka November 9–11

Boelens J. (1992a) Aerobic Biodegradation under Controlled Composting Conditions of Test Substance ZI01U. OWS Final Report.

Boelens J. (1992b) Aerobic Biodegradation under controlled Composting Conditions of Test Substance ZF03U. OWS Final Report

Degli Innocenti F., Tosin M., Bastioli C. Method for the Evaluation of Biodegradability of Packaging in Composting Conditions proposed by CEN (European Committee for Standardization): a Technical Approach. International Conference The Science of Composting, Bologna, May 30, June 2, 1995 (this volume)

De Wilde B., Boelens J. (1992) Compost Quality Tests of Test Substance ZF03U Compost. OWS Final Report

European Union. (1994) Directive 94 EC of the European Parliament and of the Council on Packaging and Packaging Waste PE-3627

Favoino E., Centemero M. (1993) Compostability Testing of injection molded items in Mater-Bi: General Information. Scuola Agraria of Parco di Monza. Final Report

Freschi G., Bastioli C., Degli Innocenti F., Gilli G., Molinari G.P. (1994) Proceedings of the Congress of Società Italiana Chimica Agraria. In press

Hauer W. (1993) Vorarbeten für die Markteinführung von Säcken aus Mater-Bi zur Samulung von biogenen Abfällen. Kornenburg, Report September 1993

Landratsamt Fürstenfeldbruck (1993) Pilot projekt Mater-Bi-Säcke. Fürstenfeldbruck , Germany

Mater-Bi Technical Bulletin (1991) Novamont: the living Chemistry, Novamont, August 1991

Mater-Bi Technical Bulletin (1992a) Mater-Bi Properties, Applications, Markets. Novamont, Italy.

Mater-Bi Technical Bulletin (1992b). Mater-Bi Converting: Blown Film Extrusion. Novamont, Italy.

Mater-Bi Technical Bulletin (1993). Mater-Bi Converting: Injection Molding. Novamont, Italy.

Molinari G.P. (1993) Saggio di biodegradabilita' aerobica secondo D.M. del 07/12/1990 di ZI01U. Università Cattolica del Sacro Cuore, Piacenza, Italy.

Molinari G.P., Freschi C. (1994) Saggio di biodegradabilita' aerobica secondo D.M. del 07/12/1990 di ZF03U. Università Cattolica del Sacro Cuore, Piacenza, Italy.

OECD (1981) Method 301B. Ready Biodegradability: Modified Sturm Test. OECD Guideline for Testing of Chemicals.

ORCA Technical Committee (1994) ORCA Compostability criteria. Brussels, Belgium.

Piccinini S., Rossi L., Degli Innocenti F., Tosin M., Bastioli C. Behaviour of biodegradable Mater-Bi ZI01U plastic layers in a composting pilot plant. International Conference The Science of Composting, Bologna, May 30, June 2, 1995 (this volume)

Silvestri S., Zorzi G., Degli Innocenti F., Bastioli C. Use of Mater-Bi ZF03U Biodegradable Bags in Source-Separated Collection and Composting of Organic Waste. International Conference The Science of Composting, Bologna, May 30, June 2, 1995 (this volume)

Yoshida Y., Tomori M. (1993) Properties and Application of Mater-Bi. 3rd International Scientific Workshop on Biodegradable Plastics and Polymers, Osaka, Japan 9-11 November, 1993, 72

Zucconi F., Forte M., Monaco A., De Bertoldi M. (1991) Biological Evaluation of Compost Maturity. Biocycle, 22, 27–29

Acknowledgment

Many thanks to Ms. Silvia Costa for her valuable help in preparing the manuscript.

Degradation of Naphthalene by Microorganisms Isolated from Compost

Civilini M. and Sebastianutto N. – Food Science Department, University of Udine, Italy

Microbial strain capable of degraded complex polycyclic aromatic hydrocarbons are enought difficult to isolate becouse their scarcity in nature. With increasing PAHs contamination is possible to find sites where microrganisms had developed the mechanisms of adaptation in confront of these compounds. This is expecially true for the community present in soils or sediments that have been chronically exposed to PAHs.

Our past experiences (1) showed also the possibility to increase the population with PAHs catabolic activity by composting. It could be done looking for to optimized the most important parameters to improve biologial activity and using an preadapted inoculum in the starting material.

Isolation of strains

We believed that organic matter from Municipal Solid Waste could be a good starting material due its high number of microbial population and ecological nick. For these reason we choosen to use this material to prime adaptetion mechanisms in microrganisms respect at napthalene and byphenil.

For several mounths about ten kilograms of this material with right moisture content was feed by air in let with naphthalene and byphenil due their volatilization propriety.

Enrichment cultures were set up involving minimal salt basal medium (2) and supplemented with naphthalene and byphenil as the sole source of carbon. The enrichments were seeded with the PAH contaminated samples mentioned above. The enrichment cultures were monitored for the presence of microorganisms and subcultured into fresh medium when significant quantities of microorganisms were obtained. After four such subcultures the cells were plated onto solid medium and the PAH was applied as an ethereal solution to the surface of the agar. Colonies appearing were purified on the same medium and then stored at –80°C.

Characteristics of strains

Representative strains resulting from these isolation experiments are listed in Table 1.

The strains 2NR, 10N1 and C4B was chosen for further study becouse of the strong growth and due their different phenotypes.

The strains were identified to the genus level using the Oxi/Ferm Tube II (Roche) system. The presuntive species name were obtained by further analysis as indicated on the system and from Bergey's manual of systematic bacteriology (vol.1). The strain 2NR was identified as *Pseudomonas aeruginosa*, 10N1 as *Pseudomonas putida* and C4B as *Pseudomonas putrefaciens* ; their characteristics are reported in Table 2.

In order to confirm that the strains could completely degrade Naphthalene, several experiments were performed with non-radiolabeled substrates following carbon dioxide evolution (3).

Respiration tests

Microbial inoculum for Biometer flasks (Figure 1) was a sospension of cells in mineral salt basal medium. The optical density at 600 nm was adjusted around the value of 0.4.

The cells of strain test were growth on suitable broth in order to produce the same definite sequential-induction pattern.

2NR and 10N1 were growth on mineral salt basal mediun plus Succinate (20 mM) and Naphthalene (2.5 mM). C4B had Biphenyl instead Naphthalene as induction conmpounds. The two water-insoluble PAH was added trought a N,N Dimethylformamide stock solution (100 mg/ml).

The cells growth at 30 °C for 36 hours were harvested by centrifugation, washed three times with Na-K buffer 50 mM, pH 7. The cells were resuspended in 300 ml of mineral salt basal medium and adjusted at the value of 0.4 of optical density (600 nm).

50 ml of this sospension was share in each of the four biometer flask; two were used as a control and the other as a repetition for degradation test. The biometer flasks were incubated in rotary shaker at 30°C. Carbon dioxide was evaluated, by titolation of the KOH with HCl normex, at different time using some attention for external CO_2 contaminations.

Table 1 Screening of the activity of the isolated strains

Isolated strains	Screening for microorganisms selection		Antibiotic resistence of selected microrganisms							Plasmid / NAH
	Growth on Agar Minimal medium plus Naphtalene	Yellow on Broth Minimal medium plus Naphtalene	Streptomycin		Carbenicillin	Tetracycline		Ampicillin		Naphtalene-Dioxigenase Activity
			50 µg/ml	200 µg/ml	100 µg/ml	10 µg/ml	50 µg/ml	20 µg/ml	50 µg/ml	
C2N	+	+	+	+/-	+	+	-	+	+	+
2NA	+	+	+	+	+	+	-	+	+	+
2NB	+	+	+	+/-	+	+	-/+	+	+	+
2NC	+	+	+/-	-	+	+	-	+	+	+
2NF	+	+	+	+/-	+	+	-	+	+	+
2NL	+	+	+	-	+	+	-/+	+	+	+
2NR	+	+	+	-/+	+	+	-/+	+	+	+
10N1	+	+	-	-	+	-	-	-/+	-	-
10N2	+	+	-	-	+	-	-	-/+	-	-
10NBD	+	-	-	-	+	-	-	-/+	-	-
10NBP	+	-	-	-	+	-	-	-/+	-	-
10NBG	+	-	-	-	+	-	-	-/+	-	-
10NG	+	-	-	-	+	-	-	-/+	-	-
10NQ	+	-	-	-	+	-	-	-	-	-
C3B	-/+	-	-	-	+	-	-	-	-	-
C4B	+	+	-	-	-	-	-	-	-	-
4B	+	+	-	-	+/-	-	-	-	-	-
4BA	+	+	-	-	+/-	-	-	-	-	-
CEFK	-	-	-	-	+	-	-	-/+	-	-
AoA3	-	-	-	-	-	-	-	-	-	-

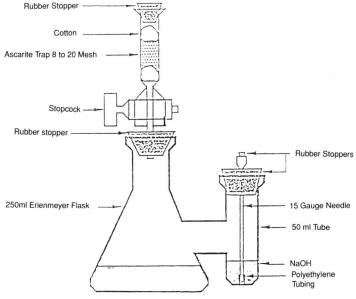

Figure 1 Biometer flask

GC mass analysis

To identified intermedies of naphthalene catabolism, 250 ml of a sospension of cells of the strain 2NR in mineral salt basal medium (O.D.600 nm: 0.4) was added with naphthalene (1 mg/ml) and incubated for 290 hours at 30°C with agitation. At different interval time, 1.5 ml of the culture was take out for samples. After centrifugation 1 ml of supernatant was acidified with HCl 1 N to pH 2 and then extracted with ethyl acetate. The extracts were evaporated to dryness. The residues were analyzed with GC-ITDMS either directly or derivatized with diazomethan to produce methylated compounds. The GC instrument was a Varian 3400 gas chromatograph associated with an ion trap mass spectrometer Varian Saturn (Ion Trap Detector ITD). Compounds separation was achieved by using a fused silica capillary column DB-1701 (J & W, Folsom, CA, USA) (30 m x 0.25 mm I.D., 0.25 μm film tickness) using the following condiction: Helium as carrier gas (1 ml/min), the injection was in split mode (1:20, v/v, ratio); the column temperature was 70 °C increased to 270°C at 10 °C/min and maintained at 270 °C for 10 min. The temperature of injector, transfer line and manifold were 270°C, 270 °C and 170 °C respectively. The filament emission current was 10 mA and an electron beam of 70 eV was used for electron impact ionization (EI).

The same method was used looking for intermedie of 2NR catabolism starting from catechol, salicylic acid and gentisic acid which important intermedies of the naphthalene degradation pathway by microorganisms (Scheme 1).

HPLC analysis

The target compounds reduction and metabolites production was follow by HPLC analysis during 2NR growth as explain for GC-ITD-MS method.. Samples were extracted with ethyl acetate after acidification to pH 2–3 with phosphoric acid. The ethyl acetate was evaporated and the residues were dissolved in methanol (4). The instrument was a liquid chromatograph Varian 5040 (CA, USA) with UV–VIS detector Varian 50 (λ of analysis was 290 nm). The column was 5µm a Spherisorb C18 (15 cm x 4.6 I.D., Phase Sep, Deeside, UK). The solvent system was of methanol (A) and a mix of water-methanol-acetic acid 80:20:3 respectively (B). Starting from 100% B to 20 % B a gradient of 1 ml/min was applied.

Results and discussion

The characteristics of the strains showed in Table 1 induce us to look for different aspects of three strain mentioned above. In particulary, the strain 2NR was positive at the naphthalene dioxigenase tests (4) and at the antibiotic test. These two results associated at the lower growth of the strain with higer concentration of Streptomycin and Tetracycline could indicate the presence of plasmid coding for naphthalene degradation as reported in bibliography for other strains (6, 7, 8 and 9). Strain 10N1 and C4B were negative for presuntive presence of plasmids coding for naphthalene degradation.

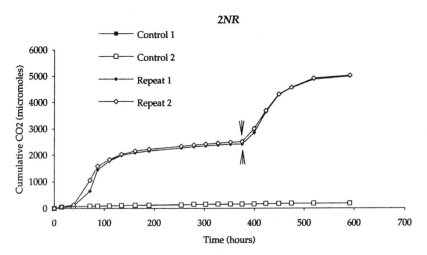

Figure 2 CO2 behaviour from 2NR growth on naphtahalene; Control: suspension of cells Control 2: suspension of cells + 500ml Dimethyl formamide Repeat 1: suspension of cells + 500 ml naphtahalene/Dimethyl formamide Repeat 2: suspension of cells + 500 ml Naphthalene/Dimethyl formamide

Naphthalene

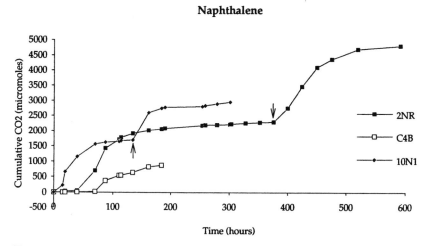

Figure 3 Comparison among CO2 behaviour of the different strains (arrows show the time where new naphtahalene was added)

The respiration results give further information on the difference among the strains; the method was applied to follow the strain's growth on naphthalene: Fig. 2 show the good reproducibility of the no-quantitative method either for the controls that for the repetition flask. A confront among the different strains behaviour on naphthalene was reported in Fig.3, where, for strain 2NR and 10N1 at time 350 (h) and 137 (h) respectively, was added new substrate to confirm its degradation. The naphthalene degradation pathway find in bibliography (Scheme 1) give us idea to use some metabolite as substrate for biometer flask method in order to know more information of degradation route of our strain. In particulary 2NR (Fig. 4) and 10N1 (Fib. 5) could be compared due the same naphthalene induction (C4B inductor was byphenil). The growth of the two strains on salicylic acid, gentisic acid and catechol was different: 2NR used all three compounds instead strain 10N1 growth only on salicylic acid. Seeing the 2NR behaviour may be hypothesize that gentisc acid pathway is different from catechol and salicylic acid pathway. Infact the initial exponential growth (quickly gentisic acid and later the other two) changed when further addition of substrates was added (quickly for catechol and salicylic acid, later for gentisic acid). This mean that gentisic acid pathway could be costitutive (chromosomal coded) whereas salicylic acid and catechol could be inductive pathway (plasmid coded).

To verify this hypothesis and to know the 2NR naphthalene pathway, studies to isolate and identify possible intermedies were performed by HPLC and GC–ITD–MS.

The disappearance of starting compounds and the production of methabolites during the growth of the microorganism was follow by HPLC analysis confirming naphthalene (non-quantitative), gentisic acid, salicylic acid and catechol degradation.

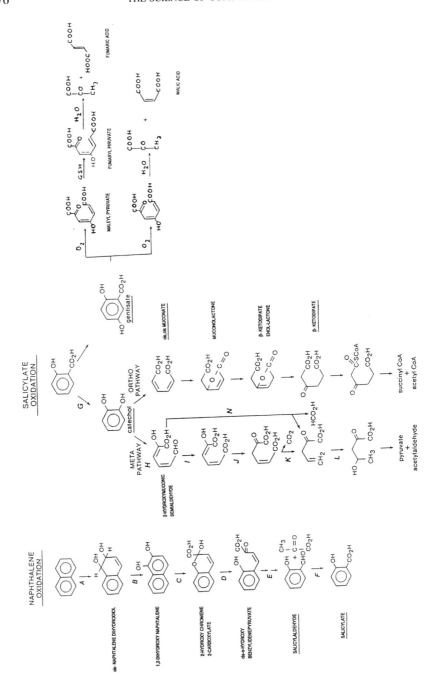

Scheme 1: Bacterial metabolism of naphthalene (underlined compounds were identified in 2NR naphthaline degradation pathway)

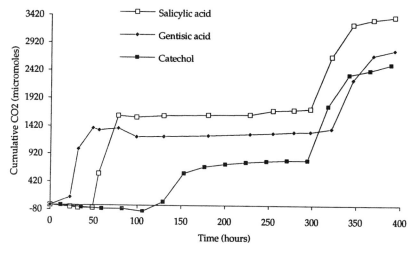

Figure 4 Comparison among CO2 behaviour of 2NR strain growth on catechol, salicylic and gentisic acid.

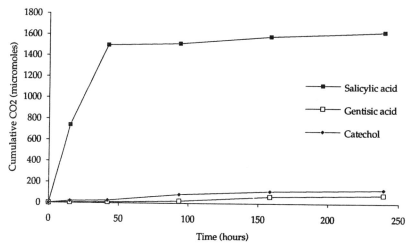

Figure 5 Comparison among CO2 behaviour of 10N1 strain growth on catechol, salicylic and gentisic acid.

The qualitative identification of intermedies was done by GC–ITD–MS analysis looking for ritention time and mass spectre of standard or only computer library mass spectre when standard of the compound was not avalaible. Starting from naphthalene as substrate, after 2 hours (Fig. 6) could be identified two compounds, salicylaldehyde (Fig. 7) and catechol (Fig. 8 and 9); after 6 hours (Fig. 10) other three compounds were found: salicylic acid (Fig. 11) and gentisic acid (Fig. 12). The presence of gentisic acid mean that at least two pathway for naphthalene degradation are involving in the same time: the ortho cleavage pathway and the gentisate route.

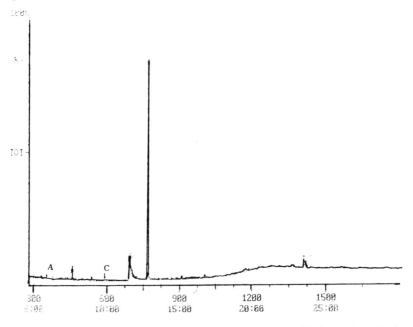

Figure 6 GC-ITDMS chromatogram obtained after two hours of 2NR growth starting from naphtahalene; A: salicylaldehyde C: catechol

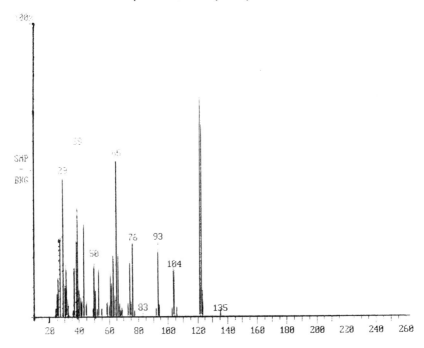

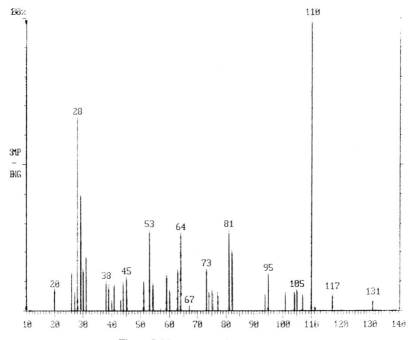

Figure 8 Mass spectra of catechol (C)

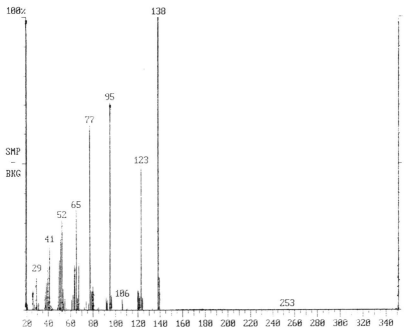

Figure 9 Mass spectra of catechol dimethyl ester (E)

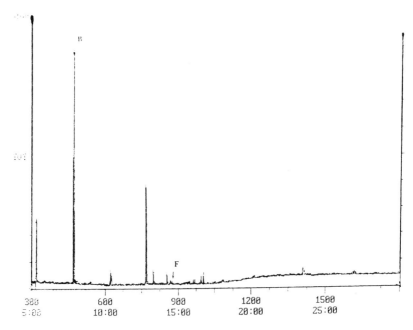

Figure 10 GC-ITDMS chromatogram obtained after six hours of 2NR growth statring from naphthalene B: salicylic acid methyl ester F: gentisic acid methyl ester

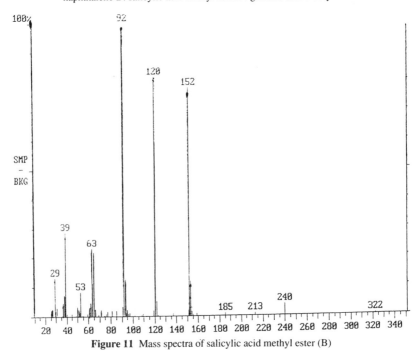

Figure 11 Mass spectra of salicylic acid methyl ester (B)

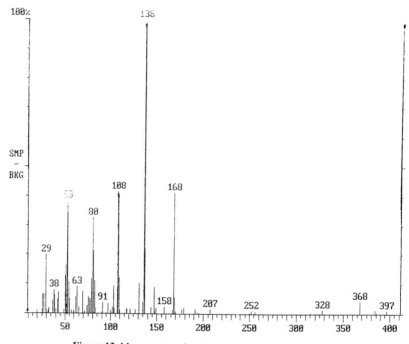

Figure 12 Mass spectra of gentisic acid methyl ester (F)

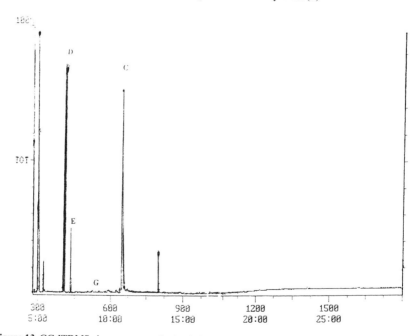

Figure 13 GC-ITDMS chromatogram obtained after eighteen hours of 2NR growth starting from catechol; D: catechol methyl ester G: cis, cis muconic acid dimethyl ester

Table 2 Morphological and biochemical characteristics of some strains

	2NR	10N	C4B
Morphology	Coccorods	Coccorods	Rods
Gram	−	−	−
Oxidase test	+	+	+
Oxi/Ferm tube			
Ana-Glucose	−	−	−
Arginine	+	−	−
Lysine	+	+	+
Lactose	−	−	−
N2	−	−	−
Sucrose	−	−	−
Indole	−	−	−
Xylose	+	+	+
Aer-Glucose	+	+	−
Maltose	−	−	−
Mannitol	+	−	−
Phenyl-Alanine	−	+	+
Urea	+	+	+
Citrate	+	+	+
Growth at 42°C	+	n.d.	n.d.
	Pseudomonas aeruginosa	*Pseudomonas putida*	*Pseudomonas putrefaciens*

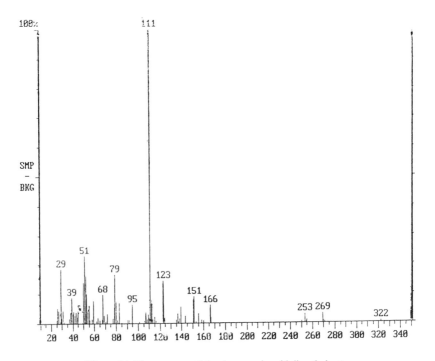

Figure 14 Mass spectra of cis, cis muconic acid dimethyl ester

Always with cultures of strain 2NR but starting with catechol was possible after 18 hours (Fig.13) point out an other compound cis,cis muconic acid (Fig. 14) to confirm the ortho cleavage pathway.

Culture of 2NR starting with salicylic acid confirmed some of the above identified compounds, instead for cultures with gentisic acid was not possible identified intermedies due their standards were not avalaible.

Conclusion

From this work is possible say that:

– the organic matter from municipal solid waste is a good source of heterogeneus microrganism which are sensible at the adptation method applied to select PAH degrading microrganisms;
– the biometer flask method has a good reproducibility when applied in pure culture and its high performance permit to have important indication on the possible metabolic pathway of the target compounds;
– the isolated strains show catabolic proprieties not only on the inductor compounds but also on other PAH;
– the strain 2NR show at least two pathway for naphthalene degradation: the ortho pathway of catechol and gentisate pathway.

Bibliography

1) Civilini M., Microbiology Europe 2, 6: 16–23 (1994)
2) Stanier r.J., Palleroni N.J. and Doudoroff, J. Gen. Micr. 43: 159–271 (1966)
3) Bartha R. and Pramer D., Soil Sci., 100: 68–70 (1965)
4) Grund E., Denecke B. and Eichenlaub R., App. Environ. Microbiol., 58, 6:1874–1877 (1992)
5) Ensley B.D. et al., Science, 22167–169 (1983)
6) Utkin et al., Folia microbiol. 35, 557–560 (1990)
7) Yen K.M. and Gunsalus I.C., Proc. Natl. Acad. Sci USA, 79: 874–878 (1982)
8) Davies J.I. and Evans W.C., Biochem. J., 91: 251–261 (1964)
9) Yen K.M. and Serdar C.M., Crit. Rev. Microbiol., 15: 247–268 (1988)

Composting and Selected Microorganisms for Bioremediation of Contaminated Materials

CIVILINI M., DOMENIS C., DE BERTOLDI M. and
SEBASTIANUTTO N – Department of Food
Science,University of Udine, Italy

Environmental contaminants such as the monocyclic aromatic group and poly-cyclic aromatic hydrocarbon (PAH) group can be found in practically all industrial areas. Pollution by these compounds are primarily associated with the processing, combustion, and disposal of fossil fuels. Industrial effluents from coal gasifica-tion and liquification processes, waste incineration, coke, carbon black, coal tar, and other petroleum-derived products are all major sources of contamination.

Bio-remediation is emerging as a proven technology to remove xenobiotic com-pounds from the environment. Advantage in using microorganisms for environ-mental clean-up include their ability to degrade a wide range of compounds, their ability to degrade those compounds to below detectable limits, the ease with wich microorganisms can be handled and applied to a contaminated sites, and their cost effectiveness.

Among the different technologies applied for bio-remediation (1), composting may be described as a system with double potentiality: treatment system in biore-actor and treatment system in-situ. The first need to move the higher quantity of contaminated materials from polluted area to a bioreactor for composting. The replacement of its original site with cleaned material, enriched with higher number of microorganisms with catabolic proprieties, can express their activity at the remained contaminated soils in the polluted area.

As regards composting as a treatment system to convert a waste to a processed residue accetable for disposal and/or productive purposes, a project was carried out to evaluate the bio-remediation of 'creosote'-contaminated materials.

The fate of 13 most abundant toxic compounds of creosote contaminated soils mixed with organic matter from municipal solid waste was followed during fif-teen days of composting (2). The laboratory apparatus consisted of three incuba-tors within three indipendent reactors having a working volume of 14 liters. The temperature set point (45°C) was regulated automatically by air flow through the mass, while samples for the chemical and microbiological analyses and the cor-rection of moisture content during the 15 days of composting were taken at days 0, 5, 10 and 15. The values of mass balance parameters (3) permit calculation of the thirteen most-important creosote compounds, herein expressed in terms of the per-

centage reduction compared with the initial concentration (Table 1). To understand the ratio between the toxic compounds elimination by biological activity,
chemical-physical trasformation and volatilization during composting the recovery
comparison among different points of the composting apparatus are showed in
Table 2. Both tables show as composting is an efficient technology for soil clean
up becouse it removes large quantities of toxic compounds in a short time and
high efficiency was also achieved for higher molecular weight compounds. From
this experience to verify the microrganism activities on creosote compounds, different microrganism groups present in organic matter (eterotrophic bacteria, gram
positive non-sporing rods, aerobic and/or facultatively anaerobic endospore-forming rods, anaerobic endospore-forming rods, gram positive cocci 'Strepto', gram
negative rods and coccorods: oxidase positive, oxidase negative lactose positive
and negative) were grown on selective and/or non-selective media. After incubation, some plates were scraped with water solution of tween 80 (2%). The microrganisms were harvested by centrifugation at 8000g for 15 minutes and three time
washed with 2.5 mM phosphate buffer, pH 7. A sospension of the cells of each
groups were inoculated in 500 ml of mineral-salt- (KH_2PO_4, 200 mg/L; K_2HPO_4,
800 mg/L; NH_4NO_3, 100 mg/L; $MgSO_4$ $7H_2O$, 100mg/L; $CaCl_2$ $2H_2O$, 100 mg/L;
$FeCl_3$, 10 mg/L) –water emulsion of aromatic compounds (33, 34, 35) and incubated in a rotary shaker (New Brunswick) at 30°C and 250 rpm. The biodegradative activities were performed by GC analysis. Target compounds were estimated
after ten days respect to the control at the same time incubated of a no-inoculated
mineral-salt-water emulsion of aromatic compounds. The microbiological results
showed in Figure 1 point out the presence of high number of different groups during composting. Their values confirm the possibility of a large biodegrading potentiality of the microbial community respect at the other elimination characteristics
of the toxic compounds by composting. The activity of some of these groups, evaluated by the above PAH-water-emulsion flask method, are rapresented in Table 3.
The disappearance of the most compounds respect at the control, may depend
from high production of enzymes as oxygenase to attach the aromatic compounds
(Scheme 1), bio-surfactant, as molecules of Scheme 2, to make more avalaible the
organic compounds for microrganisms and/or synergic activity of heterogenic
microbial population. To verify the presence in the inoculated flask of single
strains able to PAHs degradation, enriched cultures were set up involving a minimal salt basal medium (4) and supplemented with different polycyclic aromatic
hydrocarbon as the sole source of carbon. Isolation and characterization of the
strains was performed as described in reference (5). Among the numerously strains
same were tested for CO_2 evolution derived from aromatic hydrocarbons measured by a non-radioctive method (6). The results show as single strain can use
same different PAH as source of carbon and energy. Their different efficienty, valuable with the quantity of CO_2 produced, may be relatively important becouse same
time is more important produce the first attach at the aromatic molecule than to
arrive to CO_2 with a single strain. A particular organism may possess the correct
catabolic ability to catalyze the trasformation of one compound but then may not

possess the enzymatic system for further degradation. In composting populations these may be supplied by a second organism and so on, so that communities of organisms possessing complementary catalitic capabilities are established and the compound degraded.

OXYGENASES

MONOOXYGENASE: RH_2 + O_2 $\longrightarrow$ RHOH + H_2O

DIOXYGENASE: RH_2 + O_2 $\longrightarrow$ ROHOH

Scheme 1 Oxigenase mechanisms on generic substrate (R)

Growth behaviour on PAH of three out of twenty different isolated strains are represented in figure 2, 3 and 4 (for strain 2NR see also reference 5) . Naphthalene was the induction compound for 2NR and 10N1 while biphenyl for C4B. The importance of induction compound is the activation of all enzymatic pathway coded in plasmid or chromosomal genes. About naphthalene microbial degradation different pathways were establish. Organization and physical map of naphthalene genes encoding the enzymes for the 11 steps of naphthelene oxidation have been mapped on plasmid NAH 7. Looking for this plasmid in our strains, positive results were obtained for 2NR while was negative for 10N1 and C4B: indolo test (7) point out the presence of naphthalene oxigenase encoding in genes *nah ABC*. Presuntive presence of catechol 2,3 doxygenase genes (*xyl E*) was supposed be present on the three strain due the yellow color from 2 hydroxy muconic acid 6 semialdehyde production (8). About plasmid presence, only microorganism group '2' (5) confirmed the presence of R factor as antibiotic resistence plasmid encoded, but extraction procedure analysis (9 and 10) didn't show them yet.

Studies in progression on biodegradation of higher molecular weight compouds give idea that composting may be an interesting technology becouse it is a versatile process at both physical and microbial levels. Degradation of hydrophilic and hydrophobic contaminants may occur due the physical attributes of the starting materials depending from the composition of the community and its adaptative response to the presence of hydrocarbons. During composting the mechanisms of adaptetion and genetic changes resulting in a net increase in the number of hydrocarbon utilizing microorganisms and in the pool of hydrocarbon catabolizing genes within the community.

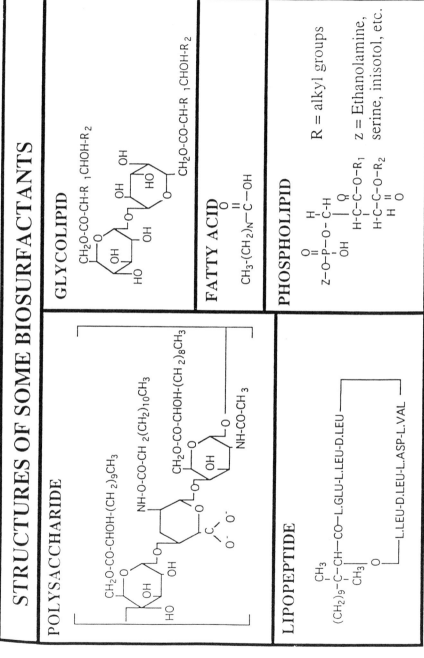

Scheme 2 Structure of some biosurfactants

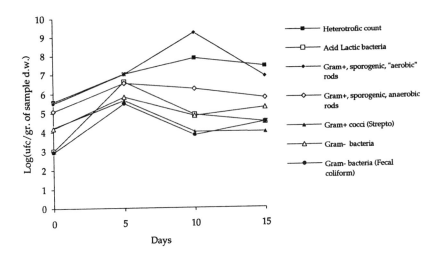

Figure 1 Microorganisms behaviour during composting

2NR

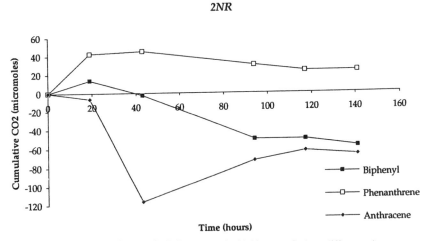

Figure 2 CO_2 evolution from strain 2NR measured with biometer flask on different substrates

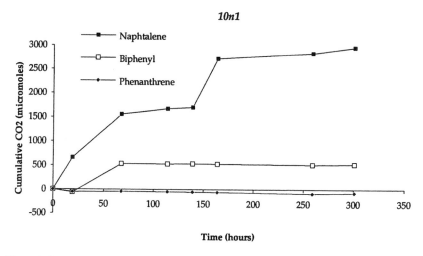

Figure 3 CO_2 evolution from strain 10N1 measured with biometer flask on different substrates

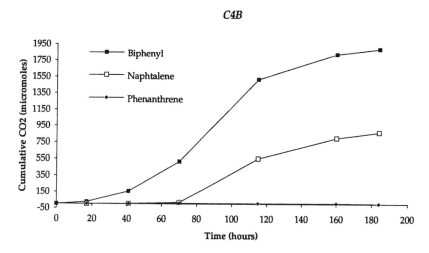

Figure 4 CO_2 evolution from strain C4B measured with biometer flask on different substrates

Table 1 Percentage reduction of each target creosote compound according to removal route (average of the three chambers)

	Recovery from Condensate plus Carbon trap	Recovery from samples	Reduction due to chemical, physical and biological biological activities	Total reduction
Naphtalene	7.990	0.142	90.40	98.53
2-Methyl Naphtalene	3.545	0.317	96.10	99.96
1-Methyl Naphtalene	5.330	0.609	94.01	99.95
Biphenyl	31.502	0.381	68.12	100
Acenaphtene	53.653	4.007	39.90	97.56
DibenzoFuran	9.012	2.502	87.86	99.38
Fluorene	6.539	2.909	89.18	98.63
Phenanthrene	0.601	2.734	94.64	97.98
Anthracene	1.525	5.184	90.98	97.69
Fluoranthene	0.114	6.797	85.76	92.67
Pyrene	0.069	5.454	90.06	95.58
Benzo(a)Anthracene	0	10.711	70.92	81.63
Chrysene	0.015	12.428	74.90	87.35

Table 2 Percentage reduction of each target creosote compound during the three composting phases (average of the three chambers)

Composting Period (day)	A–B	C–D 0–5	E–F 5–10	10–15
Naphtalene		93.33	5.23	−0.03
2-Methyl Naphtalene		94.19	4.71	1.06
1-Methyl Naphtalene		90.08	7.89	1.97
Biphenyl		65.48	33.34	1.17
Acenaphtene		17.40	69.93	10.23
DibenzoFuran		67.57	29.31	2.48
Fluorene		67.44	28.01	3.18
Phenanthrene		75.90	18.87	3.21
Anthracene		57.14	30.32	10.23
Fluoranthene		45.47	37.94	9.25
Pyrene		55.14	32.64	7.80
Benzo(a)Anthracene		34.18	22.15	25.30
Chrysene		27.39	18.80	41.16

Table 3 Percentage recovery of each target compound after 10 days of incubation (average of two tests) in Elmeyer fask at 30 °C and 150 rpm

Recovery	Recovery from control	Recovery from control	Recovery from control	Recovery from control
Naphtalene	126	n.d.	n.d.	0.5
2-Methyl Naphtalene	116	1.7	1	1.5
1-Methyl Naphtalene	119.9	n.d.	8	n.d.
Biphenyl	127	56.36	24.8	n.d.
Acenaphtene	116.9	n.d.	n.d.	n.d.
DibenzoFuran	106.7	n.d.	n.d.	n.d.
Fluorene	138.1	n.d.	n.d.	n.d.
Phenanthrene	81.3	n.d.	n.d.	n.d.
Anthracene	96.7	4.3	n.d.	n.d.
Fluoranthene	100.67	n.d.	n.d.	n.d.
Pyrene	175.8	n.d.	n.d.	n.d.

References

1) J.L. Sims, R.C. Sims and J.E. Matthews, Hazardous Waste & Hazardous Materials, 7,2, 117-149 (1990).
2) Civilini M. , Microbiology Europe, 2,6, 16–24 (1994).
3) Civilini M., ISWA Yearbook , in press.
4) Stanier R.Y., Palleroni N.J. and Doudoroff M., J. Gen. Microbiol. 43:159–271 (1966)
5) Civilini M. and Sebastianutto N., in proceeding of International Conference 'The science of composting' Ed. De Bertoldi M., Sequi P. and Lemmes B., Blackie Academic & Professional (1995)
6) Bartha and Pramer D., Soil Sci., 100,1, 68–70 (1965)
7) Ensley B.D., Ratzkin B.J., Osslund T.D., Simon M.J., Wackett L.P. and Gibson D.T., Science 222,167–169 (1983)
8) Shindo T., Ueda H., Suzuki E. and Nishimura H., Biosci. Biotech. Biochem. 59, 2, 314–315 (1995)
9) Potter A.A., in Recombinant DNA Methodology, Wiley, 147–157 (1985)
10) Wheatcroft and Williams, J.Gen. Microbiol., 124, 433–437 (1981)

Bioremediation of *PAH*-Contaminated Soil

LIS.TECH RAIMO LILJA – Engineering Office Rejlers
Ltd., Finland Ph.D. JUSSI UOTILA – JUVE-Group Ltd.,
Finland M.Sc. (Tech). HANNU SILVENNOINEN –
Ecolution Ltd., Finland

Abstract

Biodegradation of PAH-contaminated soil was studied by a slurry reactor test and
a soil column test. The initial total concentration 1000 ppm PAH soil was
decreased to 50...100 ppm in the slurry reactor in 6 weeks. In the soil columns the
concentration was decreased to 120...770 ppm in 4 weeks. Addition of coniferous
tree bark enhanced degradation significantly in the soil column tests, but not in
slurry reactor tests. Use of inoculants had only a slight positive effect in slurry
tests; the effect was even negative in the degradation of 5–6 ring compounds in
some soil column tests.

The experiences were applied to practical scale by using a composting method
where bark is used as a bulking agent. High moisture content is maintained in the
heap and the leaching water is circulated until the target concentration is reached.
The different functions of the bark addition is discussed.

Introduction

Contaminated soil remediation is a growing environmental sector also in Finland.
In a recent study by the Ministry of the Environment a total of 10,000 contaminated sites was estimated to exist in Finland. Of these 1200 are assessed to need
remediation. Biotechnical treatment is considered to be suitable for 20...25% of
the volume of the contaminated masses. (Ministry of the Environment, 1994).

A consortium of Finnish companies has established a method of biotechnical
treatment of soil contaminated by creosote-oil or other wastes containing high
concentrations of polycyclic aromatic hydrocarbons (PAHs). The method is based
on laboratory scale studies (Silvennoinen H. et.al. 1994) and it is now applied to
four practical cases in Finland.

Laboratory-scale biodegredation tests

The bioremediation technique was studied by two main methods in laboratory scale.

A *slurry reactor* test was used to study enhanced degradation of indicator PAH components in tests lasting 4–6 weeks. The method was inspired by reports by Linz D.G. et.al. (1991), Müller et. al. (1991) and others. Four 1000 ml glass reactors were equipped with aeration (1 liter/min) and mechanical agitators. The reactors were used in room temperature and protected from daylight with a aluminium foil. Neutral pH was maintained. NPK–, S–, Mg– and Ca-nutrients were added. A sample of 100 g of the contaminated soil was agitated in each reactor for 4–6 weeks. After the end of the test the residual concentrations in the slurry solids and the water phase were analyzed. Also the adsorption losses were assessed by rinsing the reactor with chloroform and analyzing the eluate.

The effect of carbon source addition was studied (10 % ground bark). Also the effect of a surface-active detergent was studied.

Another serie of tests was accomplished by using glass columns filled with the test material. The *soil column* was washed with a nutrient/buffer solution which was circulated through the soil in room temperature for 4–6 weeks.

The effect of using 20 vol–% grounded bark or acetate (3 g/ 400 g soil) as carbon source was studied. Also the effect of microbial inoculation of the soil was studied. A mixture of a pure culture growing on phenantrene and an enriched mixed-culture from creosote contaminated soil was used as inoculum.

Solid phase samples were Soxhlet-extracted by toluene and the water samples with chloroform. 21 indicator PAH-compounds were analysed using ISO-standard draft 13877 with 1,3,5–triphenylbenzene as an internal standard. Concentrations were detected by GC with a detection limit of 1 _g/kg. The absence of other PAH-compounds was verified by scanning the mass-spectrum 20–550 with a detection limit of 500 _g/kg.

Results

Slurry reactor tests

The changes in concentrations of the 21 monitored PAH compounds in one of the samples are shown in table 1.

The initial PAH concentration of the soil sample (sandy soil with 1,3% organic matter) was about 1000 mg/kg.

Of the sum concentration:

– 20% consisted of 2-ring compounds,
– 70% of 3–4 ring compounds,
– 5,4% 5–6 ring compounds and
– 4% of heterocyclic compounds.

Table 1 PAH concentrations in slurry reactor test

PAH compound	Initial conc.	Concentration after 4 weeks									
		Column no 1		Column no2		Column no3		Column no4		Column no5	
	ppb	ppb	%	ppb	%	ppb	%	ppb	%	ppb	%
2-ring PAHs											
acenaphthene	211100	137215	65	109772	52	2381	1	2368	1	33183	16
benz(a)anthracene-7,12-dione	2108	2318	110	1580	75	750	36	5380	255	1025	49
naphtalene	1994	5085	255	4247	213	906	45	4405	221	4151	208
Sum 2-ring PAHs	215202	144618	67	115599	54	4037	2	12153	6	38359	18
% of Total PAHs	21,8	19		20		3		3		8	
3-4 ring PAHs											
antanthene	<1	<1		<1		<1		<1		<1	
anthracene	103229	87745	85	67099	65	5208	5	2240	2	2936	3
1,2-benzanthracene	50019	39014	78	33012	66	1119	2	22479	45	24865	50
chrysene	61148	58885	96	37912	62	14967	24	23759	39	51439	84
4H-cyclopenta(def)phenenathrene	32087	39145	122	28236	88	22659	71	34929	109	43489	136
4,5-methylenephenanthrene	<1	<1		<1		<1		<1		<1	
fluoranthene	146763	80719	55	46964	32	15202	10	22256	15	39125	27
phenanthrene	191206	86043	45	66922	35	1418	0	21269	11	32127	17
pyrene	93197	105313	113	71762	77	26346	28	92505	99	84922	91
truxene	<1	<1		<1		<1		<1		<1	
Sum 3-4 ring PAHs	677649	496865	73	351908	52	96920	14	219438	32	278904	41
% of Total PAHs	68,5	65		61		80		61		60	

	initial	Column 1	%	Column 2	%	Column 3	%	Column 4	%	Column 5	%
5-6 ring PAHs											
coronene	1164	1583	136	2200	189	5167	444	4095	352	1603	138
benzo(ghi)perylene	7501	7276	97	7126	95	3181	42	6458	86	7098	95
benzo(a)pyrene	23240	26029	112	22311	96	5117	22	13076	56	22794	98
benzo(e)pyrene	15581	15425	99	13244	85	3363	22	16062	103	12685	81
perylene	6182	40551	656	48772	789	1531	25	77203	1249	93354	1510
Sum 5-6 ring PAHs	53668	90864	169	93653	175	18359	34	116894	218	137534	256
% of Total PAHs	5,4	12		16		15		33		30	
Heterocyclic PAHs											
5,6-benzoquinoline	6097	5829	96	4481	73	887	15	4197	69	5206	85
carbazole	16003	17923	112	8802	55	722	5	966	6	1468	9
dibenzothiophene	20709	13668	66	4556	22	115	0	4786	23	264	1
Sum heterocyclic PAHs	42809	37420	87	17839	42	1726	4	9949	23	6938	16
% of Total PAHs	4,3	5		3		1		3		2	
TOTAL PAHs	989331	769767	78	578999	59	121042	12	358434	36	461735	47

Legend:

	nutrient	inocul.	substrate.
Column 1	yes	no	no
Column 2	yes	yes	no
Column 3	yes	no	bark
Column 4	yes	yes	bark
Column 5	yes	no	bark+ acetate

In the slurry reactor tests a 90–95% reduction in the total PAH concentration from 1000 ppm to 50...100 ppm was achieved in 6 weeks in all the vessels.

Addition of ground tree bark (reactors 3 and 4) had a positive effect on the degradation rate of total PAH's, but the effect on the 5–6 ring PAH degradation was negative compared to the control (reactor 1).

The use of a detergent (reactor 2) increased degradation by about 30%. The effect was most notable on the 3–4 ring components. The best impact on the total PAH decrease was achived in the vessel with both carbon addition and detergent (reactor 4).

The degradation rate of the 5–6 ring compounds was only moderate with 40...70% of the initial concentration remaining. Addition of carbon and detergent favored the degradation of 2–4 ring compounds.

Adsorption losses were 88...1122 µg/vessel, which corresponds to 0,3...4,8% of the end content of the soil sample. Evaporation losses were not studied. They were presumed to be low on the basis of former research data (Müller 1993).

Soil column tests

The results of the soil column tests after recycling the nutrient solution for 4 weeks are shown in table 2.

With no addition of extra carbon source only a 20% reduction of the initial total PAH–concentration was achieved. Microbial inoculation had a small positive effect on the degradation of 2–4 ring PAH's.

Addition of bark had a strong positive effect with only 12% of the initial concentration left after 4 weeks. Addition of bark resulted also in degradation of the resistant 5–6 ring PAH's in contrary to the findings in slurry reactor tests. However, when adding both bark and inoculum the positive effect was modest.
Addition of an easily available carbon source (acetate) was not as effective as when using bark. Specially the microbial attack on the 5–6 ring compounds was hindered.

The residual PAH concentration in the recycled nutrient solution correlated with the concentration in the corresponding soil sample. The total amount of PAHs in the water phase was 9...28% of the remaining content in soil. This result differs notably from the finding in the slurry reactor tests, where the PAH concentration in the water phase was below detection limit.

Discussion

According to the results, PAH compounds can be degraded in a slurry reactor or by composting, which has been proved also in various other studies (Freeman, H.M., Sferra, P.R. 1991, Sayler, G.S. et.al. 1991, Pollard, S.J.R. et.al, Seman, P–O., Svedberg, R. 1990, Civilini, M. 1994 etc).

Interesting findings in our studies were that:

- coniferous tree bark enhanced the degradation in the soil leaching method, but not in the slurry tests.
- use of inoculants had only a slight positive effect in slurry tests; the effect was even negative in the degradation of 5–6 ring compounds in some soil column tests
- in the slurry reactor the remaining PAH concentration in the water phase is very small, but it can be quite significant in soil leaching procedures.

Some contradictory results were achieved in the soil column tests with negative reductions for some PAH–compounds. For 2–ring compounds the explanation is probably metabolic reactions. For the high increase in the concentration of several 5–6 ring compounds one explanation could be a change in their adsorption to the soil matrix and thus an increase in their extractability during composting.

The conclusion of the studies was that creosote can be degraded by composting with the following conditions:

- the soil must be effectively mixed with a suitable bulking agent
- a high moisture content must be maintained; this means that the leaching water must be collected and recycled or treated
- the degradability of the 5–6 ring PAHs might be enhanced by adding a detergent, but the detergent must be selected so that it does not compete as a substrate with the PAH compounds.

Bark addition seems to be very beneficial. It can have the following effects:

- it increases porosity in the mixture enhancing aerobic degradation
- it increases the moisture retention capasity of the soil
- it provides a large area for microbes to attach themselves on
- it provides a matrix for efficient PAH adsorption
- it provides the microbes with a co-substrate; not a too easily available carbon source (see eg. Keck et.al 1989).
- bark contains a high percentage of aromatic compounds some of which could act as co-substrates for PAH metabolism (eg. Fengel D., Wegener G. 1984).

Practical applications

Remediation project in Helsinki

The experiences from the laboratory studies were first applied to a full scale remediation of a contaminated site in Helsinki, Finland. The site was contaminated by a former enterprise handling tar products. A total of 1500 m^3 soil was contaminated with concentrations of 100...1200 ppm of PAHs in spot samples. Of the individual PAH-substances phenantrene, fluoranthene, anthracene, pyrene and carbazole were most abundant. The soil type was loam.

After laboratory scale pre-tests were performed successfully, the full scale project was accomplished in the following way:

1) The contaminated soil was excavated and stones were screened away. The masses were mixed during excavation so that the medium starting concentration was diluted to about 200 ppm PAH.

A composting place was prepared near the site on a plot with low permeable soil plus a HDPE liner. The area was equipped with a drainage layer and drainage pipes to collect the trickling water from the compost into a water tank. On top of the drainage layer a 20 cm layer of gravel was spread to improve aeration.

The soil mass was mixed with crushed bark and a long compost pile with a 6–8 m breadth and a 2 m highth was formed. pH was adjusted to neutral by adding lime. Nutrients were added as a solution of potassium phosphate and ammonium nitrate. The soil mass was wetted to optimal moisture. Preparations were made to cover the piles with plastic covers and to suck the evaporating hydrocarbons from the pile and through a compost filter. However the odor emissions were so low that the authorities gave up with these requirements.

The mass was aerated and mixed 3 times in 2 months with a heavy duty crushing device (ALLU SM) mounted on an excavator. Water and nutrients were added as needed. Samples were taken with an auger drill through the pile. Oxygen was measured with a probe from the pile.

After a 2 month composting period the PAH concentration dropped to 50 ppm. 200 ppm is usually regarded as the limit concentration for dumping contaminated soil at a sanitary landfill in Finland. The project is now continuing until autumn 1995. The objective is to bring the PAH concentration down to 20 ppm, which could allow the masses to be used in landscaping.

Remediation in Petäjävesi

A wood preservation facility using creosote in Central Finland had caused contamination of the soil in the 1970's. The top 5 cm layer of soil was removed and transported for incineration at the hazardous waste treatment plant. The deeper layer of sandy soil was moderately contaminated up to 50 cm.

The soil mass of 2800 m^3 was excavated and transported to the municipal landfill. There the composting/heap leaching process is to be started in summer 1995. The PAH concentration ranges from 800 to 1000 ppm. The concentration limit for landfill dumping is in this case 50 ppm. The target is expected to be reached in one or two years.

A major environmental problem is however still unsolved. Creosote oil has leaked through drainage pipes into the nearby lake. A 10–20 cm thick layer of creosote sludge was detected in the lake sediment. The contaminated sediment covers an area of about 10.000 m^2 of the lake bottom. PAH concentrations are high: 2.000 – 20.000 ppm.

Laboratory tests in a slurry reactor revealed that the PAH concentration could be decreased from 12.000 ppm to 500...1100 ppm (4...9% of original) in 4 weeks.

Luckily the concentrations of 5–6 ring PAHs were low (0,5% of total PAHs) compared to the corresponding concentration in soil (5,4%).

Pumping of the contaminated sediment would cause a major risk of contaminating the lake. The PAH compounds are very toxic to water organisms. It would also produce huge volumes of thin slurry, which should be concentrated and dewatered before treatment. The waste water should be treated before discharge back to the lake. An alternative response could be to cover the contaminated sediment with a filter cloth and a layer of sand. The problem has been disputed between the owner and the authorities for several years now.

Remediation plans at Ilmajoki

An abandoned wood preservation plant in Ilmajoki, Western Finland is ranked as potentially one of the most expensive remediation projects in Finland. The site is highly contaminated with both creosote oil and CCA-salts (chromium, copper, arsene). The soil type is loam and clay.

7000 m^3 of highly contaminated masses with PAH concentrations from 2000...20.000 ppm have been assessed. Additionally we calculated 50.000 m^3 of masses with 200...2000 ppm PAH. A part of the mass contains both PAH- and CCA- contamination. Leaching from the site causes emissions into a nearby river. The river's water is used to produce drinking water.

Our laboratory tests have shown that a concentration of 600 ppm As does not prevent biodegradation of PAH.

Thus our suggestion for remediation is:

1) sorting of the contaminated masses into different hazard classes based on concentration and soil type.
2) degradation by heap leaching of the moderately contaminated masses until the limit for landfill treatment is achieved
3) degradation by composting of the heavily contaminated masses until a residual concentration 2000 ppm or lower is reached. Special methods for attacking the 5–6 ring PAHs may be necessary.
4) stabilization of the toxic metals and As by precipitation with salts and lime to decrease solubility
5) placing the masses on site isolated with an impermeable bottom and top cover and cut-off walls.

The tendering process for contractors for this remediation project is going on in March 1995. As an alternative to composting solidification into concrete has been suggested as a solution.

Table 2 PAH conentrations in slurry reactor test

PAH compound	Initial conc.	Reactor no1		Reactor no2		Reactor no3		Reactor no4	
				Concentration after 6 weeks					
	ppb	ppb	% of init	ppb	% of init	ppb	% of init	ppb	% of init
2-ring PAHs									
acenaphthene	211100	567	0.27	949	0.45	692	0.33	552	0.26
benz(a)anthracene-7,12-dione	2108	674	31.97	1466	69.54	1004	47.63	1191	56.50
naphtalene	1994	450	22.57	107	5.37	62	3.11	72	3.61
Sum 2-ring PAHs	215202	1700	0.79	2522	1.17	1758	0.82	1815	0.84
% of Total PAHs	21.8	1.53		2.66		2.42		3.11	
3-4 ring PAHs									
antanthene	<1	<1		<1		<1		<1	
anthracene	103229	845	0.82	895	0.87	1034	1.00	696	0.67
1,2-benzanthracene	50019	17047	34.08	15872	31.73	2947	5.89	2519	5.04
chrysene	61148	20862	34.12	22552	36.88	20836	34.07	9317	15.24
4H-cyclopenta(def)phenenathrene	32087	2028	6.32	803	2.50	437	1.36	387	1.21
4,5-methylenephenanthrene	<1	<1		<1		<1		<1	
fluoranthene	146763	34283	23.36	7509	5.12	4321	2.94	2866	1.95
phenanthrene	191206	1090	0.57	914	0.48	904	0.47	755	0.39
pyrene	93197	10863	11.66	4264	4.58	2544	2.73	2610	2.80
truxene	<1	<1		<1		<1		<1	
Sum 3-4 ring PAHs	677649	87019	12.84	52810	7.79	33024	4.87	19151	2.83
% of Total PAHs	68,5	78.34		55.69		45.41		32.83	

	Initial	R1	R1 %	R2	R2 %	R3	R3 %	R4	R4 %
5-6 ring PAHs									
coronene	1164	<1	0.00	<1	0.00	<1	0.00	<1	0.00
benzo(ghi)perylene	7501	1571	20.94	2935	39.13	2845	37.93	2868	38.23
benzo(a)pyrene	23240	10258	44.14	20732	89.21	19872	85.51	19444	83.67
benzo(e)pyrene	15581	6435	41.30	10205	65.50	10020	64.31	9898	63.53
perylene	6182	3489	56.44	4975	80.48	4891	79.12	4724	76.42
Sum 5-6 ring PAHs	53668	21754	40.53	38848	72.39	37629	70.11	36935	68.82
% of Total PAHs	5,4	19.58		40.97		51.74		63.31	
Heterocyclic PAHs									
5,6-benzoquinoline	6097	270	4.43	335	5.49	45	0.74	94	1.54
carbazole	16003	223	1.39	201	1.26	188	1.17	269	1.68
dibenzothiophene	20709	110	0.53	115	0.56	78	0.38	77	0.37
Sum heterocyclic PAHs	42809	603	1.41	651	1.52	311	0.73	440	1.03
% of Total PAHs	4,3	0.54		0.69		0.43		0.75	
TOTAL PAHs	989331	111076	11.23	94831	9.59	72722	7.35	58341	5.90

Legend:	nutrient	bark	deterg.
Reactor 1	yes	no	no
Reactor 2	yes	yes	no
Reactor 3	yes	no	yes
Reactor 4	yes	yes	yes

Acknowledgements

This work was sponsored by the Finnish Technology Centre TEKES, the Finnish Wood Preservation Association and Imatran Voima Ltd.

References

Civilini, M. (1994), Fate of creosote compounds during compostin, Microbiology Europe, vol 2, no 6, 16-24.

Fengel, D., Wegener, G. (1984) Wood Chemistry, Ultrastructure reactions, Walter de Gruyter, Berlin and New York.

Freeman, H.M., Sferra, P.R. (eds),(1991), Innovative Hazardous Waste treatment technology series vol 3, Technomic Publishing Co., Pennsylvania USA.

Keck, J., Sims, R.C., Coover, M., Park, K., Symons, B. (1989), Evidence for cooxidation of polynuclear aromatic hydrocarbons in soil, Wat. Res., vol 23, no 12, 1467–1476.

Linz, D.G., Neuhauser, F., Middleton, A.C. (1991) , Perspectives on bioremediation in the gas industry, in Sayler, G.S. et.al. (1991), 25–36.

Ministry of the Environment of Finland (1994), Contaminated soil site survey and remediation project, Helsinki.

Müeller, J.G., Lantz, S.E., Blattmann, B.O., Chapman, P.J. (1991), Bench-scale evaluation of alternative biological treatment processes for the remediation of pentachlorophenol- an creosote-contaminated materials: slurry-phase bioremediation, Environ. Sci. Technol. vol 25, no 6, 1055–1061.

Müeller, J.G., Lantz, S.E., Ross, D., Colvin, R.J., Middaugh, D.P., Pritchard, P.H. (1993), Strategy using bioreactors and specially selected microorganisms for bioremediation of groundwater contaminated with creosote and pentachlorophenol, Environ. Sci. Technol. vol 27, no 4, 691...698.

Pollard, S.J.R., Hrudey, S.E., Fedorak, P.M. (1994), Bioremediation of petroleum- and creosote-contaminated soils: a review of constraints, Waste Management & Research vol 12, 173...194.

Sayler, G.S., Fox, R., Blackburn, J.W. (eds.) (1991), Environmental Biotechnology for Waste Treatment, Environmental Sci. Research vol 41, Plenum Press, New York.

Seman, P-O., Svedberg, R. (1990), Sanering av kreosotkontaminerad mark, Swedish Wood Preservation Institute, Reports no 162, Stockholm (in Swedish).

Silvennoinen, H., Uotila, J., Lilja, R., Laakso, P. (1994), Petäjäveden kreosoottipitoisen maa-aineksen ja sedimentin biohajoavuus, Final report 30.6.1994, Imatran Voima Oy, ympäristönsuojeluyksikkö, Helsinki. (in Finnish). (*Biodegradability of the PAH-contaminated soil and sediment from Petäjävesi*).

Minimum Effective Compost Addition for Remediation of Pesticide-Contaminated Soil

XIANZHONG LIU and MICHAEL A. COLE –
University of Illinois Department of Agronomy 1102
South Goodwin Urbana, IL 61801 USA

Abstract

Mature yard waste compost was mixed with pesticide-contaminated soil obtained from an agrichemical retail dealership to determine how much compost was needed to significantly stimulate plant growth, microbial activity, and pesticide degradation. The soil initially contained 1.6 mg kg^{-1} trifluralin, 1.7 mg kg^{-1} metolachlor, and 2.0 mg kg^{-1} pendimethalin. Proportions of the mixes were 0, 1, 5, 10, 20 and 40% compost (w/w). Mixtures were planted with sweet corn (*Zea mays*) and placed in a greenhouse for 4 weeks. Plant dry matter production, microbial activity (dehydrogenase), and pesticide content were determined. Maximal stimulation of plant growth was found at 20% compost, at which amount, plant growth was 154% of growth in only contaminated soil. Significant increases in soil dehydrogenase were seen at 20% and 40% compost with the activity being 18.8 times higher than it was in only contaminated soil. The soil was inhibitory to microbial activity at all rates of compost addition, with no stimulation in microbial activity at <20% compost. Degradation of trifluralin and pendamethalin was increased at compost rates of 20% or more. Only 40% compost significantly stimulated metolachlor degradation. Percentage degradation of the pesticides after 4 weeks of greenhouse incubation and 16 weeks of laboratory degradation was 85% (trifluralin), 100% (metolachlor), and 79% (pendimethalin). The results indicate that addition of relatively large amounts of compost to soils with phytotoxic and antimicrobial properties can significantly improve prospects for successful remediation, but smaller amounts have no significant benefit.

Introduction

A survey of 49 agrichemical facilities in Illinois (Krapac, *et al.*, 1993) showed that soil contamination with herbicides was very common, while contamination with insecticides was much less common. Herbicide concentrations ranged from a few ug kg^{-1} soil to several g kg^{-1} soil. Since most herbicides are effective in the mg kg^{-1} range, there were secondary problems at the sites, including off-site erosional transport of contaminated soil by wind and water, which resulted in plant death on adjacent property or pollution of surface water. Similar results were found in a survey of Wisconsin agrichemical dealerships (Habacker, 1989). Taylor (1993) found detectable pesticides in groundwater samples from wells at agrichemical retail sites, including several compounds for which drinking water standards exist. He suggested that agrichemical facilities are primary sources of groundwater contamination in Illinois. The combination of adverse environmental impacts on the earth's surface and subsurface indicates that remedial activities at these sites would be appropriate. In most cases, the pesticides detected at the agrichemical facilities were not there as the result of recent spills, but rather were the result of years of accrued contamination.

Felsot and coworkers (1988, 1990) attempted bioremediation of pesticide-contaminated soil from an inactive agrichemical facility and found that pesticide degradation occurred quite slowly, with detectable concentrations of alachlor, atrazine, metolachlor, and trifluralin still present at 380 days after land application of excavated soil. Since these authors demonstrated that freshly-added herbicides were degraded rapidly in comparison to aged materials when applied to agricultural fields, it is likely that the bioavailability of the herbicides from the agrichemical facilities was low, thereby decreasing the degradation rate. Felsot and Dzantor (1990) found that dehydrogenase activity increased only slightly in herbicide-contaminated soils amended with corn or soybean stubble, so lack of sufficient stimulation of microbial activity may also have been a contributory factor to the slow degradation they observed.

Biodegradation is the principal mechanism for destruction of trifluralin and metolachlor (Weed Science Society of America, 1989). Based on work by Nelson (1979), pendimethalin is apparently degraded by both chemical and biotic reactions, since soil sterilization reduced degradation to about 50% of the values obtained with non-sterile soil.

Because microbial degradation was the major route for environmental destruction of trifluralin and metolachlor and also contributed to pendimethalin degradation, one objective of this work was to stimulate microbial activity in the contaminated soils with the expectation that such stimulation might accelerate degradation of these pesticides. Two approaches and a combination of both approaches were tried (Cole, *et al.*, 1994; Cole, *et al.*, in press). First, an attempt was made to grow plants in the contaminated mixes, since plant growth increases microbial populations in soil, especially root-associated (rhizosphere) bacterial populations. Second, compost was added as a source of microorganisms and organic matter because addition of compost or other organic materials can stimulate soil microbial activity. Compost often has a stimulatory effect on plant growth, and therefore, a combination of planting and compost addition was tried.

Materials and Methods

Soil and Compost Samples. Contaminated matrix was obtained from a loading area at an agrichemical retail facility located in Illinois that was designated Site 20 in survey sponsored by Illinois Department of Agriculture (Krapac, *et al.*, 1993). The sampled area was used as a work area for loading and mixing of pesticides. Cores of 8.1 cm diameter were collected to a depth of 457 cm and a composite sample of all contaminated cores was used for the present work. The matrix consisted of 27% sand, 32% silt, 19% clay, and 22% gravel, had a pH of 8.4 and the electrical conductivity was 2.08 ds m^{-2}. The organic content was very low and inorganic–N (NO_3^- + NH_4^+) was < 3 mg kg^{-1}. Prior to conducting the experiments described below, the material was screened and material passing a 4 mm mesh was used.

Mature yard waste compost was obtained from DK Recycling Systems, Inc., Lake Bluff, IL. The compost was produced by a thermophilic process with average high temperatures of 60°C. Material that passed a 6 mm mesh was used.

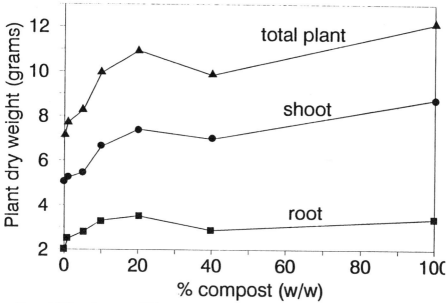

Figure 1. Effect of compost addition rate on plant dry matter production. Values are the means of three replicates of each treatment.

Physical and Chemical Analysis of Samples. Bulk density, % sand, silt, clay, and gravel, inorganic-N, pH, and electrical conductivity were determined by standard methods (Klute, 1986; Page, et al., 1982).

Plant Growth and Analysis Procedures. Blends of compost and contaminated matrix which contained 0, 1,5, 10, 20, and 40% compost (w/w) were prepared and transferred into 15 cm diameter pots for greenhouse studies. Three pots of each mixture were planted with 6 seeds of sweet corn (*Zea mays*, cv. 'Golden Beauty') and placed in a greenhouse. Plants were watered weekly with NPK fertilizer.

Plants were harvested at 30 d after planting and separated from soil. Roots and shoots were separated, and the roots were washed in tap water to remove adherent soil. Dry weights of roots and shoots were determined by drying at 90° C to a constant mass.

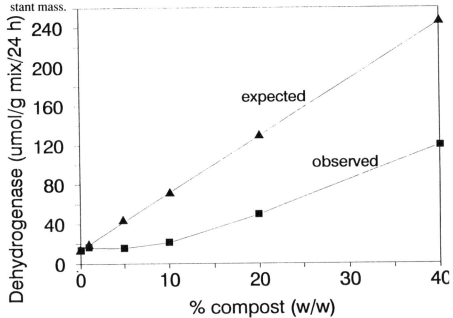

Figure 2. Effect of compost addition rate on dehydrogenase activity. Values are the means of duplicate analyses of three replicates of each treatment.

Soil Dehydrogenase Activity. Dehydrogenase activity was determined as previously described (Cole, *et al.*, 1994).

Pesticide Extraction and Analysis. Procedures were described in detail elsewhere (Cole, *et al.*, submitted for publication). Briefly, a 25 g sample was ground in a Waring blender and transferred into a rectangular 160 mL glass bottle with a PFTE cap liner. Ten milliliters of $1\underline{M}$ sodium chloride solution was added along with sufficient water to make a soil slurry, followed by 50 mL ethyl acetate and 2 mL acetone. The bottle was shaken horizontally on a rotary shaker at 150 rpm for 24 h at 20° C. The ethyl acetate (upper) layer was removed, dehydrated by passage through a column of anhydrous sodium sulfate, and reduced to 1.0 mL final volume in a Kuderna-Danish concentrator. Samples were injected without further purification into a Chrompack CP9000 gas chromatograph equipped with a nitrogen-phosphorus detector. A 50 m X 0.25 mm (i.d.) column of WCOT fused silica with CP-Sil-8 CB stationary phase (Chrompack, Inc.) was used for all analyses.

Recovery of freshly-added pesticides ranged from 70 to 100% of addition with no significant effect of compost on % recovery when compared to recovery from soil alone. Identity of the pesticides was established by GC/MS analysis of ethyl acetate extracts. Compounds with the retention times of the pesticides listed in Figures 4, 5, and 6 were not detected in pure compost, nor in uncontaminated soil. The principal contaminants found in the samples used for this investigation were trifluralin (2,6–dinitro–N,N–dipropyl–4–(trifluoromethyl)benzenamine; marketed under the US tradename, TREFLAN), metolachlor (2–chloro–N– (2–ethyl–6–methylphenyl)–N–(2–methoxy–1–methylethyl)acetamide; marketed under the tradenames DUAL or PENNANT), and pendimethalin (N–(1– ethylpropyl)–3,4,dimethyl–2,6–dinitrobenzenamine; marketed under the tradename, PROWL).

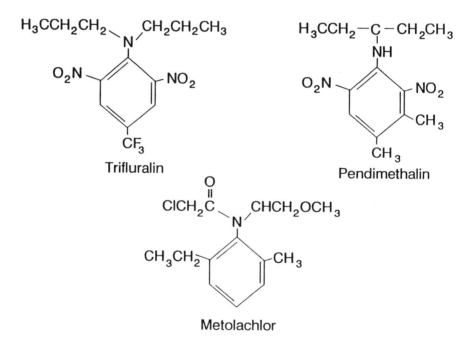

Figure 3. Structures of pesticides whose degradation was studied.

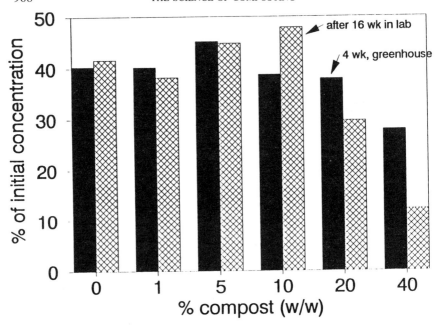

Figure 4 Trifluralin degradation as affected by % compost in mixtures. Values are the means of duplicate analyses of three replicates of each treatment.

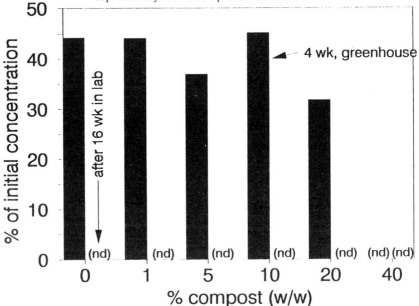

Figure 5 Metolachlor degradation as affected by % compost in mixtures. Values are the means of duplicate analyses of three replicates of each treatment.

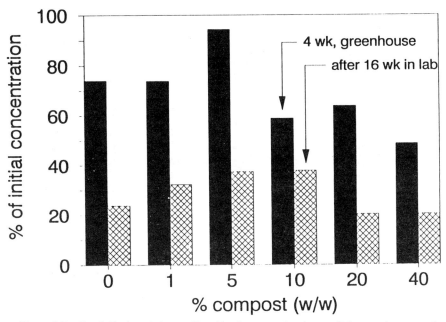

Figure 6 Pendimethalin degradation as affected by % compost in mixtures. Values are the means of duplicate analyses of three replicates of each treatment.

Results and Discussion

Plant Growth. Plants were harvested when the root systems had completely filled the pots. Maximal benefit to dry weight of roots, shoots, and total weight was obtained at 20% compost (Figure 1). The results indicate that compost has a significant protective effect against phytotoxic compounds in the contaminated matrix. The phytotoxic contaminants included herbicides (as described below) as well as other organic compounds such as phthalate esters (data not shown), which are phytotoxic (Herring and Bering, 1988). There is considerable interest in the ability of plants to stimulate xenobiotic degradation and the data in Figure 1 indicate that compost can be a significant aid in establishing plants in contaminated soils.

Microbial Activity. We used dehydrogenase activity as a broad–spectrum indicator of microbial activity in the mixes. This enzyme has been used by several investigators as an indication of overall heterotrophic activity in soil (Schaffer, 1993). There was no significant increase in dehydrogenase activity until the compost content reached 20% (Figure 2). Dehydrogenase activity was significantly lower than expected from the additive values of activity in the compost + contaminated matrix, a result that indicates that the matrix contained antimicrobial compounds as well as phytotoxins. We had reported previously that microbial activity was reduced in samples obtained from a different location at the same facility used for this work (Cole, *et al.*, in press), but there was no inhibition by samples obtained from a different facility (Cole, *et al.*, submitted for publication). Inhibition of microbial activity would be expected to increase the time required for successful remediation and the results indicate that addition of at least 20% compost is needed to overcome the inhibitory effects of the contaminated matrix.

Herbicide Degradation. The major herbicides found in contaminated matrix were trifluralin, metolachlor, and pendimethalin, whose structures are given in Figure 3. Initial concentrations in the contaminated matrix were 1.6, 1.7, and 2.0 mg kg–1 for trifluralin, metolachlor, and pendimethalin, respectively.

Trifluralin degradation after greenhouse incubation was significantly increased only at the 40% compost addition (Figure 4), but was significantly increased at 20% and 40% compost addition after a 16 week laboratory incubation. Compost had no effect on trifluralin degradation at 10% compost or less. Comparison of residual trifluralin in the 4 week greenhouse samples with residual trifluralin after 16 weeks of laboratory incubation indicates that 40% compost greatly increased the degradation rate of trifluralin during the 16 week period when compared to lower percentages of compost.

Metolachlor degradation was increased at the 20% and 40% compost rates after greenhouse incubation, with no detectable residues (< 0.1 mg kg^{-1}) in the 40% compost samples (Figure 5). Since metolachlor had degraded completely in all samples after laboratory incubation, the effects of compost could not be determined.

Mixtures containing 40% compost had less residual pendimethalin than samples with lower percentages of compost after greenhouse incubation (Figure 6). High addition rates of compost did not increase pendimethalin degradation during laboratory incubation, and the results suggest that lower rates of compost retarded degradation somewhat.

Conclusions

Plant growth has been shown to facilitate degradation of several environmental contaminants in addition to pesticides (Anderson and Coats, 1994). As described by Shann and Boyle (1994), pesticide degradation in the rhizosphere can be rapid, which may decrease the time required for remediation. An established plant cover will decrease erosional transport of contaminated soil to adjacent surface water or property. Since many pesticides are toxic to fish, decreased transport of contaminated soil to surface water would substantially diminish adverse impacts of contaminated sites. Plant growth will improve soil structure and provide organic materials which may stimulate microbial cometabolism of pesticides. The ability of compost to decrease phytotoxicity toward sensitive weed species was demonstrated previously with mixtures contained 50% w/w compost + contaminated soil

(Cole *et al.*, submitted for publication). The data presented in Figure 1 indicate that only 20% compost is needed to maximize plant growth. Regardless of any beneficial effect on soil microbial activity, the ability of compost to increase plant growth in contaminated soils is a good reason for its use. One would expect the optimal amount of compost to vary depending on site and matrix conditions, and therefore, preliminary greenhouse studies should be conducted with a specific sample to determine the best mixture of compost + contaminated soil.

The failure of compost to stimulate microbial activity in mixes containing 10% compost or less is different than usually reported. Numerous studies have demonstrated that addition of 1% to 10% (w/w) organic materials such as compost, manures, sludges, or other organic wastes will increase microbial activity in ordinary field soils and will stimulate degradation of some pesticides (Pettygrove and Naylor, 1985; Winterlin, et al., 1989). The low activity found in our samples containing 10% or less compost is probably due to the presence of antimicrobial compounds in the contaminated matrix. This inhibition may explain why Felsot and coworkers (1988) found little stimulation of degradation of herbicide wastes when applied to agricultural soils along with supplemental crop residues.

Compost, when added at less than 20% w/w, was not effective in stimulating degradation of any of the herbicides studied in this work, either when plants were growing in the soil or not. In previous studies, we had found much faster degradation of the same herbicides that were present in this study when the mixtures contained 50% compost. The difference in rate may have been the result of differences among batches of compost or differences in the contaminated matrix among different samples. We are currently investigating the impact of compost production methods and source materials on the effectiveness of compost for remediation of xenobiotic-contaminated soil.

Acknowledgments

This work was funded by a grant from the Hazardous Waste Research and Information Center, Champaign, IL and gifts from Solum Remediation Services, Lake Bluff, IL.

References

Anderson, T.A. and J.R. Coats (eds.) *Bioremediation Through Rhizosphere Technology*. Washington, DC: American Chemical Society.

Cole, M.A., X. Liu, L. Zhang. 1994. Plant and microbial establishment in pesticide-contaminated soil amended with compost, pp 210–222. IN Anderson, T.A. and J.R. Coats, eds. *Bioremediation Through Rhizosphere Technology*. Washington, DC: American Chemical Society.

Cole, M.A., X. Liu, and L. Zhang. Accelerated pesticide degradation in compost-amended, planted soils. *In In Situ and On-Site Bioreclamation: The Third International Symposium Proceedings*. In press.

Cole, M.A., L. Zhang, and X. Liu. Remediation of pesticide-contaminated soil by planting and compost addition. Submitted for publication in *Compost Science and Utilization*.

Felsot, A.S. and E.K. Dzantor. 1990. Enhancing biodegradation for detoxification of herbicide waste in soil, pp 249–268. IN K. D. Racke and J.R. Coats, eds. *Enhanced Biodegradation of Pesticides in the Environment*. Washington, DC: American Chemical Society.

Felsot, A. , R. Liebl, and T. Bicki. 1988. *Feasibility of land application of soils contaminated with pesticide waste as a remediation practice*. 55 pp. HWRIC RR–021. Urbana, IL: Hazardous Waste Research and Information Center.

Habecker, M.A. 1989. *Environmental Contamination at Wisconsin Pesticide Mixing/Loading Facilities: Case Study, Investigation and Remedial Action Evaluation*. 80 pp. Madison, WI: Wisconsin Department of Agriculture, Trade, and Consumer Protection.

Herring, R. and C.L Bering. 1988. 'Effects of phthalate esters on plant seedlings and reversal by a soil microorganism.' *Bulletin of Environmental Contamination and Toxicology*, 40, 626–632.

Klute, A., ed. 1986. *Methods of Soil Analysis, Part 1*. Madison, WI: American Society of Agronomy.

Krapac, I.G., W.R. Roy, C.A. Smyth and M.L. Barnhardt. 1993. 'Occurrence and distribution of pesticides in soil at agrichemical facilities in Illinois.' IN *Agrichemical Facility Site Contamination Study*. Springfield, IL: Illinois Department of Agriculture.

Nelson, J.E. 1979. 'Residues of pendimethalin (N-(1-ethylpropyl)–3,4-dimethyl–2,6-dinitrobenzenamine), trifluralin (alpha, alpha, alpha-trifluoro–2,6-dinitro-N,N-dipropyl-*p*-toluidine), and oryzalin (3,5-dinitro-N^4,N^4,dipropylsulfanilamide) in soil organic matter' 122 pp. Unpublished Ph.D. dissertation, East Lancing, MI: Michigan State University.

Page, A.L., R.H. Miller, and D.R. Keeney. 1982. *Methods of Soil Analysis, Part 2*. Madison, WI: American Society of Agronomy.

Pettygrove, D.R. and D.V. Naylor. 1985. Metribuzin degradation kinetics in organically amended soil. Weed Science, 33, 267–270.

Schaffer, A. 1993. 'Pesticide effects on enzyme activities in the soil ecosystem' pp. 273–340. IN Bollag, J.-M. and G. Stotzky, eds. *Soil Biochemistry, Volume 8*. New York: Marcel Dekker, Inc.

Shann, J.R. and J.J. Boyle. 1994. 'Influence of plant species on *in situ* rhizosphere degradation pp 70–81. IN Anderson, T.A. and J.R. Coats, eds, *Bioremediation Through Rhizosphere Technology*, Washington, DC: American Chemical Society.

Taylor, A.G. 'The effect of agrichemical use on water quality in Illinois.' *Abstracts American Chemical Society Annual Meetings*, Chicago, IL., August, 1993.

Weed Science Society of America. 1989. *Herbicide Handbook*. Champaign, IL: Weed Science Society of America.

Winterlin, W., J.N. Seiber, A. Craigmill, T. Baier, J. Woodrow, and G. Walker. 1989. Degradation of pesticide waste taken from a highly contaminated soil evaporation pit in California. *Archives of Environmental Contamination and Toxicology*, 18, 734–747.

Enhancement of the Biological Degradation of Contaminated Soils by Compost Addition

HUPE, K., LÜTH, J. C., HEERENKLAGE, J., STEGMANN, R. – Technical University of Hamburg-Harburg, Department of Waste Management Harburger Schlossstrasse 37, 21079 Hamburg, Germany

Abstract

Within the programme of the Research Centre SFB 188 – entitled 'Treatment of Contaminated Soils' and funded by the DFG (German Research Foundation) since 1989 – one project has been investigating the fundamental principles of the biological treatment of contaminated soils in bioreactors. The aim is to optimize the processes of biological soil treatment so that the highest possible degree of degradation is reached within the shortest possible period of time. Pre-investigations using test systems on different scales will provide information on the potential for enhancement in decomposition processes. This is dependent upon various influencing factors such as milieu conditions, additives, etc., which must be known before remedial action can be taken.

The investigations carried out so far have shown that it is beneficial to add compost during the biological treatment of oil-contaminated soils. The degradation of contaminants was enhanced by the addition of compost. This positive effect is attributed to various mechanisms. This paper presents the results from a variety of test systems on different scales.

Introduction

To apply biological soil treatment under optimum conditions and to discover the treatment limits, the specific factors of influence must be determined in advance by using a series of bench scale experiments. Since each case is different, this step is mandatory. Using these tests the following influencing factors can be optimized: oxygen supply, temperature, water content, addition of structual material (e.g. compost), microbes, nutrients, and other additives.

Within the framework of these investigations it was shown that the biological degradation of contaminated soils can be enhanced by the addition of compost. The degree of biological degradation in relation to the maturity of the compost added, and the influencing factors of the compost were investigated.

Methods Applied

For systematic investigations a model soil from an A_h–horizon, which had been contaminated artificially with diesel fuel in a defined way, was used (description of the material: Goetz et al., 1990; Miehlich & Wagner, 1990). To obtain a material as homogeneous as possible for the test series the soil was sieved to 2000 _m. The soil that was treated had a hgh content of solid ('dry process'). The optimum water content lay at 60–70% of the maximum water capacity (WC_{max}). The diesel fuel, which was analyzed in detail had the following compostion: n–alkane, 31.9%; rest aliphatics, 39.8%; monoaromatics, 16.4%; diaromatics, 8.0%; and polyaromatics, 3.9% (Steinhart et al., 1990).

The compost added was biocompost derived from kitchen and garden waste separately collected in households; it showed very low contamination. The compost was composted at a small windrow plant in Hamburg-Harburg.

Test Units

In order to be able to draw up mass balances only closed systems were used to evaluate and optimize biological soil treatment processes. Special glass vessels and respirometers were used in the experiments to optimize the milieu conditions (temperature, water content, additives). These systems have been described in numerous publications (e.g. Stegmann et al., 1991; Hupe et al., 1993; Hupe et al., 1994).

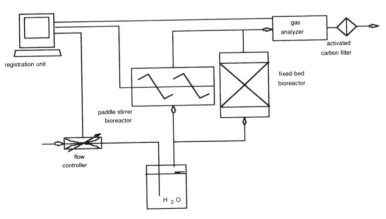

Figure 1 Principle of the bioreactor test system (schematic of a static bioreactor and a paddle-stirrer reactor)

Reactor systems were applied at different scales (volume: 3 litre, 6 litre or 90 litre) to simulate the conditions in an aerated windrow or in large-scale reactors. Aeration was conducted by compressed air and the volatiles can be easily measured in these systems. Figure 1 shows the schematic of a static and a dynamic bioreactor[1] (here: a paddle-stirrer reactor) including continuous recording and control of the measured values.

The bioreactors developed by the authors were continuously improved on the basis of the data and experience gained during the test series. In the lower part of the static bioreactor there is a sieve on which soil material is placed. Sampling pipes are located at different heights in the reactor. Controlled aeration is conducted from the bottom to the top, where the feed air is led through a wash bottle filled with water in order to avoid the soil drying out. The CO_2-content of the exhaust gas is quasicontinuously measured by infrared (IR) spectrometry and the volatile organic carbon (VOC) content by flame ionization detector (FID). Additional gas samples are taken from the reactors with a syringe through a septum and directly injected into a gas chromatograph (GC).

The paddle-stirrer reactor consists of a horizontal glass cylinder. The driving shaft is driven by an infinitely variable motor. For optimal mixing various mixing tools can be affixed to the shaft. In the middle of the glass cylinder there is a sampling pipe. The air is supplied into the reactor over the face areas. For the continuous monitoring of CO_2 and VOC in the exhaust air stream of the reactor an infrared device and a FID are used.

Investigations and Results

Carbon Balance of the Oil Degradation

Based on the results of extensive test series carried out in static bioreactors, an approach was developed to determine the carbon balance of oil degradation. During the investigations the parameters listed in Table 1 were measured.

Table 1 Investigated parameters and analytical methods for the balancing approach (Lotter et al., 1992)

parameter	analyzing method
hydrocarbon content in the soil	H18 after ultrasonic extraction
biomass	SIR-method in the respirometer
CO_2 in the exhaust air stream	gas chromatography with TCD
VOC in the exhaust air stream	gas chromatography with FID

H18: DIN 38 409 H18 (Anonymous; 1981); SIR-method: substrate-induced-respiration method (Anderson & Domsch, 1978; modified after Beck, 1984); TCD: thermal conductivity detector; VOC: volatile organic carbon; FID: flame ionization detector

Using the data of the contaminated samples and the uncontaminated control for each parameter (Table 1), an oil carbon balance can be calculated. This approach only partly describes the changing conditions in the soil after the addition of oil; however, it provides an approximation and a good overview of the decomposition of the contaminants. The C-content of the measured hydrocarbon concentrations is calculated on the basis of the carbon content of the initial oil. The analysis of the oil showed a carbon content of 86.1% for diesel fuel and 83.8% for lubricating oil (Francke, 1990).

Figure 2 shows the C-balance of the carbon content of the oil during treatment in the static bioreactor. These data are based on the results of the test series. It was observed that there was always a gap in the balance sum. The analytically measured reduction of hydrocarbon cannot be quantitatively explained by decomposition into CO_2, by volatilization or the production of biomass. In relation to this the weak and/or strong interactions of the contaminants or the metabolites with the humus matrix of the compost are of particular importance (Lotter et al., 1990; 1992). The potential interactions were investigated and described by various research groups (Gerth et al., 1990; Kästner et al., 1994). For biological remediation it is important to know whether the adsorption to the soil particles and/or to the humus matrix is reversible or irreversible.

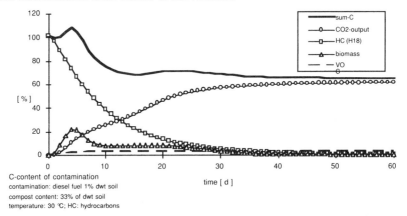

Figure 2 Model of a carbon balance of the C-content of the oil during treatment of an artificially oil-contaminated soil material with compost addition in a static bioreactor at 30 °C (Lotter et al., 1992)

In the case of a permanent gap in the C-balance it may be assumed that bound residues have formed. The analytics to determine the incorporation of the contaminant into the humus matrix are still in the process of being developed (Michaelis et al., 1991). If the balance is restored even after a period of time, it can be concluded that the adsorbed/incorporated substances were possibly set free during humus degradation and have been subsequently biologically degraded. If a C-balance is made for soil contaminated with oil without compost addition, the balance works out within the scale of measuring accuracy (100 _ 10%) (Stegmann et al., 1991). Since the humus content of the soil is clearly lower than that of the compost this is a further indication of the importance of the interactions between compost and contaminants.

Effect of Compost Addition

Several investigations have shown that the degradation of organic contaminants may be enhanced by means of the addition of biocompost. This effect was further investigated in several respirometer test series using compost at different stages of maturity. The following influencing factors may be of importance:

- compost as bulking agent to improve aeration particularly in cohesive soil materials
- compost as supplier of a great variety of microorganisms (Kästner et al., 1994)
- improvement of the pH buffer capacity (Dalyan et al., 1990) and water storage capacity due to the addition of compost
- compost as structural material to reduce pellet formation when cohesive soil materials are treated in mixing reactors (dynamic treatment)
- compost as source of nutrients and trace components (particularly nitrogen and phosphorus); compost as depot fertilizer
- compost as co-substrate
- interaction of the organic matrix of compost with the contaminant (incorporation of the contaminants in the soil/humus matrix)

In Germany, the degree of compost maturity is measured on a scale of I to V (measured as self-heating or respiration; Anonymous, 1985). The composting process itself can be divided into three stages: high-rate decomposition, stabilization and curing. During the high-rate decomposition stage, fresh organic matter is transformed into 'fresh compost'; this compost is hygienically acceptable and the most significant odour-causing organics have degraded (about 2 – 4 weeks, degree of maturity: II). In the stabilization stage the material is further decomposed and stabilized (2 – 4 weeks, degree of maturity: III–IV). The final curing formation takes place over a period of several weeks to as long as 18 to 20 weeks (degree of maturity: V; Krogmann, 1992).

In order to discover the most effective biocompost additive the following factors were investigated: compost age in relation to degradation efficiency, the relation of compost quantity to degradation and the effect of the addition of nitrogen.

Quantity of Compost Added

It could be shown that the addition of compost in a soil/compost ratio of 2:1 enhanced the degradation of oil significantly. For remedial action in practice, however, this compost ratio is too high. In Germany, when windrow techniques are applied, a volume rate of about 9:1 (soil/additive) is frequently used. For this reason soil/compost–mixtures of 2:1, 4:1, and 8:1 (refering to dry matter) were investigated. In all cases the diesel oil addition was 1% by weight of soil dry matter. The temperature was adjusted at 22° C.

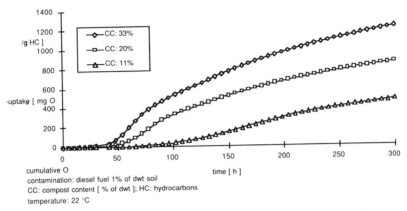

Figure 3 Cumulative O2–consumption from the degradation of the oil contaminant in the soil materials using various soil/compost-mixtures (Stegmann et al., 1991).

Figure 3 presents the cumulative oxygen consumption for the test series described above. For all set-ups the lag-phase was approximately 2 days. It was found that with decreasing compost content the cumulative O_2-consumption caused by the oil degradation decreased. Moreover, the maximum oxygen consumption in relation to the oil degradation per hour was reached when the compost addition was increased at an early stage.

Abiotic Influence of Compost

As mentioned above, a balance gap was detected when compost was added to the soil during balancing investigations with artificially contaminated soil material. No carbon balance gap was observed where compost was not added to the soil (accuracy of measurement: 100 _ 10%). The interaction between compost and the contaminants in the soil was of fundamental significance with regard to the carbon balance gap (Lotter et al., 1992). In order to find out more about the phenomenon, matured compost artificially contaminated with diesel fuel was used in tests for balancing contaminant degradation in bioreactors (Wolff, 1993). In Table 2 the different test conditions are presented.

Table 2 Test conditions for determination of the interactions between compost and the diesel fuel contamination

test	concentration of diesel fuel [% of dwt]	addition of CaCl$_2$ [% of dwt]	aeration	comment
A	1	-	air	'basic setup'
B	1	10	air	inhibition of microbial activity by CaCl$_2$ addition
C	1	10	N$_2$	inhibition of microbial activity by CaCl$_2$ addition and oppression of oxidation process by N$_2$ aeration
D	–	–	air	'control'; endogenous respiration

dwt: dry weight of compost

For test series A, B and C the compost was contaminated with diesel fuel (1% of dwt compost). The compost in test series D was left uncontaminated in order to measure the endogenous respiration ('control'). In test series A activated biomass was used to investigate the degradation of oil in the compost; in test series B and C the microbial activity was inhibited to a high degree by the addition of CaCl$_2$ (10% of dwt compost) in order to determine the abiotic interactions between compost and contaminants. The influence of oxidation processes by aeration was investigated indirectly by using nitrogen instead of air in test series C.

As is shown in Figure 4, after 21 days in test series A (basic set-up) only 8% of the carbon content of the original contamination was extractable, 59% was mineralized, 5% was stripped and about 4% of original contamination had been transformed into biomass. Thus a gap of 24% in the carbon balance was established. In comparison, only 14% of contamination was mineralized, strippped or transformed into biomass in the 'inhibited' series; a balance gap of 16% was observed and about 70% of the original contamination was extractable after a test period of 21 days.

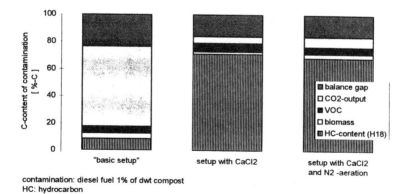

contamination: diesel fuel 1% of dwt compost
HC: hydrocarbon

Figure 4 Abiotic influence of compost on the carbon balance of diesel fuel; here: carbon balance after 21 days at 30°C

This experiment demonstrated that the incorporation of the contaminants in the compost also could be due to abiotic interactions between these compounds such as sorption and/or diffusion effects. Apparently the incorporation of contaminants in the compost matrix was not dependent upon an oxidative milieu since the incorporating effect in an oxygen atmosphere (Test A: 'basic set-up') was almost identical to that taking place in an atmosphere lacking in oxygen (test C: set-up with $CaCl_2$ and N_2–aeration).

Influence of compost maturity on the carbon balance

In order to examine the influence of the maturity of the compost on the biological degradation of hydrocarbons, soil material of an A_h-horizon was contaminated with 1% (of dwt soil) diesel fuel. 20% (of dwt soil) compost of varying ages (2 weeks, 2.5 months, 6 months, 13 months) was added in the different set-ups. In Table 3 the compost materials used are described.

Table 3 Description of the characteristics of the used compost materials employed

parameter	age of compost			
	0.5 months	2.5 month	6 months	13 months
biomass [g C/kg dwt]	17.6	14.1	1.4	0.4
TC [% dwt]	26.8	19.4	12.7	10.6
nitrogen (TKN) [% dwt]	1.2	2.4	1.4	1.0
ignition loss [% dwt]	44.8	41	24.4	20.8
degree of maturity	I	III	IV	V
reduction of hydrocarbon content after 60 days [%]	95	95	94	93

The results (Figure 5) show that the reduction in the hydrocarbon content of the diesel fuel after 60 days was more or less independent of the age of the compost; it amounted to approximately 94% of the original contamination. The established carbon balances indicate different balance gaps. The size of the balance gap decreased with increasing compost age.

In the case of fresh compost (degree of maturity < IV) it was very difficult to determine the influence of the contamination on the CO_2– and biomass-production. This was due to the high endogenous respiration of the fresh compost material. The differences of these parameters in contaminated and uncontaminated set-ups were marginal. It was possible to reduce the size of the balance gap for the set-ups with fresh compost by increasing the concentration of the diesel fuel added (to 2.5% of dwt soil). The higher quantity of oil made it easier to observe the degradation of oil into CO_2 and biomass. The results indicate that the balance system established here is not useful when fresh compost/organic matter is used as an additive.

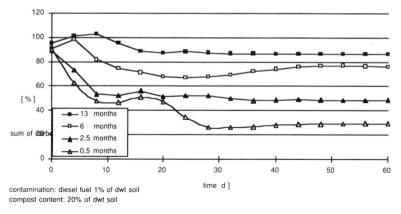

Figure 5 Influence of varying compost age on the carbon balance

Dynamic Treatment

In investigations carried out with paddle-stirrer reactors it appeared that the dynamic treatment had a positive effect on the contaminant turnover. But these tests also showed that during turning and mixing in the reactor the soil had a tendency to form pellets. It was discovered that pellet formation could be reduced if the water content was relatively low (<55% WC_{max} (maximum water capacity)), the proportion of organic structural material was high, the structural material contained only a low proportion of fines (d_p < 630-m; <10%) and the rotational speed of the mixer was low (8 rpm).

The addition of biocompost as structural material proved to be useful for the reduction of pellet formation. The addition of compost leads to a reduction of the water content and to a disintegration of the soil materials. However, it will be necessary to carry out further investigations to optimize this process.

Summary

A series of investigations has shown that the degradation of organic contaminants may be enhanced by means of the addition of biocompost. It was shown that a number of the characteristics of biocompost were important factors:

- compost as bulking agent to improve aeration particularly in cohesive soil materials
- compost as supplier of a great variety of microorganisms
- improvement of the pH buffer capacity and water storage capacity due to compost addition
- compost as structural material to reduce pellet formation when cohesive soil

materials are treated in mixing reactors (dynamic treatment)
- compost as source of nutrients and trace components; compost as depot fertilizer
- compost as co-substrate
- interaction of the organic matrix of compost with the contaminant

Moreover, it was established that the addition of compost led to a decrease in the ecotoxicology of soils contaminated with oil. This was measured by bacterial activity as well as in the results from algae and plant toxicity tests (Ahlf et al., 1993).

In addition, it was discovered that the degree of maturity of the biocompost used seems to have no influence on the degradation rate of the contaminant. The formation of non- bioavailable (non-extractable) residual contaminants remains to be investigated (interaction of the contaminants with humus / compost). Here the processes of carbon turnover in the soil have to be thoroughly examined.

References

Ahlf, W., Gunkel, J., Rönnpagel, K./Stegmann, R. eds. (1993). Toxikologische Bewertung von Sanierungen. Economica Verlag, Bonn, Germany, pp. 275–286.

Anderson, J.P.E., Domsch, K.H. (1978). A physiological method for the quantitative measure-ment of microbial biomass in soils. Soil Biology Biochemistry, 10, pp. 215–221.

Anonymous (1981). German standard methods for teh examination of water, waste water and sludge; gerneral measures of effects and substances (group H); determination of hydrocarbons (H18). DIN 38409, Beuth Verlag GmbH, Berlin, Germany.

Anonymous (1985). Qualitätskriterien und Anwendungsempfehlungen für Kompost aus Müll und Müllklärschlamm. Mitteilung der Länderarbeitsgemeinschaft Abfall 8, Merkblatt 10, Erich Schmidt Verlag, Berlin, Germany.

Beck, T. (1984). Mikrobiologische und biochemische Charakterisierung landwirtschaftlich genutzter Böden, I. Mitteilung: Die Ermittlung einer Bodenmikrobiologischen Kennzahl. Zeit–schrift für Pflanzenernährung und Bodenkunde, (1), pp. 456–467.

Dalyan, U., Harder, H., Höpner, T. (1991). Hydrocarbon biodegradation in sediments and soils, A systematic examination of physical and chemical conditions – Part II. pH values. Erdöl und Kohle – Erdgas – Petrochemie vereinigt mit Brennstoff-Chemie, 9, pp. 337–342.

Francke, W. (1990). Teilprojekt 'Mikroanalytische Untersuchungen an kontaminierten Böden: Erfassung und Quantifizierung ausgewählter Schadstoffe in unbehandeltem und behandeltem Material' im SFB 188. Mitteilung in der SFB-Arbeitsgruppe 'Analytik', Hamburg, Germany, unveröffentlicht.

Gerth, J., Förstner, U./Stegmann, R., Franzius, V., eds. (1990). Reinigung kontaminierter Böden, Economica Verlag, Bonn, Germany, pp. 13–26.

Goetz, D., Wiechmann, H., Berghausen, M. (1990). Teilprojekt 'Einfluß von Öl-kontamina-tionen auf bodenphysikalische und –mechanische Eigenschaften von kontaminierten Stand-orten' im SFB 188. Mitteilungen in der SFB-Arbeitsgruppe 'Boden', Hamburg, Germany, unveröffentlicht.

Hupe, K., Heerenklage, J., Lotter, S., Stegmann, R./Stegmann, R. eds. (1993). Anwendung von Testsystemen zur Bilanzierung und Optimierung des biologischen Schadstoffabbaus. Bodenreinigung. Economica Verlag, Bonn, Germany, pp. 97–119.

Hupe, K., Lüth, J.–C., Heerenklage, J., Stegmann, R./Alef, K., Blum, W., Schwarz, S., Riss, A., Fiedler, H., Hutzinger, O. eds. (1994). Einsatz von Testsystemen und Laborreaktoren für bilanzierende Optimierungsuntersuchungen. Eco-Informa-'94: Bodenkontamination, Boden-sanierung, Bodeninformationssysteme. Umweltbundesamt, Wien, Austria, pp. 111-121.

Kästner, M., Lotter, S., Heerenklage, J., Breuer-Jammeli, M., Stegmann, R., Mahro, B. (1994). Fate of ^{14}C-labeled Anthracene and Hexadecane in compost manured soil, Applied Micro-biology and

Biotechnology, submitted.

Krogmann, U. (1992). Proceedings of the second international Forum on Resource Recovery from Waste, September 21–24, 1992, Imola, Italy.

Lotter, S., Stegmann, R., Heerenklage, J. (1990). Proceedings of fourth international KfK/TNO Conference on Contaminated Soil, December 10–14, 1990, Karlsruhe, Germany. pp. 967–974.

Lotter, S., Stegmann, R., Heerenklage, J. (1992). Proceedings of international Symposium on Soil Decontamination Using Biological Processes, December 6–9, 1992, Karlsruhe, Germany. pp. 219–227.

Michaelis, W. (1991). Teilprojekt 'Chemische Wechselwirkung von Erdölkontaminationen und deren Abbauprodukten mit der Humusfraktion von Böden' im SFB 188. Hamburg, Germany, unveröffentlicht.

Miehlich, G., Wagner, A. (1990). Teilprojekt 'Veränderung bodenchemischer Eigenschaften durch Ölverunreinigung und –dekontamination' im SFB 188. Mitteilung in der SFB-Arbeits-gruppe 'Boden', Hamburg, Germany, unveröffentlicht.

Stegmann, R., Lotter, S., Heerenklage, J. (1991). Proceedings of the first international Symposium on In Situ and On-Site Bioreclamation, March 19-21, 1991, San Diego, USA. On-Site Bioreclamation, pp. 188-208.

Steinhart, H., Herbel, W., Bundt, J., Paschke, A. (1990). Strukturtypentrennung von Dieselöl-kraftstoff mittels Festphasenextraktion. Protokoll der 6. Arbeitsgruppensitzung 'Analytik' des SFB 188 vom 4.4.1990, Hamburg, Germany, unveröffentlicht.

Wolff, M. (1993). Experimentelle Untersuchung des Einflusses der Sorption und der Kompostreife auf die Öl-Kohlenstoffbilanz dieselkontaminierter Böden bei der Behandlung in statischen Bodenbioreaktoren. Diplomarbeit an der Fachhochschule Hamburg-Bergedorf, Germany, unveröffentlicht.

[1]dynamic bioreactors: systems in which the contaminated soil is mixed gently
static bioreactors: no mixing

Heavy Metals Removal by Clinoptilolite in Pepper Cultivation using Compost

E . G. KAPETANIOS* – Solid Waste Research Laboratories

Abstract

The compost used, was produced in a pilot plant in which household refuse from Attica Region was treated. The composting took place in piles using forced aeration with air blow and suction. In compost, soil, and manure, that were used for the cultivation of pepper plants in pots and under greenhouse experimental conditions, the following metals were determined : Cd, Cr, Ni, Pb, Cu, Zn, Mn . There were also determined the metals fraction concentrations in compost. In part of the experimental cultivation, natural zeolite was used in various compositions with compost and soil. The natural zeolite used was the clinoptilolite. The removal of metals in compost and the compost-soil system were studied, and the metal contents in the leaves, body, crops, and roots of the plants were determined .The results showed that the plants contained lower quantities of metals when cultivated with zeolite and even lower when more zeolite was used .

Introduction

The presence of heavy metals in composts derived from household refuse is a significant problem which can make prohibitive their application in agriculture when these heavy metals are found in high concentrations in the composts (Kapetanios et al 1988, Kapetanios et al 1993) . The heavy metals when contained in small quantities are valuable trace elements necessary for the growth of plants, while when contained in high quantities they become phytotoxic and toxic for the man factors (Zucconi and De Bertoldi 1986, Petruzzelli and Lubrano 1986, Petruzzelli et al, 1992).

Zeolites, synthetic or natural, have the ability to remove heavy metals by ion exchange (Loizidou and Townsend, 1987), and this characteristic ability they possess can be used to remove the heavy metals contained in composts so that they do not pass to the plants (Kapetanios, 1990) . From previous studies (Kapetanios and

*Association of Communities and Municipalities in Attica Region

Loizidou, 1992, Kapetanios and Loizidou, 1995) on tomato plant cultivation in compost with zeolite added compared to tomato plants cultivated in compost without zeolite added, a significant percentage of the heavy metals contained in the compost was removed by zeolite. In this work are reported the experimental observations when a natural zeolite, namely clinoptilolite, was used for pepper plant cultivation in compost.

Materials and methods

For the experimental work, the soil (S) used was from the Attica region. Compost (C) and stabilized natural sheep manure (K) were used as soil conditioning materials. The compost used which originated from the biodegradable fraction of household refuse from Attica region, was recovered from a pilot plant with a capacity of 2 Mt / h,and the composting was completed in piles ventilated through forced aeration with air blow and suction. The zeolite (Allison F.E., 1973) used, was clinoptilolite (CLI) and as an experimental plant, pepper was used in a 12 l pot.The proportions of compost, manure, and zeolite used, were as shown :

$C_1 = 200$ cm^3 compost $K_1 = 200$ cm^3 manure
$C_2 = 400$ cm^3 compost $K_2 = 400$ cm^3 manure
$C_3 = 800$ cm^3 compost $K_3 = 800$ cm^3 manure

$Z_1 = 15$ gr CLI $2Z_1 = 30$ gr CLI
$Z_2 = 30$ gr CLI $2Z_2 = 60$ gr CLI
$Z_3 = 60$ gr CLI $2Z_3 = 120$ gr CLI

The various combinations resulted were :

$C_1 = 200$ cm^3 compost $K_1 = 200$ cm^3 manure
$C_2 = 400$ cm^3 compost $K_2 = 400$ cm^3 manure
$C_3 = 800$ cm^3 compost $K_3 = 800$ cm^3 manure

$Z_1 = 15$ gr CLI $2Z_1 = 30$ gr CLI
$Z_2 = 30$ gr CLI $2Z_2 = 60$ gr CLI

As references, plain compost, manure and soil were used and the respective processes are denoted as OCO, OKO, SOO .

A total of nineteen combinations were performed and each one was repeated four times. Pesticides used, were selected on a basis that no metal under consideration was present in any of them. The experiments were carried out in an open experimental cage. For the determination of the total heavy metals in the compost and the various parts of the plants, the HNO_3 / $HClO_4$ method was applied and measurements were taken by atomic absorption. The metal fractions, which can be solubilized, after consecutive extraction with distilled water, KNO_3 (1M) and

EDTA, were determined again by atomic absorption. The chemical exchange capacity (CEC) was determined using the sodium acetate method At a pH of 8.2 . The organic substance was evaluated from the organic carbon (C_{org}) which was determined by oxidation with K_2CrO_4, using the formula:

$$\text{Organic substance} = K_X C_{org} \qquad K = 2$$

Experimental Results – Discussion

The most significant parameters which influence the cation exchange (Cooke G .W., 1967, Gerritse et al, 1985) of zeolites are :

– Type, size and valence of the anhydrous and hydrated cations, pH, temperature, and characteristics of the zeolite structure and ion exchange capacity .

In this work, the theoretical ion exchange capacity of the zeolite was 2.358 meq / gr obtained by chemical analysis of the mineral. From kinetic experiments it was observed that the selectivity series of clinoptilolite for the various metals is Pb > Cd > Zn > Cu > Ni > Mn > Cr .

It was also concluded that the selectivity series remains the same even when the solution concentration changes and also high quantities of metals are removed within the first minutes. There is an unstable equilibrium between the soluble and insoluble forms of heavy metals in compost and the major parameters which influence it are pH, organic substance, the presence of various anions and the redox potential. The chemical form of metals in compost depends significantly on their chemical form in refuse as well as on the composting process followed.

Table 1 Metal and metal fraction contents in compost

Metals (mg./kg. d.w)	1	2	3	4	Deduced % from 2, 3, 4
Cd	2.80	0.05	0.04	0.03	4.28
Cr	80.00	0.78	0.22	0.13	1.41
Ni	35.00	0.96	0.35	0.82	6.09
Pb	236.00	0.32	0.41	6.73	3.16
Cu	204.00	3.37	1.11	4.89	4.59
Mn	196.00	3.65	1.14	8.50	6.78
Zn	759.00	4.64	0.78	26.88	4.26

1   : Total metals in compost using the $HNO_3/HClO_4$ method
2, 3, 4: Metal fraction soluble in H_2O, KNO_3 and E.D.T.A. respectively.

Heavy metals are found in the soil in one of the following chemical forms :

– Simple or complex ionic forms.
– Exchangeable ionic forms.
– Complex organometalic or chemical forms.

– Coprecipitated (with various cations as carbonates, sulphates, etc.) forms, as well as participation of the various minerals contained in the soil.

The root system of plants takes from the soil solution not only the free ionic forms but the exchangeable ones as well as the organometalic forms. In Tables 1 and 2 are given the contents corresponding to compost and soil in Cd, Cr, Ni, Pb, Cu, Zn, and Mn . In those tables, for every metal, is given the total content in column 1, the simple and complex ionic forms in column 2, and the exchangeable forms in column 3. The main parameters that regulate the passage of heavy metals from the soil into the plant are, pH, organic substance, humidity, temperature, and ventilation (Allison F.E.,1973, Cooke G.W.,1967, Gerritse R.G et al, 1985, Smilde K.W.,1981). In Table 3 the pH, CEC, and the organic substance of the soil used, compost and manure are given. In Table 4 the concentration of total heavy metals in the soil, manure, and compost are given. In Tables 5 – 11 are presented the metal contents of the pepper plants in the roots, bodies, leaves, and crops.

Table 2 Metal and metal fraction contents in soil

Metals (mg./kg. d.w.)	1	2	3	4	Deduced % from 2, 2, 4
Cd	1.70	0.01	0.02	0.03	3.53
Cr	67.00	0.23	0.14	0.39	1.13
Ni	31.00	0.65	0.27	0.61	4.91
Pb	58.00	0.46	0.51	0.63	2.76
Cu	93.00	1.15	0.72	1.31	3.42
Mn	187.00	4.75	2.25	3.78	5.76
Zn	256.00	2.08	2.18	5.48	3.80

1 : Total metals in soil using the $HNO_3/HClO_4$ method
2, 3, 4: Metal fraction soluble in H_2O, KNO_3 and E.D.T.A. respectively.

Table 3 pH, C.E.C. and organic substance for compost, manure, and soil

Material	pH	C.E.C. (meq./200g. d.w.)	Organic substance %
Compost	7.62	54.20	45.00
Manure	9.55	64.70	51.60
Soil	7.30	16.30	1.43

Examining Table 1, it is shown that the compost has a low content in available ionic forms for all the metals. It is also shown that it has a low content in organometalic compounds. From the same Table it is shown that the percentage of the metals Cd, Cr, Ni, Pb, Cu, Zn, Mn that is available to the plants, corresponds to 4.1, 1.41, 6.09, 3.16, 4.59, 6.78, 4.26 of the respective total content in the compost. Observing Table 2 it can be seen that Cd, Cr, Ni, Pb, Cu, Zn, and Mn in soil are found in a lower content compared to their content in compost. Also, it can be seen that the available ionic and organometalic forms of the above mentioned metals are very low. The percent proportion of Cd, Cr, Ni, Pb, Cu, Zn, and Mn which is directly available to the plants corresponds to 12.73, 1.42, 1.95, 10.16, 6.56, 5.76, and 5.33 of the respective total percent proportion in the soil. In Table 3 a high CEC of compost and manure is shown in relation to what other researchers report (Guidi G.,1981, Manios V.I.,1986), a fact which resulted in a low mobility of the heavy metals. It is also observed that compost and manure have high organic matter while the soil has a very low one. The pH of the compost and soil are about neutral but that of manure is quite alkaline. Observing Table 4, it is apparent that Cd, Cr, Ni, Pb, Cu, Zn, and Mn are in a quite lower content in manure in relation to compost and soil. The metal contents in the roots, bodies, leaves, and crops of plants are shown in Tables 5–11 and the following comments can be reported for each one. Finally, in Table 12 are shown the contents in Cd, Cr, Ni, Pb, Cu, Zn, and Mn in pepper plant crops from the market of Athens. In fact, 15 samples were taken from vegetable markets and the results obtained from their mixture are shown in this Table.

Table 4 Cd, Cr, Ni, Pb, Cu, Mn, and Zn contents in compost, manure, and soil

Metals (mg./kg. d.w.)	Compost	Manure	Soil
Cd	2.80	0.76	1.65
Cr	80.00	0.94	67.42
Ni	35.00	2.30	31.76
Pb	236.00	2.46	58.25
Cu	204.00	2.10	93.23
Mn	196.00	2.71	187.20
Zn	759.00	4.73	256.32

Table 5 Cd content (mg./kg. d.w.) in pepper plants

Numb.	Process	Roots	Bodies	Leaves	Crops
1	OCO	0.143	0.111	0.130	0.112
2	OKO	0.060	0.092	0.970	—
3	SOO	0.101	0.094	0.107	0.082
4	SC_1O	0.102	0.080	0.106	0.091
5	SC_2O	0.113	0.091	0.240	0.106
6	SC_3O	0.130	0.138	0.105	0.101
7	SK_1O	0.093	0.061	0.091	0.072
8	SK_2O	0.087	0.073	0.086	0.063
9	SK_3O	0.097	0.067	0.093	0.087
10	OCZ_1	0.157	0.103	0.123	0.118
11	OCZ_2	0.130	0.110	0.127	0.125
12	OCZ_3	0.098	0.086	0.101	0.108
13	$OC2Z_3$	0.072	0.078	0.090	0.071
14	SC_1Z_1	0.113	0.092	0.139	0.103
15	SC_2Z_2	0.100	0.117	0.103	0.090
16	SC_3Z_3	0.106	0.103	0.117	0.108
17	SC_12Z_1	0.121	0.105	0.105	0.103
18	SC_22Z_2	0.095	0.109	0.095	0.097
19	SC_32Z_3	0.087	0.093	0.102	0.080

Table 6 Cr content (mg./kg. d.w.) in pepper plants

Numb.	Process	Roots	Bodies	Leaves	Crops
1	OCO	1.137	1.098	1.195	0.532
2	OKO	0.309	0.280	0.428	—
3	SOO	0.933	0.656	0.893	0.470
4	SC_1O	1.012	0.863	0.977	0.200
5	SC_2O	0.918	0.978	0.893	0.312
6	SC_3O	1.103	0.907	1.004	0.305
7	SK_1O	0.470	0.597	0.921	0.462
8	SK_2O	0.386	0.610	0.720	0.378
9	SK_3O	0.412	0.508	0.811	0.360
10	OCZ_1	1.156	0.922	1.225	0.581
11	OCZ_2	1.110	0.970	1.207	0.595
12	OCZ_3	1.126	1.056	1.111	0.463
13	$OC2Z_3$	0.900	0.712	0.904	0.428
14	SC_1Z_1	1.128	0.930	0.908	0.229
15	SC_2Z_2	1.109	0.971	0.913	0.287
16	SC_3Z_3	0.830	0.860	0.925	0.328
17	SC_12Z_1	0.941	0.875	0.890	0.320
18	SC_22Z_2	0.957	0.788	0.723	0.297
19	SC_32Z_3	0.768	0.536	0.670	0.213

Table 7 Ni content (mg./kg. d.w.) in pepper plants

Numb.	Process	Roots	Bodies	Leaves	Crops
1	OCO	1.722	1.830	1.930	0.812
2	OKO	0.315	0.400	0.411	—
3	SOO	1.605	1.513	1.722	0.690
4	SC_1O	1.780	1.722	1.768	0.650
5	SC_2O	1.590	1.791	1.813	0.687
6	SC_3O	1.613	1.710	1.700	0.592
7	SK_1O	1.628	1.431	1.840	0.813
8	SK_2O	1.645	1.370	1.892	0.708
9	SK_3O	1.830	1.455	1.714	0.740
10	OCZ_1	1.830	1.647	2.070	0.730
11	OCZ_2	1.671	1.720	1.820	0.747
12	OCZ_3	1.414	1.563	1.714	0.634
13	$OC2Z_3$	1.219	1.510	1.580	0.602
14	SC_1Z_1	1.687	1.739	1.829	0.612
15	SC_2Z_2	1.693	1.615	1.690	0.671
16	SC_3Z_3	1.580	1.417	1.735	0.593
17	SC_12Z_1	1.652	1.680	1.510	0.627
18	SC_22Z_2	1.112	1.675	1.221	0.530
19	SC_32Z_3	0.087	1.044	0.937	0.485

Table 8 Pb content (mg./kg. d.w.) in pepper plants

Numb.	Process	Roots	Bodies	Leaves	Crops
1	OCO	0.747	0.593	0.614	0.305
2	OKO	0.272	0.387	0.498	—
3	SOO	0.580	0.572	0.598	0.243
4	SC_1O	0.618	0.627	0.580	0.222
5	SC_2O	0.586	0.641	0.608	0.208
6	SC_3O	0.645	0.610	0.540	0.231
7	SK_1O	0.471	0.495	0.327	0.225
8	SK_2O	0.493	0.454	0.404	0.210
9	SK_3O	0.422	0.408	0.326	0.237
10	OCZ_1	1.695	0.611	1.602	0.281
11	OCZ_2	0.687	0.582	0.672	0.298
12	OCZ_3	0.635	0.570	0.585	0.227
13	$OC2Z_3$	0.578	0.591	0.563	0.240
14	SC_1Z_1	0.689	0.600	0.645	0.267
15	SC_2Z_2	0.622	0.623	0.570	0.245
16	SC_3Z_3	0.638	0.570	0.593	0.271
17	SC_12Z_1	0.580	0.582	0.562	0.208
18	SC_22Z_2	0.565	0.545	0.567	0.214
19	SC_32Z_3	0.476	0.507	0.522	0.202

Table 9 Cu content (mg./kg. d.w.) in pepper plants

Numb.	Process	Roots	Bodies	Leaves	Crops
1	OCO	15.950	12.410	16.740	11.970
2	OKO	5.130	4.170	7.110	—
3	SOO	8.620	9.720	10.270	7.290
4	SC_1O	13.270	12.480	15.730	12.080
5	SC_2O	12.110	11.140	16.020	11.620
6	SC_3O	14.820	11.390	15.090	11.970
7	SK_1O	7.620	9.140	10.460	11.060
8	SK_2O	7.150	8.630	9.400	11.480
9	SK_3O	7.030	8.560	8.750	10.100
10	OCZ_1	15.880	11.600	16.400	11.260
11	OCZ_2	15.240	12.130	16.720	10.950
12	OCZ_3	14.330	11.080	16.140	9.120
13	$OC2Z_3$	14.080	11.110	15.200	8.730
14	SC_1Z_1	14.360	13.270	15.830	12.180
15	SC_2Z_2	13.280	11.150	14.760	11.740
16	SC_3Z_3	13.400	12.860	14.200	11.400
17	SC_12Z_1	12.960	12.340	12.970	10.760
18	SC_22Z_2	13.070	12.570	12.700	10.240
19	SC_32Z_3	12.180	11.900	12.060	8.200

Table 10 Zn content (mg./kg. d.w.) in pepper plants

Numb.	Process	Roots	Bodies	Leaves	Crops
1	OCO	39.310	51.270	40.750	22.200
2	OKO	24.070	32.910	29.500	—
3	SOO	35.120	40.060	27.260	17.150
4	SC_1O	36.180	43.900	28.350	19.140
5	SC_2O	36.700	42.710	28.900	19.010
6	SC_3O	39.310	42.830	27.630	20.230
7	SK_1O	34.650	37.220	23.150	18.880
8	SK_2O	33.180	36.120	29.170	15.610
9	SK_3O	34.900	31.330	27.520	17.950
10	OCZ_1	40.370	30.220	26.400	23.270
11	OCZ_2	41.080	27.310	27.200	22.000
12	OCZ_3	39.700	28.290	25.760	21.630
13	$OC2Z_3$	36.170	25.120	23.600	21.720
14	SC_1Z_1	36.380	33.700	28.300	19.100
15	SC_2Z_2	35.130	35.620	29.150	19.170
16	SC_3Z_3	36.150	35.380	25.100	18.630
17	SC_12Z_1	36.720	39.270	25.960	18.070
18	SC_22Z_2	35.060	34.100	23.120	18.350
19	SC_32Z_3	33.120	33.700	23.710	17.700

Table 11 Mn content (mg./kg. d.w.) in pepper plants

Numb.	Process	Roots	Bodies	Leaves	Crops
1	OCO	20.230	22.400	27.370	11.840
2	OKO	6.700	9.820	17.120	—
3	SOO	10.280	15.430	20.100	6.120
4	SC_1O	12.720	14.180	20.550	7.100
5	SC_2O	12.940	13.920	19.700	7.560
6	SC_3O	13.800	13.670	21.070	7.920
7	SK_1O	8.470	16.710	19.400	7.600
8	SK_2O	8.650	16.020	19.100	7.710
9	SK_3O	7.260	15.130	18.770	6.600
10	OCZ_1	19.200	21.220	26.130	12.270
11	OCZ_2	19.780	20.100	24.200	10.250
12	OCZ_3	16.100	18.350	20.810	7.130
13	$OC2Z_3$	13.120	15.920	17.400	5.750
14	SC_1Z_1	13.300	14.710	19.120	10.110
15	SC_2Z_2	12.900	14.870	18.730	9.630
16	SC_3Z_3	13.370	13.250	18.810	8.900
17	SC_12Z_1	11.820	13.060	17.100	7.820
18	SC_22Z_2	12.070	11.930	16.170	7.190
19	SC_32Z_3	10.100	11.700	15.540	6.180

Cadmium

The potential properties of cadmium in their entireness show a similarity with those of zinc, but cadmium is more kinetic and found in less in soil. The use of zeolite in single proportion gave a reduction of the cadmium content, but an even higher reduction was observed using a double proportion of zeolite which in the case of a higher proportion gave significantly lower cadmium contents compared to plants cultivated in plain soil. Also, the cadmium content in pepper plant crops cultivated in a mixture of compost-soil-double proportion of zeolite was significantly lower compared to the cadmium content in pepper plant crops from the market of Athens.

Chromium

Chromium according to Leeper is considered less toxic than the other heavy metals because the hexavalent chromium which is probably found in the biodegradable fraction of refuse, with the aerobic conditions that prevail during composting, is oxidized to trivalent chromium which is much less toxic. In pH 6–8 chromium gives compounds of very low solubility which has as a result a low mobility. The addition of zeolite, mainly in double proportion, gives a chromium content for the plants much lower than the one for plants that were cultivated in plain soil. Also, the chromium content in pepper plant crops cultivated in a mixture of compost-soil-double proportion of zeolite was significantly lower compared to the content in chromium of pepper plant crops from the market of Athens.

Table 12 Metal content (mg./kg. d.w.) in pepper plant crops from the market of Athens

Cd	Cr	Ni	Pb	Cu	Zn	Mn
0.063	0.356	0.747	0.220	9.140	22.760	8.170

Nickel

Humus, according to Leeper, has the ability to remove nickel with which it forms chelates. The highest percentage of it is adsorbed from the negatively charged surfaces of humus, and the adsorption increases when the pH values get close to 7. The addition of zeolite, mainly in double proportion, gave nickel contents for the plants much lower than the ones that were cultivated in plain soil. Also, the nickel content in pepper plant crops cultivated in compost-soil-double proportion of zeolite was significantly lower compared to the content in nickel of pepper plant crops from the market of Athens.

Led

Led has been studied by many researchers (Petruzzeli, 1982, Wolnik, 1985) due to its significant mobility in the chain soil-plant-man. According to Leeper bivalent led is hydrolyzed and polymerized easily in the soil and with the increase of pH becomes inert.

From Table 1 can be seen that only 3.16 % of the total led, found in compost, is available to the plants. From Table 8 can be seen that the lower content in led is found in the crops.

The best results were obtained by a double proportion of zeolite. Indeed, the led contents in plants cultivated in soil-compost-double proportion of zeolite were lower compared to the ones in plants cultivated in plain soil.

Led content in pepper plants cultivated in a compost-zeolite mixture was lower compared to the one in plants cultivated in plain compost.

The led content in pepper plant crops cultivated in soil-compost-double proportion of zeolite was significantly lower compared to the one in crops from the market of Athens.

Copper

A significant part of the Cu^{2+} ions found in compost is combine with the organic substance in compost to form complex organic salts. In the soil-compost mixture a part of Cu^{2+} ions combine with clay, but this combination is less strong than the one with the organic substance. Also, a significant part of the Cu^{2+} ions, in the compost-soil mixture, form insoluble $CuSO_4$.

Only 4.59 % of the total Cu contained in compost is found in extractable form with H_2O, KNO_3, and E.D.T.A, and consequently can pass into the plants.

The best results in soil-compost mixtures were obtained with plants cultivated with addition of a double proportion of zeolite. But it was observed that plants cultivated in plain soil had a lower content in copper compared to every soil-compost mixture even in the cases when a double proportion of zeolite was used.

It was also observed, that the addition of zeolite in compost significantly reduced the content in copper compared to that of plants cultivated in plain compost.

The copper content in the pepper plant crops which were cultivated in a soil-compost-double proportion of zeolite mixture was significantly lower compared to the one in crops from the market of Athens.

Zinc

The Zn^{2+} ion has the tendency to combine strongly with the organic substance and form organic complex salts. In the soil-compost mixture the Zn^{2+} ions combine with the organic substance and the clay.

Only 4.26 % of the total zinc contained in compost is in extractable forms with H_2O, KNO_3, and E.D.T.A, and consequently can pass in the plants.

The best results in soil-compost mixtures were obtained with plants cultivated with addition of a double proportion of zeolite. But it was observed that plants cultivated in plain soil had a lower content in zinc compared to every soil-compost mixture even in the cases when a double proportion of zeolite was used.

It was also observed that the addition of zeolite in compost significantly reduced the content in zinc compared to that of plants cultivated in plain compost.

The content in zinc in the pepper plant crops which were cultivated in a soil-compost-double proportion of zeolite mixture was significantly lower compared to the one in crops from the market of Athens.

Manganese

The chemistry of manganese in the soil is complicated and Linday reports that this is due to the fact that manganese forms mixed oxides with different valences, and these oxides have different crystal forms or are amorphous, and it is possible for them to combine with iron or other metals oxides.

In Table 1 can be seen that 6.78 % of the total manganese contained in compost is found in extractable forms with H_2O, KNO_3 and E.D.T.A, and consequently can pass in the plants.

The best results were obtained by a double proportion of zeolite. In fact, the led contents in plants cultivated in soil-compost-double proportion of zeolite were lower compared to plants cultivated in plain soil.Manganese content in pepper plants cultivated in a compost-zeolite mixture was lower compared to plants cultivated in plain compost.

The Maganese content in pepper plant crops cultivated in soil-compost-double proportion of zeolite was significantly lower compared to crops from the market of Athens.

Conclusions

– Plants cultivated in plain compost gave the highest heavy metals content in the crops, roots, bodies, and leaves. The addition of zeolite in plain compost significantly lowered the metal content in the plants.
– Plants cultivated in plain manure gave the lowest metal content found in the, roots, bodies, and leaves.
– Plants cultivated in plain soil gave a metal content lower than that with plain compost and higher than that with plain manure.
– Plants cultivated in various compost-soil mixtures gave a metal content lower than those cultivated with plain compost, and higher or even equal to those cultivated in plain soil
– A significant reduction in metal content was observed in plants cultivated in compost-soil mixtures with zeolite added in a single proportion.
– The addition of a double proportion of zeolite contributes to the further reduction of metal content for almost all parts of the plant, and in the most of the cases, especially with the higher proportion in zeolite, it was observed a metal content even lower than that of cultivation with plain soil.
– The results concerning the metal content in the pepper plant crops, cultivated in a compost-soil mixture with a double proportion of zeolite added, and especially for the higher proportion of zeolite, compared to pepper plant crops from Athens markets the metal content was lower for all metals, except Cd,when zeolite was used.
– The observed reduction in metal contents of plants cultivated with the addition of a single and even more with the addition of a double proportion of zeolite,

showed that zeolite removes significant quantities of the metals available to the plants and prevents their intake from the plants.

References

1. Allison, F.,(1973). Soil organic matter and its role in crop production. Elsevier. Amsterdam.
2. Cooke, G.,(1967). The control 0f soil fertility. Crosby and Lockwood, London.
3. Gerritse, R., Van Driel, W., Smilde, K. and Van Luit, B. (1985). Uptake of heavy metals by crops in relation to their concentration in the soil solution. Institute for soil fertility. Haren, Netherlands.
4. Guildi, G., (1981). Relationships between organic matter of sewage sludge and physico-chemical properties of soil. 11th Europ. Symp. on characterization, treatment and use of sewage sludge. 21–23 Oct. 1981, Vienna.
5. Haghiri, F. (1974). Plant intake of cadmium as influenced by cation exchange capacity, organic matter, zinc and soil temperature. J. Envir. Qual., 3, 180–183
6. Kapetanios, E. and Loizidou, M. (1995). Influence of clinoptilolite on the heavy metals when applied to soil amended with compost. Science of the total environment. Received.
7. Kapetanios, E.,Loizidou, M. and Valkanas, G. (1993). Compost production from domestic refuse. Bioresource Technology, 44, 13–16.
8. Kapetanios, E., Loizidou, M. and Malliou, E. (1988). Heavy metal levels and their toxicity in compost from Athens household refuse.Environmental Technology Letters, 6, 799–802.
9. Kapetanios, E. (1990). Refuse derived compost production and characterization, and heavy metals removal by the use of clinoptilolite. National Technical University of Athens, Thesis in Ph.D.
10. Kapetanios, E. and Loizidou, M. (1992). Heavy metal removal by zeolites in cultivations using compost. Acta Horticulturae, 301, 63–71.
11. Leeper, G. (1978). Managing the heavy metals on the land. Dekker, New York.
12. Loizidou, M. and Townsend, R. (1987). Ion exchange properties of natural clinoptilolite,ferrierite and mordenite, Part 2: Sodium-amonium equilibria. Zeolites, 7, 153.
13. Manios, V. (1986). Organohumic material production from organic refuse and their applications in agriculture. Heraklio.
14. Petruzzelli, G. and Lubrano, L. (1986). Heavy metal extractability. Biocycle, 26(8), 46–48
15. Petruzzelli, G., Szymura, I., Lubrano, L. and Pezzarossa, B. (1992). Heavy metal speciation in compost with a view to its agricultural use. Acta Horticulturae, 301, 377–383.
16. Smilde, K., (1981). Heavy metal accumulation in crops grown on sewage sludge amended with metal salts. Plant and soil, 62, 3–14.
17. Zucconi, F. and De Bertoldi, M. (1986). Compost specification for the production and characterization of compost from municipal solid waste. Int. Symp. on : Compost production, quality and use. Udine.

B5 Composting Design

Composting Technology in the United States: Research and Practice

ROBERT J. TARDY and R. W. BECK, Denver, CO, USA

Introduction

Composting is a time-honored, worldwide practice as an inexpensive, effective means of enriching soil for crops and gardens. Over the past two decades, composting has become an increasingly common method for reducing the volume of organic materials – particularly yard waste – sent to landfills. In Europe, the entire organic portion of municipal solid waste (MSW) is often composted. Up until recently, that approach had been less common in the United States; however, as more states enact tougher landfill laws, diverting a higher percentage of organics found in MSW is attracting more interest.

Although composting has been around for centuries, we are continuing to learn more about ways to make composting a more cost-effective alternative to source reduction, methods of controlling odors, and new technologies enabling acceptance of a greater variety of organic materials.

In the United States, this knowledge is pursued through the efforts of developers, academia, environmental entities, and municipal governments.

Composting technologies

Basics of the Composting Process

In its most basic form, composting needs no technology – it is simply the natural result of microbial breakdown of organic materials in the presence of oxygen; this same process has been producing humus (organic topsoil) in forests for millennia. Technology was introduced into the composting process to regulate the methods used in order to produce certain qualities of compost, and to increase the speed of the process to enable the processing of larger volumes of materials.

Types of Technologies

In the United States, composting technologies are typically classified into three basic methods: windrow systems, aerated static pile systems, and in-vessel systems. The first two methods are relatively simple and inexpensive to develop. In-vessel systems are more costly to build and operate; however, they can be built in a smaller area and offer better process control.

Windrow Systems

In the windrow approach, the feedstock is arranged in long piles (windrows) on a gently sloping site that may be open to the air or covered. Windrows are aerated through natural convection, assisted by periodic turning with front-end loaders or special turning equipment. The turning frequency depends on the material's moisture, texture, and stability, aeration methods, and operational goals such as odor control, composting speed, or pest control. Blowers may be used to force air through the windrows for more efficient aeration and heat removal. Both windrow and aerated windrow facilities typically rely on process controls to minimize odors.

Static Pile Systems

In a static pile system, the feedstock is placed in a large pile that is not disturbed during composting. Air is introduced into the pile through duct systems installed beneath the pile. Aerated static pile operations avoid anaerobic conditions by introducing a controlled volume of air into each pile. This air can be positive, blowing up through the pile, or negative, drawing air down through the pile.

Several operators of static pile systems use negative aeration during active composting and no aeration during curing. Some operators use positive aeration during both stages, while other operators use positive aeration during active composting and no aeration during curing. One operator uses negative aeration during both phases.

Most aerated static pile facilities use process controls for managing odors (i.e., optimizing aeration, moisture control, porosity, etc.), as opposed to using more costly systems for capturing and processing odorous process air.

In general, aerated static pile systems have higher capital costs but lower operating costs than windrow systems.

In-Vessel Systems

In-vessel systems are designed to promote rapid digestion rates by careful monitoring and control of the composting process. Although these systems can produce an end-product more quickly, they are more complex and costly to build, operate, and maintain. Designs for in-vessel systems vary widely; however, they commonly use a system of fixed augers or agitated beds to promote mixing. Moisture and temperature levels must be closely monitored. Most systems feature forced aeration, as well as vessels or bays that allow new material to be introduced at one end as more mature material exits from the other.

The more technologically advanced in-vessel systems are more costly to build and operate and require more operator attention. Those systems are typically applied to more complex mixtures such as sludge MMSW; they are almost never used to compost yard waste alone.

Actual operations

General

The interest in composting continues to grow throughout the United States, despite the difficulties encountered by some facilities who are no longer operating, or are continuing to operate, but providing only a marginal financial return. Although these early projects have been unsuccessful for some, they have provided insight for others who have an interest in developing composting facilities in the future. However, mistakes continue to be made by communities who either ignore, or are not aware of basic composting principles.

Following are case histories of actual operating facilities. Each facility experienced significant operating problems. One facility, located in Sevierville, Tennessee, has overcome some of their difficulties. The other three facilities (Riedel, Oregon Compost Facility; Fembroke Pines, Florida facility; and the Agripost facility) are no longer operating due to their inability to provide the basic operating parameters needed to ensure an efficient operation.

Case Studies

Sevierville, Tennessee Co-Composting Facility. The Sevier County Solid Waste Composting Facility was designed and built, and is operated by Bedminster Bioconversion Corporation. The facility began operations in September 1992.

The facility processes a 2 to 1 ratio of unseparated MSW (150 tons per day (TPD)) and dewatered, aerated biosolids (75 TPD), along with grease trap waste. The quantity of MSW received at the facility varies with the influx of tourists, at times exceeding design capacity by 30 percent. The operators accommodate the excess tonnage by maximizing the material in each stage of the composting process.

Ram pits feed material into one of three rotating drum digesters. The drums are separated into three chambers where each chamber is separated by a bulkhead with a square door. The digesters are slightly inclined, and material is transferred from one drum to the other by gravity. The continuous tumbling of material inside the drums reduces particle size and helps provide mixing of oxygen, moisture, and nutrients required for microbial growth.

Material exiting the drums drops onto a conveyor, which feeds the material into the primary trommel screen. The screen separates the non-organics from the organics. Ferrous metal is extracted for recycling from the non-organic fraction and the residue is landfilled. The organics are composted in aerated bays, where the demand for air is governed by cooling requirements. Large, plastic hoods collect air over each bay, which is scrubbed through biofilters. The material in the bays is moved by bucket loaders about twice a week.

Processed material is typically cured for an additional 25 to 30 days.

Three biofilter areas scrub odorous compounds from process air leaving the digester, holding tank and curing building. In 1992, the facility began accepting material before it was fully enclosed, and the present biofilter system had not been installed. As a result, the facility received numerous odor complaints. The last biofilter installation was completed in August 1993, and few odor complaints were received until the peak tourist season the following spring.

Finished compost from the Sevier Solid Waste Composting Facility is utilized in a number of beneficial uses. Much of the material is given away in an effort to develop markets for the material. A small quantity of the material is taken by SSWI for use as final landfill cover. Other uses include agricultural research, landscaping, athletic fields, and parks. The price for the finished compost is at approximately $10 per ton.

Riedel Oregon Compost Facility. This facility was owned by Riedel Oregon Compost Company, Inc. (the Company). The Company offered a residential mixed waste processing and disposal contract to the Portland Metro Service District for 600 TPD, six days per week. Major elements of the facility included a tipping floor with twin apron feed conveyors, picking stations for manual recovery of recyclables and undesirables, twin DANO drums for maceration and mixing, twin covered aeration slabs, a final screen to separate contaminants and produce finished compost, and scales, conveyors, laboratory, truck wash, offices, and other support facilities.

The facility experienced problems from the time it started operating in April 1991. Limited front-end processing equipment was a key problem at the facility. The conveyor from the tip floor to the picking lines transported the waste in trickles and surges. This meant that a large portion of the waste passed by the picking stations in clumps, hampering recovery of recyclables and hazardous materials. High employee turnover and bag-breaking problems also contributed to material recovery rates that were well below the guaranteed material recovery rate of 5 percent.

The facility also experienced significant biological process control problems. Moisture content was often too low for composting to progress at a reasonable pace, clumping of material on the aeration slabs led to channeled air flow, and conditions became too hot and anaerobic in maturation piles. Together these conditions resulted in an immature compost product and additional screening problems. Odor was also a significant problem. Major odor sources were: the tipping floor, aeration bays, maturing material, and the screening operation. Other conditions that may have contributed to the facility's failure were inadequate housekeeping, resulting in pest problems, and liquids collecting in sumps and floor drains, which became septic.

Another contributing factor to the shut-down of this facility was the fact that this was Riedel's first composting project. Riedel facility staff seemed to believe that composting was a 'simple' and 'natural' process. Furthermore, Reidel was apparently not concerned over the lack of certain contingencies built into the final design of the facility. One result of this attitude was that the people running the plant had little or no compost experience; no compost 'brewmaster' was permanently on-site to draw on past experience for day-to-day operations and to develop operation practices for the new facility. The operators had no apparent appreciation for the plant as a 'manufacturing process' instead of a waste disposal operation.

The facility ceased operations in January 1992 and was purchased by the Credit Suisse bank in February 1992.

Pembroke Pines, Florida Facility. The Pembroke Pines facility, a MMSW materials recovery and composting facility, started operations in September 1991. This facility incorporated extensive front-end material recovery technology as well as process control measures such as aeration, moisture control, agitation, odor scrubbers, size reduction/screening and density separators, and oxygen monitoring equipment. The Pembroke Pines facility was operated by Reuter of Florida (Reuter) and utilized Buhler equipment. The facility processed 600 TPD of MSW and produced 200 TPD of compost.

The total cost of the project including financing was nearly $50 million. Tipping fees were between $50 to $60 per ton. The facility used a front-end processing system that removed non-processibles from the tipping floor and then fed the material through a trommel to separate three different sized fractions (smaller than 2 inches, between 2 inches to 6 inches, greater than 6 inches). The two larger fractions were hand picked to recover recyclable material and the unders were routed to mixing drums. Material not recycled at the hand picking station was shredded and then mixed with the unders from the primary trommel.

The organic material was composted using aerated windrows for six weeks. Although the composting was accomplished under a roof, no side walls were used on the composting building.

The final compost refining system used a hammermill as the first step followed by screens and density separators. All final processing equipment was located within a building to control dust and blowing material separated from the finished compost. The final compost product produced at the Pembroke Pines facility was guaranteed by Buhler to meet Florida requirements for use by commercial, agricultural, institutional or governmental operations.

Operation of the Pembroke Pines facility continued for 14 months. During this time, approximately 55,000 tons of finished compost was produced, most of which was land applied for agricultural use.

Due to the vicinity of residential homes to the facility, fugitive air emissions led to numerous odor complaints. In addition to the siting problem, the following critical problems were also experienced at the facility: undersized biofilters; an inadequate aeration system; inappropriately sized final compost screening and processing equipment; and an undersized compost hangar.

The compost building hangar was undersized by 15 percent causing overstacking of the compost rows and/or reducing the time for the compost material to decompose. Thus as the compost proceeded to final screening and processing, the material was not fully mature and could not be processed properly. The facility is currently not operating.

Agripost Facility. The Agripost facility was located in Dade County, Florida. The 800 TPD facility began operations in September 1989. The process included a tipping floor where large articles were removed for recycling or landfilling, a shredder, a covered windrow composting area using composting turning machines for aeration, and final screening equipment. The basic operational design concept for the facility was 'shred and compost.'

Problems encountered at the Agripost facility included several episodes of odors detected off-site, a severe bottleneck in its product finishing equipment, poor siting, difficulty with financing, and poor public relations. Odor problems at the Agripost facility were compounded by weather conditions including wind shifts and thermal inversions. There was also controversy over a new cell being prepared at the county's landfill next to the Agripost plant.

The plant's material processing problems were related to a bottleneck in the fine shredding and screening machinery. This bottleneck meant that while Agripost was able to receive, process and compost the incoming waste, it could finish less than a quarter of the incoming waste.

Although the facility was located adjacent to a landfill, it was also across the street from an elementary school and surrounded by a middle- to low-income, community. During the public hearings regarding the siting, concerns were raised by the neighbors about odors, noise, truck traffic, and property values. However, the site was approved for the facility.

The project's financing was based on the assumption that the process required minimal equipment and would produce virtually no residues. Unexpected cost for operations, regulations, public relations, and product marketing quickly depleted contingency funds, and refinancing became impossible due to the uncertainties associated with the project. With no ability to finance needed improvements, the facility was forced to cease operations.

Composting research

As was mentioned in the aforementioned case studies, there remains a lot to be learned about how certain mechanical devices help promote the decomposition process and how they affect certain environmental factors, such as odor levels, during the composting process. Much of the research currently being conducted is by developers trying to perfect their specific mechanical devices or systems, which will allow them to offer systems economically, while still providing the necessary environmental controls. Typically, the information these groups acquire is not available to the public unless it will help promote their products.

In an attempt to provide communities, regulators, and developers with specific guidance to help ensure successful composting projects, several initiatives have been developed through cooperative efforts with the United States Environmental Protection Agency, the Compost Council, and the National Audubon Society. Although many of these initiatives would not be considered hard research by most scientists, significant effort has been expended to determine the critical aspects of composting that will help ensure successful composting operations.

The Composting Council contracted R. W. Beck to produce the *Compost Facility Operating Guide* (the CFOG) to assist compost facility operators with:

- Developing standard operating procedures for facilities producing compost;
- Controlling the composting process in order to protect public health, safety, and the environment;
- Producing marketable compost; and
- Training and education.

The CFOG focuses on the biological needs of the microorganisms and the practical steps that facility operators can take to satisfy those biological needs. It also defines steps to prevent chemical, biological, and physical pollution of the finished product.

The CFOG is intended to have universal application to all technologies commonly used in treating organic feedstocks derived from any residential, commercial, and institutional source. It presents discussions of principles as they apply to the various steps involved in the recovery and preparation of feedstocks, composting through the high rate and stabilization, screening and refining compost, and finally, curing, storing, and packaging. It also presents principles for operating associated systems, such as odor and dust management.

The CFOG is intended for use primarily by:

- *Facility Operators and Technicians* – The operating principles are designed to assist operators and technicians as they conduct day-to-day operations at their facilities.
- *Public Planners and Decision-Makers* – As public planners and decision-makers gain a better understanding of what is required for successful composting, they will be better equipped to plan and implement compost facility projects.
- *Consultants and Designers* – The operating principles found in the CFOG should be considered when designing a facility, and it can be used to form the basis for facility-specific operating manuals.
- *Regulators and Enforcement Agencies* – The CFOG is an excellent source for

regulators and enforcement agencies who are interested in understanding basic principles of compost facility operations.

Widespread use of the CFOG within the industry is intended to promote a common understanding of composting and lend credibility and appropriate consistency to operating facilities. The CFOG is designed to be updated as innovative technologies are developed.

To date, all 200 copies of the first edition have been sold, and a second edition of the CFOG is under way.

Additionally, the Composting Council's Standards Committee has been developing product standard guidelines, which can be used throughout the industry as a basis for product quality. They are currently working with experts from academia, regulatory agencies, and the composting industry to develop stability and maturity standards.

The Water Environment Federation has also recognized the need for more operating guidance for biosolids composting. A member group is currently finalizing a publication focusing on technical issues associated with composting municipal wastewater biosolids, which also includes co-composting with MSW.

Composting Promotion Alliances

Work in the public acceptance and educational arena is being championed by a few alliances. The first, Compost...for Earth's Sake (CFES), a partnership made up of the National Audubon Society and the grocery industry (made up of groups like the Grocery Manufacturers of America, the Food Marketing Institute, and individual companies like Procter & Gamble, Quaker Oats, Nestle's, Kraft, Scott Paper, Krogers, and others) is promoting the collection of compostable organics in homes and grocery stores.

Another alliance is Food for the Earth (FFE). This partnership of the Composting Council, the National Audubon Society, and the food service industry (groups like McDonalds, Cargill, James River, and International Paper) is working with the National Restaurant Association to promote composting in the food service industry.

Both partnerships have or will implement a number of pilot programs throughout the United States. The long-term vision for both partnerships is to make source-separated composting a reality without interfering with recycling, and even improving the economics of recycling. For example, a Santa Barbara, California collection pilot program demonstrated that wet/dry bag combinations can reduce the system costs in a community doing curbside recycling. The partnerships intend to work with environmentalists, industries, and the government to accomplish this vision.

In addition, a third partnership, the Compost Action Alliance, was recently formed. It comprises the some governing bodies as the other two partnerships, plus representatives from compost businesses and the U.S. Conference of Mayors, among others. The Compost Action Alliance intends to fill in the gaps that are hindering composting growth.

A composting demonstration project is planned in Glocester City, New Jersey to research the collection and composting of source-separated organic waste from residential and commercial sources. The types of waste that will be composted include source-separated yard waste, food wastes, wet and soiled paper, diapers and sanitary products, pet waste, and dry paper packages that are not being recycled because of weak or nonexistent markets. The organic waste will be composted by Compost America, using an enclosed system, which will demonstrate a unique technology and construction methods. The developers hope to demonstrate how their system can provide lower cost systems that will gain wide acceptance in the composting industry.

Conclusion

Operators of composting facilities, regulators, and others in the composting arena have created a base of knowledge that has dramatically improved the odds for project success. Facility managers are responding more quickly and effectively when challenges arise, and customer satisfaction with the finished compost is strong.

Composting continues to grow as a technology for processing the organic fraction of a solid waste stream, a trend driven by disposal bans at landfills and a scarcity in virgin organic materials like peat, forest products, and to a certain extent, topsoil. Cities and counties continue to see higher diversion rates, as they strive to attain their waste reduction goals, produce a beneficial end-product, and minimize environmental pollution from organic solid waste. Emerging statistics are very promising, showing an increase in biosolids composting facilities and in farms using yard trimmings. Other strong indicators of composting growth can be seen in the educational area – such as hundreds of school programs placing compost bins in classrooms, adjoining the recycling containers. Further, composting continues to be a useful application for private industry seeking to manage selected organic waste streams in a more cost-effective and environmentally beneficial manner.

References

'Biosolids Composting Strengthens its Base, 1994 *BioCycle* Biosolids Survey,' *BioCycle*, pp. 48 through 57.
'Year-end Review of Recycling,' *BioCycle*, December 1994.
Goldstein, Jerome, *BioCycle*, December 1994.

Reconversion of Traditional Composting Plants for a Policy of Quality

Dr. BERNHARD RANINGER – University Lecturer,
R.A.B. Waste Management Consulting, 5322
Hof/Salzburg – Austria

Abstract

In Austria the separate collection and treatment of municipal biowaste was introduced about 1989, and is now ordered by law since January 1995. From a population of about 7,8 Million, living in 3,1 million households, the expected total quantity of biowaste, coming from municipalities, parks and gardens, and the commercial sector is about 2 million tons per year. This quantity should be collected by the year 2000. At this time about 42 % of the organic waste will be managed decentralised and composted by private people and farmers. About 58 % of the biowaste is treated in composting plants with a capacity of more than 50t/a up to 100.000t/a. On the one hand in Austria a lot of new biowaste treatment plants have to be built, on the other hand existing co-composting plants have to be reconstructed and enlarged to convert biowaste with aerobic or anaerobic technologies into high a quality compost product.

Introduction

Composting and the methods of biological waste treatment have a tradition over more than 20 years in Austria. In the eighties about 25 % of municipal solid waste (MSW) and sewage sludge was treated in co-composting plants. At least there have been 19 plants with an overall capacity of approximately 600.000 t/a.. Depending on a poor endproduct quality, containing impurities, ballast matters and heavy metals, it was the next step to force, the separate collection of the biogenic fraction out of the commercial and municipal garbage.

To date the implementation of the collection policy has progressed well. Almost 78 % of the households being connected to the separate collection system, or practising composting at their compound themselves. In the meantime recycling of biowaste and its conversion into a useful product, which can be applied without ecological risks to the soil, is an essential component of integrated waste management concepts, not only in Austria. Also in other European countries like Germany, Switzerland, Sweden and the Netherlands composting of GFT is usual. to prevent biowaste being landfilled or incinerated

The separate collection and treatment of biowaste in Austria is regulated by the 68. amendment to the waste management law, the so called „Biological Waste Decree' which came into effect on 1.1.1995. This decree orders to utilise waste with an high contend of biodegradable matter, such as natural organic waste from gardens and parks (yard waste), solid vegetable refuse from households, commercial and industrial processing and distribution of agricultural products, as well as uncoated paper which was in contact with food stuff or used for the collection of biowaste. All this wastes have to be particularly suitable for an aerob (composting) and/or an anaerob (fermentation) treatment. Commercial kitchen and food waste is only then to be collected, as far as a suitable aerobic or anaerobic plant is available. Another field to operate the methods of biological waste treatment is the pre-treatment of remaining waste after separation of the recyclable fractions, before incineration and landfilling. Using modern process controlled indoor composting systems, it will be possible to prevent biodegradable organic substances from waste disposal. Furthermore it's to minimise the capacity of methane gas production out of landfills, to reduce organic harmful substances and the quantity of leachate emissions, to ensure the hygienisation and to produce a plant tolerant humus product which can be used as a landfill construction material specially for the final cover and recultivation. If incineration of the remaining waste, which contains a high quantity of not collected biogenic waste is taken in consideration, the biological processing will be changed to grant the reduction of the water content to less than 18 % and the decrease of the total quantity before.

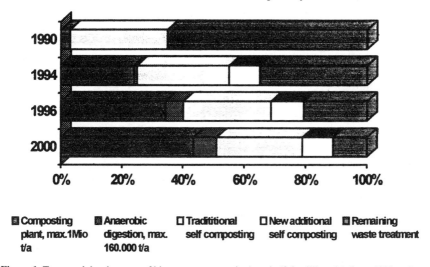

Figure 1 Types and development of biowaste treatment in Austria (2,2 million t/a), from 1990 to the year 2000

Reconversion of Co-composting plants

In Austria 12 of former 19 co-composting plants have been reconversed or supplemented to pure biowaste composting plants for a policy of quality. Different types of composting systems are implemented depending on the local conditions, like:

- the climatic situation (rainfalls, wind)
- distance to the neighbourhood (odour, noise and spores emissions control)
- requirements on the hygienisation (removal of human- and phytopathogens)
- constants of compost quality (maturation degree, plant tolerance)
- space availability
- type of biowaste (water content, density, structure stability, availability of structure material)
- quantity of biowaste
- possibilities of energy recovery (electricity, biogas, heat)

The following types of composting and digestion systems are implemented (Table 1)

Table 1 Types of composting and digestion systems used at reconverted co-composting plants in Austria

Type of treatment	Number of plants	Tot. Biowaste-capacity (t/a).
Open pile composting	3	13.500
Hangar composting (not closed, without automatically turned piles, ROTTEFILTER)	2	14.000
Hangar composting (automatically turned table piles AE&E-KOCH, WENDELIN, DYNACOMP)	2	13.000
Tunnelcomposting (BAS)	5	58.000
Anaerobic fermentation (DRANCO)	(1*)	18.000

* in connection with a Tunnelcomposting plant

Finally there will be in Austria about 50 biowaste composting plants with an specific annual capacity between 2.000 tons and 100.000 tons and an total processing capacity of 380.000 t/a. In all the expected number of composting plants (> 50 t/a) will be more than 350 with an over all capacity of 680.000 t/a.

Selected examples of reconverted composting plants in Austria

Allerheiligen

At the Allerheiligen co-composting plant a daily amount of 80 t of solid waste and 25 t of sewage sludge was treated from 1978 to 1992 by a VA – compost technology, running a forced aerated table pile at a maturation platform and a post-maturation, turned by a ENGELER windrow machine. The treatment of separate collected biowaste was started 1992 with a „ THÖNI – Silo composting plant'. This processing was not working successfully because of an insufficient oxygen supply of the rottingmaterial. The neighbourhood was complaining about massive odour emissions, threating the company to block the driveway. Finally this type of plantoperation has been stopped after a two years testrun period in 1994.

After this the plant design has been changed. In consideration of the extreme sensitive situation at Allerheiligen site a tunnelcomposting was recommended by the consultant, but not only to treat 4.000 t/a of biowaste, also the remaining waste and the sewage sludge with a total quantity of 20.000 t/a will be automatically processed in 9 composttunnels after an appropriate pre-treatment, in the future. The special requirements at this site are the minimisation of odour emissions caused by the process or by the endproducts, which are on one hand recycled (high quality compost according to the ÖNORM quality standard I A) and which are on the other hand going to be dumped at the same site after the biological stabilisation has happened. The operation times are 2 to 3 weeks in the tunnelcomposting, 5 weeks at the forced aerated maturation platform and about 12 to 16 weeks storage and stabilisation in the compost hangar.

Liezen

The co-composting plant from 1979 of *Abfallwirtschaftsverband Liezen* was expanded 1992 by a biowaste composting plant with an annual capacity of 3.000 t/a. After all, checking different available biowaste composting technologies by the consultants, the a BAS tunnelcomposting plant was selected to be built.

Before implementation of the tunnelcomposting system, a test run with Austrian biowaste in the Netherlands was done at the BAS pilotplant for biowaste treatment in Tholen. In cooperation with the Technical University of Vienna, Prof. P. Lechner, the benefits of an overall closed and process monitored composting system have been demonstrated.

After two years operation experience in Liezen the tunnelcomposting system has been completely adapted to the biowaste substrate. The major benefits of the tunnel composting are

- a batchwise operation of the waste materials with a high flexibility to the different types of waste (seasonal differences, yard waste, sewage sludge, commercial biowaste, . . .)
- an optimal process control ensures a fast microbiological degradation of bidegradable organic matter under regulated and optimised ecological conditions
- the essential process parameters are controlled and monitored by computer, like the oxygen supply (> 12 % O_2), the water contend (45 – 60 %), the energy output and the process temperature, depending on the stage of process like heating up, hygienisation (65 °C), maturation (56 °C), drying, cooling e.t.a.
- minimal odour emissions, the off air contains a relative low level of smelling units

Table 2 Co-Composting Plants in Austria (Capacity > 2000 t/a) which have been reconverted or supplemented to Biowaste-Composting plants

Location	Operator	Co-Composting		Biowaste-composting	
		1990	Technology	1995/96	Technology
		tons/a		tons/a	
AICH-ASSACH	municipality	6.000	windrow, forced aeration (Rottefilter) of presepared MSW	6.000	windrow, forced aeration (Rottefilter) of preseparated MSW (on trial same type of collection)
ALLERHEILIGEN	municipality	20.000	mixing drum, forced aerated tablepile, aerated platform and postmaturation (VA) system for Co-composting)	4.000 + 16.000 MSW	mixing unit, tunnelcomposting (BAS) and forced aerated postmaturation (same
BREITENAU / NEUNKIRCHEN	private	8.000	windrow, forced aeration (Rottefilter) of wet-waste fraction (Grüne Tonne)	8.000	windrow, forced aeration (Rottefilter) of wet-waste (Grüne Tonne)
GRAZ / LANNACH	private	4.000	municipal biowaste windrow forced aeration (Rottefilter)	15.000	tunnelcomposting, postmaturation and aerated storage in the final stage of maturation (planned)
LIEZEN	municipality	15.000	mixing drum, windrowmachine for aeration of triangle piles at postmaturation (VA)	3.000	tunnelcomposting (BAS), forced aerated postmaturation and trianglepiles
LUSTENAU	private	10.000	forced aerated indoor windrow (WENDELIN), triangle pile pos tmaturation, turning by frontendloader	10.000	planned anaerobic digestion (COMPOGAS) and aerobic stabilisation by indoor tunnelcomposting
OBERPULLEND ORF / GROßHÖFLEIN	municipality	50.000	dynamic prefermentation (DANO) forced aerated indoor table pile, windrow system (WENDELIN), postmaturation turned by frontendloader	6.000	triangle pilecomposting, operated by farmers (there is no biowaste composted at the existing co-composting plant)
PÖCHLARN	municipality	12.000	dynamic fermentation (DANO) outdoor piles	4.500	outdoor triangle windrows
ROPPEN	municipality	22.000	Fermentation drum, forced aerated indoor pile (DYNACOMP)	7.000	forced aerated indoor windrow (DYNACOMP)
SIGGERWIESEN	municipality	160.000	Fermentation drum, forced aerated table pile, outdoor postmaturation or degased landfill	18.000 +8.000	anaerobic digestion (DRANCO), yard t wasteunnel-composting (BAS), roofed postmaturation, outdoor windrows for yard waste with tablepile turn-ing machine
TRAISKIRCHEN	municipality	17.000	dynamic fermentation (DANO), outdoor piles	3.000	windrows outdoor triangle piles for yard waste
ZELL AM SEE	municipality	20.000	Fermentation drum (DANO), forced aerated table pile (VA), outdoor pile maturation	6.000	forced aerated indoor windrow, turned stack composting (AE&E-KOCH) for co-compost and biowaste compost

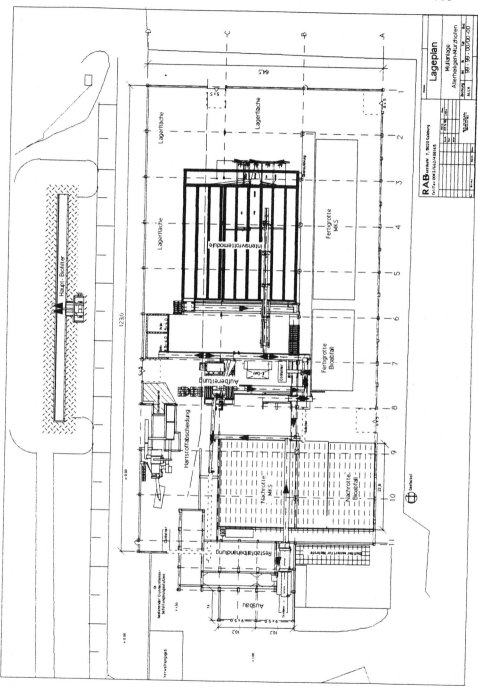

Figure 2 Layout of Allerheiligen Plant (20.000 t/a) after improvement of biological waste treatment via tunnelcomposting

- no waste water emissions, depending on the consumption on percolate water to add process water to the compost during the exothermic processphase. The process water is evaporated and discharged via the biofilter.
- short residence time to reduce the biodegradable organic matter to 50 % or 60 % DS in mass.
- low land requirement of about 0,1 to 0,2 m2/t input, concerning a residence time of 14 days
- acceptable working condition, as far as the process is automatically operated, especially during the filling and emptying procedure.
- modular design, easy to extend the capacity.
- even the central process monitoring stops, each unit is selfmanaged by a satellite processor
- a warranted hygienisation including the reused process water
- small demand of structure material or other additives
- no mechanical parts of the equipment are exposed to the corrosive atmosphere inside the operating tunnel
- an intensive air and oxygen supply ensured by an intensive air circulation in combination with the required fresh air supply, while a relatively small air volume is exhausted to the air cleaning system, which is normally a biofilter, transporting corbondioxid, water steam and energy.

Oberpullendorf

In Oberpullendorf the second largest co-composting plant of Austria is operating. permanent problems, concerning odour emissions with the DANO – WENDELIN technology have been one of the main reasons not to enlarge this plant building a biowasteline. Therefore an alternative concept was created for this mostly rural area. The biowaste of the Burgenland is treated at different decentralised composting plants, which are operated by farmers via traditional pile composting.

Roppen

Originally the THYSSEN DYNACOMP composting system was implemented for the processing of MSW. The biowaste composting is managed by the same technology. The raw compost tablepile is aerated by underpressure and the off air cleaned via biofilter. The homogenisation of the compost material is done by a mixing screw, moving through the table pile, fitted on the pile set up system.

The dynamic prefermentation in the DANO drum is not used for the biowaste processing because of the deficit of structure and pours volume of the raw compost.

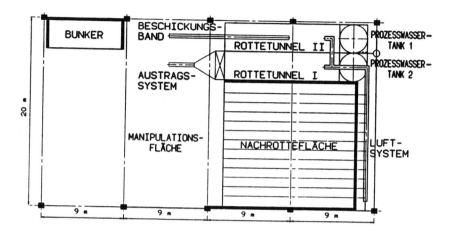

Figure 3 Basic layout of biowaste composting plant LIEZEN

Siggerwiesen

For Salzburg, a region of about 280.000 inhabitants, a biowaste plant capacity of about 15.000 to 20.000 t/a is needed. In this area two separate collection systems, one for biowaste (Biotonne) and another one to collect yard waste are existing. The mainly 120 l biowaste bins are to dispose the biowaste from households, containing mainly kitchen waste with a high water contend. Depending on former odour emissions, caused by the old co-composting plant and the Siggerwiesen landfill, before a degassing system was installed, a biowaste technology without the hazard of any production of smell has to be chosen. Depending on an anaerobic systems, which has sufficient references, available on the marked, the DRANCO fermentation system in connection with a controlled aerobic postcomposting, the BAS tunnelsystem, was favoured.

The DRANCO process is a one step, thermophilic anaerobic digestion process. After a mechanical pre-treatment the fraction < 50 mm is directly pumped into the fermentor. Depending on the type of biowaste (high moisture contend, minimal lignin contend) the total solids in the system are adjusted between 35 to 15 %. Below about 18 % DS a liquid anaerobic system should be chosen. In Salzburg the specific biogas production is about 130 – 140 Nm3 /t input material. The gas is used to run a 1,5 MW gasgenerator.

The postmaturation and hygienisation of the output of the anaerobic digester, is done after dewatering and after adding 10 – 20 % of yard waste screen overflow

(>12 mm < 50 mm) as a bulky agent in the BAS tunnelplant. The thermophilic phase is easily reached during the last period of aerobic stabilisation after the 10 th day of processing. After this time an additional maturation and material storage for the nitrification of ammonia is needed.

In a separate line the yard waste composting is done via a outdoor tablepile, regularly turned by a WILLIBALD tablepile windrow machine.

Zell am See

In Zell am See the old co-composting system from 1978 (DANO drum and forced aerated table pile) is totally be renewed to a forced aerated AE&E-KOCH System with automatic stack turning and moistening in a closed composting hangar. The pile is turned and moved by an automatic turning unit that travels across the trapezoid or table pile as well in line with the movement of the compost through the composting hall. At the end of the hall the compost is reclaimed with the turning unit itself. Depending on the process requirements, that are based on the composition of the compost, the required degree of maturity of the final compost is reached after approximately 12 weeks of biodegradation. The fields of compost, each pile formed as the turner moves across the trapezoid, are generally moved on a weekly basis corresponding to the rates of compost degradation and new material introduced to the hall as finished material is extracted. The exhaust air is treated in a biofilter.

The major advantages of the free programmable stack turning method are

- The composting hall is covered with compost and this provides the maximum amount of compost production from the smallest possible building footprint
- the turning frequency is freely selectable, for homogenisation and aeration of compost
- the air required for the composting hall is recirculated and finally washed, cooled and exhausted through the biofilter
- a heat exchanger provides the ability to regulate the temperature of the incoming or the exhaust air

Compost quality

The goal of all this collection and treatment strategies are, besides of the avoidance of landfill capacity, to improve the compost quality and to ensure a compost market on a long-term basis. Ballast matters, heavy metals, organic pollutants and a lack in plant tolerance have been the reasons that co-compost and sometimes also sludge – compost could not be successfully applied for example in the plant production, gardening , landscaping and agriculture. A clean inputmaterial and a high developed compost technology will provide that the objectives of organic waste recycling can be reached. In Table 3 the actual heavy metal restrictions are shown for co-compost and biowaste compost in Austria and other European countries.

Table 3 Limiting values for heavy metal in different types of Compost of Austria, Germany and Netherlands

	Limiting value SOIL	Limiting value COMPOST ÖNORM S 2200 and S 2202				Limiting value COMPOST Nl & Germany	
	ÖN L1075	Class I Biowaste ecompost	Class II Biowaste ecompost	Class III Co-compost	Netherlands	GERM RAL/ LAGA Kl. I	GERM LAGA M 10 Kl. II
Cr	100	70	70	150	50	100	200
Ni	60	42	60	100	20	50	100
Cu	100	70	100	400	60	100	200
Zn	300	210	400	1000	200	400	750
Cd	1	0,7	1	4	1	1,5	2,5
Hg	1	0,7	1	4	0,3	1	2
Pb	100	70	150	500	100	150	250

Conclusion

Most of the former co-composting plants in Austria have improved their facilities with an equipment appropriate for biowastematerials. Composting and especially biowaste composting is a very complex microbial oxidation process, influenced by a few parameters like moisture contend, process temperatures, oxygen supply and CO_2 removal, the concentrations and compositions of nutrients, operated by a huge number of different types of mikroorganism in a very sensitive ecological environment. The selection of an optimal biowaste treatment technology is a very important decision. Whilst for biowaste composting dynamic systems more and more loose their importance, static, frequently turning over and homogenising indoor systems, batchwise operated and process controlled are most successful. Specially for wet-biowaste mainly one step liquid anaerobic fermentation technologies are recommended for implementation. Furthermore energy recovery from biogas is a positive effect on the world wide corbondioxid economy.

References

BGBL. NR. 442 (1992), Verordnung des Bundesministerium für Umwelt Jugend und Familie über die getrennte Sammlung biogener Abfälle

BINNER, E., B. RANINGER (1995), Composting of biowaste in Austria, Biocycle

ÖNORM S2022 (1889), Quality requirements for waste-compost

ÖNORM S2201 (1993), Quality requirements for biowaste-compost

RANINGER, B. (1995) Sammlung und Verwertung kommunaler biogener Abfälle in Österreich, Studie im Auftrag des Bundesministeriums für Umwelt.

RANINGER, B.(1995) Composting and compost quality in Austria, Biological Waste Management, Bochum

Basic Processing Technologies and Composting Plant Design in Italy.

F. CONTI**, G. URBINI**, G. ZORZI*.

Summary

This study offers an up-to-date picture of the basic processing technologies and composting plant design undertaken in Italy for the treatment of unselected MSW. Present trends in the field of the high quality compost production from pre-selected organic refuse are here analyzed in order to highlight the different components and sections of the composting plants. A special care was used in analyzing technologies proposed by the different firms for the management of the first phase of the biological process. This work integrates and concludes the state of the art and perspectives about the composting theme developed in a specific report presented in this symposium.

Introduction

Composting in Italy follows two different procedures:

- compost production from unselected wastes, developed in 42 plants located all over the nation;
- recovery and utilizing of source pre-selected organic substances, a procedure used in 33 plants located in northern and central regions.

The mixed composition of unselected refuse makes it necessary to provide a series of devices capable to ensure high efficiency and effectiveness in the operation of selection and refining upstream and downstream the biological process. The biological transformation of organic substances is effected by means of different technological systems, but process may not be satisfactory and, on the other hand, recent projects cannot fill this gap. The composting of source pre-selected waste allows for a simplified technology. This last system has been widely used, recently, but sometimes it is not well managed. The analysis of the solutions applied in both composting procedures, is not complete yet, being only a first contribution to a critical knowledge of technological situation of Italian plants.

*Istituto Agrario di S.Michele all'Adige (TN), Italy.
**Hydraulic and Environmental Engineering Dept., University of Pavia, Italy

Analysis methodology

The scheduled plants were analyzed according a logical path of all the operations performed in sequence to treat refuse for composting. All operations that are not significant in the process or that are common to all solutions (storage, magnetic separation, rejected material treatment) were not taken into account. Following this path, the plants were divided into three fundamental operational blocks [Vastola and Pizzo, 1987]; inside each block the sections and the related electro-mechanical devices are highlighted.

The first block, defined as pretreatment, encompasses unit operations designed for the preparation of the organic substances for an efficient and controlled development of biological reactions.

The second block specifies the process solutions proposed for accelerated biostabilization and curing.

The third block includes a unit operation for the final refining of raw compost.

Composting plants for raw municipal solid wastes

In Table 1 technical characteristics of the scheduled plants are summarized. The planning and technological options of the different systems used in each block or section are defined below.

The Pretreatment block.

Receiving and feeding section.
In nearly all the scheduled plants the receiving and feeding sections are uniform made of three days capacity storage pits, by overhead cranes with grapple bucket and feeders. In three plants wastes are discharged on a tipping floor and daily removed.

Shredding section.
It is well known that this section is fundamental in composting plants, in view of both reliability and management, and of final product quality.
In 15 out of the total scheduled plants the shredding equipment are mills, but of different types:

- in 6 plants they are high speed rotating horizontal hammermills which pulverize the inert material (glass and ceramic) whose separation downstream becomes very difficult;
- in 3 plants they are impact breaker mills, with comb and counter comb or with tough steel bars;
- in 2 plants they are mills with two horizontal cutting shaft with blades and counter blades;

— in 4 plants they are low speed roller shredders, which simplify subsequent operation of separation.

In 9 plants the operation of size reduction is substituted or preceded by bag breaking. The operation is done by a rotary drum with sharp elements inserted inside. In 8 plants bag breaking is done on the upper part of the plate feeder.

The DANO type rotary drum operates also as bag breaker and size reducer of the organics by means of friction with hard materials in the refuse and by impacts with steel pikes inside the drum. This type of drums is proposed as biostabilizator of the organic fraction (see part 3.2.1), but since its detention time is too short (10 hours instead of 3–4 days), it can only by used to reduce the size of wastes.

The preselection with this rotary drum is more effective than the one realized with only mechanical systems, because the first biological transformations contribute to optimize efficiency in subsequent separation. On the other hand the costs of this kind of devices are considerable.

Organic fraction separation

Pre-treated organic waste have presumable a smaller size than raw refuse. The first and coarser separation of the organic fraction that feed the biological process, is mainly effected by means of rotary screens with hole diameters of 60–100 mm; only 3 of the scheduled cases use vibrating screens.

The organics separation in 9 plants is completed with density separation following the screening; the devices used are either ballistic or density separators. The former process is based on the different paths the particles follow after bouncing on a rotating surface; the latter process is based on the different deviations undergone by particles carried in an air stream.

In presence of fresh organic substances we must take into account that the process of density separation has a low efficiency, because the difference in specific weight between the undesirable materials and the organic fraction of the refuse is very small.

Biological process block

The section and the related technologies directly involved in the biological transformation of the organic substances are considered in the second block.

The whole process is usually divided into two distinct phases:

— the accelerated bio-oxidation at the beginning of the process, carried out in reactors;
— the curing at the end of the treatment, carried out in windrows.

The two steps are in general physically separated as above, but in some cases they can both take place either in reactor (4 plants) or in windrows (4 plants).

Accelerated bio-stabilization

The first phase of biological transformation is characterized by high-rate reaction; it is exothermic, it have a high oxygen uptake rate and it generate a potentially high environmental impact which can be lowered in a vessel system. Here the organic mass is mixed and aerated by means of forced ventilation systems. The accelerated bio-stabilization section characterizes the 'composting system'. The solutions used in Italy are presented below.

a) Horizontal rotating drums.

This system is widely used in Italy (in 8 plants) to start the biological processes (with the exception of what stated in capt. 3.1.2), it allows a first hygienisation of the material and facilitates the separation of the organic fraction from the inert components. The rotating drum is manufactured with welded steel section installed at regular intervals. Its length and diameter range between 25 and 3 m, and 42 and 6 m respectively. At the end of the drum, installed at 120 degrees, there are 3 fans. These fans insufflate the air against the flow of the refuse; an exhaust air fan is installed on the front of the drum.

This system was proposed by SLIA, Daneco and Snamprogetti.

b) Vertical reactor with multiple decks.

The structure is cylindrical and enclosed; material is fed at the top of the structure and the product is outfed at the bottom. In the Cuneo plant there is a group of 4 reactors, 24 m total height, 8 m diameter, 3 m height for each module. Each one of the 6 decks has rabbler arms for material movement and forced ventilation. The organic refuse drops from an upper deck to a lover one.

This system was proposed by Peabody.

c) Vertical cylindrical packed bed reactor.

This system is used in 2 plants. The material is fed at the top of the structure by means of a screw conveyor. The refuse passes into a system composed of 3 horizontal rotating disks that distribute the material uniformly on the pile below. The reactor is not mixed. The aeration system is composed of radial manifolds installed at the basis of the reactor, grouped into 8 sectors (angle of 45°) and fed in sequence. The air flows upward, counter-current to the flow of refuse, by means of an exhaust air fan installed at the top of the reactor. The outfed system is a screw conveyor installed on an adjustable cylinder inside the reactor; the discharge of the product is done at the center of the reactor.

The system was proposed by Weiss-Kneer.

d) Fixed basins with automatic turners.

The organic material, in these reactors, is moved and put forward inside rectangular or circular basins; these reactors are horizontal, nearly horizontal or with inclined bed; the organic fraction is moved by means of screws installed on traveling bridges. Continuous aeration is guaranteed by forced ventilation systems. The plans are different in number, volume and geometry of the reactors and in number of the screws, as summarized below:

— single rectangular reactor, horizontal or nearly horizontal, with traveling bridge on which screws are mounted coupled up to a number of 14; the material is loaded along one side of the reactor and is discharged at the opposite side.

The system is present in 4 plants and was proposed by Sorain Cecchini and De Bartolomeis.

— single rectangular reactor with inclined bed (13°) agitated by means of 4 screws attached on a traveling bridge; the mass is transferred perpendicularly to the upper loading side, of the reactor;

The system is used in 2 plants, proposed by Emit.

— single rectangular reactor with automatic turning of the material. This system is realized in one plant, with one basin crossed by a traveling bridge carrying a movable arm with screws and belts that collect and distribute the material; the organic fraction is deposited in one pile 3.5–4 m height, and moved from the cross-front to the back toward the extraction zone.

The system was proposed by Ferrero-Savona.

— multiple reactors side by side; the organic fraction is loaded along the short side and is moved forward by means of 4 screws installed on a traveling bridge. The extraction is done from the opposite side.

The system is used in 4 plants, proposed by Secit and Emit.

— circular shape reactor, with screws installed on a traveling bridge (Fairfield); the material is fed along the external circumference by means of a screw moving with the traveling bridge; the product is extracted from a central pit; the organic fraction is moved forward by means of 13 screws; the diameter of the bin varies from 10 to 17 m, with volumes from 200 to 450 cubic meters; the aeration system is on the bottom of the reactor, the manifold is divided into concentric rings (3 for each sector).

The system was proposed by De Bartolomeis.

e) Aerated dynamic troughs with automatic turning.

This system has two variants depending on the turning device employed:

— in the first case the material is discharged at the back of the device, along the advancing axis,
— in the second case the material is discharged sidewise in an adjacent trough.

accelerated bio-stabilisation area, in both systems, is divided into multiple trough; the length of these troughs depends on the detention time and on the number of the week turnings. The volume of the material to be treated each day determines the number of the troughs. The troughs in the system proposed by Saceccav and used in 5 plants (3 of which treat source-selected refuse) are 3 m wide and 2.13 m high. These troughs have a loading zone (each one loading 19 m^3 of material daily), different aerated zones and a discharge zone. Rails are set on top of the walls where the compost turner moves the material along all the length of the trough.

The device is made of a rotor (1 m diameter) with agitators mounted on the surface; of a conveyor at 45° that moves the material backward for 3.66 m at each passage and makes room for the next day; of a hoist for lowering and raising the rotor and the belt conveyor; and of a transferring dolly for moving automatically the turner from one trough to the other. The aeration system is composed of different fans, each one serving a part of a trough.

In 2 plants the turner is of the Siloda type, and is made of a paddle drum (3.5 m diameter and 4 m width), that while moving loads the material into a screw mounted on its shaft. The screw transfers the material to the adjacent trough. At the end of the cycle the turner returns to the initial position by means of a transfer dolly that moves it to the next trough; the system is aerated by forced ventilation. This technologies was proposed by OTV and Daneco.

f) Aerated cells.
The process develops inside primary cells with forced aeration.
This system is used in only one plant and was proposed by Voest Alpine.

g) Aerated and turned windrows.
This system is used in 2 plants. The turning of the windrows is carried with a turner provided with a thread-guide system that allows its automatic advancing; in both plants, since the material is side-discharged, a side free windrows needed, as well as the movement of all the windrows.
The system was proposed by Buehler and De Bartolomeis.

h) Aerated static windrows.
This system is used in 2 plants; the material is aerated by forced aeration and turned by means of a loading shovel.
The system was proposed by Daneco.

Curing
The biological process is completed in a suitable curing area. In this phase the oxygen demand is reduced, but the need to ensure favorable conditions for a regular transformation of the material and in particular for the organic substances humification is constant. The proposed solutions, in most cases, are very approximate and look more like a storage than like a biological process phase that must be controlled. The proposed solutions are usually limited only to unplanned movements the material with a loading shovel, and are managed with too extended windrows.

The final treatment block

The polluting elements still present in the compost are removed in the main section of this third block; the same equipment seen in the organic fraction separation is employed in this section. The final screening is made by means of a rotary screen with circular (or, in some cases, elliptical) mesh, with a diameter between 4 and 20 mm. In 6 plants this selection is made by means of an elastic vibrating screen (flip-flow type). The gravimetric selection is made in most plants by means of a vibrating inclined table with a fluidizing up flow forced ventilation. The ballistic separation is used in 5 plants (in one case it is used as a pretreatment). A zig-zag separator is used in 2 plants. The secondary shredding of raw compost is used in few cases; this practice is due to the need to have a thin final product.

Composting plants for source selected materials

The recovery of organic refuse by means of separate collection allows for the construction of plants with simplified electro-mechanical devices. However it must be taken into account that some types of refuse, as the catering food residues and the refuse derived from greengrocery markets, can be contaminated with 10–20 % of inerts (glass, plastics and ceramics). This requires the use of more complex structures, particularly when the inflow of refuse is important, as with the plants of Turin and Milan.

The fundamental blocks described in capt. 3 are present also in the simplest plants; for the analysis of technological systems these must be referred to (see table 2).

Conclusions

The examination of technological systems used in the composting plants described in this paper is not complete and can certainly be improved, but however it can give a better knowledge of the solutions employed in Italy both in the recent past and at the present; the aim of this work is to improve the planning philosophy for the future.

In most cases collecting data and information about the treatments is difficult because the planning solutions vary during execution. The number increases if we consider the great delay in plant commissioning.

The general view of the situation highlights a wide availability of systems and equipment, which, however, are not always reliable and efficient.
The modifications that can be proposed for plants treating raw MSW are the following:

- to provide the plant with receiving section for storage of source pre–selected organic fraction and the conditioning with specific equipment before composting;
- to manage the biological process of this pre-selected materials, with the aim of producing a first quality compost by means of existing devices or of up-to-date equipment;
- to remove hammermills shredding;

– to remove the thin fraction of the material very often full of heavy metals;
– to manage the bio-oxidation phase in a more controlled way;
– to equip and manage the curing phase in a suitable manner.

Modern trends, employed by plants treating source pre-selected materials, warrant the highest technological simplification and, at the same time, the production of a highly agronomic valuable compost, environmentally safe.

Bibliography

VASTOLA F., PIZZO G., (1987), 'Esame delle attuali tendenze tecnologiche nel settore degli impianti per la produzione di compost da rifiuti solidi urbani', RS – Rifiuti Solidi, vol.1 n°6, november–december, 528–538, Milan.
ZORZI G., URBINI G., (1992), 'State of the art and trends of composting in Italy and other European Countries', Proceeding of the First Italian-Brazilian Symposium of sanitary and Environmental Engineering, 216–228, Rio de Janeiro, Brazil.

Table 1 The technological solution in raw MSW composting plants.

PLANTS AND TREATMENT CAPACITY	PRETREATMENT Devices	BIOLOGICAL PROCESS Accelerated bio-oxidation	FINAL TREATMENT Curing	Detention time	Devices
Cuneo 150 t/d	Hammermill (330 kW, 1,490 rpm) Rotary screens, mesh 50x80 mm and 8 mm	Cylindrical vertical reactor, enclosed (Peabody), 6 modules, each deck 3 m height, total height 24 m, diameter 8 m	Windrows in covered area	12 days + 20 days	Rotary screen, mesh 50 x 70 mm, densimetric table, final shredding.
Novara 255 t/d	Primary rotary screen with bag breaker, secondary screen, air classification, sludge mixer	Rectangular covered reactor, inclined bed (13°), length 72 m, width 21 m, useful deep 2.5 m, bridge crane with 8 screws	In covered area	30 d+ 40 d	Air separation, cylinders-mill.
Cedrasco (SO) 100 + 50 t/d	Refuse receiving pit, bag breaker, rotary screen, mesh 250 mm.	Two Dano horizontal rotating reactors, length 42 m, diameter 4 m; the first phase is completed in a 1,500 m² area, turning windrows.	In area, not yet realized	4-5 d + 20-25 d	Screening before the curing phase, vibrating table.
Alessandria 100 + 40 t/d	Rotary screen with bag shredding.	Dano horizontal rotating reactor, length 42 m, diameter 4 m; raw screening (100 mm) in the final side.	In 2 sections with 7 windrows in each one; CNTM turning device; forced ventilation with 14 fans	72 hours + 60–70 d	Screens 50 mm and 16 mm before curing phase, final rotary screen with elliptical 6x12 mm holes, vibrating table.
Ceresara (MN) 160 t/d	Hammermill with combs and counter-combs (160 kW), rotary screens 40 and 80 mm mesh, air classification	Static aerated windrows in covered area, 1,200 m² divided in 10 spans; forced ventilation realized with 4 manifold/windrow served by 9 + 9 fans	Covered area, 1,200 m²	30 d + 90 d	Zig-zag separator, vibrating screen, 15 mm mesh.
Pieve di Coriano (MN) 210 t/d	Like above	Like above	Like above	Like above	Hammermill with 12 little hoes, Flip-flow screen, 5–12 mm mesh, vibrating table.
S.Giorgio di Nogaro (UD) 280 t/d	Hammermill with vertical axis (250 HP), rotating screen with 4 sectors with holes from 10 to 35 mm, ballistic separator.	Horizontal bioreactor, 3,000 m², mobile arm with screws installed on a traveling bridge; one pile up to 4 m height, forced aeration by means of 27 fans.	Covered area	28 d	Densimetric table, vibrating elastic screen (flip-flow) with mesh 10 mm.

Plant	Mechanical pre-treatment	Composting (primary)	Composting (secondary)	Time	Refining
Udine 280 t/d	Rotating/shredding screens with mesh 10 and 70 mm, vibrating screens with mesh 40–70 mm, hammermill	Two cylindrical bio-stabilizators then, in an area of 1,200 , 5 turned and aerated troughs with a wheel turning device installed on rails.	Static aerated windrows in an area of 1,950 m²	14 d + 30 d	Flip-flow screen, densimetric separation, compost dryer.
Villa Santina (UD), 90 t/d	Shredding cylinder with blades	Horizontal rotating reactor, DANO-type, 25 m length	Eight static aerated windrows	24 hr + 30 d	Rotating octagonal screen (mesh 30 mm) before curing, Zig-zag classification, vibrating screen (18 mm mesh).
Bolzano	Rotating screen, mesh 70x70 mm, with bag breaker, manual selection on a tape conveyor, roller shredder (160 kW), secondary screen, sludge mixer	Three primary vertical cylindrical reactors (Weiss-Kneer), 18.3 m diameter, 13 m total height, 9 m useful height, 2,400 m³ volume, bottom aeration	Three secondary reactors, as before.	14 d + 21 d	- Flip-flow screen, 8 mm mesh, densimetric table.
Bressanone (BZ) 75 t/d	Hammermill with 2 horizontal rotors, 130 kW, mixer/rotating screen 35 mm mesh.	Six turned aerated windrows in a covered area of 6,000 m², automatic turning equipment Buehler Compo-Star 4000, 52 kW, 150 m³/h.	Four windrows in covered area of 3,600 m²	42 d + 40 d	Rotating screen 10 mm mesh, densimetric table.
Pontives (BZ)	Impact breaker, rotating sludge mixer drum.	Primary aerated cells, each one with a capacity of 150 t.	Secondary cells in covered area	21 d + 60 d	Vibrating combs screen, Flip-flow screen, 10 mm mesh, air separation.
Feltre (BL) 120 t/d		Aerated area, turning with a specific turning machine			
Schio (VI) 160 t/d	Refuse receiving pit with a capacity of 1,500 m³	Horizontal rotating reactor, DANO-type, 30 m length, 3 m diameter	One windrow in a covered area, the others uncovered, turned by a loading shovel.	48 hr + 90 d	Rotating screen, 12 mm mesh, and ballistic separator before the curing. Screen, 8 mm mesh, and rolling mill.
Carpi (MO) 300 + 25 t/d	Bag breaker, knifes-mill for pre-selected organic refuse, primary selection.	Two horizontal rotating reactors DANO-type.	Turned windrows, aerated in a half of the area	4 d + 60 d	Rotating screen, air classification.
Ozzano (BO) 110 t/d	Shredding screen with 60 mm holes, air separation, knifes-mill.	Two vertical cylindrical reactors (Weiss-Kneer), 11.4 m total height, 8.2 m useful height, 1,500 m³ each one.	Uncovered aerated windrows	14 d + 30 d	Screen, 10 mm mesh, rolling-mill.

Table 1 Continued

PLANTS AND TREATMENT CAPACITY	PRETREATMENT Devices	BIOLOGICAL PROCESS Accelerated bio-oxidation	FINAL TREATMENT		
			Curing	Detention time	Devices
S.Agata bolognese (BO) 170 + 70 t/d	Knifes shredding screen with 60 mm holes, slow knifes -mill (45 kW), mixer (22 kW), rotating screen.	Eighteen windrows, in an area of 2,800, turned by a device installed on a traveling bridge; forced aeration with 2 insufflating and 2 extracting fans.	Covered area, 2,800 m², turned by CRAI device, 137 kW, 1,200–2,000 m³/hr.	21 d + 50 d	Rotating screen, ballistic separator; for the high quality line: fine screen and rolling mill
Massa Carrara 250 t/d	Hammermill, rotating screen with 20 mm mesh, sludge mixer with 6 coupled screws	Circular aerated reactor Fairfield-type, 17 m diameter, 2 m useful height, 400 m³ volume; Traveling bridge with 13 screws, 250 kW.	Static windrows, 3 m height, forced aeration in a covered area	70 hr + 40–50 d	Air separation, rotating screen 8 mm mesh.
Sesto Fiorentino (FI) 385 t/d 45 d	Three-dimensional screen, densimetric table, air separation, ballistic separation, sludge mixer. Rotating screen, air separation, final shredding.	Turned and aerated troughs (19+5), 3,920+1,200 m³ total volume, 2 Royer-type devices installed on rails.	Static windrows, turned with loading shovel.	18-30 d +	
Pistoia 210 t/d 60 d	Bag breaker with sharps.	Two horizontal rotating cylinders Dano-type, 42 m length, 6 m diameter.	Turned windrows in uncovered area.	36–48 hr +	Rotating screens, 30 and 10 mm mesh, and ballistic separation before the curing.
Ascoli Piceno 260 t/d	Bag shredding on belt conveyor, rotating screen for three outflows (organic fraction from 20 to 80 mm), sludge mixer.	Five horizontal rectangular covered reactors, 30 m length, 10 m width, 2.5 m deep, 3,750 m³ volume, with 4 turning screws installed on a traveling bridge, forced aeration by each one 5 fans.	Uncovered windrows (8), turned with a loading shovel, on 2 areas of 1,600 m²	30 d + 30 d	Screen, 20 mm mesh, vibrating . table
Fermo(AP) 210 t/d	Like above, no sludge mixer	Four reactor, like above.	Uncovered windrows	30 d+30 d	Rotating screen, densimetric table.
Pollenza (MC) 180 t/d	Hammermill, rotating screen, 40 mm mesh, sludge mixer.	Two horizontal cylinders Dano-type, 27 m length, 3.5 m diameter; the first phase is done in 20 turned and aerated troughs, 2,300 m³ volume, with a Royer turning device.	The second phase is done in 27 turned and aerated troughs with another Royer device	2–3 d 15 d + 24 d	Rotating screen, 16 mm mesh, gravimetric separator.

Plant	Pretreatment	Reactor / Composting	Storage / Curing	Days	Refining
Perugia 380 t/d	Bag breaker, rotating screen with square 80 mm holes.	Horizontal aerated reactor 70 x 21 m, turned by screws installed coupled on a traveling bridge.	Uncovered storage of large windrows.	28 d	Rotating screens, 12 and 4 mm mesh, air separation.
Foligno 110 t/d	The refuse is selected by mechanical, magnetic and by air systems	Like above		28 d	
Col S.Felice (FR) 60 t/d	Bag breaker, knife-mill, rotating screens with 35 and 20 mm mesh, sludge mixer.	Rectangular aerated reactor turned by screws.		28 d	Rotating screens, air separation.
Terracina (LT) 150 t/d	Sharps bag breaker.	Horizontal rotating reactor Dano-type	Windrows turned by a specific device	48–72hr + 20 d	Rotating screen and vibrating before curing, then air separation.
Pescara 200 t/d	Bag breaker on belt conveyor, polygonal screen with 3 flows of material, sludge mixer.	Two side-by-side inclined closed reactors, 3,000 m³ each volume, screws installed on a traveling bridge.	Windrows turned by a loading shovel	30 d + 30 d	Flip-flow screen, densimetric table.
Sulmona (AQ) 80 t/d	Rotating screen with 110 mm holes and bag breaking shafts, sludge mixer.	Two side-by-side uncovered reactors, 700 m³ each volume, 4 turning screws installed on a traveling bridge, aerated by 1 fan/reactor.	Uncovered windrows, on a 2,400 m² area, aerated by 4 fans	15 d + 20 d	Ballistic separation, hammermill, rotating screen and densimetric table.
S.Maria Capua Vetere (CE)	Two knife-mills (37 kW each one), rotating screen with 20-60 mm mesh.	Two turned aerated closed rectangular reactors, 50 m length, 8 m width, 14 coupled turning screws installed on a traveling bridge, forced aeration by means of 7 manifolds.	3+1 covered windrows, aerated and turned by a specific device.	28 d + 32 d	Rotating screen, air separation.
Vallo di Diano (SA) 60 t/d	Bag breaker, hammermill, rotating three-dimensional screen (little fraction out at 20 mm)	Circular closed reactor (Fairfield), 10 m diameter, 13 turning screws installed on a rotating bridge.	8 covered windrows, turned by a CNTM specific device (55 HP)	4 d + 60 d	Rotating screen, 12 mm mesh, densimetric table.
Matera 62 t/d	Bag breaker on a belt conveyor, knife mill (75 kW), rotating screen, sludge mixer.	Aerated reactors (4 basins), 700 m³ each volume (30 m length, 10 m width, 2.5m deep), turned by 4 screws installed on a bridge crane, forced ventilation by 1 fan.	One windrow, 4 m height, on an uncovered floor, 1,000 + 1,000 m².	28 d + 60 d	Two rotating screens, 40 and 12 mm mesh respectively, densimetric table.
Macomer (NU) 160 t/d	Rotating screen, 350 mm mesh, with shafts, secondary screen, 30 mm mesh, sludge mixer.	Aerated troughs (6 + 6), 50 m length, turned by a Siloda-type paddle wheel.	Uncovered area	28 d + 60 d	Flip-flow screen, 10 mm mesh, and ballistic separation.

Table 2 Present trends for processing technologies and composting plant design in Italy.

PLANTS AND TREATMENT CAPACITY	PRETREATMENT Devices	BIOLOGICAL PROCESS Accelerated bio-oxidation	FINAL TREATMENT		
			Detention time	Curing	Devices
Trento District 55 t/d	Sludge and vegetables received in 3 uncovered tanks, hammermill (124 kW) and knife-mill (290 kW), mixing with loading shovel.	The process is now managed in one phase, in a covered 10,000 m² area, 10 windrows turned by a Scat 4883 specific device (103 kW, 1,200–2,000 m³/hr). In a short time it will be realized a new 3,600 m² area with 4 windrows, aerated by 10 fans	now 90 d, then 33 + 57 d		Hammermill (60 kW), rotating screen, 10–25 mm mesh (Favorit max, 42 kW)
'Consorzio Bonifica Valle Scrivia' Tortona (AL) 90 t/d	Sludge receiving in 9 pits with dosing screws, vegetables storage in uncovered area (7,320 m²) and in covered area (1,800 m²), 3 mixers.	Closed aerated troughs (15), forced ventilation by 4 fans/trough, 3 turning device Royer-type moving on rails (62 kW each one).	21 d + 45 d	Uncovered area, 3,650 m², the windrows are turned by a loading shovel.	ballistic separation, rotating screen with 3 outflows (compost at 10 mm), packaging device.
Agrinord Isola della Scala (VR) 96 t/d	Sludge receiving in 2 covered tanks, vegetables storage in uncovered area; the mill is rented; turning with loading shovel.	Five closed glasshouses, 2,800 m², turning by a straw-distributing device (in a short time it will be a CRAI specific device, 137 kW 1,200–2,000 m³/hr).	30 d + 80 d	Twenty unclosed glasshouses, 11,200 m², turned windrows.	Rotating screen, 8 and 12 mm mesh, packaging device.
CE.LO Mira (VE) 57 t/d	Sludge and wet fractions receiving in a covered box, 1,500 m², vegetables storage in uncovered area, hammermill (184 kW), mixing by loading shovel.	Closed area, 4,500 m², the windrows are turned by a specific device (178 kW, 1,00–2,000 m³/hr), forced aeration.	30 d + 95 d.	Uncovered turned windrows.	Hammermill (135 kW), rotating screen (10–20–40 mm mesh), packaging devices.
Eco-pol Bagnolo Mella (BS) 68 t/d	Sludge receiving in 2 concrete uncovered tanks, vegetable storage in a covered area, hammermill (200 kW), mixing by loading shovel.	Closed tunnel, 90 x 10 m, turning and forwarding by belt conveyors installed on a bridge crane (45 kW, 400–500 m³/hr).	28 d + 60 d	Windrows turned by a Backhus specific device (136 kW, 1,00–1,500 m³/hr),	Rotating screen, 8–15 mm mesh (41 kW).
Ecopi Ghislarengo (AL) 77 t/d	Organic refuse storage in closed sectors, vegetables storage on uncovered floor, shredding, mixing.	Aerated windrows in a closed area, turned by a specific device	20 d + –	Curing on uncovered floor	Dryer, refining and packaging.

Facility	Pre-treatment	Composting process	Windrows (further)	Days	Screening
Ecopi Casalcermelli (AL) 34 t/d	Sludge receiving in 2 tanks, vegetables storage in uncovered area, hammermill, mixing by loading shovel.	Windrows (12) turned by a specific device (Morawetz, 96 kW, 400–500 m³/hr).	Windrows covered by a sheet (synthetic non woven material)	28d+ 60 d	Screen with 50 mm mesh, Hammermill, dryer, screen 2–5 mm mesh.
Maserati Piacenza 79 t/d	Vegetable storage in uncovered 60 m² area, hammermill (135 kW), mixing by loading shovel.	One phase process, 26 windrows, coupled, in 10,350 m² area, covered by a sheet (synthetic non woven material), the specific turning device (Kompatech 300, 500–700 m³/hr), is moved by a tractor (90 HP).		80 d	Hammermill and vibrating screen (50x50 and 14x14 mm mesh, 67 kW).
AMIU (MO)–CRPA (RE) Test-plant	Storage, shredding, mixing by loading shovel.	Horizontal reactor, 60 m length, 3 m width, 1 m deep. specific turning device Okada-type, forced aeration in the first 2/3 of the trough by 2 fans.	Uncovered windrows turned by a loading shovel.	30 d + 60 d	Rotating screen
AMIA Rimini 41 t/d	Four covered pits, vegetable storage in uncovered area, hammermill (300 HP), pre-mixer, rotating cylindrical mixer.			45–60 d	Rotating screen 20 mm mesh.
Bioter 25 t/d	Vegetables storage in uncovered area, hammermill (243 kW).	One phase process in uncovered area, table-windrows 10 x 20 m up to 10 x 40 m, 2.7 m height, turned with loading shovel		120 – 180 d	Rotating screen (54 kW)
Castelfranco (TV) 25 t/d	Vegetables storage in uncovered area, hammermill (271 kW).	Process in one phase in 13,000 m² area, windrows turned by a specific device.		150 – 180 d	Rotating screen 10 mm mesh.
Castiglione delle Stiviere (MN) 11 t/d	Receiving floor, hammermill, mixing by loading shovel.	Covered 2,800 m² area, 6 aerated windrows, 3 aeration manifold for each one, 3 fans.		80 d	To be defined.
Vignola (MO) 15 t/d	Three receiving boxes, mixer-shredder with 5 horizontal and 1 vertical screws.	A 1,028 m² area, 7 windrows aerated by 4 manifolds each one and 7 fans.	A 1,828 m² area, with 7 windrows.	21 d + 40 d	Rotating screen.
Alba-Brà (CN) 74 t/d	Closed 320 m² area, shredding for vegetables, screen-bag breaker, mixer with 2 dosing screws.	Closed 4,770 m² area, 9 aerated windrows, 18 fans, turning with Morawetz specific device.	Nine windrows turned a second Morawetz device.	21 d + 40 d.	Rotating screen with 10–15 mm mesh.
Vigonza (PD) 25 t/d	Three receiving tanks (160 m²), hammermill to be defined, mixing §by loading shovel.	One phase process, covered 3,178 m² area, 8 windrows turned by CRAI specific device with elevator belt (137 kW, 1,200–2,000 m³/hr, belt width 3.15 m).		90 d	Screen with 10–15 mm mesh.
Spresiano (TV) 14 t/d	Vegetables storage in 160 m² tanks, market refuse receiving in 80 m² tank, shredding, mixing by loading shovel.	Covered 1,000 m² area, 4 windrows, the specific turning device (Kompatech 300, 500–700 m³/hr), is moved by a tractor (90 HP).	Uncovered area, 5 windrows 77 m length, 4 windrows 22 m length.	28 d + 120 d	To be defined.

Table 2 Continued

PLANTS AND TREATMENT CAPACITY	PRETREATMENT Devices	BIOLOGICAL PROCESS — Accelerated bio-oxidation	FINAL TREATMENT Devices — Curing	Detention time	FINAL TREATMENT Devices
Turin AMIAT 118 t/d	Two receiving pits, bag breaker on the belt conveyor, hammermill, rotating screen with 100 mm mesh, rotating mixer.	Thirty windrows turned by a specific Royer device with rotor and redler, installed on rails, forced aeration by 2 fans for each trough.	Covered area, windrows turned by 2 screws installed on a traveling bridge.	32 d + 38 d	Rotating screen with 20 mm mesh, densimetric table.
Milan AMSA 118 t/d	Receiving pits, bag breaker on the belt conveyor, hammermill, rotating screen with 100 mm mesh, rotating mixer with 2 screws.	Twenty windrows turned by a specific Royer device with rotor and redler, installed on rails (72.6 kW), forced aeration by 2 fans for each trough.	Windrows (28) turned by 2 screws installed on a traveling bridge.	28 d + 60 d.	Rotating screen with 20 mm mesh, densimetric table.
Verbania (NO) 14 t/d	Three receiving tanks, shredder-mixer with 2 screws.	Closed 720 m² area, 3 turned windrows, forced aeration by 6 fans.	Closed 1,440 m² area, 6 windrows.	28 d + 90 d.	Rotating screen with 10x25 mm holes
Saluzzo (CN) 20 t/d	Sludge receiving in 2 closed sectors, vegetables storage in uncovered area, shredder-mixer.	Closed 1,580 m² area, 4 windrows aerated and turned by a specific device.	Five windrows in a closed 1,580 m² area, turned by a loading shovel.	24 d + 60 d.	To be defined.
Pirinoli Paper-mill Roccavione (CN) 14 t/d		Turned and aerated windrows in covered area		20 d +60 d	
Agrisesia (NO) 27 t/d		Windrows inside a tunnel, turned by a specific device (Tritter, 44–59 kW, 300–350 m³/hr) moved by a tractor	Uncovered area.	120–150 d.	Shredding and screening with 15 mm mesh.
Italconcimi (TO) 27 t/d		Uncovered 15,000 m² area, windrows turned by a loading shovel.			
S.Carlo fertilizers (TO) 27 t/d		Uncovered 25,000 m² area, windrows turned by a loading shovel			
Faenza (RA) 55 t/d		Nine turned and aerated troughs, 1,800 m² area.	The plant is described in a specific relation presented in this Congress.		
Senigallia (AN) 151 t/d	Sludge and saw dust storage in closed tanks, 2,000 m³ volume, with screw extraction, mixer.	BAV vertical closed reactor, 300 m³ volume, fed on the top, outfed in the bottom, forced aeration in the bottom	Windrows 1.5–2 m height	14 d + 42–56 d	

Design of Passively Aerated Compost Piles: Vertical Air Velocities between the Pipes

Nancy J. Lynch and Robert S. Cherry – Idaho National Engineering Laboratory P.O. Box 1625, Idaho Falls, ID 83415–2203

Abstract

Our goal is to develop design procedures for passively aerated compost piles. In this system, piles are built on a base of porous material such as straw or peat moss. Perforated pipes are placed on this base, and the compost pile built over the base and pipes. The pipes are open to the atmosphere on the ends. As the warm moist air in the compost piles rises, the pipes increase the amount of fresh air drawn into the pile by free convection. The piles are not turned, nor is forced aeration equipment used, which significantly reduces the operating and capital expenses associated with these piles. The specific geometry considered here is that of a long windrow with pipes embedded in the porous base perpendicular to the length of the windrow. Currently, mathematical models are not available that describe these piles to aid in design, and optimum pile configurations and materials are worked out by trial and error.

Models are difficult to develop for passively aerated piles because the air flow rate is not explicitly known, but coupled with pile temperature. In this paper, we develop a mathematical model to calculate the upward air flow velocity over a pipe, v_{yo}, and the fraction of that velocity $\dfrac{v_y}{v_{yo}}$, that is present as one moves away from the pipe. The temperature difference between ambient air and the pile is used to calculate the driving pressure drop. This model can be solved analytically, and contains one dimensionless number, $$\zeta = \frac{Kd^2\rho}{K_b HH_b \rho_b}$$, which characterizes $\dfrac{v_y}{v_{yo}}$, where K is the permeability of the composting material, d is the distance between the pipes, K_b is the permeability of the porous base, H is the height of the pile, and H_b is the height of the porous base. The results of these air flow calculations will serve as starting point to design passively aerated piles.

Introduction

In a compost windrow, air is heated due to microbial metabolism and rises, pulling in fresh air via a chimney effect. This passive aeration can be enhanced by strategically placing perforated pipes in the pile to increase airflow through the pile. Such an approach is attractive when the capital and operating costs associated with frequent turnings or forced aeration equipment cannot be justified. These piles would also be attractive in bioremediation technologies to minimize handling and potential spreading of contaminated wastes. Although several workers (Mathur, Daigle et al. 1988; Mathur, Patni et al. 1990; Mathur 1991; Patni, Fernandes et al. 1992) have illustrated the feasibility of these systems, design procedures are not readily available to extend these results to different pile configurations or feed materials. This is because, unlike in forced aeration systems, air flow, substrate degradation and heat removal are coupled. Consequently, it is difficult to accurately predict how pile dimensions, the number and placement of perforated pipes, and the feed composition will affect pile performance.

The air flow rate is a key parameter in any compost model because it determines whether the pile will be aerobic or anaerobic, and the rate of heat removal by water evaporation. Current models for static pile forced aeration processes are possible because the air flow rate can be controlled by the operator and is therefore known. In a passively aerated pile, the air flow rate intrinsically is coupled with temperature. This coupling prevents direct analytical solutions, and requires rigorous numerical solutions involving finite element or finite difference methods. This problem is even beyond the scope of current computational fluid dynamics models because the problem includes several difficult aspects, such as free convection in porous medium, reaction on the media surface, a phase change (vaporization of water), and the need to track various species, especially oxygen.

We seek to develop a simpler, mathematically tractable model that still embodies the important aspects of a composting process. A simpler model should emphasize the relationships between design variables and process variables that may be lost in all of the detail provided by a numerical model. This paper represents a first attempt at developing correlations for the air flow rate in a passively aerated pile.

Problem Definition

Figure 1 illustrates the windrow geometry considered in this paper. The windrow is built upon a base of porous material, such as straw or wood chips, having a height H_b and a permeability K_b. The permeability is a measure of how freely air can flow through the material. Pipes are embedded in the porous base across the pile's width, with a distance d separating them. The air here has a temperature T_b and density ρ_b, which are at ambient conditions (T_o and ρ_o) since no composting reaction occurs in the base. The compostable material (permeability K) is piled to a height H on this porous material. We neglect temperature gradients, and assume that the temperature T and air density ρ in the pile are constant. Typically, the pile is covered with a material like peat moss, straw or finished compost to provide insulation and act as a deodorizer. We assume that this material had the same physical characteristics as the compostable material, and this layer was included in the pile height H. However, the model can easily be modified to include a covering layer. In the chosen coordinate system, x is in the horizontal direction perpendicular to the pipe length, y is the vertical direction, and z is the horizontal direction parallel to the pipes.

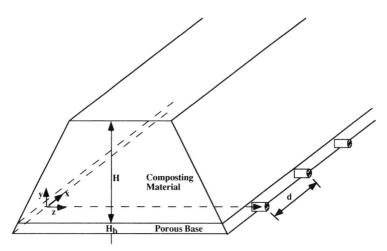

Figure 1 Compost Pile Geometry

Air flows through this pile via free convection, as the warm air in the pile rises and pulls in fresh air. In the considered problem, air enters the pipes and travels down the pipe length. At some point, the air will exit the pipe, and travel upward and outward through the compost pile, eventually exiting through the pile surface. It is this part of the journey, after the air exits the pipe and begins to travel through the pile itself until the air exits the pile, that we will concern ourselves with in this paper. Pressure losses and air flow rates in the pipe will be addressed in a later paper.

Several simplifying assumptions were be made in the following calculations:

1. Air can exit the pipe at any point along the pipe. This effectively neglects the effects of hole spacing, and allows the following assumption.
2. Once air exits the pipe at a particular value of z, it no longer can move in the z direction; but is limited to the x–y plane at that particular z, until the air exits

the pile. This implicitly assumes that pressure drop along the length of pipe and across the exit holes, ΔP_{pipe}, is small, and confines us to regions where the pile height is constant. This assumption allows us to reduce air flow through the pile to a two dimensional problem.

3. Air can only travel horizontally through the porous base in the x direction, and only vertically (y direction) through the compostable material. While we might realistically expect the air to follow a curved streamline, identifying these streamlines requires a rigorous numerical finite difference or finite element model. This assumptions allows the problem to be solved analytically by reducing the two dimensional problem even further to two one-dimensional problems.

4. The fluid, air, can be treated as an incompressible fluid in the composting and base material since the pressure drops encountered in this process are small. We assume a uniform temperature in the composting material, and allow a density change between the porous bed and the composting material.

5. The base material has a higher permeability than the composting material ($K_b >$ K).

6. The mass of air entering the pile through the porous base and exiting the pile surface are the same. This neglects the change in mass caused by oxygen consumption, CO_2 and NH_3 generation, and water vaporization.

The specific goals of this part of the model are:

1. Determine v_{yo}, the upward air velocity at x = 0.

2. Determine $\Phi = \dfrac{v_y}{v_{yo}}$ from x = 0 to x = d/2.

Darcy's Law

Flow through porous media is governed by Darcy's Law (Nield and Bejan 1992), which is written as

$$v = -\frac{K}{\mu}\left(\frac{\Delta P}{L}\right)$$

(1)

where v is the fluid (in this case, air) velocity; μ is the fluid viscosity, K is the permeability of the porous medium, and ΔP is the pressure drop across a length L of the medium. The permeability K, has units of area; as K increases, air can flow more easily through the medium. Darcy's Law applies only for low flow rates, and requires that

$$Re_K = \frac{\rho v K^{1/2}}{\mu} < 1$$

(2)

where Re_K is the Reynolds number.

The pressure difference driving passive aeration can be calculated as

$$\Delta P = \int_0^H (\rho_b - \rho(z))g\,dz$$

where g is the gravitational constant. This becomes

$$\Delta P = (\rho_b - \rho)gH \tag{3}$$

with the assumption that T and consequently ρ are constant throughout the pile. When calculating ρ, the effect of the water vapor must be included as this can contribute significantly to the density difference between ambient air and air within the pile (Haug 1980). ΔP represents the pressure difference driving force from the point where air first enters one of the pipes until it exits from the surface of the pile.

As stated in assumption 3 and illustrated in figure 2, in this model the air is constrained to flow only in the z direction through the pipe, then to flow only in the x direction through the porous base beneath the pile, and finally flow vertically upward in the y direction only through the pile. The entire pressure drop of equation (3) can be thought of as the sum of three components, namely, the pressure drop in the pipe and its exit holes, the base, and through the pile,

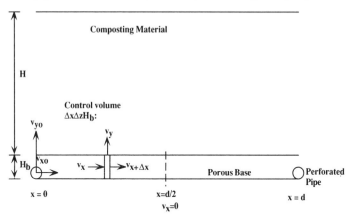

Figure 2 An interior slab of the compost pile. Air exiting the pipe can either travel straight up , or travel horizontally through the pile base before traveling straight up, and exiting the pile surface.

$$\Delta P = \Delta P_{pipe} + \Delta P_{base} + \Delta P_{pile} \tag{4}$$

At the present time, calculating the pressure drop as the air moves through the pipe, ΔP_{pipe}, will be neglected here. Thus, let us approximate equation (4) as

$$\Delta P \approx \Delta P_{base} + \Delta P_{pile} \tag{5}$$

Both ΔP_{base} and ΔP_{pile} will change as a function of x, but the sum of these pressure drops, ΔP, will remain constant. At x = 0, $\Delta P = \Delta P_{pile}$, since air traveling upward at that point does not have to travel (much) through the porous base. For simplicity in the equations which follow, let us define

$$Y = \Delta P_{pile} \qquad (6)$$

$$Y_o = \Delta P \qquad (7)$$

and

where Y is a function of x and Y_o is a constant.

At any point x, the vertical velocity of air traveling through the pile in the y-direction is

$$v_y = \frac{KY}{\mu H}, \qquad (8)$$

and let us define

$$\Phi = \frac{v_y}{v_{yo}} = \frac{Y}{Y_o}. \qquad (9)$$

Our challenge, then, is to determine Y as a function of x. To do this, consider the control element $\Delta x \Delta z H_b$ shown in the porous base in Figure 2. Applying a mass balance and using the constant density assumption, we find

$$v_x H_b \rho \Delta z = v_{yb} \rho \Delta z \Delta x + v_{x+\Delta x} H_b \rho \Delta z \qquad (10)$$

which simplifies to

$$\frac{v_x - v_{x+\Delta x}}{\Delta x} = \frac{v_{yb}}{H_b}. \qquad (11)$$

The quantity v_{yb} in (10) and (11) is the velocity the air has when exiting the porous base, where it has a density ρ_b. Considering an arbitrary area $\Delta x \Delta z$ and assumptions 6 (constant mass), 2 and 3 (fluid flows only vertically in pile), we can write

$$\rho_b v_{yb} \Delta x \Delta z = \rho v_y \Delta x \Delta z \qquad (12)$$

which leads to

$$v_{yb} = \frac{\rho}{\rho_b} v_y \qquad (13)$$

Taking the limit as $\Delta x \to 0$ in equation (11), and combining with equations (8) and (13), we find

$$\frac{dv_x}{dx} = -\frac{K\rho Y}{\mu \rho_b H H_b}. \qquad (14)$$

Applying Darcy's Law to this control volume yields

$$\frac{dY}{dx} = -\frac{v_x \mu}{K_b}. \qquad (15)$$

Equations (14) and (15) will be subject to the following boundary conditions:

$$\text{At } x = 0, Y = Y_o; \tag{16}$$

$$\text{At } x = d/2, v_x = 0. \tag{17}$$

By applying the chain rule, x can be eliminated from (14) and (15) to obtain

$$\frac{dY}{dv_x} = \frac{Av_x}{Y} \tag{18}$$

where

$$A = \frac{\mu^2 H H_b \rho_b}{K_b K \rho} \tag{19}$$

The boundary condition will be

$$\text{At } Y = Y_o, v_x = v_{xo} \tag{20}$$

where v_{xo} denotes the velocity at $x = 0$, which is an unknown that must be determined.

Integrating with boundary condition (20), we recover

$$Y^2 - Y_o^2 = A\left(v_x^2 - v_{xo}^2\right). \tag{21}$$

To determine v_{xo}, equation (21) is solved for Y, and the result inserted into equation (14). Integrating yields

$$v_x \sqrt{A} + \sqrt{Y_o^2 - A\left(v_{xo}^2 - v_x^2\right)} = \left(v_{xo}\sqrt{A} + Y_o\right)e^{-\sqrt{\zeta}\left(\frac{x}{d}\right)} \tag{22}$$

where

$$\zeta = \frac{K\rho d^2}{K_b H H_b \rho_b} \tag{23}$$

Now we can apply the boundary condition in equation (17) and solve for v_{xo}:

$$v_{xo} = \frac{Y_o}{\sqrt{A}}\left(\frac{1 - e^{-\sqrt{\zeta}}}{1 + e^{-\sqrt{\zeta}}}\right). \tag{24}$$

and using equation (21), can also determine Y at $x = d/2$.

Next, equation (21) is solved for v_x, and the result inserted into equation (15). Integrating again with boundary condition (16), we recover

$$Y + \sqrt{Y^2 - Y_o^2 + Av_{xo}^2} = \left(v_{xo}\sqrt{A} + Y_o\right)e^{-\sqrt{\zeta}\left(\frac{x}{d}\right)} \tag{25}$$

This equation can be simplified considerably by using the following substitution

$$a = e^{-\sqrt{\zeta}}, \tag{26}$$

and combining equation (25) with equations (24) and (9) to recover

$$\Phi = \frac{a^{(x/d)} + a^{(1-x/d)}}{1+a}$$

(27)

We have plotted Φ as a function of (x/d) for various values of a in Figure 3. The parameter a must be between 0 and 1, and can be increased by decreasing the distance between pipes, d, decreasing K, or by increasing H, H_b, or K_b.

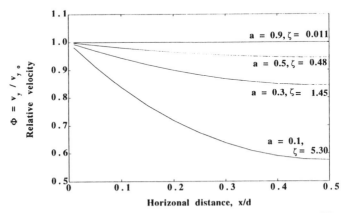

Figure 3 F/Fo as a function of x/d for several values of $a = \exp(-\sqrt{\zeta})$.

Results and Discussion

There are several interesting things to note about this model. As we might intuitively expect, the highest air flow rate will be v_{yo}, which is directly over the perforated pipes. That velocity is given by

$$v_{yo} = \frac{KY_o}{\mu H}.$$

(28)

Note that changing the properties of the porous base or the pipe spacing, d, does not affect v_{yo}, but only $\dfrac{v_y}{v_{yo}}$. This result will only be valid if ΔP_{pipe} is negligible; otherwise, the distance down the length of the pipe must be included. Also, if the assumption of constant T and ρ applies, incorporating equation (3) yields

$$v_{yo} = \frac{K(\rho_b - \rho)g}{\mu}$$

(29)

This result implies that v_{yo} can only be manipulated by changine K, for instance by adding a bulking agent.

The dimensionless number ζ occurs throughout the model and can be written as

$$\zeta = \frac{K\rho d^2}{K_b H H_b \rho_b} = \frac{\dfrac{Qd}{K_b \rho_b (H_b \Delta z)}}{\dfrac{QH}{K\rho(d\Delta z)}} \tag{30}$$

Q represents the volumetric flow rate, which is the same order of magnitude in the porous base and the pile. Darcy's Law, equation (1), can be rearranged to

$$\Delta P = \frac{\mu v L}{K} = \frac{\mu Q L}{\rho A K} \tag{31}$$

and the two equations combined to find

$$\zeta = \frac{\Delta P_{base}}{\Delta P_{pile}} . \tag{32}$$

For small values of ζ, the air flows relatively freely through the porous base, and the profile of $\dfrac{v_y}{v_{yo}}$ versus x/d is fairly flat (Figure 3). As ζ increases, the air finds it easier to move upward through the pile than to travel through the porous base, and $\dfrac{v_y}{v_{yo}}$ drops off quickly as x/d increases to 0.5. This effect becomes significant when ΔP_{base} is comparable or larger than ΔP_{pile}, and ζ approaches or exceeds one.

Permeabilities for compost materials have not been commonly reported in the literature. We have measured the permeabilities of several compost mixtures, and report the values in Table 1 only to help determine reasonable values for ζ. The materials were tightly packed into a 30.5 cm inner diameter reactor to a nominal height of 30 cm. The reactor ends were sealed except for air inlet and outlet lines and pressure taps. The pressure drop across the bed was measured using an inclined manometer (Cole-Parmer) with a range of 0– Pa. Flow rates were measured using a bubble flow meter (Cole-Parmer) of the appropriate size. The permeability was determined from equation (1).

Consider a 1 meter tall pile constructed on a 15 cm bed of wood chips with perforated pipe every 1 m, where the inside temperature is 60°C and the ambient air is dry at 20°C. Here, $\rho_b = 1.21$ kg m^{-3}, and $\rho = 0.94$ kg m^{-3} at 760 mmHg, and $\mu = 2.04 \times 10^{-5}$ kg m^{-1} sec^{-1} at 60°C.. If the pile is composed of the most porous compost feed tested (K = 2.4×10^{-5} cm^2), we find $\zeta = 2.26$, a = 0.22, and $v_{yo} = 0.031$ cm sec^{-1}. If the pile permeability K were instead 1.0×10^{-6} cm^2, $\zeta = 0.094$, a = 0.73 and $v_{yo} = 0.0013$ cm sec^{-1}. In the second case, the lower K leads to a flatter

$$\frac{v_y}{v_{yo}}$$

profile, but v_{yo} is also considerably lower. These numbers can also be used to check that Darcy's law applies, as stipulated by equation (2). Using the largest compost permeability, 2.4×10^{-5} cm^2, and the corresponding velocity, $Re_K = 0.00090$ in the compost bed, so Darcy's Law applies. In the porous bed, v_{xo} will have its largest value as $= \zeta$ tends to ∞. For this system, $v_{xo} = 0.107$ cm sec^{-1}, and the corresponding $Re_K = 0.0047$.

The next steps in this analysis will include accounting for ΔP_{pipe}, removing the assumption that ρ and T are constant in the pile, and experimentally verifying the results presented here.

Table 1 Permeabilities for several compost materials.

Description of Compost Material	Permeability, cm^2
Wood chips	5.5×10^{-5}
Sheep Manure / Straw Mixture (predominantly straw)	1.0×10^{-5}
Equal volumes of cow manure, straw, horse manure and wood chips	8.1×10^{-6}
One volume straw, one volume cow manure, two volumes horse manure, two volumes wood chips	2.4×10^{-5}
Finished cow manure and straw compost (small particle size, no bulking agents)	$\approx 2.1 \times 10^{-6}$

References

Haug, R. T. (1980). *Compost Engineering: Principles and Practice.* Ann Arbor, Lewis Publishers (out of print).

Mathur, S. P. (1991). Composting Processes. *Bioconversion of Waste Materials to Industrial Products* Ed. Martin. London, Elsevier. 147–183.

Mathur, S. P., J. Y. Daigle, et al. (1988). Composting seafood wastes. BioCycle, 29(8 (September)): 44–48.

Mathur, S. P., N. K. Patni, et al. (1990). Static pile, passive aeration composting of manure slurries using peat as a bulking agent. Biological Wastes, 34: 323–333.

Nield, D. A. and A. Bejan (1992). *Convection in Porous Media.* New York, Springer-Verlag.

Patni, N. K., L. Fernandes, et al. (1992). Passively aerated composting of manure slurry. 1992 International Winter Meeting of The American Society of Agricultural Engineers, Nashville, TN, American Society of Agricultural Engineers.

A Review of Features, Benefits and Costs of Tunnel Composting Systems in Europe and in the USA

KEITH PANTER, – Thames Water International, UK,
RICHARD DE GARMO – PWT Waste Solutions USA,
and DAVID BORDER DBBC Consultancy, UK

Abstract:

Odour control is identified as the critical technical issue for the success of composting projects. Tunnel composting has the advantages that the compost can be kept aerobic, all the excess air is collected for treatment and that building corrosion, that leads to fugitive emissions, is prevented. Data is given for the number of installations in Holland and USA. Rough costs indicate that tunnel composting has comparable costs to other s mechanised systems. Two types of tunnel system are described: batch and plug flow and their features compared.

All compost projects are subject to the same pressures to a lesser or greater degree depending on the difficulty of composting the source material and local sensitivities. Compost projects fail for 3 major reasons: lack of proper project finance; poor community acceptance normally due to poor odour control and third, lack of product market. The first and the last come down to good management, but some systems are more prone to odour problems than others. Table 1 is a summary of technical factors to consider when choosing a composting system.

There has been a gradual evolution and a wider range of wastes composted. In the US there was a large increase in composting following the USDA work at Beltsville in the late 70s. There was a dramatic increase in aerated static pile systems. Many of these early systems have had to be retrofitted with expensive leachate collection and odour control systems. The researchers did not take into account community acceptance as a key issue. This then led to upsurge so called invessel systems in the mid 80s. Some being true invessel built around a silo configuration and others being based on shed or hangar systems where aerated static pile, windrows or a combination has been carried out in a building. Many of the early silo systems had problems, primarily due to poor porosity and mechanical failure. The longitudinal or agitated bin system has proved to be the most robust system of the shed systems and has been widely adopted. There are still some inherent weaknesses due to the possibility of fugitive emissions and building corrosion for the more putrescible wastes which is coming to light for some of the earlier systems.

Table 1

	heap	windrow	aerated static pile	shed hangar systems *	vertical enclosed vessel	tunnel systems
Odour control	low	low	low/med	med	variable	high
Temp control	low	med	med	med/high	variable	high
Pathogen kill	low	low	med	med/high	variable	high
Weather tolerance	low	low	med	high	high	high
Ease of operation	med	med	med	med	med/low	med/high
O & M staffing	med	med	med	med/low	med/low	low
Refits	low	med	high	med	high	low
O & M costs	med	med	med	med	med/low	med/low
Operator envionment	low	low	med/low	med/high	high	high
Capital costs	low	low	med/high	high	high	high
Appearance	low	low	low	med/high	med	med/high
Noise	high	high	med/high	low	low	low
Process reliability	low	med	low	med/high	low	high
Feedstock versatility	low	low	low	med/high	low/high	high
Process tolerance	low	med	med	med/high	low	high
Product reliability	low	med	med	med/high	low/high	high

In Europe the development of composting has varied from country to country. The most rapid rise in composting has been in Holland where the problem of ground-water contamination has driven a movement to take organic waste out of landfills. The development of Biowaste composting systems is more recent than the USA developments and there has been a more rapid move to tunnel systems. Statistics are given in Table 2 for the USA and Holland

Table 2

Holland 1993 1		USA 1993 2	
Biowaste		Biosolids + MSW	
Shed systems	12	In vessel and Shed	66
Tunnel systems	15	Agitated Bed	19
Other Aerobic	5	Tunnel	7
Anaerobic	3		

Whilst the use of open systems is generally appropriate for garden waste, more in-vessel systems are being used for sludge and for MSW their use is virtually universal.

Composting tunnels are large rectangular boxes which tightly control the composting environment of the organic material by means of a computer controlled supply of forced air through a perforated floor structure. Two main forms of tunnel are available :batch tunnels and plug flow tunnels.

Batch tunnels have been used in the mushroom industry for 30 years to produce high quality compost which acts as the growing medium for the white mushroom, *Agaricus bisporus*. A great number of these tunnels are in use around the world to produce many hundreds of thousands of tonnes of compost each year. They work on the principle of composting one batch of waste at a time under highly controlled conditions. The control of the composting environment is brought about by means of recirculated forced air supply system under computer control. This allows temperatures throughout the composting mass, often up to 400 tonnes, to be kept within $2-3^0$ C. of that required for the different stages of the composting process. This in turn allows a fine manipulation of the microbiology of the composting process to bring about a near optimum composting regime and to produce a tightly specified compost product. Composting is normally complete within 14 days.

The plug-flow tunnel system has been designed with similar principles in mind but operates in a continuous rather than in a batch mode. This system allows the continuous feeding of a wide range of organic wastes into the composting tunnel by means of a supply conveyor and a hydraulically driven ram plate at one end of the tunnel. This has the advantage that the material can be kept fully enclosed during feeding the conveyor and that there is no chance of fugitive emissions. The composting material within the tunnel is compressed enabling considerable quantities of waste to be composted in a small area. As the composting mass passes along the tunnel it is taken through the various composting stages by means of differential temperature control brought about by the supply of air in the various zones in the tunnel. The completed compost leaves the tunnel at the other end of the tunnel and is removed by a second conveyor. Once again the residence time in the tunnel is about 14 days. The process can be totally automated from the beginning to the end .

The computerised regulation of temperature within both systems by the supply of forced air allows an unrivalled control over the composting process. Further control is enabled by the strict regulation of the nature and consistency of the organic feedstock. Many of the problems that have been encountered in the composting of organic wastes, such as the generation of offensive odours, are caused by the lack of sufficient control leading to anaerobic conditions.

Other papers describe the virtues of tunnel systems most notably that of Loikin and Oorthuys [1]. They summarised the benefits as follows: substrate flexibility; optimum process control; acceptable working conditions; low maintenance; short retention time; small footprint; no leachate. The same authors reviewed operating costs which are allustrated in Table 3.

Table 3

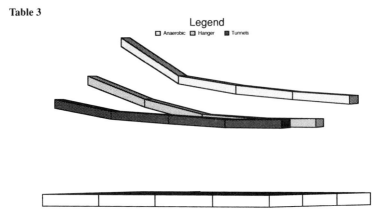

Batch tunnel systems have had a rapid growth in Europe. In Holland there are now more tunnel systems than other system for composting Biowaste(source separated MSW). In the US the development of tunnel systems has been led by the Tunnel Reactor[R] system which is a plug flow system.

The main advantage of this configuration is that odour is completely contained not just during the composting phase but during the materials handling prior to loading. The other benefits are that it has lower capital cost for small systems due to low number of tunnels to be constructed (sometimes only one) and thus also a very small footprint. There are several reference facilities in the US – the longest running one is near Cincinnati at Hamilton OHIO.

Tunnel systems are the next generation composting system. Their use is growing rapidly in Europe and USA. The preference is for the tunnel reactor [R] in the USA because of the very high community reaction to odour and a preference for systems that do not expose employees to composting atmospheres. Costs are similar for both systems and comparable to other mechanical systems.

Waste generators need the right technology coupled to effective management. For instance in MSW composting over half the cost and skill is in the operation of MRFs to be able to produce a reliable feedstock. There is now a growing trend for technology vendors to develop other skills: financing, operations and product marketing so that customers can be offered the certainty of a gate fee with guaranteed compliance and recycling, product marketing and odour control. Waste management companies are in the USA and Europe are exploiting the benefits of Tunnel composting to reduce landfill and promote beneficial use, to the extent that some proprietary systems are now owned or operated by waste companies.

1. P Loikin and T Oorthuys, Major Benefits from Tunnel Composting for Production of Biowaste and Dried Sewage Sludge. European Conference on Sludge and Organic Waste, University of Leeds 1994.
2. Biocycle November & December 1993

B6 Marketing and Economy

Compost Marketing Trends in the United States

LINDA L. EGGERTH – CalRecovery, Inc.Hercules, California USA

Abstract

As programs are being implemented throughout the United States to divert wastes from landfill disposal, composting is continuing to increase in popularity as a means of treating organic wastes. As the number of composting facilities increases, the competition for markets and the importance of producing a high quality compost also increase. This paper provides an overview of composting in the United States, a discussion of compost markets and marketing efforts, and a presentation of factors that affect compost marketing. Trends in composting and compost marketing are discussed, including the use of source separated feedstocks, composting of segregated wastes, and governmental policies that affect composting.

Introduction

As communities in the United States implement new programs and expand existing ones in order to divert larger quantities of materials from landfill disposal, composting is playing an increasingly important role as a waste management tool. However, as the number of composting facilities grows, the quantity of compost product available on the market increases. As a result, more attention is being given to product quality and market development. Composting projects that produce inferior quality material or that do not take the steps necessary to develop markets will find it difficult to compete.

As background to the subject of compost marketing, this paper begins with a brief review of the current role of composting as a waste management tool and of the types of wastes being composted in the United States. Compost uses and markets, marketing practices, and factors affecting compost marketing are then discussed.

Composting as a Waste Management Tool

In 1993, 207 million tons of municipal solid waste (MSW) were generated in the United States, or 4.4 lb per capita-day (U.S. EPA, 1995). As shown in Table 1, paper and paperboard make up the largest component of MSW, at 77.8 million tons annually or 37.6% of the waste stream. At 15.9% of the MSW generated in the nation, yard trimmings also represent a substantial fraction of the waste stream. Nationally, 21.7% of the MSW was recovered (18.6% through recycling and 3.1% through composting), 15.9% was incinerated, and 62.3% was landfilled in 1993 (U.S. EPA, 1995). Although still relatively small, the role of composting as a waste management tool has increased substantially during recent years (i.e, from less than 1% of MSW generated during the mid 1980s, to more than 3% in 1993).

Table 1 Quantity and Composition of MSW Generated in the United States in 1993

Component	Million Tons	% by Weight
Paper and paperboard	77.8	37.6
Yard trimmings	32.8	15.9
Plastics	19.3	9.3
Metals	17.1	8.3
Food	13.8	6.7
Wood	13.7	6.6
Glass	13.7	6.6
Other	18.7	9.0
Totals	206.9	100.0

Types of Wastes Being Composted

Composting facilities that utilize biosolids, mixed municipal solid waste (MSW), or yard trimmings have been operating in the United States for many years. More recently, facilities using source separated MSW or segregated fractions of MSW other than yard trimmings (e.g., grocery waste, cafeteria food waste) as a feedstock have been gaining popularity.

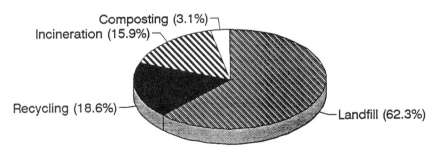

Figure 1 Disposition of MSW Generated in the United States in 1993

Municipal Solid Waste

As early as the 1950s, MSW composting was being considered as a solid waste management process in the United States. However, in the 1960s the prospects for MSW composting decreased, largely due to a lack of a market for compost, the low cost of landfilling, and the high carbon-to-nitrogen ratio of MSW.

In the late 1980s, MSW composting began once again to gain in popularity. As shown in Figure 2, the number of operational facilities increased substantially from 5 in 1988 to 21 in 1992 (Eggerth, 1992; Goldstein, 1994B). The interest in MSW composting was the result of a number of factors including the closure of substandard landfills, strong anti-incineration sentiments, the introduction of high technology systems to process a mixed waste stream, a growing confidence in composting as an option to handle municipal waste, and economics that would allow composting to compete with incinerators and landfills.

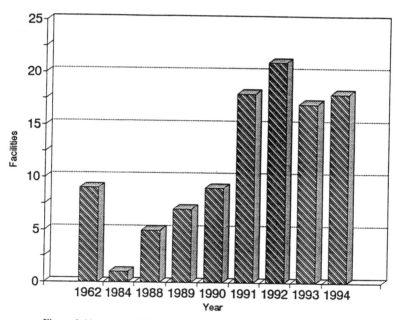

Figure 2 Number of MSW Composting Facilities in the United States

In the early 1990s, trends emerged that had a negative impact on project development. Cutbacks in state and municipal budgets restricted the funds available for projects, the implementation of recycling and yard trimmings composting programs resulted in sufficient levels of diversion to meet short-term mandates, the ability of a project to control the flow of waste to a facility was challenged in the court system, some of the larger mixed MSW composting plants were closed, and large landfills were opened making landfill disposal more cost effective. The number of MSW composting facilities in 1994 was 18.

In the 1990s, mixed MSW project failures, combined with skepticism about the quality of a mixed waste compost product sparked interest in composting source separated organics. Currently, 4 of the 18 operating MSW composting projects process a source separated stream. In these communities, residential waste is separated into three fractions – compostables, recyclables, and trash.

Segregated Fractions of MSW

Yard Trimmings

Although composting of yard trimmings has been practiced for many years, it was not until the late 1980s that this practice began to attain widespread application in the United States. As shown in Figure 3, the number of yard trimmings composting facilities increased from about 650 in 1988 to about 2200 in 1991, and to 3014 in 1993 (Eggerth, 1992; Steuteville, 1994).

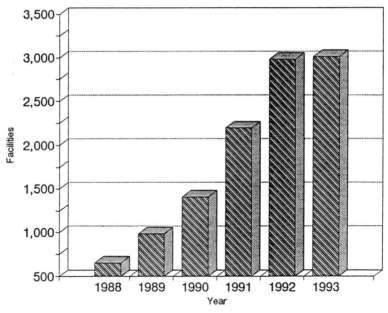

Figure 3 Number of Yard Trimmisting Facilities in the United States

A number of factors have been responsible for the growing interest in yard trimmings composting: 1) the potential for diversion of substantial quantities of organic materials from landfill; 2) implementation of regulations by some local governments to ban yard trimmings from landfills; 3) the compostability of the material; 4) lower technology process requirements than MSW composting; 5) less stringent regulatory requirements than MSW composting; and 6) the high quality of the end product.

Approximately two-thirds of the yard trimmings composting programs in the United States involve the composting of leaves collected in the autumn. These types of programs are more common in the northeastern portions of the country where greater quantities of leaves are generated. Leaf composting programs generally utilize low-technology composting processes.

The implementation of programs that process a mixture of yard trimmings materials (e.g., leaves, grass clippings, and brush) throughout the year are becoming more common, particularly in the western portion of the United States. These yard trimmings composting programs generally require higher levels of processing technology and more comprehensive marketing programs than do leaf composting programs. Mixed yard trimmings are generated year-round, and consequently the facility must be capable of processing the materials continually. In addition, some portions of the feedstock (e.g., brush) should be size reduced prior to composting.

Other

In 1993, 18 pilot programs were being conducted to evaluate the technical feasibility of composting segregated components (other than yard trimmings) of the solid waste stream. The projects primarily use windrow technology and range in size from 1 ton/week to 100 tons/day (Martin, 1993). Most of these programs process some type of food waste that had been separately collected from residential, commercial, or industrial sectors. Sources for the food waste vary, e.g., commercial food processing, restaurants, grocery stores, residences, and college cafeterias. Other types of segregated waste fractions that are being tested in pilot programs are chicken litter, paper, and combinations of separately collected organic wastes.

The pilot programs have demonstrated the technical feasibility of the composting process itself. As they consider full-scale implementation, communities are dealing with questions regarding collection methodology, economics, and markets.

Biosolids

Although biosolids composting has been practiced for decades in the United States, it wasn't until the 1980s that it became a preferred method for managing the material. The interest in biosolids composting, as demonstrated by the number of operational facilities, has continued to grow during recent years – from 159 operations in 1992, to 186 in 1993, to 201 in 1994 (Goldstein, 1994A).

During the 1990s, a few significant trends have developed related to biosolids composting, including an emphasis on producing an exceptional quality of product (i.e., one that meets federal requirements for unrestricted distribution); an increase in the use of yard trimmings (primarily leaves and woody material) as a bulking material; and private ownership and operation of composting plants.

Compost as a Product – Its Uses and Markets

The benefits of using compost as a soil amendment have been known for many years. When incorporated into the soil, compost increases the organic content of the soil and can improve its texture, its nutrient content, and its water retention and aeration capacities. Because of its beneficial characteristics, the material can be used in a variety of applications. Six major market segments for compost in the United States have been identified (U.S. EPA, 1993):

— Agriculture – food and nonfood crops and sod farms
— Landscaping – industrial and commercial properties; grounds maintenance (golf courses, cemeteries, and athletic fields)
— Nurseries – potted plants, bare root planting, and forest seedling crops
— Public agencies – highway landscaping and maintenance, parks, recreational areas, and other public property
— Residents – home landscaping and gardening
— Other – land reclamation, reforestation projects, landfill cover, hydromulching

The agricultural industry is the largest potential market for compost in the United States, although it is also the most difficult to penetrate. In addition to product quality, cost, ease of application, and availability are extremely important to farmers.

Marketing Practices

Various approaches are used to market finished compost products. These approaches can be generally categorized either as: 1) direct marketing (e.g., to landscapers, nurseries, homeowners, product blenders); or 2) brokering (i.e., marketing through a single distributor or a series of distributors who sell the product on a retail or wholesale basis). Strategies used for developing markets include providing information on the product (data and instructions), demonstration plots, giving samples of the product away, participation in trade shows, and advertising.

The majority of composting operations in the United States market the compost directly, primarily to professional customers, in bulk form, at a relatively low cost. Many facilities are recognizing the need of having a diversified customer base.

A recent survey of 60 composting facilities (Alexander, 1994) indicated that approximately 70% have a marketing or product distribution program. Of those with a program, 70% sell the compost in bulk form only, and 30% sell in both bulk and bags. None of the facilities sell only in bags. Facilities that do not have a marketing program either distribute the material at no cost or use it on municipal projects, as landfill cover, or in land application projects.

Municipal Solid Waste

Due to a lack of a significant track record, generalizations regarding markets and marketing practices for MSW composting facilities are difficult to make. With a few exceptions, the facilities have not established long-term, paying markets for their products, for a variety of reasons. Some facilities have chosen to utilize the compost for municipal projects (e.g., landfill cover, public works projects) or to give the product away. Others have experienced difficulty in meeting state regulations or in producing a quality compost.

Currently, of the 18 MSW composting facilities in operation, only one (Lakeside, Arizona) reports that it sells all of the compost it produces. The compost is sold to a soil blender, for use by nurseries and landscapers.

Several other facilities sell a portion of total production. Other uses for the finished compost are landfill cover, public works projects, or application to farmland owned by the public entity. Some facilities give material to farmers, soil blenders, or landscapers.

Yard Trimmings

During the late 1980s, the majority of the composting facilities that processed yard trimmings (primarily leaves) were low technology, produced a medium-grade compost product, and either used the compost for municipal projects or distributed it at little or no cost to homeowners. More recently, a number of factors have resulted in an increasing emphasis being placed on the production of a high quality compost. These factors include: 1) the implementation of programs for composting a mixture of yard trimming materials (e.g., leaves, grass clippings, and brush); 2) the higher cost associated with these programs (i.e., due to size reduction, aeration, post-processing, etc.); 3) competition from other composting programs; 4) market demand for a higher quality product; and 5) privatization of some of the programs.

Direct bulk sale to the landscape and nursery industry is the primary market for compost produced from yard trimmings. Homeowners also represent a significant market, both for bulk product (either directly from the facility or through topsoil dealers) and bagged product (usually through garden centers). Markets for lower quality compost include land application, land reclamation, and landfill cover or closure. A number of facilities are having some success at marketing to agriculture.

Biosolids

At the present time, the most popular markets for biosolids compost are landscapers, topsoil dealers, and garden centers and nurseries. Biosolids compost is also frequently used for landfill closure and land reclamation, and by public agencies in parks, for highway maintenance, on building grounds, etc. Neither residential use nor agricultural use represent significant markets for the product.

Biosolids composting facilities typically manage their own marketing programs rather than utilizing the services of a broker for product distribution, and have a combination of public and private sector outlets for the compost. Most of the compost produced by facilities with an aggressive marketing program meets federal (Part 503 Class A) requirements for unrestricted distribution. According to facility operators, customer satisfaction with finished compost is strong.

Factors That Affect Demand

A large number of factors affect the demand for compost. A few of the important factors will be discussed here, including product quality, policies/regulations, competing and complementary products, and transportation.

Product Quality

Marketing studies conducted throughout the United States have identified quality and consistency of the product as key elements in the utility and marketability of the products (CalRecovery, 1988, 1992, 1994; U.S. EPA, 1993). Consequently, an increasing number of facilities are placing emphasis on producing a quality product.

High-quality, mature compost has a dark color, uniform particle size, and a pleasant earthy odor. It should not contain visually identifiable contaminants (such as bits of glass, metal, and plastic). In addition, the compost should contain minimal levels of chemical residues, heavy metals (such as cadmium, lead, and mercury), herbicides, pesticides, and other potential toxic compounds. The compost also should have a high concentration of organic matter, contain nutrients, be free from pathogenic organisms, and contain no viable weed seeds.

Policies/Regulations

A wide range of policies and regulations affect composting, including quality standards, process requirements, and procurement policies. In addition, flow control policies are also having an effect on the implementation of composting projects, particularly MSW composting projects that require a large throughput to be cost effective. The two policies that are most likely to affect compost marketing efforts are product quality standards and procurement policies.

Quality Standards

While product quality is very important to the marketability of compost, with the exception of biosolids, uniform specifications have not been developed nationwide for compost. Product quality standards for solid waste composts have been implemented in a number of states. These standards vary in terms of terminology, feedstocks regulated, parameters monitored, allowable limits, and allowable uses. The discrepancies among the states, and the evolving nature of the regulations, can negatively affect the implementation of composting programs and marketing efforts for the finished products. Efforts are currently underway by private and public agencies to develop national product quality standards for solid waste composts.

Procurement Policies

Public procurement policies in effect in many parts of the country historically have essentially served to discourage the use of composts produced from waste materials. Most often, this is the result of specifications for soil amendment products that stipulate the source material for the product. A number of state and local agencies have revised or are in the process of revising their specifications to allow or encourage the use of waste-derived compost by public agencies.

Competing and Complementary Products

Another factor that affects demand is the quality, availability, and cost of compost compared to its competing and complementary products in the marketplace. Many of these product (e.g., fill dirt, topsoils, silt, potting soils, custom soil mixes, bark mulch and wood chips, manure, peat moss, mushroom compost) have a long history of consistency, availability, reliability, acceptance, and use in agriculture, horticulture, public and private landscaping projects, and residential gardening. Development of a market niche for compost requires an aggressive and consistent marketing program, including giveaways, demonstrations, advertising, and promotional efforts.

Transportation

Distance also is a factor affecting demand. One reason is that proximity to compost facilities promotes product acceptance and recognition. Thus, a potential user is more likely to know of a product that is produced nearby, and may feel compelled to support the project. Also, if the composting facility is located a long distance (e.g, more than 50 to 100 miles) from compost markets, the cost of transport may be prohibitive, particularly for bulk product which has a relatively low product value. Strategies that can be employed to mitigate high transportation costs include siting the facility near markets, developing local markets for compost, backhauling, and establishing a network of distribution centers.

Conclusion

Composting seems to have a bright future in the United States. As land suitable for landfilling continues to decrease and regulatory mandates require the diversion of wastes from the landfill, composting will play a major role in the treatment of the wastes.

However, as the number of composting facilities increases, competition for markets will also increase. Competition will more than likely lead to stringent requirements for high quality products, which must be able to meet the demands of knowledgeable users, and will be able to command high prices.

References

Alexander, R. (1994). Compost Market Programs at 60 Facilities. BioCycle, 1, 34–36.

CalRecovery, Inc. (1988). Portland Area Compost Products Market Study. 107 pp.

CalRecovery, Inc. (1992). Feasibility Study for MSW Composting in Kane County, Illinois. 127 pp.

CalRecovery, Inc. (1994). Biosolids Compost Market Assessment. 43 pp.

Eggerth, L.L., and Diaz, L.F. (1992). Compost Marketing in the United States. Presented at the BIOWASTE '92 Conference, organized by the Danish Waste Management Association (DAKOFA), sponsored by ISWA and IAWPRC, Herning, Denmark.

Goldstein, N., Riggle, D. and Steuteville, R. (1994). Biosolids Composting Strengthens its Base. BioCycle, 12, 48–57.

Goldstein, N., and Steuteville, R. (1994). Solid Waste Composting Seeks its Niche. BioCycle, 11, 30–35.

Martin, R. (1993). Composting Facilities. Waste Age, 8, 101–108.

Steuteville, R. (1994). The State of Garbage in America. BioCycle, 4, 46–52.

U.S. Environmental Protection Agency (1993). Markets for Compost. Prepared by CalRecovery, Inc. and Franklin Associates. EPA/530–SW–90–073A, 179 pp.

U.S. Environmental Protection Agency (1995). MSW Recovery Rate Surpasses 20 Percent. Reusable News, EPA530–N–95–001, 1–2.

The Natural Markets for Compost*
RODNEY W. TYLER, Browning Ferris Industries, USA

'Every man owes a part of his time and money to the business or industry in which he is engaged. No man has a moral right to withhold his support from an organization that is striving to improve conditions within his sphere. – Lawn & Landscape Maintenance Magazine, 1993, after President Theodore Roosevelt, 1908.

In every major industry, the learning curve associated with the identification, quantification, and understanding of the markets is possibly the most limiting factor to total industry development. The faster the markets are identified, quantified, and understood, the faster the industry will grow. Where little information on the market is available, equally small market development seems to follow.

Financially oriented companies use profitability, cash flow, return on investment, and ratios of liquidity as rulers of measurement to compare success within their industry. Although all of these standard measurements are important in determining performance of compost producers today, so are identification, quantification and understanding of available compost markets. After all, if markets are not yet identified, how can total profitability be calculated? Moreover, if accurate quantification of markets are not possible, how can reliable, accurate financial forecasts be developed? If salespeople do not understand customer needs, how can they satisfy the customer?

Since no thumb rules have been established for the measuring of markets involved in composting, some limitations in the industry's growth have occurred. This is not entirely unusual. Historically, the compost industry has not considered the production of compost at each 'compost factory' a manufacturing process at all (Tyler, 1993). Compost plants have been built without adequate market assessments. Why else are there so many noted facility failures which point to excess inventory and odor as main problems? When market development does not keep up with products manufactured from the 'compost factory', a market surplus in eminent, making the tidal wave of compost feared by the year 2000 quite feasible (Tyler, 1993).

Thinking of a composting facility as a manufacturing plant identifies critical components of a successful operation. Like any manufacturing operation, there is a need for consistency in product quality, which is dependent on the incoming raw materials and the production process (Tyler, 1993). Without control over the inflow

*This is an excerpt from Chapter One of the book, 'Winning at the Organic Game'– The official Compost Marketers Handbook', by Rodney W. Tyler, available late 1995 from GIE Publishers, (publishers of Lawn & Landscape Maintenance Magazine), Cleveland, Ohio.

of materials, the factory cannot promise that high quality products will be produced. However, some compost factories have become quite advanced, timing production with sales and resulting in just in time inventory programs that satisfy market demands.

The Compost Factory Model

The compost factory model is a set of charts, theories and diagrams which help structure the thought processes of decision makers to help them become successful. The model includes a compost factory diagram, natural hierarchy flow chart, charts for uses of compost within each major market, and economic graphs indicating various example results of actions taken. The combination of these items should link the individual compost use within each market to the big picture. This is necessary for compost marketers who want to be sure to maintain the highest return for their products marketed while not creating an inventory problem. The model structures markets so simple measurement of each area can be communicated accurately, to the industry or within a company, about a broad range of issues.

The Compost Factory model is based on two major theories: 1). There are a number of dollar markets close to cities that will currently purchase all of the compost produced and 2). The historical development of compost markets in every major city of the United States has shown a *natural market hierarchy* which identifies *dollar markets* as those developing first and *volume markets* developing last. Dollar markets are described as such because they have the ability to generate revenue for organic materials by either reselling them directly or charging appropriately for their costs within current service businesses. Many dollar markets, like garden centers, actually act as middlemen in the marketing process. The Volume markets have little ability to pay middlemen and these markets are more challenging to develop without some type of built in economic incentive.

The difference between dollar and volume markets is subtle yet enormous. Collectively, the dollar markets represent the majority of the revenue potential of all markets, while massing only a minor amount of total volume. Individually, with all of the markets listed together, it is hard to say exactly where dollar markets and volume markets meet, especially considering local and regional market conditions. Therefore an arbitrary line exists at the point where the natural market development (for profit) ends and subsidized marketing takes over.

There will be over 200 million tons of urban waste generated in the U.S. yearly by the year 2000 (USEPA, 1994). Estimates vary, but most experts agree that about 60% of the waste stream is compostable, which may result in about 120 million tons of compost produced yearly. These figures generally are expected to increase proportionately as population increases. In year 3000, there will be more compost available from recovered organics than in year 2000, assuming we do not destroy the planet before then.

For the general populous reading this book, 120 million tons of compost produced is a good round number to focus on for available compost in the future. Where will all this compost go? The Battelle study (Slivka, et all.,1992) indicates that a market for 518 million tons per year exists, but does not identify time lines for development, nor does it focus on standard use guidelines within each market (Tyler, 1993). It is interesting to note that the majority of the Battelle study indicates that compost will be marketed to the volume markets. This has not yet happened.

Volume markets are the largest markets but compost sold in these markets must be offered at a low cost (USEPA, 1993). Further, the largest barrier to agriculture market development for compost is price (Jones, 1993). Over the last five years, the dollar markets have proven to be large enough to absorb most compost produced in large urban areas (see 'evidence' later in the chapter). The data search performed to validate the theories for this book was exhausting and show that the theories are indeed reality in most large cities where the majority of the organic residuals are generated..

In every city, a natural market structure, or hierarchy, has evolved. Why? Because most companies making and marketing compost products are trying to be profitable. They want to sell products for as much as possible, because that is what a good business does. Compost products of high quality have been reasonably priced compared to alternatives like peat moss and have sold well. Composts of lesser quality have been lower in price, free, or negative in value.

Sales of compost has revolved, at least initially around 'dollar markets' rather than 'volume' markets. The split is identified in the figures below and validated by the amount of evidence supporting this claim. Since a natural split already exists, it seems logical to measure the markets in the same way on a regular basis. Subsequently, yearly sales and marketing reports can be compared to understand increases or decreases in various market sectors.

The Natural Market Hierarchy

Local supply and demand helps determine a natural economic hierarchy for local markets. Major metropolitan areas will fall prey to the Dollar markets because there are enough people creating the demand for high quality compost and willing to buy the product. Volume markets will probably be left vacant for those companies entering the industry later or for lower quality products.

Business economics suggest that compost effectively marketed will find it's way to the highest dollar market. In resource management, this is referred to as highest and best use. 'Excess' compost will naturally gravitate to the next highest paying market. Unfortunately, using this method alone for market development may lead to a market surplus due to lack of diversification during initial market development. Ironically, the highest dollar markets also have correspondingly low total volume potential. It appears that these two measurements (volume and revenue potential) are inversely related. Evidence supporting this theory is plentiful. Another example of the hierarchy is listed in figure 1.

Evidence of natural hierarchy within dollar markets

1). A study of 35 compost facilities in New England by Lang and Jager concluded that only 13 of the total had developed more than one market for the final product. This is usually the case for marketing programs which involve low yearly volumes. They further showed that market development followed a somewhat logical pathway of landscape contractors, public agencies, nurseries, groundskeepers, homeowners, and finally agriculture (Lang and Jaeger, 1993). Due to many of the facilities giving the product away, the natural hierarchy was somewhat skewed, but Agriculture was noted to be underdeveloped as a market because of so many other uses in dollar markets were available within acceptable price ranges (Lang and Jaeger, 1993). This study clearly shows dollar markets develop first.

2). Malcolm Beck has been building the compost market in Texas for many years and has done it based on returns from some of the higher dollar markets identified in the hierarchy (Goldstein, 1993; Tyler, 1993). In Beck's case, retail and wholesale markets equal 40% and 60% of sales respectively. Although it is always hard to determine if the local retail market is saturated, it is obvious that dependence upon other markets is eminent as production increases. Malcolm Beck's market is entirely focused upon the dollar markets (Goldstein, 1993).

3). Collins, Fritsch and Diener (1993) identified three basic market areas at increasing distances from Washington D.C. as high value, medium value and low value markets, respectively. For low value markets at a far distance, product was planned to be given away. Markets closer to the Baltimore-Washington D. C. area commanded $14.00 per cubic yard in the plan (Collins et. all, 1993). They went one step further to suggest the market development plan be segmented by increasing the program's % of compost sold (vs. given away) from 25% to 75% over three years. This allowed for the give away market to be shifted slowly to the dollar markets over time (Collins et. all, 1993). In this particular case, the free compost actually found a home in dollar markets. This is often done in an effort to introduce new products and establish utilization habits. Danger exists when customers begin to rely on the free compost and price it accordingly in their work. When the cost of compost suddenly increases, they are unable to use it unless they take a loss.

4). American Soil inc., in New Jersey has established an extensive network of buyers including Garden Centers, Landscapers, Horticultural and Green Retail outlets, Homeowners, Urban Gardeners and Soil Blenders (Young, 1993). All of these markets represent dollar markets available between the Philadelphia-Newark swath of urbanization that have customers willing to pay top dollar for high quality products (Young, 1993).

5). Data collected on 290 Biosolids composting projects in development or operation showed the most widely developed markets were: 1) Landscapers, 2) Nurseries, 3) Public works agencies, 4) Topsoil blenders, and 5) Golf courses (Goldstein et. all, 1992). All of the options for high volume markets were listed last. It is interesting to note even in this example that the retail market, although it offers high potential returns, has rarely been targeted exclusively for any compost factory's main marketing strategy. This is because in most cases, the volume associated with the retail sector cannot completely absorb all production from the factory and other markets must therefore be relied upon.

6). In yet another review of 60 marketing programs at various facilities, Landscapers, Garden Centers, and Topsoil dealers were the top three markets identified for Biosolids or yard debris composts (Alexander, 1994). The natural market hierarchy definitely exists, at least for high quality products. In the same study, MSW compost claimed Farmers, Landscapers, Landfills and local Government as the top four markets (Alexander, 1994). This data suggests that MSW historically has not been as suitable or successful penetrating dollar markets and has developed a low value stigma.

7). A recent study conducted by USEPA focused on 30 compost factories which showed the majority of users were from dollar markets (see table 1.1).
Ironic as it may be, the majority of the volume potential for all markets lies in the volume market, and economic potential for the majority of all markets lies in the dollar markets. These names do not suggest that money cannot be made *selling* compost to the turnip growers or that lots of compost cannot be *given* to the retail sector. Indeed, special exceptions exist, but when comparing both dollar and volume potential to the total market, the titles adequately describe natural market segmentation that exists today.

8). A recent market assessment in California identified 13 separate potential markets for compost products with the dollar markets indicated as those willing to pay the most money for products (Shiralipour and Zachary, 1994). Additionally, all of these dollar markets identified quality as a key barrier to market development. (See chart below...Dollar markets are in bold).

9). Even in Florida, where composting research is quite diverse, especially in the Agriculture sector, products produced find their way first to Dollar markets. Undoubtedly, Florida is one of the largest potential Agriculture markets due to the sandy nature of soils, amount of high dollar Agriculture, citrus and greenhouse crops. Regardless of this vast market, the primary markets for compost include Landscapers, Nurseries and Soil blenders (Kelly, 1994).

10). A marketing study conducted by E & A Environmental Consultants for Hamilton County, Ohio indicated initial high end markets to be local Topsoil blenders, Brokers, Landscape Contractors, and Nurseries (Shiralapour and Zachary, 1993). This study was conducted in the early 90's and was strictly for composted Biosolids, but indicates that any organic material of high quality will first be sold to the Dollar markets.

Table 1 Compost/Mulch End Products

Community	Compost or Mulch End Product	Compost or Mulch End User	Sale Price ($)
Austin, TX	Compost	Landscapers, Retailers	Marketed as 'Dillo Dirt'
Berkeley, CA	Compost, Mulch	Wholesalers, Nurseries, Residents	$7 – $15/cy
Berlin Township, NJ	Compost	Residents	$0
Boulder, CO	Mulch	Residents, Public Facilities	$0
Bowdingham, ME	Compost	Residents	$0
Columbia, MO	Mulch, Wildlife Habitat	Residents, Landscapers	$0
Dakota County, MN	Compost, Mulch	Residents, Landscapers	$0 to $8/cy
Fennimore, WI	Compost, Farm Application	Farmers	used in City
King Co., WA	Compost, Mulch	Privately marketed	NA
La Crescent, MN	Compost	Residents	$0
Lafayette, LA	Compost for Public Facilities	Public Facilities	Not sold
Lincoln, NE	Compost for Landfill, Mulch	Landfill, Landscapers	Mulch $3–$8/cy
Lincoln Park, NJ	Compost, Mulch	NA	NA
Mecklenburg Co., NC	Compost, Mulch	Residents, Landscapers Compost $10/, Mulch $4–$6/cy	
Monroe, WI	Compost	Residents, Public Facilities	
Naperville, IL	Compost, Mulch	NA	$0 for Mulch
Newark, NJ	Compost, Mulch	Rutgers U.Urban Gardening, Businesses	$0 to $2/cy
Perkasle, PA	Farm Application, Mulch	Landscapers, Farm	$0
Peterborough, NH	–	–	–
Philadelphia, PA	Compost, Mulch	Residents, Landscapers, Community Gardens	$0
Portland, OR	Compost, Mulch	Residents, Landscapers, Nurseries Varies	
Providence, RI	–	–	–
San Francisco, CA	Compost, Mulch	Retail and Residents	NA, Mulch $0
Seattle, WA	Compost	Retail and Wholesale Outlets $6/cy, $3/cubic-foot bag Sonoma Co., CA	
Compost		Landscapers, Farmers, Residents $15–$25/cy	$0
		Residents, Garden Shops	
Tacoma Park, MD	Compost, Mulch	Wildlife Habitat, County, Residents Compost $7/cy Mulch	
Upper Township, NJ	Compost, Mulch,	Residents, Farmers $0	
Wapakoneta, OH	Farm Application		
West Linn, OR	Compost, Mulch	Residents, Public Facilities $5/cy or $3/3 ft bag	
West Palm Beach, FL	Mulch	Residents (Mulch), Landscaping at Landfill $0	

Key: cy = cubic yard   NA = Not Available   – = Not applicable

(Source: Adapted from EPA530–R–92–015, 1994)

Table 2 Market Assessment Summary

Compost User Group	Potential Demand for Compost			Willingness to Pay ($/ton)	Willingness Key User Specifications	Potential Barriers to Compost Use
	Low	Medium	High			
Avocados and Lemons	3,953	11,293	16,942	1–10	Salt content and pH	Cost, quality, availability, field accessibility
Vineyards	1,030	3,433	5,149	1–10	Salt content, pH, nutrient content	Cost, availability, product consistency
Vegetable Crops	8,961	29,871	44,807	1–10	Salt content, pH, moisture	Cost, availability, quality and nutrient content
Strawberry Crops	375	1,250	1,875	1–10	Salt content, pH, moisture	Cost, availability, quality
Miscellaneous Crops	—	—	—	—	Odor, salt content, pH, nutrient content	Cost, quality, availability content
Organic Farms	4,085	16,341	24,512	6–10	Nutrient content, salt content, moisture, pH	Cost, availability
Ornamental	345	862	1,725	6–10	Salt content, pH, consistency, odor, & moisture content	Cost, quality
Landscape Companies	2,593	6,483	12,967	20–40	Odor, salt content, pH, nutrient content, consistency	Cost, quality, availability, product consistency
Nurseries	931	2,327	4,655	20–40	Salt content, pH, nutrient content, odor, consistency	Quality
Mining Industry	700	2,100	3,500	1–5	Nutrient content, pH, salt content	Quality
City and County Parks and Recreation	46	117	233	20–40	Salt content, odor, product consistency	Cost, timing of reclamation operations
Public Works Landfill Cover	1,900	4,750	9,500	0	Moisture content	Cost
Caltrans District 05	0	0	0	0	—	Uncertainty of potential need
Total	24,919	78,827	125,865			Uncertainty of potential need

Notes: 1) — = data not available.
(Source: Shiralipour and Zachary, 1994)

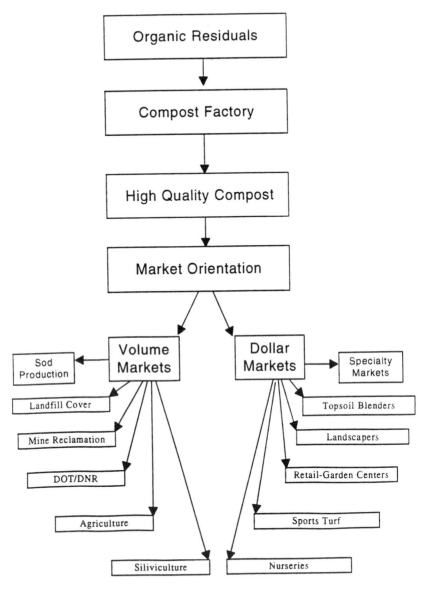

Figure 1 The compost Factory and Natural Market Hierarchy

11). In the early 80's, D. D. Southgate from Ohio State University determined that the greatest demand for compost produced in the future would be in markets willing to pay the most money, especially the horticulture and landscape industries (Southgate, 1981). Although compost was not as commonly accepted in the green industry in the early 80's as it is today, this study shows that the dollar markets were proving to be the greatest attraction for quality products.

12). A 'Market Status Report' in 1992 was written by the California Integrated Waste Management Board (CIWMB) and indicated the 'most established markets in California for organic soil amendments are Landscaping, Nursery, Gardening and Soil blending sectors'. They also point out that soil blenders are not actual end users but are middlemen, using compost as an ingredient in final products produced from a variety of ingredients (Jones, 1992).

13). AgRecycle, a leading composting company in Pittsburgh, Pennsylvania, has a diversified marketing program targeting 100% of all production to landscapers and other high end markets (Goldstein, 1994). DK Recycling Systems in Lake Bluff, Illinois processes large amounts of yard trimmings (125,000 cubic yards per year) and markets about half of the final product to landscape contractors while the rest is sold to nurseries, golf courses, and garden centers (Goldstein, 1994).

This book does not suggest pricing nor guarantee accuracy of figures listed as example prices. However, prices are listed to give examples credibility and to reflect general marketing experience within the industry at this point in time.

Benefits of the Compost Factory Model

1). Allows the industry to be identified, segmented, quantified, and understood based on natural division in the marketplace.
2). Increases research and development projects which will be based on the needs of the marketplace.
3). Improves communication about markets, uses and the difference between them.
4). Helps educate decision makers so correct compost facilities *are built after* appropriate market assessment.
5). Helps marketers keep profits in mind which should positively affect their bottom line.
6). Provides a mode of interaction between the marketplace and manufacturers of compost products to increase awareness of needs on both ends.

Applying our history lessons to future challenges

Although the composting process is centuries old, it's full development into a successful organic recovery option in a myriad of waste management choices is underutilized. The model described in the following pages helps all those associated with composting by using a standard: The identification, quantification and understanding of markets based on common use guidelines. Although the industry is far from mature, enough information is available about compost utilization in the major markets to form a measurement system.

The composting business is a relatively new industry struggling to communicate between researchers, market developers, engineers, contractors, and the green industry. They all look for common ground which they can agree upon long enough to at least present their point of view. In order for composting to reach it's full potential as a organic recovery option, markets for the final product must be identified and developed. Without adequate market development, other options for waste management may be chosen in place of composting.

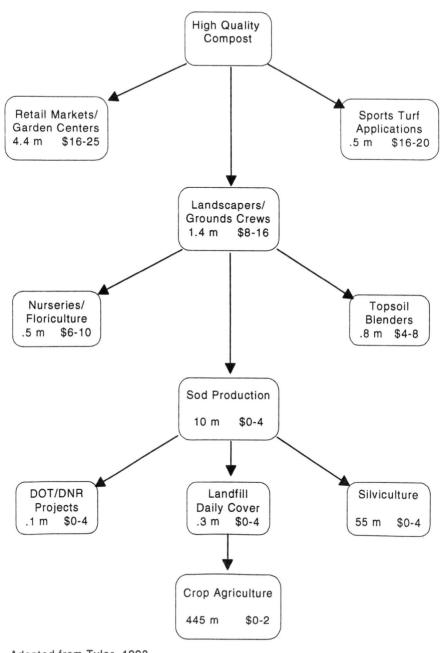

Adapted from Tyler, 1993.

Figure 2 Market Development Hierarchy with market size and value

The examples in this book set forth a measurement system of currently identified *uses for compost within each market*. The natural balance of 'dollar' and 'volume' markets is easily measurable and therefore fulfills the vital objective of quantification of each market *based on use*. As time goes by, measuring the industry using the same template, year after year, will increase the accuracy of the information. The model is therefore justified *from* the market *for* the market yet provides extremely useful information to the *entire* industry. Some key definitions follow.

Market: Defines the customer base ie., Chicago compost market

Market Sector: Customer bases within markets, ie., Landscape market segment in Chicago

Use: Defines how the product is utilized, independent of customer base ie., Topdressing can be done by homeowners, golf courses, & landscapers (each are market segments).

The method of breaking down the above definitions into organizational and informational forms can be structured using an outline format:

   I. Market (Usually geographical area)
      A. Market Sector (Usually customer base)
         1. Use (How they use it)
            a. Thumb rule guidelines

Do not confuse uses with markets when measuring market potential (because we measure what customers buy, not how they use it...yet). Compost sales and marketing programs *are based on popular uses* and within the markets affected most. A topdressing program (the use) is promoted to Golf Courses, athletic fields, and grounds managers (market sectors).

Conclusions

The objective of the Compost Factory model is to help determine potential consequences of decisions before they are actually made. Playing the 'what if' game in business is probably one of the oldest management tools around, yet a format to do this in the compost business has not been established until now. With the help of the Compost Factory model, industry personnel can determine the appropriate estimates for each market area and estimate the bottom line with some simple calculations on a computer. Of course, this is assuming that only high quality, source separated composts will be made that will be welcome in any market.

   Markets can be determined chiefly by supply and demand, product quality, feedstock materials, technology available, potential volume and product sales price (Tyler, 1993). The degree of both market development and manufacturing diversification at compost factories is crucial to long term success of any marketing program. Obtaining reliable dollar markets are worth the wait, providing profitable returns to successful marketers (Tyler, 1993). The dollar markets may be adequate in size to satisfy production from urban compost factories in the future if products are of acceptable quality. For many companies, research and development involving compost use in the dollar markets is easier to show an economic return than research relating to volume markets. Therefore, it is expected that dollar markets will naturally develop first around each urban center (Tyler, 1994).

If private or public decision makers are considering building a compost factory, using the 'what if' situation is a powerful tool when used with the compost factory model. Initial planning involving engineers, bankers, plant pathologists, marketers and the markets can be successfully accomplished by using the compost factory model as a central discussion theme. The result should be compost factories manufacturing products to meet specified demands, impacting the industry, markets, and environment only in a positive way (Tyler, 1993).

References

Tyler, Rod. 'Diversification At The Compost Factory', BioCycle, August 1993, Pgs 50, 51.

Slivka, Donald C., Thomas A. McClure, Ann R. Buhr and Ron Albrecht. 'Potential U.S.Applications for Compost', Application Study Commissioned by The Proctor & Gamble Company for The Solid Waste Composting Council, January 1992.

USEPA. 'Markets for Compost', #EPA/530–SW–90–073A, November, 1993. pgs.2.13, 2.17, 4.1,4.5, A–20.

Jones, Pat. 'Market Research into the use of organic soil amendments by Agriculture in California', A Report by Emcon Associates, August 9, 1993. pg. 10.

Lang, Mark E. and Ronald A. Jager. 'Compost Marketing InNew England', BioCycle, August 1993. pgs. 78, 79, 80.

Goldstein, Nora and Robert Steuteville. 'Biosolids Composting Makes Healthy Progress', BioCycle 1993, pgs.. 56, 57.

Collins, Alan R.; David A. Fritsch; and, Robert Diener. 'Expanding Uses For Poustry Litter', BioCycle, January 1993, pg. 66.

Young, Robert F. 'Colection And Composting Of Commercial Organics', BioCycle, March 1993, pg. 50.

Goldstein, Nora; David Riggle; and Rob Steutevile. 'Sludge Composting Maintains Growth', BioCycle, December 1992, Pgs. 49, 50, 51, 52, 53, 54, 55, 56, 83.

Alexander, Ronald. 'Compost Market Programs At 60 Facilities', BioCycle, January 1994, pgs 34, 35, 36.

USEPA. 'Waste Prevention, Recycling, and Composting options: Lessons form 30 Communities', #EPA530–R–92–015, February, 1994. pg. 34.

Shiralapour, Aziz. 'Santa Barbara County Preliminary Compost Market Assessment', prepared   for the Santa Barbara County Solid Waste Management Division and the California Integrated Waste Management Board by the Community Environmental Council, April, 1994. pgs. 3.

Kelly, Scott D. 'Sites, Plastic Bags and Compost Markets', BioCycle, September, 1994. pg. 53.

Shiralipour, Aziz. 'Compost market development: a literature review', prepared for the Santa Barbara County Solid Waste Management Division and the California Waste Management Board by the Communtiy Environmental Council, October, 1993. pg. 27.

Southgate, D. D. 'Potential markets for Akron sludge–derived compost', Unpublished research, 1981.

Jones, Patricia. 'Market Status Report Compost', California IntegratedWaste Management Board Staff Report, August 1992, pg. 11, 15.

Goldstein, Nora. 'Compost companies with green roots', BioCycle, May, 1994. pgs.42, 46, 47.

Tyler, Rod. 'Diversification At The Compost Factory', BioCycle, August 1993, Pgs 50, 51.

Tyler, Rod. 'Fine-tuning compost markets', BioCycle, August, 1994, p 41–48.

Monitoring Strategies and Safeguarding of Quality Standards for Compost

J. BARTH Dipl.Ing. – Infoservice Bundesgütegemeinschaft Kompost e.V., Oelde, Germany

Summary

The qualified separated collection of organic residues from households, garden- and park areas is the actual technical standard in Germany and obligatory as to the regulations of the waste law. The former composting municipal solid wastes has not been proved to be successful.

The German Ferderal Compost Quality Assurance Organisation (FCQAO) defined a general quality standard (RAL compost quality sign) and established a nation-wide system for external monitoring of composting plants and compost products. Today approximately 170 compost plants participate in the quality assurances of the FCQAO.

The quality assurance programme contains the definition and continuation of quality requirements, the organisation and enforcement of quality monitoring, the punishment of regulation omissions and violations and the labelling of the quality standard.

The type, extend and frequency of evaluations depends on the capacity of the composting plant. In order to guarantee an identical standard for the monitoring all over Germany, FCQAO established a central office where all results originated from external monitoring are evaluated.

Introduction

Biocomposts are composts originating solely from organic residues. These residues consist mainly of biowaste (source separated organic household waste) and residues from garden and park areas (generally municipal yard waste).

The qualified separated collection of organic residues from housholds, garden- and park areas is the actual technical standard in Germany and obligatory as to the regulations of the waste law and the TA Siedlungsabfall. The former composting municipal solid wastes has not proved to be successful because of lack of utilization. Today it is no longer permitted in Germany.

The evaluation of organic residues to be utilised in compost production has to take into account value adding contents and properties as well as those which diminish the value of the product.

Depending on the kind of use, properties such as nutrient and organic matter contents, the level of substances showing alkaline reactions or the magnitude of active substances which, for example are able to suppress plant pathogens are seen as positive with respect to compost quality.

With regard to product properties which are value diminishing, the content of undesirable materials and contaminants as well as hygienic aspects have to be considered.

Therefore, compost quality is determined by an array of different contents and properties, rendering the objective evaluation of *the* compost quality to be a difficult task. An additional problem poses the fact that the various quality parameter are evaluated quiet differently according to the type of use, the objective of the discussed regulation or to ones individual prejudice.

This was the situation when the Federal Compost Quality Assurance Organisation (FCQAO), in association with affected industries, i.e. mainly compost users, defined a quality standard for compost which ensures that the product meets the highest quality requirements posed by the industry.

This step made not only compost quality objectively determinable and comparable but also created a defined product standard which was then available on the market. To identify products of this standard, FCQAO created a quality sign for compost (RAL-GZ 251) (Figure 1). The German Federal Compost Quality Assurance Organisation is recognised by the RAL-Institute as the organisation to handle monitoring and safeguarding of compost quality in Germany.

Figure 1 RAL compost quality sign

During procedural steps to obtain RAL recognition, relevant industries and compost users as well as involved authorities were consulted on the issue of quality requirements to be contained in the compost quality sign.

The compost quality sign provides evidence for quality assurance. It has been recognised by RAL on 28. 1. 1992 and was subsequently published by the Minister for Trade and Commerce in the Federal Legal Gazette and also registered in the trade mark register with the Federal Patent Office.

During the last two years the Federal Compost Quality Assurance Organisation (registered society) established a nation-wide system for external monitoring of composting plants.

Today approximately 170 compost plants in Germany with a yearly input of 1,5 Mio. tons participate in the quality assurances of the German Compost Quality Assurance Organisation.

According to RAL principles governing a compost quality sign, a quality assurance programme needs to contain the following:

– *Definition and continuation of quality requirements*
  – regulations concerning quality and its evaluation (e.g. product diversification)
  – definition of analytical methodologies (handbook)
  – criteria and provisions for internal evaluations of a   composting plant (e.g. guarantee that the finished product is hygienically quite safe)
– *Organisation and enforcement of quality monitoring*
  – evaluation and designation of authorised laboratories
  – handling of forms, data and statistics
  – independent assessment of examination results
– *Punishment of regulation omissions and violations*
  – complaint about missing examination results
  – definition and fixing of required repeated examinations
  – decision on specific cases
– *Labelling of the quality standard*
  – award and certification of the compost quality sign
  – issuing of the test certificate containing details on
  – product and producer
  – declaration in accordance with the rules
  – appropriate application rates
  – analytical results and potential differences

Organisation of the quality assurance programme

The way to obtain the compost quality sign is described in detail in our brochure of the same title. Further information (application form for the quality assurance programme, model contract with laboratories, list of authorised laboratories, model sampling report, examination report, control of sanitation) is provided by the FCQAO office.

Fundamentally, the quality assurance programme represents a triangular setup comprising the operator of a composting plant, the Federal Compost Quality Assurance Organisation and one of the laboratories which are recognised by FCQAO. Recognition and authorization of laboratories follows established criteria.

The plant operator is a member of a regional association for compost quality. Through the regional association he files an application with FCQAO to use the compost quality sign and at the same time commits himself to obey the regulations governing the quality assurance programme.

The plant operator chooses one of the registered laboratories to carry out external monitoring of the plant, i.e. to take the required annual number of samples and to analyse them. Since the laboratory is recognised by FCQAO it is obliged to obey the regulations governing the quality assurance programme.

The laboratory conveys the examination results to the plant operator and FCQAO at the same time. This way it is guaranteed that all external examinations are used for the assessment and that external monitoring really takes place.

With respect to the award and use of the compost quality sign, the so-called recognition procedure needs to be differentiated from the monitoring procedure. The recognition procedure is carried out in order to obtain the compost quality sign and the monitoring procedure in order to check subsequent obedience with the quality assurance programme regulations. (Figure 2)

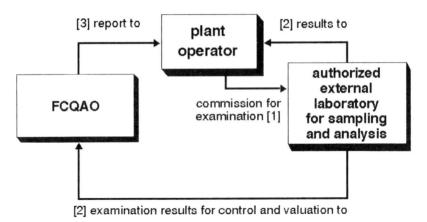

Figure 2 The monitoring procedure of the RAL quality sign compost

If the results of the one year recognition procedure comply with the requirements, FCQAO awards the RAL compost quality sign on the basis of a recommendation made by the FCQAO quality committee. Subsequently, the evaluated composting plant is transferred from the recognition to the monitoring procedure.

Assessment of external monitoring within the monitoring procedure is carried out once a year by the FCQAO quality committee. For the monitoring procedure, it is recommendable for the plant operator to sign a new contract with a recognised laboratory or to prolong the current contract. This guarantees that the sampling frequency and analysis are correct and that the tasks are carried out regularly and timely. Results of the monitoring procedure provide the basis for FCQAO´s annual confirmation that a composting plant is entitled to carry the compost quality sign. In order to guarantee an identical standard for quality monitoring all over

Germany, FCQAO established a central office where all results originating from external monitoring of composting plants are evaluated.

The establishment of this office made it possible to issue uniform certificates to composting plants throughout Germany. This certificate is updated annually and contains the following:

- acknowledgement of obeying quality requirements
- summary and assessment of analytical results, including average, median and variation of the specific parameter as well as a comparison with recommended levels
- information concerning declaration, suitability and utiisation
- information concerning the application rates based on proper agricultural and horticultural hus bandry (legal requirements)
- information concerning limited compost application rates due to potential contaminants (LAGA– information sheet M10)

Type, extend, frequency and time of evaluations

The type, extend and frequency of evaluations to be carried out according to FCQAO depends on the capacity of the composting plant, either on the authorised capacity or the actual input (Fig. 3). The type of licensed products is also of importance in this context.

The total number of evaluations has to be spread over the year in such a way that at least one sample is taken every quarter and that sampling dates are at least four weeks apart. However, it is possible to take two samples at one particular sampling date, provided it is ensured that samples are taken from different batches (proof through sampling record).

Figure 3 Frequency of quality testing within the recognition and monitoring procedure   required by the RAL compost quality sign

Input	Recognition	Monitoring
< 2.000	4	4
2.000 - 5.999	6	4
6.000 – 11.999	8	6
12.000 – 19.999	12	8
20.000 – 25.000	12	10
> 25.000	12	12

Status of external monitoring

Presently, FCQAO carries out external monitoring in some 170 composting plants *throughout Ger many.*

Total capacity of all composting plants looked after by FCQAO amounts to an input of 1,5 Mio. t. 71 % of this compost material carries the RAL compost quality sign when it is put on the market (Figure 4).

Figure 4 German composting plants participating in the quality assurance programme[1]

Characteristic value	Year
Number of plants participating in a quality assurance programme	170
Number of plants carrying the RAL-compost quality sign	130
Number of samples in 1994	900
Total authorised plant capacity (Mio. t)	1,80
Amount of processed organic residues (Mio. t)	1,50
Produced compost (total, Mio. t)	0,75

1) Quality assurance programme according to RAL GZ 251

Quality properties and characteristics of compost

Quality characteristics of compost from separately collected original materials vary in a typical range. (Fig. 5). The definition of the individual quality parameters is determined by different influences.

The inputmaterials which are used for composting have to be appropriate in structure and composition for composting. Furthermore they may only include a slight contents of impurities and harmless contents of harmful substances. These are the basic requirements for the production of usable composts in the sense of the governmental regulation TA Siedlungsabfall and the qualification for the RAL quality recognition. The producer has to guarantee these requirements by means of proper technical and organising measures.

Apart from the influences on the quality which are caused by technical/organising measures, influences on the quality of compost are to be considered which usually cannot be foreseen, such as:

– regional characteristics of geogen origin,
– regional characteristics of antropogen origin.

Considering the individual compost plant, a difference in quality will be caused respectively shown by further influences:

– Variation of produced composts on account of differences caused by seasonal conditions in character and composition of the delivered compost rawmaterial (up to approx. 30 % discrepancy of the plant average value of single quality parameters.
– Tolerances that result from the statistical possible faults of sample–taking, sample treatment, analysing and differences between authorised laboratories (up to approx. 50 % deviation from the statistical 'true value').

Finally german peculiarities are to be named such as e.g. the wide-spread conversion of the content of heavy metals in compost on a unified reference to a defined standardised compost with 30% organic substances. This means that compost having a measured value of e.g. 250 mg Zn/kg dry substance and 65% organic substances is rated in such a way as if it had 500 mg Zn/kg dry substance; (after a calculated reduction of the organic substances on 30% dry substance). Such a product would not meet the approximate value of the RAL quality sign on account of its high content of organic substances but not on account of a real high content of Zinc.

Figure 5 Valuable characteristics and constituents of biocompost

Parameter	Dimension	Average value	Range
Loss of organic matter due to burning	% DS	35	25–45
Water content	%	36	30 – 50
Net weight or volume	g/l	680	550 – 850
Soluble salt content	g/l	4	2 – 8
pH value		7,6	7,0 – 8,3
Main nutrients:			
Nitrogen (N–total)	% DS	1,1	0,8 – 1,5
Nitrogen (N–min.)	mg/l FS	150	100 – 400
Phosphor (P–total)	% DS	0,3	0,2 – 0,4
Phosphor (P–soluble)	mg/l FS	520	200 – 900
Potassium (K–total)	% DS	1,0	0,5 – 1,2
Potassium (K–soluble)	mg/l FS	2000	1200 – 4000
Magnesium (Mg–total)	% DS	0,4	0,2 – 0,7

DS = Dry Substances   FS = Fresh Substances

Hygienic requirements

Composts have to comply with hygienic requirements. This is achieved if it is assured that patho–gens are inactivated during the composting process. Many research projects investigating hygienic aspects of composting have shown that sanitation achieved during the composting process is sufficient to comply with the requirements, as long as the composting plant is managed properly.

Micro biological methods are used to evaluate elimination of pathogens in various composting systems. This requires that typical and representative microorganisms are used as test organisms which pass through the entire composting process (evaluation of the composting process). Additionally, chemical and physical parameter of the various composting processes as well as the processes themselves have to be included in the examination.

The level of sanitation depends mainly on the duration of the composting process and the reached temperatures. In order to obtain favourable conditions for progression of the composting process and associated sanitation, input materials need to be mixed and processed with special reference to water content, proportion of pore-volume, nutrient supply etc.

Due to the magnitude of pathogens potentially contained in the input material, it is not possible to meet hygienic requirements through an evaluation of the cured compost (evaluation of the compost). Only an evaluation of the entire operating composting plant (evaluation of the composting process) is able to provide reliable data. Such an evaluation of sanitary effects of the composting process involves the passage of typical pathogens through the entire composting process, followed by their isolation from the finished compost and a virulence test. It is required that used test organisms are inactivated sufficiently during the composting process.

With respect to survival of micro organisms during composting, used test organisms have to be selected in such a way that the following deduction can be assumed: If test organisms are killed, other pathogens will be also sufficiently inactivated. Therefore, it does not make sense to use many different test organisms. It should be adequate to use only one pathogen, provided its hardiness infers that the obtained results apply also to other potential test organisms.

FCQAO regulations determine

- that each composting plant utilising a specific composting system has to record operational composting conditions relevant to the sanitary status of their entire compost production (e.g. temperature level). These records are checked regularly as part of the external plant evaluation and have to be made available to FCQAO, if required.
- that a systems examination of the applied composting system is a prerequisite for the above procedure. Within the framework of this examination, relevant composting conditions (e.g. range of favourable water contents of input materials, temperature regime of the composting process, turning interval) have to be determined and recorded.

If the required records can not be provided, the produced compost can not be labelled as quality controlled.

In case no systems examination has been carried out yet for a comparable type of composting plant, it is required to take such action when a new plant is put into operation. Even if a certain type of composting system has been evaluated already, a new composting plant of the same type has to be examined when it is put into operation, if the input materials or the composting process itself vary from the examined type of plant with respect to sanitary effects.

In response to the fact that many new composting systems have been introduced over the last couple of years, FCQAO is presently preparing a list of available types of composting systems which passed the systems examination successfully, either through a so-called prototype examination or when the first plant of a certain type was put into operation. This list will also contain FCQAO requirements regarding checkable documentation of long-term operational composting conditions which affect compost sanitation.

LAGA information sheet No. 10 stipulates requirements concerning a systems examination for composting plants. In this context the representative test organisms are 'Plasmodiophora brassicae' and 'Salmonella senftenberg'.

In certain cases it might be wise to examine the finished compost for pathogens (e.g. check for club root if the compost will be used in potting soils) but generally it is not useful to postulate a product examination as obligatory.

From a quality control and quality assurance point of view documentation of long-term operational composting conditions is a lot more important. However, this approach requires that the specific composting plant either went through a systems examination when the plant was put into operation or that a prototype examination was provided when the composting operation applied for building permission.

References

1. RAL Deutsches Institut für Gütesicherung und Kennzeichnung e.V.: Gütesicherung Kompost, RAL-GZ 251, Fassung Januar 1992, Bonn
2. Bundesgütegemeinschaft Kompost e.V.: Der Weg zum RAL-Gütezeichen, Köln, Mai 1994
3. Bundesgütegemeinschaft Kompost e.V.: Qualitätskriterien und Güterichtlinien für das RAL-Gütezei chen Kompost, Köln, Mai 1994
4. Bundesgütegemeinschaft Kompost e.V.: Methodenbuch zur Analyse von Kompost, Köln, Nov. 1994
5. Bundesgütegemeinschaft Kompost e.V.: Ergebnisbericht Ringversuch 1993 zur Analyse von Kom post, Köln, April 1994

English References of the FCQAO (in printing)
No. 201e: How to obtain the RAL Compost Quality Sign
No. 202e: Quality Criteria for the RAL Compost Quality Sign
No. 222e: Handbook for analysing methods of compost (together with a summary with the essential results of a parallel interlaboratory test with 100 laboratories)

Minimizing the Cost of Compost Production Through Facility Design and Process Control

HAROLD M. KEENER, DAVID L. ELWELL, K.C. DAS, and ROBERT C. HANSEN – Department of Agricultural Engineering Ohio Agricultural Research and Development Center The Ohio State University, Wooster, Ohio

Keywords: Compost(ing), Design, Process control, Waste treatment, Aeration, Remix, Compost stability, MSW, Biosolids, Yardwaste, Kitchen waste.
Introduction

Abstract

Excessive cost of producing compost at many facilities through out the world continues to burden taxpayers. Observations on failed designs and management strategies in large scale facilities demonstrate operators have not clearly understood the interrelationship of the biological, physical and economic variables on the cost of production. This paper identifies the major variables under management control and their relationship on the cost of producing compost. Governing equations, based on energy and mass balances and the kinetics of composting, are presented and used to quantify the effects on cost and throughput capacity of in-vessel composting systems as operating practices are changed. Factors studied include air recycling, ambient air conditions, remix frequencies and moisture control, bed depths, fan management strategies (on/off, staged levels), temperature set points, and airflow paths. Data used in the analysis are from both pilot and full scale composting studies.

The composting process transforms biodegradables into nonodorous, more stabilized materials. Today it is accomplished using systems identified as windrow with turning, static pile with forced aeration and in-vessel (Anomymous,1991). For the process be used commercially, systems must be properly designed and operated (Hoitink et al., 1993).

The process is governed by the basic principles of heat and mass transfer and biological constraints of living organisms (Keener et al., 1993a). Many researchers have studied such systems and determined operational data for the success of their particular configuration (Kuter et al., 1985; Nakasaki et al., 1987). Haug's (1980,

1993) books details governing principles and reports on some of the systems that have been used for composting. Keener et al. (1993a) presented analytic expressions showing the interdependence between biological and physical factors. They used these expressions to derive the governing equations for optimizing the efficiency of the composting process. Later, Keener et al., (1993b, 1994) expanded those equations to aid in the efficient design and operation of the air handling system for compost facilities , to estimate remix times and to evaluate water requirements of a composting system.

This paper presents data on composting parameters, as derived in controlled experiments, and uses that data to analyze the operation of the composting system.. The reader is referred to Keener et al. (1993a, 1993b, 1994, 1995) for specific details on development of the derived equations used in this paper.

Theory

Kinetics of the Process

The active stage of composting is the time where substrate is readily available for microbial decomposition. From a design standpoint, this first stage of composting is most important for proper sizing of equipment. During that time, the rate of disappearance of biomass can be written as a first order reaction (Marugg et al., 1993). As the composting process approaches the curing stage, this assumption may be wrong, but it is then of lesser practical significance. The rate of disappearance of biomass, written as a first order reaction term, is

$$\frac{dm_c(\theta)}{d\theta} = -k(m_c(\theta) - m_e) \tag{1}$$

If k is constant, the solution to the equation is mR, the compost mass ratio.

$$mR = \frac{m_c(\theta) - m_e}{m_o - m_e} = e^{-k(\theta)} \tag{2}$$

The compost mass ratio, a dimensionless number, is a useful way to describe how far the process has advanced (Marugg et al., 1993; Keener et al., 1993a). It always ranges from 1 to zero. For $\theta=0$, mR=1 and for $\theta=\infty$, mR=0.

Equilibrium Mass (me) – Evaluation

The use of the term m_e in Eq. (1) assumes that decomposition of the organic matter by microorganisms is limited. The expression m_e/m_o (called β) represents that fraction of materials which remains after a long period of composting (6 months to 1 year). It has a value greater than the ash content since some organic fractions will remain in the stabilized end product. β should not be confused with non-volatiles, a proximate analysis measurement evaluated under high temperature[1] conditions. Keener et al. (1993a, 1995) discusses evaluation of β for a large number of materials. Evaluation of k and β using pilot scale data (Keener et al., 1995) are given in table 1.

Using the results for k and β, airflow, time fan could be off, and time to first remix were evaluated and included in Table 1. See appendix for the equations used. Airflows ranged from 0.5 m^3_a/kg_{ds}-day for MSW to 3.5 m^3_a/kg_{ds}-day for high grass yardwaste and separated dairy waste. Allowable fan time off ranged from 4 to 27 minutes. Time to first remix, based on moistures of 65% wb, was from 2–18 days (See Appendix for equations).

The equation for airflow,* is based on the fact that temperature control normally governs the airflow (rate) requirement rather than oxygen levels within a compost system (Finstein et al., 1986). Use of the equation with specific data for yard waste composting shows airflow requirements decreases 50 fold from day 0 to day 21 for this material (Hoitink et al., 1993)). Other materials, such as MSW, have a lower k value and therefore require less airflow.

Fan Power

Keener et al. (1993b) derived fan power per unit area of vessel size as

$$P'(\theta) = \sum_{\ell=1}^{2} \left[\frac{-h_C \rho_{©} k\, d\, e^{-k\theta}}{\rho_{a*}[HAO - HAI]} \right]^{n_\ell + 1} \frac{a_\ell d^{j_\ell}}{\alpha \varepsilon} \tag{3}$$

where $\ell = \{1 \Rightarrow$ piping and $2 \Rightarrow$ compost$\}$. For the piping system it is assumed $j_1 = 0$ and $n_1 = 2$.

The fan power required for unit weight of compostable material can be calculated from

$$P''(\theta) = \sum_{\ell=1}^{2} \left[\frac{-h_C k\, e^{-k\theta}}{\rho_{a*}[HAO - HAI]} \right]^{n_\ell + 1} \frac{a_\ell \rho_{©}^{n_\ell} d^{n_\ell + j_\ell}}{\alpha \varepsilon} \tag{4}$$

This expression can be used to investigate the effects of factors, such as n, $\rho_{©}$, h_C, [HAO–HAI] , ... , on fan power requirements. The general relationship is based on air pressure drop through porous media and in piping(Steele and Shove, 1969; Higgins et al., 1982). Equations 3 and 4 do not account for the effect of moisture on air porosity. (See Das and Keener in this proceedings) Table 2 list values for j and n for some specific composts.

*STM calls for 950°C; American Society of Agronomy uses 550°C.

Table 1 Evaluations of k, β, Airflow Requirements, Allowable Fan Off Time, and Time to First Remix Based on Moisture Loss.[a]

Compost Mix[b]	k (1/day) ave	k (1/day) sd	β ave	β sd	temp C	H₂O %	C/N	rho kg/m³	inert dec	re-mix[c] days	airflow m³/kg da	time off min.	re-mix1 days
g33/b/l33	.201	.009	.751	.011	60	62.0	27.2	286		4	2.002	5.93	4.43
g50/b/l17	.178	.024	.511	.038	60	63.0	20.4	223		3,4	3.482	4.38	2.01
g50/b/c17	.185	.008	.512	.022	60	63.0	21.3	223		3,4	3.611	4.22	1.93
yw/kw	.230		.770		57	62.1	16.7	175		3,7	2.381	9.17	4.55
yw/kw	.156		.781	.017	72	62.1	16.7	172		3,7	.683	14.45	6.32
yw/cm	.172		.663		65	43.2	22.0	345	.040	3,4	1.739	4.25	3.19
yw/cm/fw	.089	.023	.630		65	43.5	16.0	386	.040	3,4	.988	6.68	5.46
yw/cm/fw	.176		.664		65	41.5	25.0	365		3,4	1.774	3.93	3.13
yw/wc/s	.153		.701		60	63.1	19.4	209		3,4	1.830	8.89	4.41
yw2/wc/s	.132		.694		60	55.8	12.8	272		3,4	1.616	7.73	4.95
yw/wc/s	.092		.680		60	58.8	18.3	204		3,4	1.178	14.14	6.67
120/wc/s	.248	.016	.842	.015	60	58.6	13.3	309		7	1.567	7.01	10.65
wc/s	.176	.038	.800	.032	60	61.2	14.4	283		7	1.408	8.53	7.52
wc/s	.137		.709		60	65.0	11.4	211		7	1.595	10.09	5.12
wc2/s	.149		.739		60	57.4	19.4	219		7	1.556	9.97	5.54
p/s	.143		.679		60	63.6	20.6	220		7	1.836	8.41	4.27
p2/s	.097		.586		60	64.0	14.8	23		7	1.606	8.89	4.51
b/s	.169		.718		45	61.0	46.8	224	.197	7	1.387	27.33	17.97
msw	.083		.735		62	61.0	46.8	212	.197	7	.836	18.21	9.57
msw	.024	.003	.393	.023	55	56.4	33.2	226	.083	7	.758	25.79	11.98
msw	.032	.007	.594	.045	60	39.7	31.1	386	.374	3,4	.520	16.93	14.02
msw/clm	.033	.004	.349	.126	60	43.5	19.3	376	.190	3,4	.859	10.51	7.74
sdw	.215	.031	.541	.021	60	71.4		114		3,4	3.947	7.55	1.79

[a] Temperature, moisture, C/N, and compost density, dry basis, are shown for test conditions.

[b] g–grass, b–brush, l–leaves, c–cardboard, yw–yardwaste, kw–kitchen waste, cm–chicken manure, fw–food waste s–biosolids, yw2– 2nd use yardwaste, p–particle board, p2– 2nd use particle board, b–bark, wc–woochips wc2– 2nd use woodchips, msw–municipial solid waste, clm–caged layer manure, sdw–separated dairy waste.

[c] Remix days were intervals used during experimental studies.

Table 2 Pressure drop parametersa for two cereal grains, poultry manure mixes, MSW, and sludges.

MATERIAL(s)[b]	Material Ratio	Moisture % w.b.	Pressure Drop ....Parameters.......		
			a	j	n
WC/sludge[c]	2:1 v		1.02×10^{-1}	1.05	1.61
WC/sludge[c]	3:2 v		1.59×10^{-1}	1.30	1.63
WC/sludge[c]	1:1 v		5.03×10^{-1}	1.47	1.47
WC/sludge[c]	1:2 v		1.43×10^{0}	1.41	1.48
Fresh wood chips[c]			3.13×10^{-1}	1.08	1.74
Recycled screened wood chips[c]			9.74×10^{-1}	1.54	1.39
Final compost material[c]			3.61×10^{-1}	1.66	1.47
CLM/sawdust[d]	1:2 w	60	4.71×10^{-2}	1.0	1.17
CLM/corncobs[d]	1:2 w	60	3.72×10^{-2}	1.0	1.60
WC/RC/Leav/sludge[d]	2:0:2:1 v	50	1.37×10^{-1}	1.0	1.56
WC/RC/Leav/sludge[d]	3:0.5:1.5:1 v	49	9.49×10^{-2}	1.0	1.73
WC/RC/Leav/sludge[d]	3:0:1:1 v	49	1.00×10^{-1}	1.0	1.82
WC/RC/Leav/sludge[d]	3:1:0:1 v	52	6.45×10^{-2}	1.0	2.02
MSW[d]			7.82×10^{-1}	1.0	1.23

[a] Pressure drop(cm H_2O); depth(m); velocity (m/min)
[b] CLM=cage layer manure. WC=wood chips. RC=recycle. Leav=leaves. MSW=municipal solid waste.
[c]. Higgins et al., 1982. [d] Keener et al., 1993

Variable cost of composting

Composting economically requires efficient design and operation of the compost facilities. In particular, since many large compost facilities are designed to operate using forced ventillation, it is important that power requirements be low while still allowing efficient operation of the facility. For example, a facility composting 80 ton_{ds}/day may require as much as 225 kW of fan capacity. [Based on 20.3 cm static pressure, 1.34 m^3/kg_{ds}–day, 22 day compost cycle] If these fans have a 50 % duty cycle, total power cost is about \$4/ton_{ds} composted or \$88,000/yr, when electricity cost is \$0.08/kWh. Although not the major cost of producing compost, it is one which the compost system designer and/or operator can control.
Energy for composting

Energy for composting per unit compostable material assuming a variable speed fan and inclusion of piping losses is (with * signifying desired level of compost maturity)

$$E = \sum_{\ell=1}^{2} \frac{a_\ell \rho_{\copyright}{}^{n_\ell} k^{n_\ell} d^{n_\ell + j_\ell}}{\alpha \varepsilon} \left[\frac{-h_C}{\rho_{a*}[HAO - HAI]} \right]^{n_\ell + 1} \left[\frac{1 - mR^{*n\ell+1}}{n_\ell + 1} \right] \tag{5}$$

Variable cost per unit compostable, C_v, is
$$C_v = c_v E \tag{6}$$

The above expression (Keener et al., 1993b) shows the interdependence between biological and physical factors and can be used to optimize the efficiency of the composting process. (See appendix for nomenclature).

Fan Sizing and Variable Cost

Selection of fan size is governed by the airflow requirement and will be determined by (a) maximum airflow which generally occurs within the first 36 hours of the process and (b) the way the total composting operation is managed. For example, consider a compost operation with three windrows. Each windrow has a 3–week ventilation schedule and is turned in place with water addition based on moisture levels (Keener, et al. 1994). The windrows are started 1 week apart. Two fan arrangements are considered for this example.

Case 1: Each fan delivering air to only one windrow or section of a windrow. This scenario requires the largest fan power and shows the least efficient duty cycle. It generally uses some type of two stage fan control (high/low or on/off) using compost temperature for feedback control.

Case 2: One fan delivering air to all three windrows. This arrangement requires the least fan power and maximum duty cycle. It requires air dampers on feedback control regulating airflow to each windrow.

Table 3 shows the fan sizes required for case 1 and case 2 if pressure drop is similar to a 2:1 wood chip/ sludge mixture. To solve this problem, parameters associated with pile size along with the compost material have to be specified. In this example, compost depth is 3 m, compost density is 290 kg_{ds}/m^3 and pressure drop in the aeration duct is limited to 7 cm H_2O at maximum flow to the windrow. For case 1 power requirements are .0702 watts/kg_{ds} while for case 2 the requirements are .0345 watts/kg_{ds}. This example suggest that by sharing fans across windrows, fan power requirement can be cut in half. This leads to lower fixed cost and ventilation cost to ventilate the compost piles.

Some compost systems mechanically move the compost from one section to the next at regular intervals, eventually moving the compost out of the system. For these systems , sizing the fan for each section based on days into the composting will achieve minimal power requirements.

Table 3 Fan power requirements to meet temperature control in sludge composting system. Case 1 unshared fan; Case 2 fan shared across windrows.

Time	Airflow[a]	Velocity (compost)	Δp[b] (compost)	Fan Power unshared	shared
Days	$m^3/min \cdot Mg_{ds}$	m/min	cm H_2O	watts/Mg_{ds}	watts/Mg_{ds}
Case 1					
0	2.54	2.54	1.46	70.2	
Case 2					
0	2.54	2.54	1.46		34.9
7	1.04	1.04	0.34		34.9
14	0.16	0.16	0.02		34.9

[a] Mg_{ds} = 1000 kg dry solids
[b] Δp_{total} = pressure drop (duct + damper) + pressure drop (compost). For case 2 with shared fan, static pressure fan works against is maximum 8.46 cm H_2O while airflow is 1.25 $m^3/min \cdot Mg_{ds}$.

Fan sizing based on an on/off mode (signified by ') leads to higher cost. Keener et al. (1993) found a one fan size system has variable energy cost given by

$$C'_v = c_v \sum_{\ell=1}^{2} \frac{a_\ell \rho_{\copyright}{}^{n_i} k^{n_i} d^{n_i+j_i}}{\alpha \varepsilon} \left[\frac{-h_C}{\rho_{a*}[HAO-HAI]} \right]^{n_i+1} [1-mR]$$

(7)

The cost ratio between an on/off fan and a continuously variable fan speed (Eqs. 6 and 7) when $n_1=n_2$ is given by

$$\frac{C'_v}{C_v} = \frac{(n+1)[1-mR]}{[1-mR^{n+1}]}$$

(8)

Fan energy cost ratio for the on/off fan is given in Table 4 . A similar analysis can be done for a fan with various levels of control. For a mR value of 50 percent, variable cost of composting is 71 percent higher using an on/off fan system rather than a variable fan approach. Note that the earlier analysis of sharing the fan across a number of windrows approaches the variable fan concept in terms of fan sizing/utilization.

Effect of mR on Variable Cost

Table 5 shows how C_v varies with mR for the case $n_1=n_2=2.0$ and, where $C_{v\infty}$ equals the maximum cost which occurs when mR $\rightarrow 0$ (i.e., time $\rightarrow \infty$). For such compost mixtures the results showed that $C_v \backslash O(\sim,\_) 0.88 \, C_{v\infty}$ when mR=0.5. These results apply when the fan power changes with ventilation needs of the composting system and the assumption that the coefficient a and depth d are fixed. Although this is an approximation, since $n_2 < 2$ for most compost mixes, the results illustrate the importance of specifying mR when calculating system efficiency. It also indicates the relative savings on power which can be achieved by moving compost from stage one composting to an unventilated operation as soon as possible.

Table 4 Effect of fan operation on the relativecost[a] of ventilating a compost pile $n_1=n_2=2$.

mR	$C\backslash O(`_v)/C_v$
1.00	1.00
90	1.11
80	1.23
70	1.37
60	1.53
50	1.71
40	1.92
30	2.16
20	2.42
10	2.70
00	3.00

[a] $C\backslash O(`_v)$ is cost of on/off fan; C_v is variable speed fan

Table 5 Effect of final maturity ratio on the relativecosta of ventilating a compost pile for case n1=n2=2.

mR	$C_v/C_{v\infty}$
1.0	.00
9	.27
8	.49
7	.66
6	.78
5	.88
4	.94
3	.97
2	.99
0.0	1.00

[a]C_v• = ventilation cost when time goes to •.

Effect of Depth on Variable Cost

Variable cost per unit weight to be composted is proportional to d^{n+j}. For poultry manure plus corncobs doubling of the depth increases C_v four to six times ($n_1+0=2$ and $n_2+1=2.6$; see Table 2). In the case of sewage sludge and wood chips, doubling of depth would increase cost four to eight times since n_2+j_2 is as high as 3.02.

Effect of Air Enthalpies on Variable Cost

The effect of [HAO – HAI] cannot be evaluated independent of k, given current information that rate of composting may increase significantly (Q_{10} effect) as temperature increases up to a value of 60°C (Schulze, 1962; Haug,1993 ;Snell, 1991). Table 6 shows how C_v varies with the temperature of the compost pile using $C_v(60)$ as a reference value and $n_1=n_2=2$. Results show that minimum cost occurred at 60°C and would be 2/3 the cost of operating at 50°C. If k should be constant between 50°C and 60°C, the operating cost at 60°C would be only 17%

of the cost at 50°C, i.e. it would cost 5.8 times more to compost at 50°C than at 60°C.

Table 6 Effect of compost temperature on the relative cost of ventilating[a] a compost pile for the case $n_1=n_2=2$.

		k = fct(T)		k = constant	
T°C	[HAO–HAI]	k_T/k_{60}	$C_v(T)/C_v(60)$	k_T/k_{60}	$C_v(T)/C_v(60)$
30	50.6	0.125	9.59		
40	113.8	0.25	3.04		
50	215.6	0.5	1.61	1.0	5.8
60	387.7	1.0	1.00	1.0	1.0

[a] Based on incoming air, $T_{ai}=20°C$, $w_{ai}=0.107$.

Effect of Air Recycling on Variable Cost

Use of recycle air increases incoming air temperature and potentially increases decomposition rates during composting if the Q_{10} effect applies to the composting rate. Analysis of the effect of incoming air temperature on variable cost was done using equation 6 for the cases (1)decomposition rate is uniform from inlet to exit and is equal to k at the exit temperature, and (2)decomposition rate is determined by the compost temperature and is linearly increasing from inlet to exit. Evaluation of ρ_c, [HAO–HAI] and average k^2 was required. Table 7 shows the effect of raising inlet temperature from 20°C to 30, 40, or 50°C (using recycled air) and exiting the system at 50 or 60°C. Results showed for a linear k profile and 10°C difference across the bed, variable cost increased 32 to 34 time, compared to the reference conditions. The potential cost savings due to an increase in throughput caused by the uniformly higher temperatures were not evaluated here.

Table 7 Effect of Recycle Air on Variable Cost of Ventillating Compost.

T_{in} °C	T_{out} °C	[HAO–HAI]	Uniform $(k)^2$	Linear $(k)^2$	Uniform $(C_v/C_{v,ref})$	Linear $(C_v/C_{v,ref})$
20	60	387.7	1.00	0.36	1.00	1.00
30	50	165.0	0.25	0.11	3.79	4.67
40	50	101.8	0.25	0.15	19.6	32.2
30	60	337.1	1.00	0.38	1.78	1.90
40	60	273.9	1.00	0.44	4.03	4.95
50	60	172.1	1.00	0.58	21.2	34.8

$(k)^2$ represents $(k/k_{ref})^2$; Uniform represents condition where k is constant over the bed; Linear represents k increasing linearly through the bed; $C_{v,ref}$ represents C_v for the first row in table (i.e. for 20 –60°C air temperatures).

Fixed Cost and Optimizing Depth

Fixed cost of a composting facility represents land, buildings and equipment. Minimum operating cost, as a function of depth, must occur between 0 and ∞. At zero depth, infinite land area is needed while at infinite depth, vessel construction cost and power cost for fans become infinite. This minimum cost can be solved by using

$$\frac{\partial C_A}{\partial d} = 0$$

(8)

where C_A is annual operating cost and d is the depth of the bed. To solve for optimum depth requires evaluating only those cost factors which are a function of compost depth. Keener et al. (1993a, 1993b) has solved this problem for composting systems with fixed area per day (IPS system) and the fixed depth system. Because the cost function for machine size is a discontinuous function it was not included in their analysis. Solutions for optimum depth were as follows.

Case 1 – Analysis of Fixed Area Per Day

For the case of fixed daily land area, A_o (IPS system), daily feed rate, m_o, and total composting time, s, optimum depth is

$$d_{OA}^* = {}^{n_2+j_2+1}\sqrt{\frac{\Phi_c s}{(n_1 \Phi_{A1} d_o^{n_1-n_2-j_2} + (n_2 + j_2)\Phi_{a2}\Phi_{A2})}}$$

(9)

where * signifies optimum. See nomenclature for definitions of Φ_c, Φ_{a1}, Φ_{a2}, Φ_{A1} and Φ_{A2}.

Case 2 – Fixed depth (variable floor area per day)

For the case of fixed depth and daily feed rate, m_o, optimum depth is

$$d_{od}^* = {}^{n_2+j_2+1}\sqrt{\frac{\Phi_c \Phi_{vs}}{(n_1 \Phi_{a1}\Phi_{d1} d_o^{n_1-n_2-j_2} + (n_2 + j_2)\Phi_{a2}\Phi_{d2})}}$$

(10)

Eqs. 9 and 10 can be used to estimate optimum depths for the two types of system analyzed. They also allow one to estimate the effect of changing parameters on composting system efficiency.

Estimates of parameters required in the design analysis of compost systems are given in Table 8.

In this table, cost term c_l, c_f and c_v are assumed. Using these table values, and assuming Φ_c ranges from 2–32, solutions of optimum depth for a reactive compost mix (k=0.1/day) were solved and are given in Table 9. For n=2.0 and the parameters selected, optimum depth d\S(*,oA) ranged from 1.8 to 3.6 m while d\S(*,od) ranged from 1.5 to 3.1m.

Summary

This paper presents analytic expressions which show the relationship between compost properties and maturity. Specifically it analyzed the effect of pressure drop parameters on the variable cost of composting and presents an optimization analysis for composting in a system with uniform temperature across the bed. The analysis assumed continuous remixing and constant bed density during composting. The analysis of optimum depth for fixed and variable bed depth systems was performed. Although most results presented are restricted (uniform temperature), the procedure outlined (1)does serve as a focal point for optimizing composting systems, (2)has identified groups of parameters which allow results to be applied to many different types of wastes and (3)should allow improved design of in-vessel composting systems.

Table 8 Parameters and range of values for use in calculating optimum depths of compost systems. (No piping losses, $a_1=0$)

Parameters				Calculated Variables		
	Range	Value Used				
n	1.2 to 2.0	(1.5, 2.0)		$n=$	1.5	2
a	2.02×10^{-7}	.0000002				
α	8816	8816				
ε	.5 to .6	.5		$f=$	4.54e–11	
h_C	20000	–20000	J/kg			
ρ_c	200–600	500	kg/m^3			
ρ_a	1.29 to .85	1.29	kg/m^3			
ΔHA^a	330 to 410	400	kJ/kg$_a$	$D=$	19380.	
k	.01 to .2	.1	1/day	$\Phi_{a2}=$	.007501	.330247
β	.25 to .75	.5				
θ	10 to 20	15	day	$\Phi vs=$	10.67	
s	$\theta+1$	16		$\Phi_{A2}=$	3.7	3.01
r	.20 to .30	.3		$\Phi_{d2}=$	4.56	3.91
c_1	10–100	20	$/m^2$			
c_f	150×10^{-3}	.15	$/watt	$\Phi c=$	12.4	
c_v	$.05 \times 10^{-3}$	.00005	$/watt–hr			

^{a} $T_{ai}=20°C$; $T_c=60°C$

Table 9 Optimum depth for composting^a for the case $n_2=(1.5, 2.0)$, $j_2=1$, and $\Phi_c = (2,4,8,16,32)$ and no piping losses ($a_1=0$).

$n=$	1.5	2	1.5	2
Φ_c	$dS(*,oA)$		$dS(*,od)$	
	m	m	m	m
2.0	5.8	1.8	4.8	1.5
4.0	7.0	2.2	5.9	1.8
8.0	8.6	2.6	7.2	2.2
16.0	10.5	3.0	8.8	2.6
32.0	12.7	3.6	10.7	3.1

^{a} $T_{ai}=20°C$; $T_c=60°C$

Appendix

Airflow Requirements

The fundamental equation for design of the aeration requirements for a compost system (Keener et al., 1993a) is (note: h_c is negative):

$$Q(\theta) = \frac{(m - m_e)}{\rho_a} \frac{-kh_C}{[HAO - HAI]}$$

(A1)

Values for $\dfrac{-h_C}{\rho_a[HAO-HAI]}$ are given in Table A1.

Table A1 Effect of Compost Temperature on the Airflow Requirements During Composting.

T °C	[HAO–HAI] kJ/kg_a	$-h_c/\rho_a[HAO–HAI]$ m^3_a/kg_c	Φ
30	50.6	0.306	0.306
40	113.8	0.136	0.136
50	215.6	0.072	0.072
60	387.7	0.040	0.040
70	678.0	0.023	0.023

Based on incoming air, $T_{ai}=20°C$, $w_{ai}=0.0107$, $\rho_{a\,=\,1.29\,kg/m}^3$, $h_c = -20kJ/kg_c$

Maximum Time Fan is Off

The maximum time the fan can be off is given by

$$\Delta\theta^* = \frac{\varepsilon\rho_a\,(C_{O2} - C_{O2}{}^*)}{G\,(1-\beta)\,k\,\rho_{co}e^{-k\theta}}$$

(A2)

For analysis $\varepsilon = 0.4$, $\rho_a = 1.29$, $(C_{O2} - C_{O2}{}^*) = 0.16$, $G = 1.37$, and $\theta = 0$.

Remix Time and Water Addition

Keener et al. (1994) presents the equations on remix times. Remix occurs at time θ^* when moisture drops below $w_{c,L}$.

$$\theta^* = \frac{-\ln(z1/z2)}{k} = e^{-k\theta}$$

(A3)

where

$$z1 = w_{c,o}M_{c,o} + \Phi - \beta_o(w_{c,L} + \Phi)$$

(A4)

$$z2 = (1 - \beta_o)\,(w_{c,L} + \Phi)$$

(A5)

Values for Φ (Keener, et al.,1994) are given in Table A1.

Nomenclature

Letters

a_ℓ = coefficient–pressure drop airflow relationship

c_f = variable fan cost, \$/W

c_L = variable land cost, $/m^2

c_v = variable cost coefficient, $/W · hr

C_A = annual operating cost, $

C_v = variable cost, $/ kg

C_v = variable cost, $

C_{O2} = oxygen concentration of ambient air,

C_{O2}^*= lower limit for oxygen concentration of compost

d   = compost depth, m

$D = \dfrac{-h_c \rho_c}{\rho_a [HAO - HAI]}$, energy density number

E   = energy, W hr/kg

f_ℓ = $\backslash F(a_\ell, \alpha\varepsilon)$, pipe or fan factor

G   = mass rate of oxygen uptake per unit of dry matter disappearance

h_c = heat of combustion at T, MJ/kg

[HAO–HAI] = enthalpy difference air out – air in for compost system, KJ/kg$_a$

j   = exponent–pressure drop, compost pile height

k   = rate of disappearance of dry matter, day^{-1}. It is a function of substrate compounds, microbial populations, temperature, moisture content, surface exposed, interstitial atmosphere (oxygen, NH_3 ...)

m or m_c = compost-dry mass, kg

m_o = initial dry mass in composter, kg

m_e = equilibrium mass, non-compostable (dry basis), kg; $m_e = \beta_o m_o$

mR = compost mass ratio, $\dfrac{-m(\theta) - m_e}{m_o - m_e}$

M_c = m_c/m_e

n_e = exponent-pressure drop, airflow relationship

P'   = power per unit area, watts/m^2

P''   = power per kg$_©$, watts/kg (kg$_©$ implies kilograms compostable)

Δp = pressure drop, cm H_2O

q   = airflow per unit dry matter of compost, m^3/kg$_c$ day

Q   = airflow, m^3/day

r   = annual rate for fixed cost

s   = index number for total number of days of composting

T   = temperature, °C

v   = air velocity, m/min

V_c = composting vessel volume, m^3

w_c = moisture content-compost, dry basis, kg/kg

Greek Letters

α = conversion coefficient, $0.612 \dfrac{m^3 air. \, cm \, H_2O}{watts . min}$

β = compost equilibrium value, kg/kg. (determined by   level of sugar, cellulose, hemi-cellulose etc}

ε = efficiency of fan system

    = air porosity of compost

θ = time, day

ρ_a = dry air density, kg/m^3

ρ_c = dry matter density of compost, kg/m^3

ρ = average density, kg/m^3

$\rho_{©}$ = dry matter density of compostable portion; $\rho_{©}=$

$$\frac{m - m_e}{V} = \rho_c - \frac{m_e}{V}, kg/m^3$$

Φ = moisture loss term, aeration factor

$\Phi_{vi} = \beta + (1-\beta)\,e^{-k(i-1)}$, daily volume number

$$\Phi_{vs} = \sum_{i=1}^{s} \Phi_{vi} \quad , \text{ cumulative volume number}$$

$\Phi_{a\ell} = f_\ell\,(Dk)^{(n/+1)}$, aeration factor

$$\Phi A_{\ell=\backslash} = \sum_{i=1}^{s} (e^{-k(i-1)})^{n,+1}\Phi_{vi}^{\,n,+j,+2} \quad , \text{ fixed floor area number}$$

$$\Phi d_{\ell} = \sum_{i=1}^{s} (e^{-k(i-1)})^{n,+1}\Phi_{vi} \quad , \text{ fixed bed depth number}$$

$$\Phi_c = \frac{^{rc}L}{(^{rc}f + 365.24\,c_v)} \quad , \text{ cost ratio number (fixed to variable cost)}$$

Subscripts

i   = ambient condition

i   = running index associated with compost vessel length

ℓ = index, 1=pipe, 2=compost

o   = initial value at time zero

L   =lower value

Acknowledgements

Salaries and research support were provided by a grant from Columbus Division Sewer and Drainage as well as State and Federal funds appropriated to the Ohio Agricultural Research and Development Center, The Ohio State University.

Literature Cited

Anonymous. (1991). Sludge composting projects in the United States. In: The BioCycle Guide to The Art and Science of Composting. Ed. Staff of BioCycle. p 56–66. Emmaus, PA: The JG Press, Inc.

Das, K. and Keener, H.M. (1995). Process control based on dynamic properties of composting: Moisture and compaction considerations. (In this Proceedings)

Finstein, M.S.,Miller, F.C., and Strom, P.F. (1986). Waste treatment composting as a controlled system. *Biotechnology*. eds. H.–J. Rehm and G. Reed. Volume 8 (Biodegradations), Chapter 10, pp. 363–398. VCH Verlagsgesellschaft, Weinheim, FRG.

Haug, R.T. (1980). *Compost Engineering, Principles and Practices*. Ann Arbor, MI: Ann Arbor Sciences Publishers, Inc.

Haug, R.T. (1993). *The Practical Handbook of Compost Engineering*. Boca Raton, FL: Lewis Publishers.

Higgins, A.J.,Chen, S., and Singley, M.E. (1982). Airflow resistance in sewage sludge compostiing aeration systems. Transactions of the ASAE 25(4):1010–1014,1018

Hoitink, H.A.J.,Keener, H.M. and Krause, C.R. (1993). Key steps to successful composting. *BioCycle* 34(8): 30–33. August.

Keener, H.M.,Marugg,C., Hansen,R.C, and Hoitink,H.A.J..(1993a). Optimizing the efficiency of the composting process. In: *Science and Engineering of Composting: Design, Environmental, Microbiological and Utilization Aspects*. Ed. H.A.J. Hoitink and H.M. Keener. pp. 59–94. Columbus, OH: Renaissance Publications.

Keener, H.M.,Hansen,R.C. and Elwell,D.L.. (1993b). Pressure drop through compost: implications for design. Presented at Spokane, WA, June. ASAE Paper No. 934032. St. Joseph, MI: American Society of Agricultural Engineers.

Keener, H.M.,Elwell,D.L., Das,K. and Hansen,R.C. (1994). Remix frequency of compost mixes based on moisture control. Presented at Kansas City, MO, June. ASAE Paper No. 944066. St. Joseph, MI: American Society of Agricultural Engineers.

Keener, H.M.,Elwell,D.L., Das,K. and Hansen,R.C. (1995). Specifying Design/Operation of Composting Systems Using Pilot Scale Data. 7th International Symposium on Animal and Food Processing Waste, St Joseph, MI: ASAE.

Kuter, G.A.,Hoitink,H.A.J., and Rossman,L.A.. (1985). Effects of aeration and temperature on composting of municipal sludge in a full-scale vessel system. *J. WPCF* 57(4): 309–315.

Marugg, C., Grebus,M., Hansen,R.C., Keener,H.M., and Hoitink,H.A.J.. (1993). A kinetic model of the yard waste composting process. *Compost Science & Utilization* 1(1): 38–51.

Nakasaki, K.,Kato,J.,Akiyama,T., and Kubota,H. (1987). A new composting model and assessment of optimum operation for effective drying of composting material. *J. Ferment. Technol.* 65(4): 441–447.

Schulze, K.L. (1962). Continuous thermophilic composting. *Compost Science:* 22–34.

Snell, J.R. (1991). Role of temperature in garbage composting. In: The BioCycle Guide to the Art and Science of Composting. Ed. Staff of BioCycle. p 254–256. Emmaus, PA: The JG Press, Inc.

Steele, J.L. and Shove,G.C.. (1969). Design charts for flow and pressure distribution in perforated air ducts. Transactions of the ASAE 12 (2):220–224

Final Reports

A1 Composting Process

Prof E I STENTIFORD

In all a total of 12 papers were presented in this session covering the process aspects of composting on both the macro and the micro scale, and addressing some of the more frequently occurring process problems. The core feature which ran through these papers was the important of a relatively few process characteristics during composting itself. Essentially these are:

- the amount of aeration;
- the process operating temperature(s); and
- the moisture content of the composting mass.

The Macro Scale

The three characteristics are very much interactive and the changing of one generally affects the other two. It was pointed out that when considering the process itself it is important to understand something of the many types of mass transfer taking place within the system boundary. For example in most properly run systems:

- the greatest moving mass in many composting plants during the overall process period is air;
- for many waste combinations to maintain the water balance more water is added than substrate processed; and
- the key to understanding these mass transfers is a knowledge of the energy balance with the system boundary.

The papers confirmed the importance of separating heat generation from temperature with one particular example showing that a compost mass operating at 70°C showed the lowest volatile solids loss and heat output.

The Micro Scale

The use of inocula in general waste composting was shown to have no significant effect on the composting process with the organisms present in the waste being more than adequate to ensure effective composting.

Two papers dealt with organisms filling two specific niches, those of nitrogen fixation and functioning at high temperatures. In the later case some novel groups of bacteria were identified which were operating in the 75 – 80°C range. However the general consensus was that although the organisms were present at these temperatures their general level of activity was very low. There is a general consensus that the maximum rate of degradation takes place in the 45 – 55°C range and conditions outside this range are sub-optimal.

Process Problems

Composting plants around the world are experiencing local residence due to the generation of obnoxious odours to the extent that many of them have been forced to close. It is important to realise that all composting processes produce odour, what we have to ensure is that the odour is either inoffensive or at such a low level in sensitive areas to make it undetectable. The major factors to consider in relation to compost plant odour:

– the different quantities, strengths and flow rates from various parts of the process installation;
– the correct siting of the installation to minimise impact; and
– developing an odour 'footprint' for any site to anticipate the potential impact and thus minimise the chance of complaints.

The other major problems covered during the meeting involved the following:

– leachate – difficulties in this area almost totally related to open systems, covered installations only had to cope with relatively 'clean' surface run-off;
– airborne microorganisms – the difficulty here was largely related to the high levels of organisms in the immediate vicinity of the composting operation, and this was normally addressed by the use of suitable face masks; and
– compaction of the composting mass – this affects air flow rates, cooling, and thermal properties but with biosolids provided the moisture content was < 55% and the bed depths were , 2.5m compaction was not a process limiting factor.

The overall conclusion was that the majority of process problems arose either as a result of:

– bad design of the composting installation; or
– poor management and operation; or in many cases
– both of these.

B1 Starting Materials

Dr J LOPEZ-REAL and Dr J MERILLOT

A wide range of waste materials used for composting – organic and inorganic – was reported in the fourteen paper presented in this session. From lignin containing materials through agricultural wastes including composting of whole pig and poultry mortalities), grass, sludges and even diapers from the urban sector to the use of bauwte and coal ash. Solid wastes such as these to the more intractable and problematical organic liquid wastes, such as cattie slurry and olive-mill wastewater were discussed. A session that proved once and all that if the substrate/waste is organic it can be composted; if the waste is inorganic it can be incorporated into the process.

Many of the staring material covered, and mentioned above, represent wastes with problem of disposal, storage and treatment, whose previous route have become less attractive or constrained because of environmental concerns, environmental legislation and of course increasing econonuc cost.

The composting technologyapplied to these wastes covered the whole range of system avalaible from windrowing to tunnel reactors. In the agricultural sector the simpler more easily adapted windrowing technique, utilising appropriate turning regimes, predominates. The application of composting technology to agricultural wastes – the largest waste producing sector of society – is hampered by ignorance, increased costs and a reluctance to shift from traditional methods of treatmenindisposal. Alternative to composting such wastes may be ecologically unsound if this is resulting in increased levels of important greenhouse gases such as methane - an area requiring further urgent investigation.

It is self evident that the starting material determines the greater part of the quality criteria of the final product. Moreover the eonomical and even the social status of the startmxg material (urban waste, sewage sludges) may determine the rationale for its treatment by composting. Co- composting can answer these questions as a part of an integrated system. When refering to starting materials it mast be clearly stated why it is being composted as well as how.

B2 The State of the Art of Composting and Perspectives

Summary Report Presented by JEROME GOLDSTEIN – *BioCycle*

The specific challenges and operational experiences were reported by the presenters who described the status of composting in Austria, Brazil, Denmark, Finland, France, Italy, Japan, Spain, and the United States of America. The specific challenges included the following themes which were cited by most presenters:

– The need to overcome legislative barriers and enhance economic incentives so private companies will assume a greater role in composting;
– The need to integrate waste management solutions into the framework of societal issues – and to explore the links between quality of life and recovery of organic residuals;
– The need to build on the awareness of the relevance of composting through research on critical issues, and to identify those issues and use research centers more effectively to develop solutions;
– The need to form new partnerships among the private sector, government agencies, nonprofits in order to expand the impact of composting projects. To achieve successful partnerships, it is necessary to aggressively show the common interests of environmentalists, industry associations, private corporations and public policy-makers;
– The need to develop the diversity of approaches in managing and recycling organic residuals. These approaches range from decentralized, backyard programs using home compost bins and regional 'self-composting' to curbside collection of source-separated residential organics and residuals from commercial generators where they are composted at centralized facilities. The diversity also relates to combinations of aerobic and anaerobic systems to produce energy in the form of biogas as well as soil-improving materials for application to farms and gardens.

Representatives of different nations provided data of relevant conditions which directly affect the role of composting in waste management. Waste audits of organic residuals were cited for each country, indicating the percentage and volume of biosolids, paper and wood products, food waste, etc. which are generated from the commercial and residential sectors. Mention was made, for example, of national laws in Finland that require a 50 percent reduction in solid waste diversion by the year 2000. Many states in the United States of America have passed laws against the disposal of yard trimmings in landfills and incinerators. These states have also set recycling goals that encourage the greater use of composting. France has set a national policy for waste management in 1992 that requires regional plans and should foster composting of source separated organics.

In Denmark, there is a growing emphasis to reduce the use of commercial nitrogen as part of a program to save energy and protect streams. The economic rationale for new recovery plants for centralized digestion of animal waste slurries is being recognized.

Several speakers – including those from Italy and France – noted the failure of past compost plants to achieve objectives and perform successfully. A specific common problem cited was the poor quality of compost produced for agricultural markets. The presenters reported on the great potential which exists in countries like Italy and France to use existing facilities to process a wide range of source separated organic feedstocks.

For example, future developments in France have been simulated and the potential for different kinds of composts was calculated as follows: six million tons from source-separated household wastes; one million tons of yard waste composts; four to five million of sludge composts (mixed with sawdust, bark, etc.). While yard waste composting is expected to increase in France in the coming years, it is expected that composting of source-separated household organics will remain low, with only five to seven experiments. A 1993 survey of compost plants by Ademe (the French Environment and Energy Management Agency) indicated the following data: 73 municipal solid waste composting plants (no source selection); 13 sludge composting plants; 30 yard waste composting plants (plus 21 projects); 16 farmyard manure composting plants; 10 mixed organics composting plants; and five experiments for source-separated household wastes.

By comparison, data reported for the United States of America indicate that there are 3,202 yard trimmings composting facilities; 318 biosolids composting facilities (including 118 in construction, permitting, design or planning stages as of December, 1994); 17 operational mixed or source separated municipal solid waste composting facilities (and an additional 34 in various stages of development).

Several speakers also stressed how existing laws needed to be changed since they were outdated and prevented progress.

The paper concerning composting experiences and perspectives in Brazil specifically pointed out the special problems facing developing countries, where practices had to reflect the links of the compost method to food production, public health, environment and social aspects. It was noted that 80 percent of the 190,000 tons per day of refuse in Brazil are disposed of in open dumps. A previous 'rush' to use composting led to use of inappropriate technologies, systems based on wrong concepts that neglected biological principles, and poor quality products. Presently, Brazil has 74 composting plants – 20 percent are fully mechanized systems and 80 percent are described as 'simplified systems' using locally adapted technologies. Most plants have problems of strong odors, vectors and leachate production. Concluded Prof. Neto of the University of Vicosa in Brazil: 'The potential for the use of composting in Brazil can be promising if a great effort is made, mainly by the government, with policy and legislation to cover the control of the systems, in operation and its final products.'

The perspective on the state of composting in the nations represented in this session at the International Symposium is that a great effort is needed now to fulfill the potential of composting. There is a sense that in each nation, there is an emerging infrastructure of dedicated researchers, policy makers, companies, project managers and individuals to achieve the full potential of composting.

B3 Composting as an Integrated System of Waste Management

BERT LEMMES – Managing Director ORCA

The intention behind the session was to avoid the possible overfocussing on the specific composting / biogasification problem, and to avoid the kind of composting euphoria or megalomania that has caused very serious harm to the acceptance of biological treatment in the past.

We only have to think back about the experiments with total MSW-composting ,that produced an end-product that one could never call compost.Some people still advocate this strategy as a pure reduction and stabilisation-process ,with an end-product called 'STABILAT' (in German) or stabilised matter.

But apart from that, the general tendency goes in the direction of source-separated collection, and the biological treatment of the separated organic fraction.

To oversimplify, it boils down to the question of principle whether we should try to compost as much as possible or *impose* a sufficient level of self-discipline on the sector to treat only that part of the organic resources,that will lead ultimately to the production of a valuable product 'compost' and to achieve this result in an *eco-efficient* manner.

We are faced in politics and legislation with the choice between waste and non-waste and a lot will depend on the final decision concerning this issue. But for the sake of this discussion ,we will continue to use the definition of recyclables and organic matter as waste for the time being.Nevertheless I want to stress already now that a basic change should be made in the thinking and therefor in the wording and definitions.

We should start calling 'waste' what is definitively 'wasted' and should adopt the positive philosophy of 'RESOURCES-MANAGEMENT' instead of 'waste-management' for all the fractions that can eco-efficiently be recovered.

We are faced with choices in resource-management about how to deal with the organic fraction. Biological processes are one of several means of dealing with this part of the secondary resources , but there are obviously others.

The integration of composting, anaerobic digestion or biogasification and stabilisation into an overall recovery strategy, not only seems the logical way to go but will avail itself the only solution for specific fractions of the waste stream if we want to achieve the recovery targets that will be imposed.

To help the European legislators prepare decisions and mandates to implement biological treatment as a solution, it is necessary to provide a rational framework and a matrix to judge the opportunity of the solution.

This framework will have to take into account:

1. The specific objectives of the waste management (local-regional constraints, etc.)
2. The broad environmental merit via the LCI's and LCA's
3. The economic merit: impact on the primary production
                            social acceptability and ease of implementation

This should enable the decisionmakers to determine the eco-efficiency of their strategy.

Eco-efficient would mean in this case: *eco*-logically sound
                                              eco-nomically viable

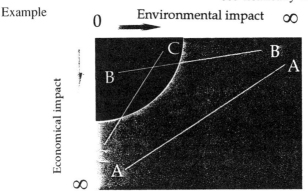

Figure 1 Operating Eco-efficiency

Legend: Project A   Big ecological impact but very expensive
                 Project B   Big Ecological impact and low price
                 Project C   Small ecological impact and very expensive

            Black section is the operating window that can assure eco-
            efficient sustainable activity

We acknowledge that in the end the decision will remain political, but we can proactively create an *operating widow* to point out to the authorities that any decision outside this operating widow would be unreasonable and certainly not sustainable.

A lot will depend on the definition of the waste and its separation in different fractions.

Depending on the collection system, the source separation system, the fraction of *grey waste could go from 100%* (as in some areas using mass burn-technology) (*to only 10% after source separation and pre-treatment.*

The more efficient the separation, the smaller this grey fraction will become and the more important the diversion from landfill will be. But, whatever we do, we admit that there will always *be* a grey fraction. But this has to be as small as possible.

Dealing with that issue, Paul Bardos exposed a possibility to implement the *co-treatment of this grey fraction with industrial wastes in order to reduce the volume and stabilise and sanitise the waste to be landfilled*. At the same time the possibility to recover biogas, and the detoxification of hazardous substances.

Two other aspects of well managing the landfill, where it avails itself unavoidable, were addressed by Prof. Cossu and Dr. Muntani. The former addressed the problems with the containment in landfilling and the treatment of the percolate.

This discussion put the finger on every sore spot in the whole waste management strategy.

> *'How do we estimate the long-term environmental impact of a landfill and the economical implications of its sanitation afterwards.'*

> *'How do we reflect this in the actual landfill-levies and do the existing or revised landfill levies such as in the UK reflect the real cost for providing an acceptable solution.'*

Dr. Muntani addressed a different issue of landfilling by advocating the use of compost to avoid some of the engineering problems in well-managed landfills. According to him compost can function as a filter, a buffer, an absorbent in landfills to avoid the problems with leachate (among others) by binding toxic ingredients.

From the landfills we turned our attention to the solutions that are provided respectively by recycling of dry matters and the use of R.D.F. and P.D.F. in co-combustion processes.

Julia Hummel presented the pioneering work done by ERRA in setting up a database to a collect data in a dozen pilot cities and to design a number of analyse-tools to interpret those data.

The motivation and participation ratio's for the source separation schemes and *the cost for collection and separation are especially interesting* for our sector as we are confronted with the same problems.

For this reason ORCA and ERRA will join forces in the very near future and combine their information and the processing of the data with this data-base to achieve a global view on the recovery-issue.and to put these data at the disposal of the decision-makers. The collaboration with a university and a number of motivated researchers will guarantee the academic and scientific accuracy of these data and make this a tool of high value to the EU , OECD and other authorities,faced with the elaboration of mandates in this complex sector.

Martin Frankenhaeuser of Borealis advocated the reduction of the grey fraction by extracting the recyclables first, splitting the waste into a dry and a wet fraction. This proves once more that the different technologies can be complementary with a common denominator : good source separation.

The colours of state of the art composting were defended by: George Savage, of California Recovery ,who advocated the use of the C/N ratio as a possible tool to judge the efficiency of process design.

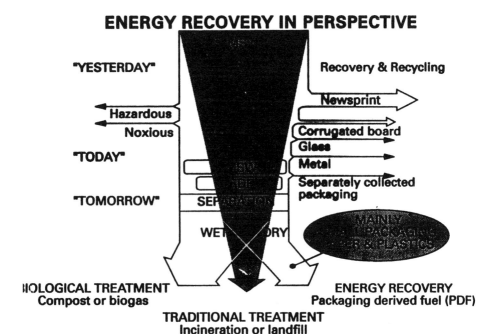

Figure 2 Vision for Municipal Solid Waste Management

At the same time he focused the attention once more on the *importance of the waste characterisation on the quality of the final compost.*

Going deeper into the problem of the definition of biowaste and broadening that definition by adding a paper fraction, Bruno De Wilde of Organic Waste Systems N.V. (OWS), Belgium, provided us with a clear insight in the advantages of paper inclusion in composting. As there are (among many others):

– reduction in salts
– more organic content
– better compost

The complete text of this research-project is available in the ORCA Technical Publications.

To close this compost section, Karel Mesuere commented and analysed the compostability criteria elaborated by ORCA as a framework to define the acceptable feedstock for state-of-the-art composting, and in order to make a quality compost.

These criteria create the link between the biodegradability (as tested in the labs) and the practice of composting, by integrating parameters that are relevant to the operations of composting/biogasification-plants. At the same time these criteria are scientifically correct and take into account the technical, biological and economical factors of the different processes. This very important work has been reviewed by a Peer-Review-Committee consisting of the leading authorities in the academic world and has therefore also been accepted by different authorities to form the framework for future legislation.

This publication is also available in the ORCA Technical Publications.

Last but not least, Peter White, Procter & Gamble Ltd., UK, author of the book 'Integrated solid waste management' showed us in his exposé, apart from the well-known advantages of biological treatment, the necessity to optimise biological treatment, to optimise source separation and to optimise the whole system, in order to reduce overall environmental impacts and to make integrated waste management economically sustainable.

Through LCI and LCA ,we can create tools to model the options ,even if questions will always remain concerning the validity of socio-political parameters and their evaluation in the overall LCI. At least we will have a more comprehensive view on the problem that will help us design a possible solution ,that will always have to be a custom made ,local or regional solution.

He warned us once more for the danger of isolating systems or technologies that will distort the total picture and shift the problems and advocated *the integrated waste management approach.*

Conclusion

I just want to close this report, by urging all you ladies and gentlemen to do the same. Biological treatment, our technology (including the correct applications of biodegradable materials and products) has so much to offer ,that we should not try to pretend it has no limitations

We most certainly have an *economical, ecological and socio-psychological advantage* over the other waste-treatment options , so why avoid the challenge to situate our technologies in an overall sustainable integrated waste-management system.

- ◆ Taking into consideration a similar environmental impact
- ◆ landfill
- ◆ incineration with energy recovery
- ◆ biological treatment
  relate as shown here

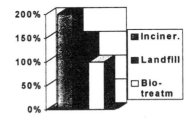

Figure 3 Cost-relation Waste-treatment

Lets sit together with the representatives of the other resource-recovery technologies, and discover where we are complementary, where competitive, to ensure an overall better result and the recovery of valuable resources and energy.

B4 Bioremediation

LF. DIAZ – Moderator CalRecovery, Inc. Hercules, California, USA

In this session, a total of seven presentations were made. Five presentations dealt with bioremediation of contaminated materials, one covered the "removal" of heavy metals by the addition of a natural zeolite, and one discussed the application of starch-based biodegradable materials in the separate collection and composting of organic waste.

The papers dealing with bioremediation reported the results of experiments and tests conducted to break down primarily hydrocarbons and pesticides. One paper described the methodology for conducting biotreatment analyses and presented the results of laboratory-scale and pilot-scale studies. The results of the studies indicated that existing technology can be applied to the composting of hazardous biodegradable wastes through suitable moditications. Another paper reported the results of composting creosote in reactors (under mesophilic temperatures). The results of these tests indicated that composting can be applied for the treatment of soils polluted by creosote. The results of experiments performed on the bioremediation of soils contaminated by PAII indicate that biotreatment of the contaminated soils in slurry form is too expensive under tull- scale conditions and that heat treatment and solidilication may be more cost-effective. Another presentation dealt with the use of compost in the treatment of soils contaminated with pesticides. The results of this work demonstrated that the addition of 20% to 40% of compost to contaminated soils stimulated the degradation of pesticides. Another presentation described the use of closed systems to evaluate biological degradation of hydrocarbons in soil. The results of the work indicated that the application of compost had a positive effect on the degradation of FAIls. In general, the results of all of the studies discussed in this portion of the session showed that hydrocarbons and pesticides can be effectively degraded biologically. The results also indicated that degradation increases as the quantity of composts added to the contaminated materials increases. Presenters stressed the importance of conducting laboratory-scale tests to assess the impact of the addition of compost to the haaardous wastes before large-scale projects are perlbrmed.

With regard to using natural zeolites, results of tests conducted cultivating peppers in compost showed that plants contained lower quantities of metals when cultivated with zeolite. Lowerconcentrations of metals were obtained when higher concentrations of zeolite were used.

Finally, it was reported that starched-based biodegradable materials compatible with composting and having good mechanical properties currently are available. Starch-based materials can be used as a replacement for LDPE in the manufacture of products such as trash bags, packaging materials, and diapers. In addition, the results of experiments of separate collection and composting of organic wastes conducted in Germany, Austria, Italy, Sweden, and Finland using starch-based materials indicated that these materials have acceptable mechanical properties and they are compatible with the process of composting.

B5 Composting Design

ROGER T. HAUG, PhD, PE – Session Moderator

Six papers were presented in Session 5B, each addressing aspects of composting plant design. A brief summary of each paper is presented along with the moderator's closing comments.

Summary of papers

Mr. Robert Tardy, Director of Compost Projects for R.W. Beck, presented an overview of U.S. composting practice and experiences, with particular emphasis on recent experiences with municipal solid waste (MSW). MSW composting in the U.S. has had some successful facilities, but also a number of high profile failures. Significant progress has been made in understanding the design criteria necessary for successful operation. Despite this better understanding and the fact that some MSW facilities are still being planned, Mr. Tardy felt that the industry was moving away from MSW composting toward composting of source separated organics.

Dr. Bernhard Raninger, Unversity of Salzburg, presented an overview of composting practices in Austria. 12 of 19 former co-composting plants (MSW plus biosolids) are being converted to source separated biowaste. Over 78% of households are required to source separate and this is expected to increase to 90% by the year 2000. Dr. Raninger felt that future facilities, particularly those composting substrates with high odor potential, should be 'industrial' types with enclosed processes and odor control. Several examples of 'industrial' conversions were presented, including composting plants at Salzburg and Allerheilegen. The Salzburg facility has added anaerobic composting of biowaste followed by enclosed aerobic curing.

Mr. Fabio Conti, Research Engineer at the University of Pavia, presented an overview of Italian experiences and plants. Photos and descriptions were presented of a number of plants of historic significance. Many of these are being modified to more modern standards. Mr. Conti mentioned that some drum systems did not provide sufficient residence time to start the compost process.

Mr. Mogens Hedegaard, Project Manager for Kruger A/S, presented new ideas on biosolids stabilization and removal of heavy metals. Some laboratory work has been conducted in support of the concept, but there are no commercial installations at present.

Dr. Nancy Lynch, Research Engineer with the Idaho National Engineering

Laboratory, presented the results of simulation modeling to develop design guidelines for passively aerated compost piles. The latter rely on the 'chimney effect' for pile aeration, enhanced by perforated aeration pipes, open at each end, and arranged perpendicular to the pile or windrow length. This work is a good example of basic research designed to improve understanding of the physics of air movement in such piles. It is expected that guidelines will be developed on pile dimensions and the distance between aeration pipes.

Mr. Keith Panter, Product Development Manager for Thames Water International, presented an overview of tunnel systems in the U.S. and Europe. A new tunnel system at Camden, New Jersey, began commissioning in April, 1995. Designed for 50 dtpd of biosolids, it is one of the largest in the U.S.

Moderator's comments

Today's composting market is characterized by a wide variety of commercially available systems being applied to a variety of substrates. Many systems are enjoying commercial success, whereas some formerly popular systems are less prominent. Systems with more advantages and improvements are replacing less successful designs. The composting industry is maturing and design critieria for successful facilities are understood better by today's design professionals than at any time before. Many of the recent improvements focus on the needs for containment and treatment of odors, particularly for substrates with 'high odor potential.'

This is a time of great optimism in the composting industry for a number of reasons, including the following.

- More substrates are being considered for composting than at any previous time. In the U.S. the 1950's and 60's saw composting applied to MSW with only limited success. In the 1970's composting was first applied to biosolids with the 80's devoted to expansion and problem solving. Yard waste started to be a significant composting substrate in the late 80's and has grown explosively since then. The 90's will be the decade in which composting is applied to almost all degradable organics using latest generation technologies.
- There is general acceptance that composting is part of an integrated approach to solid waste management.
- There is a return to source separation of urban organics, begun in Europe and now spreading in the U.S. This will be a tremendous help to composting because once wastes have been mixed together, such as with MSW, it is very difficult to separate the clean compostables. The engineer can design separation equipment, but no separation process is ever 100% efficient. Separation efficiency can be increased, but only at the expense of producing more rejects. It is becoming generally accepted that producing a high quality compost with an acceptable level of rejects is not possible with MSW. Simply stated, the quality

of the compost product depends on the quality of the feedstocks.
- Support has grown to make the difficult political decisions to support reuse of waste materials, such as by composting. In addition, favorable regulations have been adopted by many governments which support composting and attempt to reduce the regulatory burden associated with permitting new facilities.
- Research in support of composting is being conducted in many related fields, such as plant physiology and disease suppression. Such research is advancing to the point where the information may soon influence both process design and operation.

While optimistic about the current state-of-the-industry, I am concerned about certain clouds that we must jointly work to dispell. First, the history of the composting industry is not well documented. Many current mistakes, such as recent MSW composting failures in the U.S., may have been avoided if the lessons of history were better known. Many of our early pioneers are aging and we risk further loss of our history with their eventual passing. Second, significant barriers exist to the transfer of information across different parts of the composting industry. In the U.S. for example, lessons learned from biosolids composting appear to be little applied by those engaged in MSW composting. This is in part caused by the fact that professionals active in the biosolids industry in the U.S. are generally not the same ones active in the solid waste industry. Also, the MSW composting industry in the U.S. seems to be dominated by entrepeneurial ventures. The wealth of talent developed in the biosolids industry is not well used ourside that industry. Third, high profile failures of several MSW composting facilities have occured in recent years, mainly due to odor control issues. This has had a cooling effect toward composting in many circles. While it is true that composting can be applied to a variety of substrates, it is equally true that composting is not a panacea for all wastes and circumstances. Using the wrong substrate or not building odor control adequate to the site and substrates is a disservice to the composting industry. Eagerness to 'do the right thing' must be tempered by the cool eye of engineering analysis, field experience, and historical perspective.

B6 Marketing and Economic

L.L. EGGERTH – Moderator CalRecovery, Inc Bercules, California, USA

In the session on marketing and economy, live presentations were given: two dealing with the importance of marketing efforts and markets for composts; one dealing with the principles of the German quality assurance program; one on the activities of the Italian Composting Council; and one on economic modeling.

The presentations on marketing focused on the status of compost marketing in the United Statesof America. Important points that were addressed include:

- marketing is ofien forgotten, but needs to be included early in the project development process
- there is a sillR in mentality in the USA from considering composting strictly as a waste
- disposal process to consideration as a product manufacturing process
- production of a quality product is critical to market development
- public education must be an integral part of the composting program

The presentation on the principals of the German quality assurance program stressed the fact that marketing depends on user conlidence – in the product, in the producer, and in the compost quality. In Germany approwmately 170 composting plants representing 1.5 million tons of organic materialparticipate in the quality assurance program. Of the participants, 130 plants, producing 71% of the product in the country, meet the standards and carry the compost quality standard emblem.

The Italian Composting Council is a teclmical national scientific organization designed to promote compost use. The objectives of the Council are to:

- disseminate information by organizing meetings to increase awareness
- work with agriculture to develop conlidence in compost
- implement a certification process
- create markets

The speaker also stressed the importance of producing a Ngh quality compost to agriculturalmarket development.

The presentation on economics stressed the fact that composung currenlly is a subsidized industry and that there is a critical need to reduce costs. The author presented a series of mathematical relationstups to describe the composting process and thus use the mathematical models to optimize the system through process control, producing a stabilized product, and actueving the lowest cost.

The pervading theme of the session was on the need to utilize feedstock selection, process control, product quality standards, and public education efforts to ensure the availability of Ngfi quality composts and to develop markets for the products.

Posters

Thermophilic Pilot Scale Composting of Olive Cake

BACA, M.T[1].; BELLVER, R.[1]; DE NOBILI, M.[2] and
SANCHEZ-RAYA, A.J.[1]

Summary

Olive press cake are solid agroindustrial residues produced during the extraction of olive oil, disposal of wastes produced during this process is a particularly pressing problem in Southern Spain and other mediterranean countries. We developed a pilot-scale windrow composting process for optimal transformation of this kind of substrate, characterized by a high lignocellulose content. We found that the 82% of lignocellulose degradation took place when temperatures were above 50°C. After three months temperature was still relatively high; this notwithstanding the phytotoxicity of the compost was almost nil as demosntrated by biological tests.

Introduction

Decomposition of biodegradable organic wastes is a continuous and natural process developed in the soil by its biological community especially at the soil surface. But the soil has a limited resources to adsorb, absorb and transform this organic matter. Debris inputs and soil biomass are in dinamic equilibria influenced by many environmental factors. The elimination of the orgánic wastes generated by the human activities without breaking the equilibria could request a hight soil surface which makes unviable the process.

With composting, we reproduce in industrial scale these natural degradative processes, obtaining a stable organic product with similar characteristics to soil organic matter in a short time. Olive press cakes are solid agroindustrial residues produced during the extraction of olive oil, disposal of wastes produced during this process is a particularly pressing problem in southern Spain and other mediterranean countries. Residues are produced for 4–5 months in a high rate producing serious problems of storage, transport and disposal. Composting it, we eliminate those problems obtaining paralelely an organic matter which can be applied as ammendment on several crops and soils. The aim of this work was to develop in

1 Estación Exp. del Zaidin, CSIC. Apdo. 419. E18080, Granada, Spain
2 Dep. Produzione Vegetale e Tec. Agr.; Fac. Agraria.; Univ. Udine. I33100, Udine, Italia

pilot scale, a composting process to transform this lignocellulosic byproduct in a substrate rich in nutrient and organic matter suitable as soil ammendment.

Material and methods

We developed a windrow composting process to transform pressed olive cake. The characteristics of the product during composting are in Table I. We prepared a windrow of 5000 Kg of dry mater and we controled its physico-chemical characteristics during 3 months. Initial C/N and C/P ratios were adjusted to 33 and 240 respectively by adding urea and superphosphate. The composting mass was turned periodically to homogenize the compost and to control humidity, temperature and aeration. Humidity was mantained above 40–50% to promove fungal growth. O2 concentration was measured by gas chromatography at 20 cm of deep where the temperature was higher. Humification was determined according to Sequi et al. (1986), which separate polyphenolic (humic like) and non polyphenolic (labile) substances extracted in alkaline solution by adsorption chromatography on PVP resin. Biological tests were realized to determine the evolution of the phytotoxicity during the proces; we determined the shoot dry weight of sunflower plants grown for 21 days on soil:compost (1/1 w:v) mixtures at 0, 15, 30, 60 and 90 days of compost transformation. Nitrate content was determined using the phenol-disulphonic method proposed by Lachica et al. (1972).

Table 1 Physicochemical characteristics of the compost at the beginning, and at the end of the thermophillic phase (60 d) at the end of the experiment (90 d).

	Days		
	1	60	90
pH	6.60	7.70	7.80
E.C. (dS/cm)	3.50	3.50	2.50
WHC	21.50	84.30	92.10
%H	60.00	35.00	44.00
Density (g/l)	0.48	–	0.58
Volatil solids (%)	90.30	86.90	85.80
TOC stractable (mgC/gds)	87.90	54.50	59.00
HS mgC/gds	50.40	36.20	42.90
NHS mgC/gds	40.70	12.00	12.90
HI	0.80	0.30	0.30
C/N	33.00	19.20	17.00
C/P	240.00	180.00	128.00
Lignin	31.20	22.30	19.90
Cell + hemicell.	29.80	16.60	9.60
NO3 %	0.04	0.02	0.01

Result and discussion

Temperature reached thermophilic values after several hours and remained elevated throughout the process. The thermophilic phase lasted for 55 days with an average T^a of 55°C and recovered fastly after turning (Fig. 1) . After 55 days, the temperature decreased slowly to mesophylic values above 40°C. At the end of the experiment after 90 days of composting, it remained at mesophylic values.

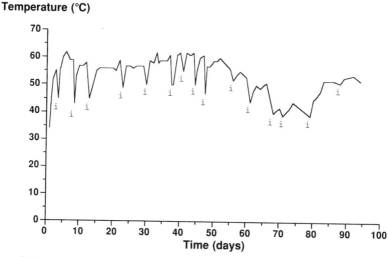

Figure 1 Temperature curve during composting determined at 20cm deep. Arrows indicate turning points.

Aeration was carried out by turning to assure an adecuate oxigen level into the compost mass that required over the 5% O2 to assure the aerobicity. Only during the first days, when the biological activity was highest, the O2 level decreased tol 2% (Fig. 2).

During the thermophilic phase, lignin, cellulose and hemicellulose were hardly mineralized by the microorganisms, which performed above 82% of the total degradation in these phase and the remaining 18% during the last month at temperatures around 40°C (mesophilic phase)

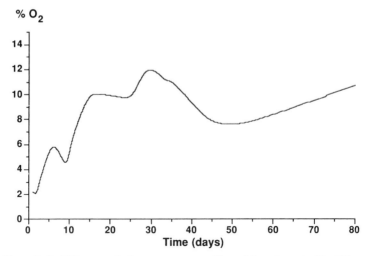

Figure 2 % of O2 content in the compost mass at 20cm of deep determined by GC.

Stuctural changes were observed during composting, the density increased from 0.48 to 0.58 g/cm^3 with a paralell loss of 2/3 of the initial volume. WHC changed in the same time from 21.6% to 92.1% (Table I).

At the end of the experiment, the temperature was still quite high (above 40°C). Nevertheless, the phytotoxicity of the compost was almost nil as demosntrated by biological tests carried out periodically with sunflower plants grown on soil-composts mixtures. After 90 days of composting, the plant development were similar in the controls and in the mixture (Fig. 3).

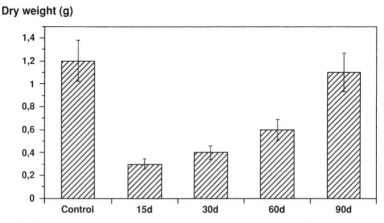

Figure 3 Shoots dry weight of the plants grown on the control and on the soil-compost mixture at 15, 30, 60 and 90 days of transformation (any growth was found at 1 day of transformation

In a previous work, nitrificant activity was found to be as an index of the aerobicity and maturity of these processess (Baca et al. 1995), nevertheless, the presence and activity of the nitrificant bacteria diminished with temperature and it could´n be used in this experiment as an indicator of compost stabilization complementary to the biological test.

Total extractable organic carbon (TOC) decreased during the composting passing from 87.9 to 59.0 mgC/g ds. In the same time, extractable polyphenolic (HS) substances decreased from 50.4 to 42.9 mgC/gds while non polyphenolic (NHS) were strongly consumed by microorganisms as a carbon source, passing from 40.7 to 12.9 during the process (Table I). Humification index, which is the quotient between NHS and HS decreased from 1.1 at the beginning to 0.3 after 60 and 90 days indicating that the chemical stability reached before the disapearance of the phytotoxicity.

Discussion

The composting process proposed in this works suitable for transforming by-products with a very homogeneous composition and high content of polyphenolic (lignin) and polysaccharidic (cellulose and hemicellulose) substances. The phytotoxicity was almost nil even before the maturation phase, probably because the biological activity was very restricted and specialized due to the composition of the by-product with more than 60% of dry weight constitued by polysaccharides and lignin. The later compound, rich in polyphenolic and polycarboxilic acids could be transformed by the biological degradation and posterior synthesis and diagenesis in a stable organic matter rich in humic-like substances. The disappearance of the toxicity was a slow process throughout the experiment, while the chemical stabilization of the orgánic matter seemed to be related to HI values of 0.3 found at the end of the thermophylic phase.

The compost obtained by these method can be used as a soil ammendment and even also as an organic fertilizer in ecological agro-culture (Guía de productos utilizables en agricultura ecológica, 1993).

References

Baca, M.T. 1988. Fertilizantes orgánicos: Estudio y preparación de compost para su utilización en agricultura. PhD. Thesis. Univ. Granada. Spain. pp:341.

Baca, M.T.; Delgado, I.C.; De Nobili, M.; Esteba, E. and Sanchez Raya, A.J. 1995. Influence of compost maturity on nutrient status of sunflower. Commun. Soil Sci. Plant Anal. 26: 169–181.

Baca, M.T.; Estaban, E.; Almendros, G. and Sanchez Raya, A.J. 1993. Changes in the gas phase of compost during solid state fermentation of sugarcane bagasse. Bioresource Technology. 44: 5–8.

Baca, M.T.; Fornasier, F. and De Nobili, M. 1992. Mineralization and humification pathways in two composting processes applied to cotton wastes. J. Fermentation and Bioengineering. 74: 179–184.

De Bertoldi, M.; Filipi, C. and Picci, G. 1986. Olive residue composting and land utilization. In: Olive By-products valorization. INIA (Eds). pp: 307–319.

Guía de Productos utilizables en Agricultura Ecologica. 1993. Servicio de Investigación y Desarrollo Tecnológico (colección monografías). Conserjería de Agricultura y comercio, Junta de Extremadura (Eds.). pp: 21

Lachica, M., Aguilar, A. and Yañez, J. 1972. Analisis foliar. Métodos utilizados en la Estación Experimental del Zaidín (I). Anal. Edaf. Agrobiol. 32: 821–830.

Sequi, P.; De Nobili, M.; Leita, L. and Cercignani, G. 1969. A new index of humification. Agrochimica. 30: 175–183.

Changes in the Amino Acid Composition of Grass Cuttings During Turned Pile Composting

M.T. BACA*, I. FERNANDEZ-FIGARES*, C. MONDINI** and M. DE NOBILI***

Abstract

The amino acid composition of composts derives in part from the original amino-acid pool of the substrate and in part from extracellular enzymes and cells of the dead and living micro-organisms. Because of their high biodegradability, the relative molar distribution (RMD) of amino-acids in compost depends on a balance between re-synthesis and decomposition and should be related to changes in microbial community. The total amino-acid content increased sharply during the first days of composting fairly as a consequence of weight loss and remained fairly constant up to the 75th day of composting, decreasing thereafter during the curing period to about one half of the maximum value obtained (139 mmol/100g). The RMD shows a decrease in acidic amino-acids and particularly in the content of glutamic acid (11.19% at 7 days and 7.75% at 100 days), with a concomitant increase in cystine (6.85% at 7 days and 10.07% at 100 days) and methionine (8.45% at 7 days and 11.95% at 75 days). Comparison with previous work suggests that changes in RMD are related to process strategies.

Introduction

Good quality compost could represent a used alternative to peat as growing substrate and organic ammendant. The quality of compost is related to the quantity and the quality of humified fraction of organic matter, but also to the presence of a biologically active fraction that acts as a readily available source of nutrients. Amino acids (AA) represent an important part of this fraction. Recently, relative molar distribution (RMD) of AA was related with changes in soil organic matter quality (Campbell et al., 1991), but information on the composition of the

*Estacion E~erunental del Zaidin, CSIC, -of Albareda 1, E- 18008 Granada, Spain.
**Istituto Sperimentale per 1a Nut-one delle ~ante, Se~one di Goriiia. Via Trieste 23, 1-34170 Go~a, Italy.
***Dipartimento di ~produzione Vegetale e Tecnologie Agrarie, Via delle Scienze 208, I- 33100 Udine, Italy.

aminoacidic fraction of compost are very scarce. The AA composition, in soils as in compost, depends on a balance between re-synthesis and decomposition: if the former is prevalent, changes in the amino-acids composition should reflect changes in the composition of microbial population.

In this work we studied the variations in AA composition of grass cuttings during composting and its relation with the degree of stabilization of compost.

Material and methods

In the compost process grass cuttings (50 Kg) were piled and manually turned, at regular intervals, for a period of 100 days. Humification index was determined according to Sequi et al. (1986). Samples for amino-acid analysis were hydrolyzed with HCl 6M and 1% phenol for 24 h at 110 $\times$ C. Aminoacids were analyzed quantitatively by high performance liquid chromatography (HPLC)using the Waters Pico-Tag method and Nor-leucine as internal standard.

Results and discussion

The regular trend of 111 (table 1) indicates the progress of humification processes with the increasing time of composting. Nevertheless HI already at 7 days shows values which are typical of well humified materials. This could be explained by the presence in the starting substrate of a high content of phenolic (humus-like) substances that are extractable with alkali and contribute to the amount of C in the B fraction (humic). In fact the B fraction decrease during composting up 61% of the initial value. Total extractable C decreased as well during composting.

Table 1 Extractable organic carbon and humification index of grass cuttings during composting

Days		O.C. (mg/g)		H.I.
	TE	A	0	
7	182.70	48.76	131.82	0.37
21	129.05	25.63	99.73	0.26
41	107.34	18.34	86.81	0.21
75	120.49	17.93	103.11	0.17
100	60.63	8.93	50.13	0.18

The total AA content (table 2) increased sharply during the first day of composting, mainly as a consequence of weight loss and remained fairly constant up to 75th day of composting, decreasing afterwards during curing period to about one half of the maximum value obtained (139 mmol/100g). This would indicate that with advancing humification, decomposition overcomes re-synthesis. The prevalence of mineralization processes at the end of composting period could be demonstrate by the increasing content of Nh4 (tab. 1).

Table 2 Total aminoacid acid and NH4 content of grass cuttings during composting

Days of composting (mmol/100g)	Total aa content (mmol/100g)	NH4
7	130	13
21	139	13
75	125	18
100	71	17

Table 3 Amino acid composition of grass cuttings during composting

Amino acid	7 days	21 days	75 days mmol/100g	loo days
Acidic				
Aspartic a	6.235	4.806	4.392	2.022
Glutamic a., a.	14.531	12.507	11.784	5.478
Total	~.788	17.313	18.178	7500
Basic				
Hi~histidine	2.382	2.430	2.301	1.332
Lysine	3.987	4.219	3.723	1.949
Arginine	4.949	5.687	4.504	2.473
Total	11.318	15338	1O.6~	5.754
Neutral				
Threonine	6.529	7.142	5.796	3.685
Serine	8.290	9.792	8.786	5.324
Prone	4.210	5.676	3.730	2.470
Glycine	14.133	15.995	13.731	8.736
Alanine	14.555	15.572	13.242	7.554
Vane	7.990	9.606	6.907	4.020
l~leucine	4.662	5.093	4.076	2.558
Leucine	9.437	10.627	8.319	4.821
Tyrosine	3.809	4.054	3.343	2.044
Phenilalanine	4.305	5.001	3.984	2.333
Total	77.9~	88.55	71.914	43.545
Sulphur c				
Methionine	10.968	10.718	14.988	6.746
Cystine	8.902	9.700	11.765	7.113
Total	19.870	418	~753	13.859

The amino acid composition is shown in table 3. Total acidic AA is the class of AA that shows the higher decrement respect to initial value (63.9%). This trend is confirmed by the 1RMD of total acidic AA (fig. 1) showing a regular decrease, particularly in the content of glutamic.c acid (11.19% at 7 days and 7.75% at 100 days). This is in agreement with the trend found during humification of spruce and aspen leaf litter (Ladesmaki and Phspanene,1989). RMD of sulphur-containing AA shows a marked increase (28.2%). In particular cystine increased from 6.85% at 7 days to 10.07% at 100 days and methionine from 8.45% at the beginning to 11.95% towards the end of composting. Change in RMD of sulphur-containing AA is particularly evident between 21 and 75 days, suggesting a change in the composition of microbial population from thermophilic (50-60 xC) to mesophilic (30-40 0C) microorganisms of total basic and neutral AA remained practically constant. of total acidic and sulphur containing is quite different from that of a well stabilized population such as that of an organic matter rich soil (Campbell et al., 1991). Comparison with previous work (Baca et al., 1992), suggests that changes in RMD are related to process strategies. Therefore changes in AA composition can be a useful parameter to characterize the compost, but -her works is necessary to use AAcomposition to assess compost maturity.

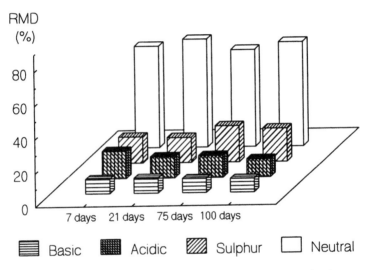

Figure 1 RMD of amino acid of grass cuttings at different composting times

References

Baca, M.T. ,Fernandez-figares, 1. and De Nobili M. (1992). Amino acid composition of composting cotton waste. J. Ferment. Bioeng., 74,179-174.

Campbell, C.A., Schnitzer, M., Lafond, G.P., Zentner, RP. and Knipfel, J.E. (1991). Thirty-year crop rotations and management practices effects on soil and amino nitrogen. Soil Sci. Soc. Aba. 1., 55,739-745.

Ladesmaki~ P. and Phspanen, R (1989). Changes in concentration of flee amino acids during humification of spruce and aspen leaf litter. Soil Biol. Biochem, 21,975-978.

Sequi, P., De Nobili li, M. Leita, L. and Cercignani G. (1986). A new index of humification. Agrochimica, 30, 175-178.

Composition and Chemical Characteristics of the Main Compostable Organic Wastes

BARBERIS RENZO, PANZIA OGLIETTI ALDO and
CONSIGLIO MICHELE – Istituto per le Piante da
Legno e l'Ambiente I.P.L.A. – Torino

Introduction

In order to produce high quality composts it is necessary to start from selected organic wastes, non polluted by heavy metals and with limited inerts and unwanted materials. The source separation of municipal solid wastes and especially the separate collection of wastes from specific categories of producers, can help to supply good quality waste materials that can be composted together with yard wastes to advantage. This poster reports the most important results of a wide range of experiments, carried in Piedmont (Northern Italy), on separate collection of organics from specific categories of producers such as restaurants, canteens, and vegetable markets, foodstuff stores, and so on. Chemical characteristics and composition of these wastes were compared with those of mixed municipal wastes selected after collection.

Discussion

Solid wastes contain high percentages of organic compostable materials (food and yard wastes, paper, wood, ..); in particular wastes coming from specific producers are prevalently made of compostable materials. The pictures in annexe show the main components of MSW and of wastes from specific producers (restaurants, canteens, foodstuff stores, etc.), the latters contain very high percentages of compostable materials which reach 90% of the waste when separate collection programs are implemented on purpose for the abovesaid categories. The table report the results of chemical analyses carried out on the organic fraction of wastes collected from the above–mentioned producers, the results are then compared with those of chemical analyses of the organic fraction of MSW separated after collection. The organic materials coming from selected producers contain vary low percentages of heavy metals, this is particularly evident when the data are compared

Table 1 Chemical analyses of compostable materials (dry weight except for pH and moisture content)

Parameter	Domestic Wastes (1)	Foodstuff Store Wastes	Restaurant Wastes	Canteen Wastes	Yard Wastes	MSW (2)
Moisture (%)	66,5	40,5	61,5	78,2	33,0	42
pH	5,3	5,2	4,9	4,6	5,8	7,1
Conductivity (mS/cm)	3,0	3,6	2,7	2,9	2,0	2,4
Salinity (meq/100g)	80,1	98,7	71,2	77,3	58,6	64
Ash (%)	23,6	33,8	9,6	5,8	17,7	50,9
Organic carbon (%)	35,1	31,8	41,40	40,83	41,6	22,7
Total Nitrogen (%)	2,1	1,6	4,1	2,8	1,3	1,3
C/N Ratio	16,9	19,5	10,0	14,8	40,4	17,4
Phosphorus (%)	0,42	0,27	0,25	0,196	0,17	0,37
Potassium (%)	1,30	1,15	0,60	1,26	1,15	1,03
Calcium (%)	4,10	2,32	2,69	1,45	2,41	1,28
Magnesium (%)	0,07	0,05	0,01	0,01	0,61	0,76
Sodium (%)	0,35	0,58	0,53	0,68	0,57	0,60
Lead (mg/kg)	50	41	9	10	183	891
Nickel (mg/kg)	23	97	6	19	62	48
Manganese (mg/kg)	53	213	11	24	124	568
Zinc (mg/kg)	284	197	80	187	111	822
Copper (mg/kg)	31	41	8	18	13	302
Chromium (mg/kg)	41	58	6	27	41	71
Cadmium (mg/kg)	1,2	0,5	0,4	1	0,3	5

(1) Source separated organic wastes from the homes
(2) Organic fraction selected from Municipal Solid Wastes after collection

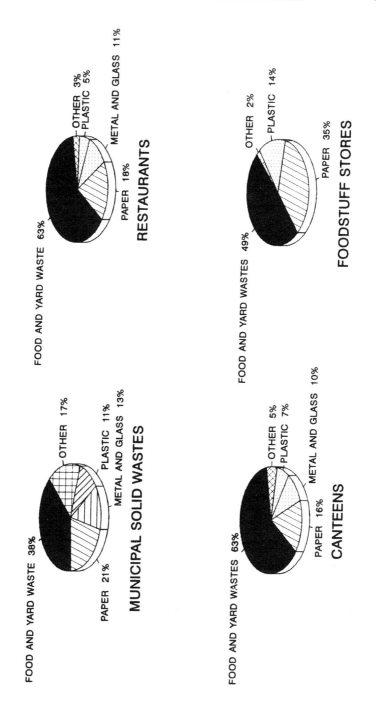

Figure 1 Comparison of Waste Composition

with those of the organic fraction selected from MSW alter collection. The latter in fact, being contaminated by heavy metals, is not suitable for high quality compost production.

Organic materials from selected producers, preferably bed with sludges of the food industry, wood debris and agricultural residues, can coy make high quality composts. The results prove that an effectively carried out separate collection, both from specific waste producers and from homes, is able to convey to the composting facility little polluted materials (low heavy metal concentration) containing easy-to-remove quantities of glass and plastic. Thus the obtained product will comply with the most severe qualitative standards.

Impacts of Separation on Compost Quality

SAMIRA BEN AMMAR (INBST*, Tunisia)

Composting treatment reduce the volume of refuse to be dumped in landfills since household waste will thus lose a large portion of its organic components. Correctly implemented, composting is a reliable method of transforming the putrescible organic matter of wastes in to a soil conditioner.

Throughout the fermentation period, it is extremely important to avoid contact between organic and inert matters contained in household refuse, to prevent contamination from heavy metals.

Heavy metals are used in many household products (soap, detergents, cosmetics, packaging, leather and batteries). They are concentrated during the composting process as the organic matter which dilutes them gradually degrades. The metals of greatest concern are those which tend to bioaccumulate, causing short or long term toxic effects to organisms in the environment. Those most commonly regulated include cadmium (Cd), chromium (Cr), copper (Cu), mercury (Hg), lead (Pb), nickel (Ni) and zinc (Zn).

Separation at the source seems to be the best way to reduce inert matter contained in household refuse thereby provides a compost with low heavy metal content.

Source Separated Program

Concerned about the improvements and the quality of the Environment, the Tunisian Ministry of Environment has decided to implement a collection pilot project involving source separated organics. The household separation project for collection and composting started operating in 1994 January in a city of 6,000 inhabitants. The aim of the project is a reduction of the total amount of wastes and the production of compost of better quality as compared with the previously produced municipal waste compost. Each household is given two collection bins. One is green for the organic fraction (food waste, garden refuse, etc...); the other is blue for the inert fraction (plastic, metal, glass, etc...).

Alter daily collection, 'blue waste' is sorted and either recycled or landfilled. 'Green waste' – 65 percent of the total domestic waste by weight – is transported to an experimental composting plant, designed to process 30 Ton/day. Experiments are performed by research workers of the 'Institut National de Recherche Scientifique et Technique'.

*Institut National de Recherche Scientifique et Technique – 13P.1 5, Cité Mahrajène 1082 Tunis-Tunisie.

Experimentation

Two composts were tested in this research. The first of the two – referred as C 1 – was obtained with the usual process: sorting, trituration, screening, fermentation and maturation.. The second of the two composts – referred as C2 – was obtained with a new process: sorting, prefermentation, screening, fermentation and maturation.

The putrescible organic matter obtained through household separation is characterized by a high moisture content wish causes – after trituration – compacting during the first 3 weeks of the process. Thereby compromising ventilation inside the biomass and increasing the risk of anaerobic conditions in windrows. To avoid these anaerobic conditions 'C2' was composted without trituration.

Composting was done in windrows that are turned once every week or two. Turning was carried during the entire thermophilic phase to dissipate excess heat. Throughout the fermentation period, temperatures are recorded daily and moisture levels checked weekly.

Compost Product quality

The finished compost produced by this project was analysed for both its fertilizer value as well as any potential contamination that might limit the product's use. Main Characteristics of the compost issued from the source separated household waste compared with the previously municipal waste compost are listed in Table. 1 and Table.2.

Table 1 Compost quality Data

Parameters	Organic waste compost	Municipal waste compost
pH	8,5	8,1
Organic matter (%)	52	40
Moisture content (%)	40	36
% dry matter:		
Carbon	23	17
Nitrogen	2,2	1,0
Phosphorus	0,85	0,6
Potash	0,9	0,7
Calcium	7,5	10
Magnesium	0,43	0,8
Iron	0,4	0,8

Table 2 Metal contaminants in composts

Heavy Metals	Organic waste compost	Municipal waste compost	Guide values for compost in Germany
mg/kg dry matter			
Zinc	440	605	400
Copper	195	346	100
Lead	182	545	150
Cadmium	4,2	4,5	1,5
Mercury	0,9	2,0	1,0
Nickel	7,7	36	50
Chromium	31	52	100

This study demonstrates that contents of total N, P2O5 and K$_2$O are significantly higher in organic waste compost as compared to Municipal waste compost. Thereby indicate that this material should prove a high quality soil amendment for landscape use.

The results of the metal analysis – standardized to an organic matter of 30% – show that the heavy metal concentration found in the organic waste compost is considerably lower as compared to municipal waste compost.

Considering the trituration parameter, results indicate that it increase significantly heavy metals content in organic waste compost (table. 3).

Table 3 Metal contaminants in organic waste compost

Heavy Metals	Organic waste compost (C1)	Municipal waste compost (C2)
mg/kg dry matter		
Zinc	556	325
Copper	256	135
Lead	178	176
Cadmium	4,7	3,7
Mercury	1,0	0,8
Nickel	10	5,0
Chromium	33	30

Conclusion

Results demonstrate that under practical conditions, source separation composting will lead to a final product of better quality, with regard to the contents of undesirable components.

Nitrogen in Composting: Relevance of the Material and the System Used

M.P. BERNAL, A. ROIG, M.A. SÁNCHEZ-MONEDERO, C. PAREDES, D. GARCIA – Department of Soil and Water Conservation and Organic Waste Management, Centro de Edafología y Biología Aplicada del Segura, CSIC. P.O. Box 4195, 30080 Murcia, Spain.

Introduction

The composting of organic wastes rich in easily biodegradable nitrogen compounds leads to the formation, accumulation and subsequent loss of nitrogen mostly through ammonia volatilization (Witter, 1986). Mixing these organic wastes rich in nitrogen with lignocellulosic wastes which have high C/N ratio can result in partial incorporation of N into the organic fractions through immobilization (Bernal et al., 1993). Therefore, selecting the appropriate waste mixture may be an appropriate way of reducing ammonia losses during composting. Frequent turning of the pile may facilitate this NH3-volatilization (De Bertoldi et al., 1982). It might be possible to control NH3-volatilization by using the Rutgers static pile system which can also regulate the organic-N mineralization through controlling the pile's ceiling temperature during composting.

Materials and methods

Five composting piles were prepared with the following organic wastes mixtures (fresh weight):

Pile 1: 95 household biowaste 1 5% sweet sorghum bagasse (turned pile).

Pile 2: 95 household biowaste 1 5% sweet sorghum bagasse (static pile).

Pile 3: 34.6 poultry manure 1 65.4% cotton waste 1 1.93 1/kg olive-mill wastewater.

Pile 4: 32.1 sewage sludge 1 67.9% cotton waste 1 0.94 1/kg olive-mill wastewater.

Pile 5: 47.2 sewage sludge 1 52.8% maize straw 1 1.76 1/kg olive-mill wastewater.

About 1500 kg of the mixture were placed in trapezoidal piles of 1–1.5 m high with a 2 × 3 m base. Only pile 1 was composted by turning every two days during the first week, twice a week during the second week, and once a week during the rest of the biooxidative phase. The Rutgers static pile composting system was used in the rest of the piles involving on-demand ventilation through temperature feedback control (Finstein et al., 1985). The air was blown from the base of the pile, the timer was set for 30 s. ventilation every 15 min. and the ceiling temperature for continuous air blowing was 55°C. The biooxidative phase of composting lasted for 77,77,49,84 and 63 days in the piles described above, respectively. The air-blowing was then stopped to allow the compost to mature over a period of two months. The piles were sampled weekly during the biooxidative phase. Moisture content was ascertained by drying at 105°C, pH in water soluble extract 1:10 (w/v), organic matter (OM) by loss-on ignition at 430°C during 24 h. Inorganic-N was extracted with 2M KCl, NH4–N was determined by a colorimetric method based on Berthelot's reaction and NO3–N by the ultraviolet technique. Total nitrogen and organic carbon were determined by automatic microanalysis. Losses of organic matter were determined from the initial (X1) and final (X2) ash contents according to the equation (1). Similar equations were used to calculate total and organic–N losses.

$$OM-loss\ (\%) = 100-100\ [X, (100–X2)]/[X2\ (100–X,)] \tag{1}$$

Results and discussion

In pile 1 a substantial amount of NH4–N was produced during the first week of composting (Fig. 1a) and its concentrations were higher than those in pile 2 during the following weeks. At the same time pH values increased in pile 1. These two facts resulted from the faster organic–N mineralization in pile 1 than in 2, as it is showed by the losses of organic– N (56.7 of initial organic–N in pile 1 and 48.1 in pile 2). The total–N concentration decreased in pile 1 indicating great N–losses during composting. These losses are mainly due to NH3–volatilization (Bishop and Godfrey, 1983). N–losses by denitrification were probably very low, since oxygenated condition prevailed in the composting mass during the whole process. Also, very low lixiviation was observed from the piles mainly in those which the Rutgers state pile system was used, therefore NO3–lixiviation was presumably negligible. Losses of N through NH3–volatilization accounted for 59 in pile 1 and 50 in pile 2. The NH4–N nitrification started after 49 days of composting increasing the NO3–N concentration.

Using the Rutgers static pile system instead of the traditional turned pile reduces the organic–N degradation rate as well as the losses through NH3–volatilization. The higher N–losses from the turned pile were due to greater pH values, higher NH4–N production through organic N mineralization and the frequent turning. In the static pile the air was blown from the bottom to the external part of the pile.

The NH_3 and H_2O produced during OM decomposition would be diffused with the air through the pile. Because the surface of the pile was cooler than the interior, the H_2O may condense and NH_3–N be absorbed by the material, helping to prevent N–losses as NH_3–N could be further immobilized or nitrified by the microorganisms.

In pile 4 with sewage sludge the fermentation phase of composting lasted almost double than in pile 3 with poultry manure (Fig 1c,d) indicating higher microbial activity in the former, although both piles had similar organic matter degradation rate (data not shown). Changing cotton waste by maize straw reduced the rate of OM decomposition (69.2 to 54.2 %) indicating the lower biodegradability of the maize straw compared with the cotton waste. Total and organic–N increased with composting in piles 3,4 and 5 due to the concentration effect of the degrading organic–C compounds. Pile 3 had a high initial total–N concentration, 12.5 of which was NH_4–N, this was lost at high rate during the first 14 days as shown by the decrease of NH_4–N concentration and the total–N losses (14 %). The increase in pH during the same period strongly favoured these N–losses. However, in pile 4 only 5 of total–N was lost during 14 days, both NH_4–N and pH values increased until day 49 resulting from the organic–N mineralization. Greater proportion of initial N was susceptible to volatilization in pile 3 than in 4, due to the high initial NH_4–N concentration, high pH and extra ventilation in pile 3, conditions which favoured NH_3 losses (Bhoyar et al., 1979).

Pile 5 had the lowest initial total–N concentration, the greatest C/N ratio and initial pH <7.0. There was a sharp reduction of NH_4–N concentration during the first week, which may be due to N–immobilization since total–N loss was very low and negative organic–N loss occurred (–12.6 %). The mature compost prepared with sewage sludge and maize straw (pile 5) had a total–N concentration similar as those in piles 3 and 4 although the initial mixture had the lowest value. The use of maize straw with a high C/N ratio as a bulking agent was more effective in reducing N–losses during composting than the cotton waste and so enriched the final product. On the other hand, the great amount of NH_4–N brought about by the poultry manure caused high NH_3–volatilization at the beginning of the fermentation process.

References

Bernal, M.P., Lopez-Real, J.M. and Scott, K.M. 1993. Bioresource Technol. 43, 35–39.
Bhoyar, R.V., Olaniya, M.S. and Bhide, A.D. 1979. Indian I. Environ. Mlth., 21, 23–34.
Bishop, P.L. and Godfrey, C. 1983. BioCycle, **24**, 34–39.
De Bertoldi, M., Vallini, G., Pera, A. and Zucconi, F. 1982. BioCycle, **23**, 45–50.
Finstein, M.S., Miller, F.C., MacGregor, S.T. and Psaranos, K.M. 1985. EPA Project Summary (EPA/600/52–851059) U.S. EPA, Washington.
Witter, E. 1986. Ph.D. thesis. Wye College, University of London.

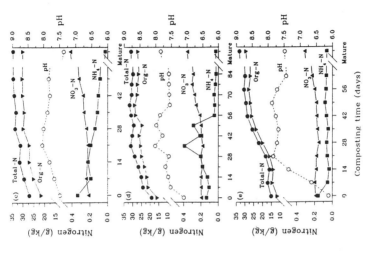

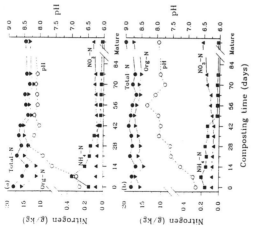

Figure 1 Changes in the nitrogen forms and pH values during composting of pile 1(a), 2(b), 3(c), 4(d) and 5(e).

Estimation of N-release and N-mineralization of Garden Waste Composts by the Mean of Easily Analysed Parameters

A. BERNER, I. WULLSCHLEGER and T. ALFÖLDI

Introduction

Determination of the biomaturity of a compost is a critical point. It can be estimated by chemical analysis or optical measurements (Mathur et a*l*. 1993). N-mineralization and N– immobilisation of composts can be assessed by the method of incubation and by germination tests. However, these methods are time consuming. We therefore compared facile and quick tests with those standard methods.

Material and methods

Samples from 38 garden waste composts from different compost plants were taken at the time as they were given to the farms.

For the incubation test a mixture of 10 vol % compost, 40 vol % soil (loamy loess) and 50 vol % Perlit (vulcanic rock stone) was used. The incubation period was 12 weeks at field capacity and 20 °C. Table 1 shows the analysed parameters.

Results and discussion

N-Releare

As shown in figure 1, the 38 garden waste composts release only a small amount of plant available nitrogen. Half of the composts immobilise nitrogen. during the 12 week incubation period. The values of the Nmi_n incubation are in the average at 2.1% of the total nitrogen content. 13% of the composts would cause N-blockage after application. The values of the Nmin incubation test are varying between + 8% and − 11% of the total nitrogen content of the composts. The short term and long term availability of nitrogen in the compost is small.

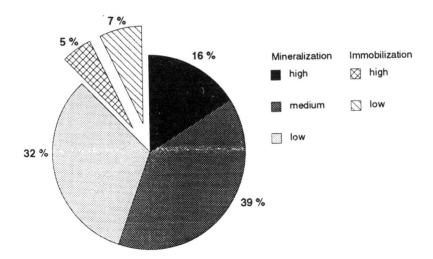

Figure 1 Distribution of the nitrogen mineralization potential of 38 different garden waste composts. Low mineralization: 0–2% of the compost nitrogen has been mineralized, medium: 2–4%, high: 4–6%: low immobilization: 0 till–2%, high: –2 till–11% of the compost nitrogen has been immobilized.

Correlations

Simple correlations show low and only partly significant correlations (p ≤ 0.05) to the aiming points Nmin incubation and N-mineralization (table 2). The highest correlations were found between Nmin incubation and respiration, C and $W_{40°}$ with r2-values between 0.38 and 0.45. The r2 -values between N-mineralization and W40°, closed cress test and respiration respectively were 0.41, 0.52 and 0.59. The closed cress test shows clearly better correlations to both aiming points than the open cress tests. C/N ratio is correlating to the N-mineralization with a r2-value of only 0.21.

Multiple correlations between quick tests, cress tests and carbon content (No. 1–5) are reaching R2-values up to 0.66. Similar correlation coefficients can be calculated with chemical analyses (No. 6–9). The optical tests are reaching the highest correlation coefficient. Determining optical parameters is easy and fast, results are available after one week.

Table 1 Parameters to characterize biomaturity of garden waste composts

Parameter	Definition/Method	Remarks
N_{min} incubation	Mineralised nitrogen of the compost $NO3$ $+NH_4$) during incubation period in % of the N_{kj} of the compost.	Plant available nitrogen at short term
N-mineralization	N_{min} incubation minus N_{min}-content of the compost at the start of the incubation test.	Plant available nitrogen at longer term
C	Carbon content (%) in the dry matter (DM). loss by dry ashing * 0.47.	
N_{kj}	Total nitrogen content (%) in the dry matter (Kjeldalil).	
C_{sol}, $N_{tot\ sol.}$ $N_{org\ sol}$	C, N_{tot}, N_{org} soluble in 0,01 M $CaCl_2$ extraction of the samples dried at 40 °C (Houba et al. 1986).	Fast mineralizable carbon and nitrogen from the microbial pool.
Respiration	Oxygene consumption of composts at 20 °C in mg O_2/mg C*day.	Microorganisms in immature composts show increased respiration.
Ammonium test Nitrate test	Quick test with Merckoquant Test sticks in mg/kg DM	Immature composts contain more ammonium, mature composts contain more nitrate.
NH_4	Ammonium content of the DM analysed by a Ion chromatogragh in mg/kg DM.	
Cress test open 100%, Cress test open 50%	Cress test on 100% compost resp. 50% compost/soil-mixture in open seed pots in g FM/pot	In open pots cress is growing on nearly every substrate.
Cress test closed	Cress test on 100% compost in closed pots in g FM/pot.	Gaseous emission of composts increase differentiation in closed pots (Nasliowski und Grantzau 1992).
$W_{40°}$	Water content of compost samples dried at 40 °C during 48h.	Composts with a high organic matter content have an increased water holding capacity.
Wf_{210}, Wf_{280}, Wf_{600}, Wf_{700}	Optical density of the water extract of a compost at a fixed wavelength at 210 nm 280 nm, 600 nm, 700 nm in absorbance/g $C_{compost}$.	Mature composts are less water soluble. More complex compounds absorb light at higher wavelengths (Mathur et al. 1993).
Ws_{280}	Absorption of water extracts stored for 7 days at 20 °C compost/water mixture at 280 nm in absorbance/g $C_{compost}$.	Storagc increases decomposing processes.
ED_{400}	Absorption of fresh EDTA-extracts from composts at 400 nm in absorbance/g $C_{compost}$.	EDTA is also solving huinic acids from the compost.
T_{20}	Temperature of the compost heap at sampling time at a depth of 20 cm.	Compost heaps show higher temperature during decomposition process.

Table 2 Simple (r^2) and multiple (R^2) correlations between simple tests, specific chemical analyses and N_{min} incubation, N-mineralization; estimation of the analytical expenses and duration

	aiming point			
	Nmin	N-Minera-	Analyucal	Duration
impact value	Incubation	lization	expenses[1]	[2]
Simple correlations	r^2		r^2	
Nitrate test	0.28	0.20	*	*
Ammonium test	n.s.	n.s.	*	*
Cress test (closed)	0.22	0.52	*	**
Cress test (open 100 %)	n.s.	0.17	*	**
Cress test (open 50 %)	n.s.	0.22	*	**
C	0.39	0.33	**	*
N	n.s.	n.s.	**	*
C/N	0.34	0.21	**	*
$W_{40°}$	0.45	0.41	*	*
ED_{400}	0.25	0.22	*	*
Respiration	0.38	0.59	**	*
Multiple correlations	R^2		R^2	
1 Nitrate test * Ammonium test	0.33	0.22	*	*
2 Nitrate test * Cress test (closed)	0.38	0.57	*	**
3 Nitrate test * Ammonium test * Cress test (closed)	0.48	0.66	*	**
4 Nitrate test * C	0.57	0.45	**	*
5 Nitrate test $W_{40°}$	0.64	0.55	*	*
6 %C * %N	0.53	0.39	**	*
7 C/N * $W_{40°}$	0.56	0.46	**	*
8 T20 $W_{40°}$ Respiration	0.59	0.77	**	*
9 C * $N_{tot sol}$ * C_{sol}	0.69	0.55	***	*
10 C * $N_{org sol}$ C_{sol}	0.63	0.64	***	*
11 Wg_{280} * Wf_{700} * Wf_{210}/Wf_{280} * Wf_{280}/Wf_{600} *Ws_{280}/Wf_{280} * $W_{40°}$	0.82	_3)	*	**
12 $W_{40°}$ NH_4 Ws_{280}/Wf_{280}	_3)	0.81	**	**
13 ED_{400} * Wf_{700} * Ws_{280}/Wf_{280} * $W_{40°}$	_3)	0.75	*	**
14 $W_{40°}$ * Ws_{280}/Wf_{280} * Cress test (closed)	_3)	0.82	*	**

$p \leq 0.05$ n.s. = not significant

1)*	low, quick test, simple analyse	2)* 1–2 days	3) missing values: impact values
**	medium, routine analyse	2)** 1 week	depending from each other.
***	high, specific chemical analyse		

Conclusions

The N_{m}in incubation method showed that the 38 garden waste composts released only little amount of nitrogen. 12% of the composts would cause nitrogen blockage in the field.

None of the tests was reliable enough to substitute the Nmin incubation method. Reliability was clearly improved by combining several parameters.

The preliminary results indicate a better prediction of N-mineralization and N-release with optical parameters than with chemical analyses and cress test.

Literature

Houba, V.J.G., I. Novozamsky, A.W.M. Huybregts und J.J. Van der Lee, 1986: Comparison of soil extractions by 0,0lM CaCl$_2$, by EUF and by some conventional extraction procedures. Plant and Soil 96,433–437.

Mathur, S.P., H. Dinel, G. Owen, M. Schnitzer und J. Dugan, 1993: Determination of Compost Biomaturity. II. Optical Density of Water Extracts of Compost as a Reflection of their Maturity. Biological Agriculture and Horticulture, Vol. 10, 87–108.

Nasliowski, K. und E. Grantzau, 1992: Kompostqualität durch Keimlingspflanzen testen. Gartenbau Magazin 12/92.

Performance Prediction of Composting Processes Using Fuzzy Cognitive Maps

C. BHURTUN and R. MOHEE – Faculty of Engineering, University of Mauritius, Reduit, Mauritius

Introduction

Composting is a process whereby organic matter is decomposed by a mixed population of microorganisms in a warm, moist, and aerobic environment. The process involves the interaction of several parameters which do not behave linearly with time and which have time constants varying between days and weeks. The complexity of the interactions between parameters produces several ecosystems which prohibit accurate modeling of the process in the form of differential mathematical equations. Physical modeling of composting [5] has been done but this imposes several constraints. Thermal balance analysis [6] may give a better insight of the process but performance prediction would not be possible because the equations might not be valid if the operating conditions change as is always the case in practice. On the other hand, engineers and scientists know the causes and effects of many parameters involve in composting. Since we know beforehand what effect (large, small, positive, negative etc.) different parameters have on one another, composting process can be treated as being 'fuzzy'.

Fit to be fuzzy

Fuzziness describes 'event ambiguity' and it measures the degree to which an event occurs, not whether it occurs. Fuzzy logic is a method of easily representing analog processes on a digital computer. It is appropriate to use fuzzy logic when one or more of the control variable are continuous, when a mathematical model of the process does not exist, or exists but is too difficult to encode, or is too complex to be evaluated fast enough for real operation and perhaps above all, when an expert is available who can specify the rules underlying the system behaviour and the fuzzy sets that represent the characteristics of each variable. Composting processes are continuous phenomena that are not easily broken down into discrete segments, and the concepts involved are difficult to model along mathematically or rulebased lines, and as such, fit appropriately into the framework of fuzziness.

Most knowledge is specification of classification and causes. In general the classes and causes are uncertain (fuzzy or random), usually fuzzy. This fuzziness passes into knowledge representations and on into knowledge bases where it leads to a 'knowledge acquisition/processing trade off ' [1]. The fuzzier the knowledge representation, the easier the knowledge acquisition and the greater the source occurrence. But the fuzzier the knowledge, the harder the (symbolic) knowledge processing. Fuzzy Cognitive Maps (FCMs) circumvent the tradeoff and have been used to model continuous phenomena [2, 3].

Fuzzy Cognitive Maps

A FCM is a causal network showing domain knowledge in the nodes and edges of a neural network whose architecture is based on the model outlined in Fig1. FCMs are fuzzy signed directed graphs with feedback. The directed edge eij from causal concept Ci to concept Cj measures how much Ci causes Cj (Fig 2). The time varying concept function Ci(t) measures the non negative occurrence of some fuzzy event. The edges eij take in the fuzzy causal interval [–1,1]. eij=0 indicates no causality. eij=–1 indicates causal decrease or negative causality: Cj decreases as Ci increases and vice versa. eij=1 indicates causal increase: Cj decreases as Ci decreases, or Cj increases as Ci increases

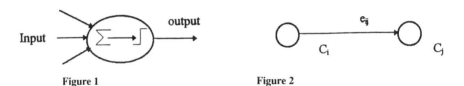

Figure 1   Figure 2

We reason with FCMs as we recall with Temporal Associative Memories (TAMs) [4].We pass state vectors X repeatedly through the FCM connection matrix E, thresholding or nonlinear transforming the result after each pass. Independent of the FCM's size, it quickly settles down to a TAM limit cycle or 'hidden pattern'. The limit cycle inference summarises the joint effects of all the intermediate fuzzy knowledge.
The algorithm is as follows:

Step 1 – Develop the FCM of the process and generate the causal connection matrix E.
Step 2 – Choose the state vector according to the effect of causal concept we want to study.
Step 3 – Pass the state vector through the FCM matrix to obtain the output.
Step 4 – Threshold the output and clamp the concept under study.
Step 5 – Check if the vector is a fixed point of the FCM. Otherwise repeat steps 3 and 4.

Performance prediction and results

As a case study the effect of forced aeration is investigated using the FCM approach. The FCM of the composting process is shown in Fig 3 where the node Ci, for i =1 to 8, represents the following: C1 = Moisture Content, C2 = Temperature, C3 = Heat Generated, C4 = Aeration, C5 = Particle Size, C6 = Microorganisms, C7 = Respiration Rate, C8 = C/N Ratio. The resulting FCM connection matrix E is the 8x8 matrix given in Fig 4.

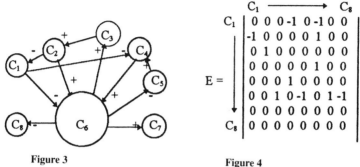

$$E = \begin{array}{c|cccccccc} & C_1 & & & & & & & C_8 \\ \hline C_1 & 0 & 0 & 0 & -1 & 0 & -1 & 0 & 0 \\ & -1 & 0 & 0 & 0 & 0 & 1 & 0 & 0 \\ & 0 & 1 & 0 & 0 & 0 & 0 & 0 & 0 \\ & 0 & 0 & 0 & 0 & 0 & 1 & 0 & 0 \\ & 0 & 0 & 0 & 1 & 0 & 0 & 0 & 0 \\ & 0 & 0 & 1 & 0 & -1 & 0 & 1 & -1 \\ & 0 & 0 & 0 & 0 & 0 & 0 & 0 & 0 \\ C_8 & 0 & 0 & 0 & 0 & 0 & 0 & 0 & 0 \end{array}$$

Figure 3   Figure 4

For forced aeration, the causal state vector is defined as $X1=[0\ 0\ 0\ 1\ 0\ 0\ 0\ 0]$. $X1*E = [0\ 0\ 0\ 0\ 0\ 1\ 0\ 0]$, thresholding and clamping the concept C4 the new state vector X2 is given by $[0\ 0\ 0\ 1\ 0\ 1\ 0\ 0]$. Using an arrow to indicate thresholding and clamping, the different steps are executed as follows to reach a fixed point in the FCM:

$X1*E = [0\ 0\ 0\ 0\ 0\ 1\ 0\ 0]$ $[0\ 0\ 0\ 1\ 0\ 1\ 0\ 0] = X2$
$X2*E = [0\ 0\ 1\ 0\ -1\ 1\ 1\ -1]$ $[0\ 0\ 1\ 1\ -1\ 1\ 1\ -1] = X3$
$X3*E = [0\ 1\ 1\ -1\ -1\ 1\ 1\ -1]$ $[0\ 1\ 1\ 1\ -1\ 1\ 1\ -1] = X4$
$X4*E = [-\ 1\ 1\ 1\ -1\ -1\ 2\ 1\ -1]$ $[-1\ 1\ 1\ 1\ -1\ 1\ 1\ -1] = X5$
$X5*E = [\ -1\ 1\ 1\ 0\ -1\ 3\ 1\ -1]$ $[\ -1\ 1\ 1\ 1\ -1\ 1\ 1\ -1] = X6$

Since X5 and X6 are identical, X5 is a fixed point of the FCM and the procedure stops. From this we can infer that forced aeration has reduced moisture content (since the first element of vector X5, which is C1, is equal to –1), raised temperature (C2=1) due to the production of heat (C3=1) resulting from the multiplication of microorganisms (C6=1) causing a rise in respiration rate (C7=1), a decrease in C/N ratio (C8=–1) and a reduction in particle size (C5 =–1). This result has been verified experimentally in [7] whereby the effects of forced aeration is analysed experimentally. Using the FCM approach, however, more inferences could be made. Like, before the thresholding and clamping procedure, it is seen that forced aeration causes microorganisms to multiply (C6 increases to reach a value 3) and if constant supply of air is not maintained the process becomes anaerobic (C4 reduces). Although this approach gives the final steady state result of the effect of some causal concept, it is evident that certain inferences could be made beforehand as the process evolves. This helps to decide what parameters need to be controlled to prevent degradation of the process.

Conclusion

An FCM model of composting is developed by treating composting processes as fuzzy systems. The model reflects the designer view of the process and it depicts the various causes and effects as available to the designer. The model can be used to predict the performance of composting processes subject to some causes. The effects of forced aeration is studied and the results obtained has been verified in practice thus confirming the validity of the approach used. The technique is simple and the designer does not need to know the dynamics of the process to develop the model. Moreover the approach gives a better insight of the process by depicting the the different interactions that might exist among the states.

References

[1] Kosko B ,'Fuzzy cognitive maps', Int. J.Man-machine studies, (1986) 24, pp. 65–75.
[2] Taber W. R, 'Estimation of Expert Weights with Fuzzy Cognitive Maps', Proceedings of the first IEEE international Conference in Neural Networks, Vol. II, June 1987.
[3] Taber, W. R,'Knowledge Processing with Fuzzy Cognitive Maps', Expert System with Applications, Vol. 2, No 1, pp. 83–87, 1991.
[4] Kosko B., 'Bi-directional Associative Memories,' IEEE Transaction on Systems, Man and Cybernetics, Vol. 18, 1988.
[5] Hogan, J.A, et al, 'Physical Modeling of the Composting Ecosystem', Applied and Environmental Microbiology, May 1989, pp 1082–1092.
[6] Bach, P.D, et al, 'Thermal Balance in Composting Operations', J. Ferment. Techno., Vol. 65, No. 2, 1987, pp.199–209.
[7] Mohee, R., et al, 'Composting of solid Organic Waste in Subtropical Regions', Proceedings of European and Asean Conference on Combustion of Products and Treatment of Products, Thailand, February 1995.

Justification and usefulness of support

Research is being carried out in composting both at the university and at national level. The difficulty in modeling composting processes poses serious problems as to the prediction of its performance under various conditions. Analyzing such a process on a large scale is even more complicated because of the absence of a mathematical model. Consequently, applications of certain control strategies are done on a trial basis. This conference will enable us to find out what other researchers are doing to optimise composting processes and how they are controlling such a complex process. Their techniques can eventually be adapted to works carried by our local researchers.

Biodiversity of Thermophilic Bacteria Isolated from Hot Compost piles

MICHEL BLANC, TRELLO BEFFA, and MICHEL ARAGNO – Microbiology laboratory, Utuversite of Neucbttel, Ne Enu.le-AIgand 11, GH-2007 Ncuchatel, Switzerland

Introduction

Composting is a self-heating, aerobic solid phase biodegradative process of organic waste materials, making possible its return to the environment as soil fertilizer and conditioner (Finstein & Morris, 1975; de Bertoldi & Zucconi, 1987; Finstein, 1992). Temperature increase involves a rapid transition from a mesophilic to a thermophilic microflora (65–75 °C) within a few hours, providing the pile is regularly aerated or frequently turned.

The present state of knowledge on microbial diversity during the thermogenic phase is surprisingly poor: only a few heterotrophic strains related to Bacillus stearothermophilus were identified at temperatures above 60 °C (Strom, 1985a–b).

However, microbial diversity is a prerequisite for the degradation and mineralization of complex organic waste materials (Finstein & Morris, 1975; de Bertoldi & Zucconi, 1987; Finstein, 1992). Active degradation was also observed at temperatures above 60 °C (Schulze, 1962). Therefore, microbial diversity should be expected during this phase.

In this work we tried to isolate thermophilic bacteria by Using a basal mineral medium (Aragno, 1991) supplemented as to isolate bacteria showing three different metabolic types:

Chemolitho-autotrophic, aerobic bacteria:

CO_2-fixing bacteria, using hydrogen or sulfur reduced compounds (thiosulfate) as energy source, and oxygen as respiratory substrate.

Beterotrophic, aerobic bacteria:

using organic compounds such as sugars, organic acids or amino-acids broth as carbon and energy source, and oxygen as respiratory substrate.

Beterotrophic, anaerobic bacteria:

using organic compounds such as organic acids or amino-acids broth as carbon and energy source, and sulfate or nitrate as respiratory substrate in absence of oxygen.

Material and Methods

Organic material from hot (60–75 °C) compost was suspended in sterile basal mineral medium and shaken. Enrichments were performed by parallel serial dilutions (10212102l0) in the media and under gas phase as described above. These were incubated for 1 to 14 days at 70 °C for autotrophic cultures and at 65 °C for heterotrophic cultures. Pure colonies were isolated by successive plating on the same media solidified with agar-agar, and then routinely cultivated in liquid cultures.

Germ numbers and types were estimated by phase contrast microscope examination of enrichment cultures performed from serial dilutions and by the ability to form colonies on solid media.

DNA:DNA homologies were measured spectrophotometrically by following the renaturation rates at Tm 220 °C according to De Ley et al. (1970).

Results

We studied hot compost from 10 composting facilities throughout Switzerland: windows, garden piles, and a bioreactor.

We report the minimal and maximal values for germ numbers observed:
Hydrogen-oxidizing autotrophic bacteria: 105 2107
Sulfur-and thiosulfate-oxidizing autotrophic bacteria: 104 2106
Heterotrophic, aerobic bacteria: 107 21010
Heterotrophic, aerobic (sulfate-reducing or denitrifying) bacteria: 102 2107
Actinomycetes: 103 2104

Hydrogen-, sulfur- and thiosulfate-oxidizing autotrophic bacteria

Five strains were Gram-negative rods able to grow with either hydrogen, sulfur or thiosulfate as sole energy and electron donor, and with CO_2 as carbon source. They were not able to grow with the organic substrates tested. Thus, we regard them as obligate autotrophic strains.

These strains had DNA G l C content similar to those published for the strains related to *Hydrogenobacter* isolated from geothermal areas and showed high DNA:DNA homology with one of these strains isolated from Tuscany, Italy (Aragno, 1991). Soluble proteins electrophoresis profiles confirmed these results.

Unlike the above mentioned strains, strains THS-4 and TH-102 were able to grow on acetate or pyruvate as sole energy and carbon source, but they were unable to grow with reduced sulfur compounds such as thiosulfate or elemental sulfur as sole energy and electron source. They showed similar mol G 1 C and high DNA:DNA homology (74–84%) with *Bacillus schlegelii* (Aragno, 1991).

Heterotrophic, aerobic strains

A. Most of the strains growing at 65 °C on nutrient broth under air or on basal mineral medium with pyruvate were 4–7 mm-long rods forming terminal, oval endospores. Estimations gave as many as 109 cells per gram dry weight of compost in some samples. DNA:DNA and growth characteristics gave strong evidences for these strains belonging to Bacillus st*earother*mophi*lus* complex.
B. Three strains were nonmotile, non-sporeforming rods, 3–6 mm long. They grew well at 70 °C on various organic substrates and showed high DNA:DNA homology with the reference strain of ***Thermus aquaticus*** (65–79%). 12 newly isolated strains are under investigation in our laboratory: estimated numbers reached as many as 10^{10} cells per gram dry weight (cf Beffa *et al.*, this issue).
C. Three strains growing at 65 °C on basal mineral medium supplemented with pyruvate were motile, short rods (0.8 3 2–3 mm) forming central to subterminal, oval endospores. They have not been further characterized yet.

Heterotrophic, anaerobic strains

A. Six strains growing on basal medium supplemented as described above were nonmotile rods, with differences in size and spore-forming ability. Estimated numbers of colony forming units rose to 10^5 per g dry weight of compost. The medium containing sulfate and no fermentable sugar suggests a respiratory metabolism with sulfate as the final electron acceptor. No growth occurred under 20% oxygen (air).
B. Three strains were isolated on nutrient broth supplemented with nitrate. Growth on this medium in absence of oxygen is thought to be due to nitrate reduction (final electron acceptor). The strains resembled those cultivated on nutrient broth under air and one of them proved facultatively aerobic when grown on nutrient broth under air. This is good evidence for these strains belonging to ***Bacillus stearothermophilus*** complex.

Actinomycetes strains

Two strains were isolated on amino-acids broth at 50 °C and tested for growth on antibiotics according to Amner et al. (1988): both strains were insensitive to novobiocin, but growth was inhibited by streptomycin and kanamycin. Along with morphological features, this pattern suggests a close relation with ***Thermoactinomyces spp.***

Conclusions

This study presents the first evidence that highly thermophilic chemolitho-autotrophic Bacteria related to *Hyd*rogenobacter *spp* and to Bacillus schlegelii are not confined to the highly specialized and sparse geothermal habitats but also occur in compost piles, which are short-term high-temperature habitats. Furthermore, heterotrophic bacteria thriving above 60 °C are not restricted to the single species Bacillus stearothermophilus, but include highly thermophilic strains related to The*rm*us a*qu*atic*u*s. Besides, micro-environments deprived of oxygen also permit growth of a diversified anaerobic microflora.

We aim to better understand the functional diversity during the thermogenic phase, suggesting that it is possible to compost at higher temperatures (60–75 °C) for a longer period of time.

Acknowledgments

This research was supported by grants 31–28597.90 and 5002–038921 of the Swiss National Science Foundation.

Literature cited

Amner, W., McCarthy, A.J. & Edwards, C. Appl. Environ. Microbiol. 54, 3107–3112 (1988).

Aragno, M. in The Procaryotes (eds. Balows, A., Trüper, H.G., Dworkin, M., Harder, W., Schleifer, K.-H.) 3917–3933 (2nd ed., Springer-Verlag, New York, 1991).

de Bertoldi, M. & Zucconi, F. in Bioenvironmentals systems (ed Wise, D.) 95–141 (CRC Press Inc., Boca Raton Florida, 1987).

De Ley, J., Cattoir, H. & Reynaerts, A. Europ. J. Biochem. 12, 133–142 (1970).

Finstein, M.S. & Morris, M.L. Adv. Appl. Microbiol. 19, 113–151 (1975).

Finstein, M.S. in Environmental Microbiology (ed Mitchell, R.) 355–374 (Wiley-Liss Inc., New York, 1992).

Schulze, K.L. Appl. Microbiol. 10, 108–122 (1962).

Strom, P.F. Appl. Environ. Microbiol. 50, 899–905 (1985a). Strom, P.F. Appl. Environ. Microbiol. 50, 906–913 (1985b).

Effect of Humic Matters Extracted From Compost and From Leonardite on P Nutrition of Rye-grass

BRUN G., EZELIN K., KAEMMERER M. and
REVEL J.C. – Equipe Pédologie et Environnement,
Ecole Nationale Superieure Agronomique de Toulouse,
145 Avenue de Muret 31076 Toulouse Cedex France.

Introduction

Commercial humic solutions are used in agriculture in order to enhance mineral nutrition of higher plants (Garcia et al. 1993), and especially P nutrition (Gaur 1969). The decisive effect on P dynamics (Rouquet 1988) has often been correlated with the complexing power towards di or trivalent ions, like calcium ions in calcareous soil. In the investigation reported, the effects of two commercial humic solutions on the phophorus exported by pot grown Italian rye-grass were compared after determining their caracteristics.

Material and methods

Humic solutions

Two commercial solutions were used (table I), one issued from compost and one from leonardite after treatment with KOH 1.0M

Table 1 Characteristics of humic solutions used

Humic solution % total HS (b)	Humic substances (HS) content Complexing power towards Ca at various pH mmolc.kg–1 [c]			Molecular Weight (MW)	
	g.dm–3 [a]	MW<14000	MW>14000	pH 5.0	pH 8.0
Compost	256 (1	63 (2	37 (2	200 (25	1320 (41
Leonardite	264 (1	45 (2	55 (2	680 (28	2000 (39

a: determined by alkaline $KMnO_4$ oxidation
b: determined by dialyse with Spectra Por membrane Type A MWCO (Molecular Weight Cut Off) 14000
c: assessed by potentiometric and conductimetric measurement (Brun et al. 1994)

Pot experiment

Italian rye-grass (*Lolium italicum cv Barspectra*) was grown in a glass house (photoperiod: 16 h, temperatures: 24/19, irradiance: 3000 lux) in pots containing 400g of a calcareous soil (pH H_2O: 8.3, pH KCl: 7.5, available Truog P: 12mg.kg-1) according to the method discribed by Chaminade (1960). Each pot was sown with 200 seeds. Plants were watered alternatively every two days with 20cm^3 of distilled water or 20cm^3 of 20mg.dm-3 humic solution. Soil moisture was maintained constant at two thirds of the water holding capacity with distilled water. There were four replications per treatment. The first cut took place four weeks after starting the pot experiment and the second cut three weeks after the first. Biomass produced has been dried and analysed in order to determine the yield and phophorus content.

Results and discussion

There was no significant effect of the treatments on the plant biomass produced wich amounted to ca. 600mg per pot for the first cut and 180mg per pot for the second cut. The amount of P exported was expressed in mg per pot (fig 1)

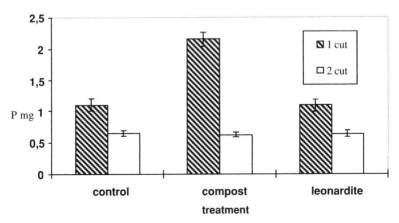

Fig 1 Amount of P exported expressed in mg per pot (average of four replications)

For the two cuts, when leonardite humic solution was used, the P exported by rye-grass was no significantly different from that of the control plants in spite of a higher complexing power towards calcium ions.

For the compost product, the amount of P exported was significantly higher than that of the control plants for the frist cut. For the second cut, no more difference was observed.

Under the experimental conditions, we can notice that:

– some humic solutions have no effect on P nutrition in calcareous soil in spite of

- there is no direct relationship between complexing power towards calcium ions and increase in phosphorus uptake by rye-grass.
- low molecular weight molecules seem to be more effective on P nutrition than high molecular weight ones.

Conclusion

The above investigations suggested that the complexing power towards calcium ions is not the only parameter involved in the enhanced P nutrition observed. Humic substances may have some effects on the root system as the uptake of some nutrients can be significantly enhanced. This could explain the effect only on the frist cut.

We notice that low molecular weight molecules appear more effective than high molecular ones.

We have also notice that during the complexation of calcium, the amount of H+ released is variable. For compost solution, this amount is highest than for the leonardite one: at pH8 (nearly soil pH) one mole of calcium complexed releases 0.5 mole of H+ for compost solution and only 0.2 mole for leonardite solution. This local acidification could perhaps enhance solubilization of insoluble phosphorus forms, like apatite, and than enhance phosphorus uptake by plants.

Bibliography

BRUN G., SAYAG d. and andre l. in ' Humic substances in the global environment and implications in human health ' Senesi and Miano eds., (1994), 192–198.
chaminade r. Ann. Agron., 11/2, (1960), 121–123.
garcia d., cegarra j., abad m. and fornes f. Bioressource Techn., 43, (1993), 221–225.
gaur a.c., Agrochimica, 14/1, (1969), 62–65.
rouquet n., C. R. Acad. Sc. Paris, 307–II, (1988), 1419–1421.

The Suppressive Effects of Composted Seperately Collected Organic Waste and Yard Waste Compost on Two Important Soilborne Plant Pathogens

C. BRUNS, S. AHLERS, A. GATTINGER, C. SCHÜLER, H. VOGTMANN* and G. WOLF** – University of Kassel, Division of Ecological Agriculture, Nordbahnhofstr. 1a, 37213 Witzenhausen, Germany

Summary

Each year during the period 1991 to 1994, three different types of input material (yard waste, biowaste and cattle manure) were composted and subsequently evaluated for their suppressiveness towards *Pythium ultimum* and *Phytophtora parasitica* spp.. This was achieved by means of bioassays conducted in a sterilised substratum (sand) employing the host-pathogen systems peas – *P. ultimum* and in non-sterilised substratum (commercial peat based potting mix) applying the host-pathogen systems peas or cucumber – *P. ultimum* and tomato – *P. parasitica nicotianae*.

Amendment of P. ultimum inoculated sterilised sand with yard waste– and biowaste compost resulted in a significantly increased yield (fm). This was mainly due to pathogen suppressive biological mechanisms induced by compost. Composted cattle manure showed such an effect only occasionally. In experiments where 90 % of control treatment plants were damped off, compost addition, apart from cattle manure compost, resulted in a reduction of the disease incidence by 30 to 50 %. In bioassays with soils naturally infected with root-rot pathogens of red beet, however, we also observed some suppressive effects of cattle manure compost.

Contrary to cattle manure compost, both yard waste- and biowaste compost show a high level of specific microbial activity, i.e. a high rate of Fluorescein diacetate-hydrolysis per unit microbial biomass. Therefore, micro-organisms originating from both waste composts are better competitors for easily available carbon

*Hesse State Board for Regional Development and Agriculture, Kölnische Straße 48/50, 34117 Kassel, Germany
**University of Göttingen, Institute for Plant Pathology, Griesebachstr. 6, 37777 Göttingen, Germany

sources. This results in a fungistatic effect towards pathogens since these organisms depend on the same carbon sources. This has been shown by the suppressive effects towards P. ultimum of a 50 % yard waste compost amendment of a peat growing media in comparison to non-amended pure peat growing media. The two waste composts have been evaluated in non-sterilised container media in all other above mentioned host-pathogen systems. Yard waste compost application rates between 30 % and 50 % (v/v) resulted in a significant reduction of the disease incidence of at least 45 % for cucumber, peas and also tomatoes. Due to a higher nutrient and salt content of composted biowaste, the application rate of this material was limited to 30 % (v/v), resulting in a reduction of the disease incidence by 10 – 20 %.

With a St. Paulia – P.parasitica bioassay similar results were obtained during a cultivation period of 14 weeks. Compost amendment of 30 and 50 % yard waste compost, respectively, to a non-sterilised peat growing media significantely prolonged the time until plants died.

Legislative and Scientific Aspects of the Production and use of Vermicompost from Biological Sludges

CECCANTI B.*, MASCIANDARO G.*,
DELL'ORFANELLO C.**, LUCHERINI M.** and
GARCIA C.***

Summary

The vermicomposting is a process widely employed for ecological recovery and transformation of biological sludges into 'casting' with a high fertilising value. While practical and scientific aspects of the process have been already studied and, for most part, understood both in laboratory and in pilot experiments, the legislative aspect is, still nowadays, very confused and lacunous. In fact, there are many national (Italian) laws which regulate the composting and vermicomposting and which overlap to regional legislation. Contrasts in defining the wastewaters origin (industrial, urban, municipal or assimilable to the municipal ones), in technical aspects of sludge processing, in final classification and use of 'stabilised' product, are frequently encountered. In this work, attention has been focused on the technical aspects of sludge vermicomposting and on the legislative aspects to classify the product as a 'compost' according to the DPR 915/82.

Introduction

The composting of municipal sludges is possible if a bulk agent (wood chips, straw, etc.) which reduces the water content and improve aeration is added; the process requires energy to turn the organic mass and to finish the product. Composting may be conducted also economically without employing bulk agents, expensive plants and energy, but only using adapted earthworms (i.e. Lombricus terrestris, Eisenia foetida, etc.), which may growth under a wide range of temperature (12–35 °C), humidity (35–65 %) (Hartenstein, 1981) and organo-mineral feeding (aerobic and anaerobic municipal or paper mill sludges).

* Institute of Soil Chemistry-CNR, Pisa -Italy
** Airfoical s.r.l., Lucca-Italy
*** Centro de Edafologia y Biologia Aplicada del Segura, CEBAS-CSIC, Murcia -Spain

The action of earthworms is to transform the sludge organic fraction into a 'casting' with a high fertilising value. The fertilising value of earthworm casting and the beneficial effects on crops, have been related to the presence of active mineral nutrients (Ceccanti et al., 1994) and plant growth regulators with phytohormonal action (Tomati et al., 1985).

The aim of this paper is to study the feasibility of sludge vermicomposting, dealing with three main aspects:

1. adaptation of earthworms and process efficiency
2. vermicompost quality and effects on soil and plant
3. legal aspects of sanitarization

Results and discussion

The experiment has been conducted in pilot-plant with mixtures of aerobic and anaerobic sludges, these last added at growing quantities (0, 25, 50, 100% v/v); Eisenia foetida was used for operating the vermicomposting process. It has been observed that the adaptation of Eisenia foetida to the sludges mixtures was quite rapid, even in the presence of relatively high concentration of ammonia (about 170–180 ppm) arising from the anaerobic sludges. Laboratory experiments of vermicomposting, to assess in a quantitative way the adaptation of Eisenia foetida to the sludges, have been carried out. These experiments have been conducted stratifying the sludges (mixed in the same proportion as in the pilot-plant) on a sandy loam soil poor in organic matter (< 1.5%) (Peccioli, Pisa – Italy), followed by the addition of earthworms to promote the vermicomposting directly in the soil (in situ).

In figure 1 are reported, in relation to the sludge mixtures (0, 25, 50, 75, 100% v/v of anaerobic sludge), a) the distribution of Eisenia foetida in soil-sludge system after one week from the addition of earthworms, b) the residual weight of the sludges which has not been incorporated by the earthworms into the soil after 8 months, c) the metabolic activity (dehydrogenase activity) of the soil at the end of 8 months vermicomposting, and d) the growth test with Lepidum sativum on soil samples kept 8 months under vermicomposting 'in situ'.

It is well evident that with the increase of anaerobic sludge concentrations, the number of earthworms increased in the soil, leaving on its surface growing quantities (residual weight) of not-digested sludge. The low microbiological activity measured in the soil populated by a relatively high number of worms, might indicate an acceleration of the composting process. The growth test with Lepidum sativum showed that the soils with 25–50% of anaerobic sludge were regenerated in the biological, chemico-physical and nutritional components, while appeared negatively affected by the concentrations of 75–100%.

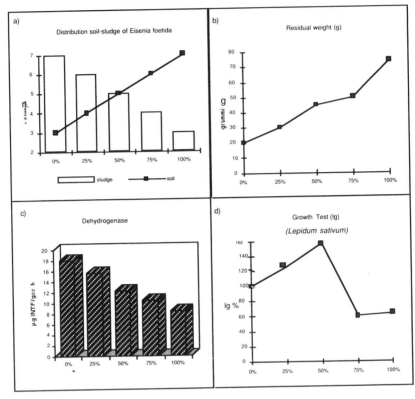

Figure 1.

Once assessed the feasibility of vermicomposting and its positive effects on soil-plant properties, it remains to define better the legislative aspect of the process. In our experiments, the final vermicompost obtained after 8 months of sludge treatment in pilot-plant with Eisenia foetida presented chemico-physical and microbiological (absence of pathogenic microorganisms, data not reported) properties which were very similar to or better than those reported by Italian law-standards for composted urban organic fraction (DPR/915/82), vermicompost from animal manures (D.L. 748/84) and for the use of biostabilised sludges (L. 99/92) in soil amendment. No one laws mentioned above, contemplate the 'vermicompost from biological sludges', so that our vermicompost is illegal for two reasons: a) it does not respect the process of composting for the termophilic phase, as stated by DPR 915/82 and b) it is not clearly stated by Italian (n. 690/86) and Regional laws (Tuscany n. 5/86) the origin of wastewaters and sludges to be processed to produce compost or vermicompost. However, there is a possibility to classify as a 'compost', the vermicompost produced from biological sludges, only if a thermosanitarization phase is activated, after worms remotion (Polperio et al., 1994). Thermosanitarization of final vermicompost is possible using ligno-cellulosic materials (20% and 25% v/v) which started bioxidative reactions so permitting to achieve temperatures over 55°C for at least three days, as reported in the DPR/915/82 (figure 2).

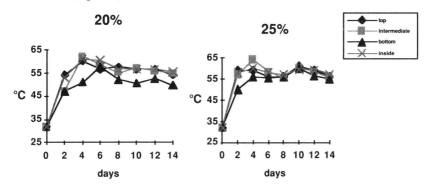

Figure 2 Temperatures measured in different position of piles under hygienization.

The vermicompost after the thermosanitarization can be considered a 'compost' according to the DPR/82 for its toxicological, agronomical and hygienic characteristics; but further studies are necessary to verify the eventual presence and the effects of phytotoxic substances on plant growth and to evaluate the economic aspects for marketing. Clearly, the 'legal suitability' of wastewaters and sludges must be re-defined and the compost must be classified on the basis of its chemico-phisical, agronomical and microbiological properties.

References

Ceccanti B., Masciandaro G. and Garcia C. (1994). Vermicompost and humic substances management for sustained productivity. Ordine Nazionale dei Biologi, Vieste, september 22–25 (in press).

Hartenstein R. (1981). Use of Eisenia foetida in organic recycling based on laboratory experiments. Wokshop on the role of earthworms in the stabilization of organic residues,1:155–165.

Polperio S., Masciandaro G., Ceccanti B. and Garcia C. (1994). Fanghi biologici: rifiuto o compost? Aspetti normativi e criteri di qualità. XII Congresso nazionale della Società Italiana di Chimica Agraria (SICA), Piacenza 19–21 settembre (in press).

Tomati U., Grappelli A. e Galli E. (1985) The alternative 'Earthworm' in the organic wastes recycle. In: Processing and use of organic sludge and liquid agricultural wastes, ed. P. L'Hermite, 510–516.

Composting of Fresh Olive-Mill Wastewater Added to Plant Residues

I. CEGARRA, C. PAREDES, A. ROIG, M.P. BERNAL, A.F. NAVARRO – Department of Soil and Water Conservation and Organic Waste Management, Centro de Edafología y Biología Aplicada del Segura. CSIC. P.O. Box 4195, 30080 Murcia, Spain.

Introduction

Vast amounts of olive-mill wastewaters (OMW) are produced in Mediterranean countries causing very serious environmental problems either through spreading on agricultural land or storage in ponds. A wide variety of technological treatments are available to reduce their polluting effects and for their transformation into valuable products (Fiestas Ros de Ursinos and Borja Padilla,1992). However, very few data, if any, have been published on OMW composting. This paper reports a methodological approach for transforming these wastewaters into organic fertilizers by letting them be absorbed by plant waste materials which were later composted.

Materials and methods

Fresh OMW, were absorbed by cotton waste (C) and maize straw (M), to which fresh poultry manure (P) and sewage sludge (S) were added to supply the necessary nitrogen. Three composts (POC, SCO and SMO) were prepared respectively by mixing 65.4% of C with 34.6% of P (fresh weight), with 1930 1/tonne OMW added to the solid mixture during the first 18 days of composting; by mixing the same C carrier plus S (67.9 and 32.1%, respectively), with a lower amount of OMW (943 1/tonne) being added on the first day of composting; and by mixing the M carrier (52.9%) with S (47.1%) to which 1765 1/torne of OMW were added also on the first day of the process.

Three trapezoidal piles 1–1.5m high with a 2 3 3 m base containing about 1500 kg of every mixture were composted in a pilot plant based on the Rutgers static pile system (Finstein et al., 1985). The air was blown from the base of the pile, the timer was set for 30s. ventilation every 15 minutes and the ceiling temperature for continuous air blowing was 55 °C. After the biooxidative phase of composting, the air-blowing was stopped to allow the compost to mature over a period of two months. The piles were sampled weekly till the end of the biooxidative phase and once again after the maturation period.

Moisture content was ascertained by drying at 105 °C, organic matter (OM) by loss-on ignition at 430 °C during 24h. Losses of organic matter were determined from the initial and final ash contents according to the equation of Viel et al. (1987). Lignin and cellulose were determined by the American National Standard Methods (ANSI/ASTM, 1977a and b), holocellulose according to Browning (1967), cation exchange capacity (CEC) by Lax et al. (1986) and germination index (GI) by Zucconi et al. (1981). Humic-like substances were isolated by treating compost samples with 0.1 M NaOH, later separating humic (HA) from fulvic (FA) acids by acid precipitation and centrifugation. Samples of whole extracts and FA solutions were analyzed for organic C by automatic microanalysis and C in HA was calculated by subtraction of the above values.

Table 1 Changes observed in some parameters during composting (% d.w.)

Sampling time (days)	Organic-C (%)			CEC (me/100g OM)			GI (%)		
	PCO	SCO	SMO	PCO	SCO	SMO	PCO	SCO	SMO
1	40.7	40.5	47.2	—	85	—	13	75	19
7	39.2	36.6	46.8	104	—	—	—	—	—
14	37.1	35.1	43.9	—	—	27	—	—	—
21	36.0	33.3	43.5	121	—	44	30	—	—
28	35.3	33.0	40.9	—	—	57	—	—	105
35	34.1	33.7	43.3	142	—	—	—	—	—
42	34.6	33.3	43.0	—	140	77	—	78	—
49	33.4	31.7	41.0	149	—	—	45	—	—
56	—	32.1	41.4	—	—	—	—	—	—
63	—	33.7	41.5	—	—	86	—	—	84
Maturity	33.7	29.4	39.4	156	195	110	77	94	91

Results and discussion

As shown in Fig. 1, the fermentation period was shortest in PCO and longest in SCO, the most extended thermophilic phase being observed in the latter compost. The C content decreased as composting progressed in all cases, the highest decrease being in the SCO compost (Table 1). The OM content decreased during composting in accordance with the fall in C, and was responsible for the important weight losses observed (Fig. 2), which reached a maximum value in the SCO compost. Thus, this compost underwent the greatest degree of mineralization, which might be related either to the lower toxic effect on the microorganisms because the quantity of OMW added was the lowest or to a greater microbial activity due to the important load of microorganisms from the sewage sludge. The N content increased in all composts (Fig. 3), particularly in SMO (N content in the 'mature' compost was almost 120 of the initial value) suggesting some biological fixation at least in this compost. The initial values of the C/N ratio were 16.5 (PCO) 22.4 (SCO) and 33.7 (SMO), although these different values became very similar at the end of composting (9.8, 9.5 and 12.1, respectively).

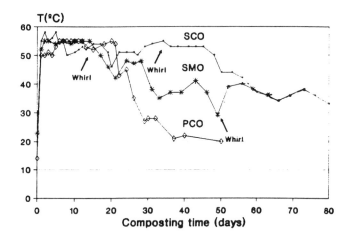

Figure 1 Changes in temperature during composting

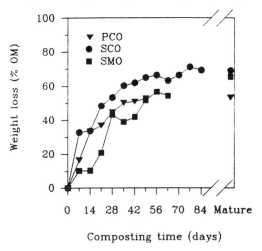

Composting time (days)

Fig.ure 2 Weight losses during composting

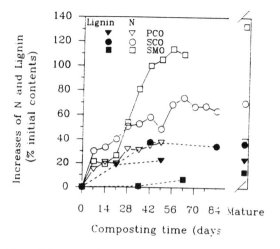

Figure 3 Increases of N and lignin contents during composting

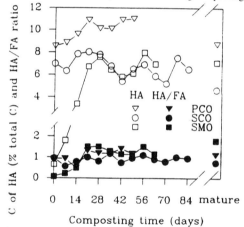

Figure 4 Changes in contents of fulvic (FA) and humic (HA) acids during composting

The lignin content increased during composting in all the cases, the highest increase being detected in SCO and the lowest in SMO (Fig. 3). At the same time, both alphacellulose and hemicellulose contents decreased, the reduction being greatest in SCO and smallest in SMO (data not shown). These results are in good agreement with the already mentioned losses of OM during composting and suggest that lignin is clearly resistant to biodegradation while cellulose was degraded to different degrees by the composting microorganisms, this phenomenon depending on factors such as the type of plant waste, the diverse rate of addition of OMW or the kind of N-source employed to prepare composts (poultry manure and sewage sludge). Because the CEC is related to compost 'maturity', this parameter was determined during composting (Table 1). The highest value was found in the 'mature' SCO compost and the lowest in SMO. However, the greatest CEC increase during composting was observed in the latter compost, as this parameter increased four times, from 27.3 me/lOOg after two weeks of composting to 110.0 me/lOOg in the 'mature' compost. Increases were less apparent in the SCO compost and were still lower in PCO. The GI generally increased as composting progressed (Table 1), its values being small at the beginning of the process. This was particularly evident in PCO whose GI remained quite low during the first three weeks of composting, probably as a result of the successive additions of OMW to this mixture during the first 18 days of the experiment. However, rather high values of the GI were recorded in SCO from the beginning of composting, which may be related to the low quantity of OMW added to the raw mixture. An intermediate pattern was observed for SMO.

The highest HA content was found in PCO and the lowest in SCO (Fig. 4), this content generally increasing during the first 2–3 weeks of composting. This latter trend was particularly marked in SMO, whose HA content increased from less than 1 at the beginning of the process to 7–8 after week 3. HA content was generally lower in the 'mature' composts compared with the values reached during the active phase of composting. Lastly, the AH/FA ratio remained practically constant during composting, with the exception of SMO which clearly increased, suggesting that more polymerized humic substances were produced during the process.

References

ANSI/ASTM. 1977a. D 1106. American National Standard.
ANSI/ASTM. 1977b. D 1103. American National Standard.
Browning, B.L. 1967. Interscience Publ., New York, 395 p.
Fiestas Ros de Ursinos, J.A. and Borja Padilla, R. 1992. Grasas y Aceites, **43**, 101–106.
Finstein, M.S., Miller, F.C., MacGregor, S.T. and Psaranos, K.M. 1985. EPA Proyect Summary (EPA/600/S2–851059) U.S. EPA. Washington.
Lax, A., Roig, A. and Costa, F. 1986. Plant and Soil, **94**, 349–355.
Viel, M., Sayag, D., Peyre, A. and André, L. 1987. Biol. Wastes, **20**, 167–185.
Zucconi, F., Pera, A., Forte, M. and De Bertoldi, M. 1981. BioCycle, **22**, 54–57.

Municipal Solid Waste Composting: Chemical and Biological Analysis of the Process

BENNY CHEFETZ, AND YONA CHEN – Dept. of Soil and Water Science, and YITZHAK HADAR, Dep. of Plant Pathology and Microbiology, Faculty of Agriculture, The Hebrew University of Jerusalem, Rehovot 76100, Israel.

Abstract

Composting of municipal solid waste (MSW) was studied in an attempt to better understand the composting process and define parameters of maturity. Composting was performed in 1–m3 plastic boxes and the following parameters were measured: temperature, C/N ratio in solid and liquid phase (C/N(s) and C/N(w) respectively), humic substance fractions and contents, dissolved organic carbon (DOC). Spectroscopic method (DRIFT) was used to study the chemical composition of the bulk organic matter (OM). A bioassay based on cucumber plants growth was correlated to other parameters. C/N(w), C/N(s) and DOC showed high rates of change during the first 60 days, then stabilized. Humic acid (HA) content increased to a maximum at 110 days, corresponding to the highest plant dry weight and the highest 1650/1560 (cm–1/ cm–1) peak ratios from DRIFT spectra. DRIFT spectra showed that the OM transformed to a more aromatic structure.

Introduction

Problems caused by municipal solid waste (MSW) in modern society have become more severe over the last decade due to increasing amounts of waste and decreasing availability of landfill space (Alter, 1991; Finstein, 1992). Recycling has become an attractive solution for the waste management. For the organic fraction composting seems to be a desirable option having the capacity of reducing the volume and weight by approximately 50% and resulting a product which can be useful for agriculture (He, 1992). One of the main obstacle to successful utilization

Corresponding Author

of MSW compost in agriculture is the lack of reliable quality criteria and understanding the OM transformations throughout the process. Proper evaluation of compost maturity is essential for the establishment of such criteria. Therefore, the objective of this paper is to correlate chemical analyses and DRIFT spectra, with plant performance bioassay.

Resultsand Discussion

Various parameters were followed during composting of MSW. The C/N(s) ratio decreased rapidly from an initial value of 28 in the raw material to 18 after only 20 days. The ratio continued to decrease, albeit less sharply, to 12.2 after 60 days. From this point on the C/N ratio stabilized at a value of about 12 (11.8–12.8) for the remainder of process. The C/N(w) ratio followed a similar trend, exhibiting three phases: (i) rapid decrease from 35 to 8 during the first 20 d; (ii) slower decrease to a value of 6, lasting till 70 d; (iii) days 70 to the end of the experiment, when it stabilizes at value between 5 to 6. The C/N ratio (C/N(w) and C/N(s) as well) in itself appeared to be a non reliable indicator of compost maturity. Although it changed dramatically during the second phase of composting, it did not change significantly during the curing phase.

The compost bacterial population is active only in the liquid phase, thus the DOC concentration during composting can indicate the biodegradability of the OM. The changes in DOC concentration followed a trend similar to that of the C/N ratio, exhibiting same three phases: (i) rapidly decrease from initial concentration of 28.30 g/Kg to 7.40 g/Kg over the first 20; (ii) days 20–60, when the DOC continued to decrease, less sharply, to 2.25 g/Kg; (iii) days 60 to the end of the experiment, when the DOC decrease moderately to 1.00 g/Kg.

HA content increased during composting, reaching a stable value after 110 days at 13.5% of the OM while the fulvic fraction (FF) content, including the fulvic acid and the non humic fraction (FA and NHF respectively), decreased. The increasing level of HA may indicate the degree of humification and the maturity of the compost. The humification index (HI=HA/FA) increased to a ratio of 3, and the humification ratio (HR=HA/FF) increased to 1.35. These values differ in other, but in general fresh compost contains low levels of HA and higher levels of FA (Saviozzi et al., 1988; Gonzalez, 1993; Ciavatta et al., 1993). During the composting process the FF decomposed whereas the HA level remained stable.

The distinct differences in the bulk DRIFT spectra resulting from composting, was a reduction of the 1560 cm–1 peak (amide II) with time probably due to the relatively rapid biodegradabtion of the amino chain. The aromatic region (1650 cm–1) became sharper during composting. Peaks in the aliphatic region at 2930 and 2850 cm–1 decreased, while the 1450 cm–1 peak, which represents C–H deformation, increased. Another method of monitoring changes during the composting process is the measurement of the intensity of major peaks and the ratios between them (Inbar et al., 1989). The peaks at 2930 cm–1, 2850 cm–1, 1650 cm–1, 1560 cm–1 and 1050 cm–1 were chosen for these calculations. The ratio 1650/2930 (aromatic C / aliphatic C) increased from 0.88 to 1.10, the ratio 1650/2850 (aromatic C / aliphatic C) increased from 0.79 to 1.54, the 1650/1050 ratio (aromatic C / polysaccharide) increased from 2.39 to 2.80, and the 1650/1560 ratio (aromatic C / amide II bond) increased from 0.94 to 1.52. These increases represent a decrease in polysaccharides, aliphatic and amide components, and an increase in the aromatic structure in the mature compost. The linear correlation between the aromatic to aliphatic peak ratio (1650/2930 cm–1/ cm–1) and the C/N ratio had a R2=0.936 calculated for the equation $Y = 1.263 - 0.014X$. This correlation indicate that the DRIFT spectra is a useful and reliable tool in the analysis the composting process.

It was hypothesized that plant growth may serve as an integrative single test for compost maturity (Chen and Inbar, 1993, Inbar et al., 1993). Cucumber plants grown in media containing fresh (14 d) compost (50% v/v) exhibited inhibited growth as compared to plants grown on older compost. The dry weight of plants grown in 110 and 132 d old composts were significantly higher than all the others. These results show that as OM decomposed the compost became a better substrate for plant growth. Only highly mature compost supported better plant growth. A correlation's between plants dry weight and other chemical analysis were correlated. A linear correlation between the 1650/1560 (cm–1/cm–1) DRIFT peak ratio and plant dry weight had R2=0.81 calculated for the equation $Y = 0.81 + 0.54X$. The linear correlation between HA content in OM and plant dry weight had R2=0.70 calculated for the equation $Y = 3.59 + 7.78X$.

Conclusions

Due to materials and processes complexity the determination of compost maturity is very difficult. Therefore, several parameters are needed to be crossbred. This article proves that DRIFT spectroscopy, together with data on HS provide useful information about the OM transformations occurring during the composting process of MSW. Plant bioassay, level of HA and the 1650/1560 DRIFT peak ratio showed the same trend during composting and can therefor be used as maturity indexes. All those parameters exhibited three distinct phases: (i) rapid decomposition during the first 30 d; (ii) stabilization till day 90; (iii) maturation from day 90 and on. The MSW compost, in our experimental system, was mature and ready to be used as an agricultural substrate after about 110 days of composting.

References

Alter, H. 1991. The future course of Solid Waste Management in the US. Waste Management & Research 3:3–20.

Chen, Y., and Y. Inbar. 1993. Chemical and spectroscopic analysis of organic matter transformations during composting in relation to compost maturity. pp. 551–600. In H.A.J., Hoitink (ed.) Science and engineering of composting: Design, environmental, microbiological and utilization aspects. Renaissance Publications, Worthington, OH.

Ciavatta, C., Govi, M., Pasotti, L., and P. Sequi. 1993. Changes in organic matter during stabilization of compost from Municipal Solid Waste. Bioresource Tech. 43:141–145.

Finstein, M.S. 1992. Composting in the context of Municipal Solid Waste management. pp. 355–374. In R. Mitcell (ed.) Environmental Microbiology. Wiley-Liss, New-York.

Gonzalez, S.J., Carballas, M., Villar, M.C., Beloso, M.C., Cabaneiro, A., and T. Carballas. 1993. Carbon-and Nitrogen-containing compounds in composted urban refuses. Bioresource Tech. 45:115–121.

He, X.T., Traina, S.J., and T., Logan. 1992. Chemical Properties of Municipal Solid Waste Composts. J. Environ. Qual. 21:318–329.

Inbar, Y., Chen, Y., and Y., Hadar. 1989. Solid state Carbon-13 Nuclear Magnetic Resonance Infrared spectroscopy of Composted Organic matter. Soil Sci. Am. J. 53:1695–1701.

Inbar, Y., Hadar, Y., and Y., Chen. 1993. Recycling of Cattle Manure: The composting Process and Characterization of Maturity. J. Environ. Qual. 22:857–863.

Saviozzi, A., Levi-Minzi, R., and R., Riffaldi. 1988. Maturity evaluation of organic waste. BioCycle 29:54–56.

Chemical Parameters to Evaluate the Stabilization Level of the Organic Matter During Composting

C. CIAVATTA, B. MANUNZA*, D. MONTECCHIO, M. GOVI and C. GESSA – Institute of Agricultural Chemistry, University of Bologna, Italy

Abstract

The stabilization level of the organic matter of compost from MSW has been monitored using the degree of humification (DH), the isoelectric focusing (IEF) and a Bimodal Gaussian Distribution (BGD) of pK of –COOH and phenolic –OH groups of HA. The results obtained show that these methods are able to monitor the processes.

Introduction

Before the application to the soil, the organic matter from composts must be sufficiently stabilized, i.e. decomposable organic compounds must be completely transformed to stabilized substances [11]. Many methods have been proposed to control the quality of the organic matter during composting processes, such as C/N ratio [1], C.E.C. [8], E4/E6 ratio [10], gel chromatography [7], and selected chemical components in water extracts [2]. The determination of the amount of humic substances extracted from samples of different materials [5, 3] and a Bimodal Gaussian Distribution (BGD) model applied to the potentiometric titrations of (HA) have also been proposed [9]. The present work shows the results obtained following the evolution of the organic matter from MSW in piles of compost by using the degree of humification (DH), the isoelectric focusing (IEF) and a study of potentiometric titrations by a BGD model.

Materials and methods

The samples from municipal solid wastes (MSW) were taken from a composting plant near Mantova (Italy) from a static pile of compost stabilized during winter

*Di.S.A.A.B.A., University of Sassari, Italy

for 55 days under a forced-pressure ventilation composting system. The extraction and fractionation of the organic carbon from the were carried out using the methods reported by Ciavatta and Govi [3]. The potentiometric titrations on HA have been performed using the method suggested by Manunza et al. [9]. The degree of humification, i.e. DH% = [(HA+FA)/TEC] x 100, was used as humification parameter. Further details on analytical characteristics of the samples have been reported in a previous paper [4].

Results and discussion

The total organic carbon content (TOC) of the samples composted steadily decreased from 25.0% C at the beginning until 17.1% C at the end of the process (the losses of organic C suggest the occurrence of high microbiological activity). The DH values started at 43% and sharply increased at the beginning of the process, then at the end tended asymptotically to stabilize around 60% (Fig. 1). These values are slightly lower than those found for organic matter extracted from soils, humified peats and well-matured composts [3]. The high DH values at the beginning of the composting process depend on the occurrence during this phase, of humic-like substances that interfere in the separation method of humic substances [3].

In figure 2 are shown the IEF patterns of the organic extracts from compost at the beginning of the stabilization processes and after 55 days. The two patterns show a group of bands in the region from pH 4 to pH 5.3 and a group of well resolved bands in the region from pH 5.3 to pH 6.0 in both samples. Before composting in the region over pH 6.0 there are only two weak bands in the IEF profile of the raw sample (zero days), while in the composted sample (55 days) there are a group of five well resolved bands over pH 6.0. Our results confirm what has been shown by other authors [5], i.e. that the evolution of the organic matter during the composting process leads to the formation of organic compounds which can be identified using the IEF and are characterized by both high molecular weights and isoelectric points.

Figure 3 shows the distribution of pK of humic acids extracted at the beginning (a) and at the end (b) of the composting process. The data obtained from the titration curves have been processed using a BGD model of binding sites. Both curves (a, b) were characterized by two Gaussian distribution of pK of –COOH and phenolic –OH groups. At the beginning of the humification processes (a) the mean pK of the –COOH and phenolic –OH groups were pK = 3.91 and pK = 9.33, respectively, while the total acidity of the –COOH was higher than that found for phenolic –OH groups (198 vs. 120 cmolc 100g–1). After composting the mean pK of the –COOH and phenolic –OH groups were pK = 4.56 and pK = 10.52, respectively and the two areas are of the same order of magnitude (203 vs. 198 cmolc 100g–1). The last pattern is very similar to that found for soil HA.

In conclusion, all the three approaches here used have shown a progressive humification of the organic matter during composting. This phenomenon can be resumed with i) increasing of the DH values; ii) by a formation of a new group of well resolved bands over pH 6 in the IEF profiles; iii) and finally by a mean pK values of –COOH and phenolic –OH groups very close to those found for soil HA. The stabilization of the organic matter in piles of compost from MSW can be monitored using the DH, the IEF and by a study of potentiometric titrations of HA It seems reasonable to suggest the use of these chemical methods for an adequate control of the composting processes.

References

1. Chanyasak V. and Kubota H. (1981). J. Ferm. Technol. 59(3), 215–19.
2. Chanyasak V., Hirai M. and Kubota H. (1982). J. Ferm. Technol., 60(5), 439–46.
3. Ciavatta C. and Govi M. (1993). J. Chromat. 643, 261–70.
4. Ciavatta C., Govi M., Pasotti L. and Sequi P. (1993). Biores. Technol. 43, 141–45.
5. De Nobili M. and Petrussi F. (1988). J. Ferm. Technol. 66, 577–82.
6. De Nobili M., Ciavatta C. and Sequi P. (1989). In: Proc. Int. Symp. on Compost: Production and Use, pp. 328–42. San Michele all'Adige, Italy.
7. Giusquiani P.L., Patumi M. and Businelli M. (1989). Plant Soil 116, 278–82.
8. Harada Y. and Inoko A. (1980). Soil Sci. Plant Nutr. 26, 127–34.
9. Manunza B., Gessa C., Deiana S. and Rausa R. (1992). J. Soil Sci. 43, 127–31.
10. Riffaldi R., Levi-Minzi R. and Saviozzi A. (1983). Agric., Ecosyst. Environ., 10, 353–59.
11. Zucconi F. and de Bertoldi M. (1987). In: Compost: Production, Quality and Use (M. De Bertoldi, M.P. Ferranti, P. L'Hermite and F. Zucconi, Eds.), pp. 30–50, Elsevier, London, GB.

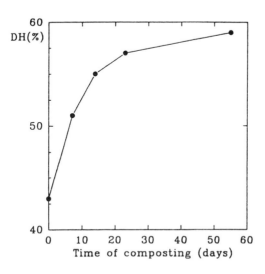

Figure1. Trend of the degree of humification (DH) during the organic matter stabilization processes in composts from municipal solid wastes (MSW).

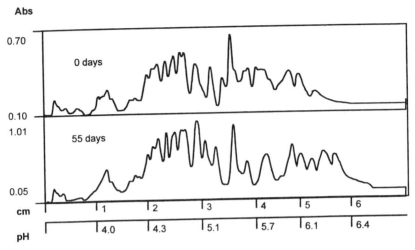

Figure 2. Isoelectric focusing (IEF) profiles of 0.5 M NaOH extracts of municipal solid wastes (MSW) stabilized during the winter season after zero and 55 days of organic matter stabilization processes.

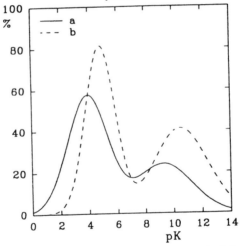

Figure 3 pK distribution of HA extracted from compost samples before (a) and after (b) composting.

Response of Three Compost-based Substrates to Different Irrigation and Fertilization Regimes in Poinsettia (*Euphorbia pulcherrima Willd.*)

G. D'ANGELO – Fondazione Minoprio Vertemate con Minoprio (Como) Italy

Introduction

Since 1985, the research section of Minoprio Foundation has been testing different composted materials of vegetable origin as components of growing media for ornamental pot plants. The study of the relationship between physical and chemical characteristics of composts and plant quality gave the opportunity to obtain good quality plants in cultivation trials, even by reducing the quantity of peat in the substrate until 1/3 of the total (Lamanna et al., 1991).

The different composition and structure of compost-based substrates, compared with peat-based ones, induced to modify irrigation and fertilization techniques so as to obtain better quality plants; compost-based media generally have a lower water retention capacity; in some cases, they showed greater requirements for nitrogen (N) and a high mineralization rate (Lemaire et al., 1989).

Materials and methods

In 1994 a trial on poinsettia plants was planned (table 1). Rooted cuttings of cvs. 'Peterstar' and 'Supjibi' were potted in four media, respectively a peat-based substrate (control) and three compost-based ones, and plants were cultivated using an ebb-flow irrigation system.

For the trial the following substrates were used:

– a commercial peat-based substrate (T);
– a mixture of a vegetable compost (n. 1) + white peat + perlite (3:2:1 ratio) (V1);
– a mixture of a second vegetable compost (n. 2) + white peat + perlite (3:2:1 ratio) (V2);
– a mixture of a conifers bark compost + white peat + perlite (3:2:1 ratio) (C).

The four media were treated in a greenhouse in following ways:

Table 1 Cultural technique used on Poinsettia cvs. 'Peterstar' and 'Supjibi' during the trial.

	August, 5.
Potting:	
Pot diameter:	16 cm.
Basic fertilization:	None in peat-based substrate; 140 mg/l N, 160 mg/l P_2O_5 and 180 mg/l K_2O in compost-based substrates.
Pinching:	August, 25
Spacing:	6.25 p./m2
Minimum and ventilation temperature:	18 – 22 °C
EC nutrient solution:	1.2 mS/cm from August, 25 to September, 14
	1.8 mS/cm from September, 15 to October, 9
	1.4 mS/cm from October, 10 to November, 20
	0.9 mS/cm from November, 21 to December, 16
Data collecting:	December, 16

— high irrigation frequency (IF) and N/K_2O ratio = 1:1;
— high irrigation frequency and N/K_2O ratio = 1:1.6;
— low irrigation frequency and N/K_2O ratio = 1:1;
— low irrigation frequency and N/K_2O ratio = 1:1.6;

Irrigation time of plants receiving high IF has been fixed by tensiometers: irrigations began when water tension of the most moistened substrate reached the threshold of 40 hPa.

Plants with high IF were irrigated twice, compared to plants with low IF.

Because of the rich K content of vegetable composts, two different N/K_2O ratios were tested.

Physical and chemical characteristics of the four substrates are reported in table 2.

Table 2 Physical and chemical characteristics of peat-based (T) and compost-based (V1, V2, C) substrates a b.

Characteristics	T	V1	V2	C
Bulk density (g/cm3)	0.139	0.296	0.306	0.186
Total porosity (% v/v)	91.2	83.7	85.0	89.0
Air capacity at pF 1 (% v/v)	23.3	33.2	27.9	42.7
Available water between pF 1 and pF 2 (%v/v)	27.5	18.4	23.3	17.8
pH	6.0	6.8	6.2	4.7
EC (mS/cm)	1.25	2.19	1.42	0.97
N (mg/l)	165	146	147	106
P2O5 (mg/l)	70	54	89	100
K (mg/l)	123	440	208	119
NDI index	1	0.7	0.3	0.2

a Water extraction with 1:1.5 (v/v) ratio.
b With basic fertilization.

The electric conductivity (EC) of the nutrient solution was the same for all substrates and treatments and varied, during cultivation, from 0.5 to 1.8 mS/cm.

At the same time, at the beginning of the trial, in a laboratory the three compost-based substrates were incubated at 21 °C for 4 days, to control nitrogen immobilization by the Nitrogen Drawdown Index (NDI) test (Handreck, 1992). At the end of cultivation, data on plant height, plant diameter and bracts development were collected and evaluated by analysis of variance. At the same time, foliar analysis for N, P, K, Ca and Mg on bracts of plants receiving high IF were carried out.

Results and discussion

NDI test

Table 2 reports the result of the NDI test on the four substrates. Regarding the peat-based substrate (T) the Nitrogen Drawdown Index was 1 (absence of immobilization). On the contrary, the compost-based substrates revealed different levels of nitrogen immobilization (lower in substrate V1, higher in substrates V2 and C).

Plant growth

During cultivation plants grown on substrate C showed symptoms of N-deficiency and a lower growth in comparison to those cultivated on the other substrates. At the same time, tensiometers revealed that substrates with a higher water retention capacity in laboratory analysis, did not have an analogous behaviour in cultivation trial: the peat-based substrate T, that showed a water capacity between pF 1 and pF 2 much greater than other substrates, during the trial revealed the same irrigation needs as substrate V1 but greater ones compared to the other two substrates.

Tables 3 and 4 show the effects of substrate, irrigation and fertilization on plant height, plant diameter, and bract diameter in cvs. 'Peterstar' and 'Supjibi' at the end of cultivation.

Table 3 Average plant height, plant diameter and bract diameter of Poinsettia cv. 'Peterstar' as affected by substrate, irrigation frequency and N/K2O ratio.

	Plant height (cm)	Plant diameter (cm)	Bract diameter (cm)
Substrate:			
T	29.9 a	47.9 a	23.7 a
V1	25.6 c	43.2 c	22.5 a
V2	27.9 b	45.9 b	23.3 a
C	24.2 d	41.9 d	21.8 a
Irrigation frequency (IF):			
High IF	30.4 a	49.8 a	24.2 a
Low IF	23.8 b	40.3 b	21.6 b
N/K2O ratio:			
1:1	27.0 a	45.2 a	23.5 a
1:1.6	27.0 a	44.4 a	22.2 b

Mean separation within columns by LSD test (P £ 0.05).

Table 4 Average plant height, plant diameter and bract diameter of Poinsettia cv. 'Supjibi' as affected by substrate, irrigation frequency and N/K$_2$O ratio.

	Plant height (cm)	Plant diameter (cm)	Bract diameter (cm)
Substrate:			
T	26.3 b	41.2 b	23.5 a
V1	28.4 a	44.3 a	24.1 a
V2	27.2 b	42.9 a	24.0 a
C	24.1 c	37.9 c	22.5 a
Irrigation frequency (IF):			
High IF	31.4 a	50.2 a	26.4 a
Low IF	21.5 b	32.6 b	20.5 b
N/K2O ratio:			
1:1	25.3 b	39.2 b	22.6 b
1:1.6	27.7 a	43.8 a	24.4 a

Mean separation within columns by LSD test (P £ 0.05).

Substrates T and V1 allowed us to obtain higher and larger plants, respectively of cv. 'Peterstar' and 'Supjibi'. The influence of substrates T, V1 and V2 on plant growth was different according to varieties; plants of both varieties cultivated on substrate C showed the lowest growth. Bract diameter was not affected by substrate. While substrate C revealed to be subject to N immobilization, substrates V1 and V2, even with a NDI index lower than 1, did not present a remarkable N-deficiency during cultivation.

Plants of both cultivars more frequently watered, showed a better growth and a greater bract diameter in comparison with plants receiving low IF.

N/K$_2$O ratio had no effect on cv. 'Peterstar', except for bract diameter, which was slightly larger with the 1:1 ratio, compared with the 1:1.6 ratio. In cv. 'Supjibi', on the contrary, a N/K$_2$O ratio = 1:1.6 led to greater plants and a larger bract diameter than with the 1:1 ratio.

Foliar analysis

Bract analysis on plants receiving high IF (table 5) showed no appreciable differences among the treatments in mineral contents, with the exception of K-content in plants receiving different N/K$_2$O ratios: on both cultivars there was an unexpected higher K-content in bracts of plants fertilized with the highest N/K2O ratio (3.4 % d.w. vs. 2.8 % in cv. 'Peterstar' and 3.0 vs. 2.8 in cv. 'Supjibi').

Conclusions

The comparison between laboratory analysis and cultivation trial pointed out that not always there is correspondence between physical characteristics of a substrate and its behaviour in practice.

At the same time, also the NDI test for measuring N immobilization does not seem efficient to describe the nitrogen drawdown for a period longer than 4 days.

Table 5 Bract analysis (% on dry weight) of Poinsettia cvs. 'Peterstar' and 'Supjibi' as affected by substrate and N/K2O ratio (only plants receiving high IF).

	N	P	K	Ca	Mg
Cv. 'Peterstar'					
Substrate:					
T	2.7	0.5	2.7	0.3	0.3
V1	3.1	0.6	3.4	0.4	0.3
V2	2.6	0.6	3.2	0.4	0.3
C	2.6	0.6	3.1	0.3	0.3
N/K2O ratio:					
1:1	2.7	0.6	3.4	0.4	0.3
1:1.6	2.7	0.5	2.8	0.3	0.3
Cv. 'Supjibi'					
Substrate:					
T	2.7	0.6	3.0	0.4	0.3
V1	2.7	0.6	2.7	0.4	0.3
V2	2.7	0.6	3.0	0.5	0.4
C	2.7	0.6	2.8	0.3	0.3
N/K2O ratio:					
1:1	2.6	0.6	3.0	0.4	0.3
1:1.6	2.8	0.6	2.8	0.4	0.3

For the above-mentioned reasons it would be better, with regard to physical characteristics of substrates, to control irrigation of each individual substrate by tensiometers, according to water tension measurements. Regarding the tendency to immobilize nitrogen by compost-based substrates, it is necessary to test substrates for a longer period of time, at values of temperature similar to those maintained during cultivation.

References

Handreck, K.A., 1993. Rapid assessment of the rate of nitrogen immobilisation in organic components of potting media. Commun. Soil Sci. Plant Anal. 23(3–4): 201–215.

Lamanna, D., Castelnuovo, M., and D'Angelo, G., 1991. Compost-based media as alternative to peat on ten pot ornamentals. Acta Horticulturae 294: 125–129.

Lemaire, F., Dartigues, A., Rivière, L.M., and Charpentier, S., 1989. Cultures en pots et conteneurs. INRA – PHM Revue Horticole, Paris – Limoges: 76–80.

Method for the Evaluation of Biodegradability of Packaging in Composting Conditions Proposed by CEN (European Eommittee for Standardization): a Technical Approach.

F. DEGLI INNOCENTI, M. TOSIN, C. BASTIOLI – NOVA-
MONT, via Fauser 8, 28100 Novara

Introduction

The European Directive on Packaging and Packaging Waste, recently promulgated, considers composting as a form of recycling of biodegradable packagings. To become effective the Directive must be endowed with technical tools such as definitions, criteria, and test methods. CEN (Comité Européen de Normalisation), the European Standardisation Committee has established on 1991 a subcommittee (TC261/SC4) named 'Packaging and the Environment' to provide the technical and standardization support to the Directive. Within this subcommittee the Working Group 2 (WG2: 'Degradability') is defining a test method to measure the biodegradability of packaging in composting conditions. Both this test and ASTM D5338–92 are derived from a proposal of Organic Waste Systems (Gent, B) and based on the measurement of CO_2 evolved by the test substance in conditions simulating the composting environment: aerobic fermentation, solid state, high temperature, and inoculum particularly rich in thermophilic microorganisms (mature compost). In order to verify the accuracy and precision of the method and gather useful data for the final standardization, the CEN TC261/SC4/WG2 has organized a ringtest at European level using cellulose, paper and Biopol, a polyhydroxyalcanoate, as test materials. In Italy NOVAMONT, a Company actively involved in the development of biodegradable materials, has taken part to the ringtest. A comprehensive description of the results of this testing activity will be soon published by the WG–2. In this presentation we describe the technical approach followed by us to set-up the test procedures and test apparatus.

Materials and methods

Apparatus. Compressed air pressure is reduced by a pressure reduction device to about 0.5 atm. The air flow is then dehydrated by a silica gel (Fluka, Ch) column and decarbonated by a soda lime (Ohmeda, UK) column. The flow is then split into 12 lines, corresponding to 12 composting reactors. The air flow rate of each line is measured by a rotameter and adjusted by a valve (Flow-Meter srl, I), and leaded to the reactor by means of natural rubber thick tubing (Ascenso, I). The air is fluxed upwards in each reactor (3 litres glass bottles) via an hollow shaft finishing with a 'T'. The reactors are heated in a thermostatic water bath. The exit gas is passed, before measurement, through two vapour traps. The first is a glass bottle at ambient temperature; the second is a water-cooled glass coil. Twice a day, the air flow rate of each line is measured with a precision rotameter (Flow-Meter srl, I) and the CO_2 concentration is measured by an infrared CO_2 detector (Gas Monitor ADC 2000 Series The Analytical Development Company, UK). This instrument makes autozero every ten minutes. The span is checked every 2 weeks using a 1% CO_2 in He calibration mixture (SIAD, I).

Test substances. Cellulose microcrystalline 'Avicel' by Merck. Packaging paper, made with virgin kraft unbleached softwood pulp, by Centre Technique du Papier (France). Biopol, grade D800P, obtained by Zeneca Bio Products (UK). The compost used as inoculum was obtained by a composting plant located in Trento (I).

Test procedure. A mixture of mature compost (600 g, dry weight) and test material (100 g) is introduced in a reactor (about 3 liters) maintained in thermophilic conditions for 45 days and fluxed with a continuous air stream. In these conditions the mixture will produce, by respiration, carbon dioxide. If the test material is metabolized there will be an extra production in comparison with control blank reactors. The mineralisation percentage is the ratio between net CO_2 produced by sample and the amount which could be produced in case of a complete transformation of its carbon into CO_2. The temperature in this experiment followed a profile, according to ASTM D5338–92 test procedures; the final CEN draft prescribes a fixed 58°C temperature. Three replicates for test materials and blanks reactors were used.

Data treatment. The row data collected during each sampling are the following: date and hour of sampling; reactor identification number; air flow rate (AFR = L/h); CO_2 concentration (% v/v). A home prepared program on '123–Lotus' processes these row data and gives the cumulative CO_2 production (grams) and the biodegradation (% $CO_2/ThCO_2$) of each reactor. It calculates: the elapsed time from test start; elapsed time from last measurement. It converts the CO_2 concentration from % (A) to g/L (B) by:

$$B (g/L) = A/22.263*44/100$$

and works out the CO2 evolution rate (CER):

$$CER (g/h) = B \times AFR.$$

The amount of CO_2 produced during the time interval within two measurements is estimated by multiplying the CER by the elapsed time from last measurement.

The sum of these CO_2 amounts is the total cumulative CO_2 production. The net production is determined subtracting from each total value the corresponding mean CO_2 production of blanks. Finally, the net values are divided by the theoretical CO_2 production and multiplied by 100 to obtain the % biodegradation.

Results & discussion

The total cumulative CO_2 productions of replicates were very similar for each test material. At test termination (46 days) the CO_2 production (+ standard deviation) and biodegradation values (in parenthesis) were: Blanks 35.6+2.3 g; Cellulose 187.8+3.9 g (97.68%); Paper 173.4+4.9 g (88.29%); Biopol 247.3+6.9 g (102.10%). These results indicate that the procedure in spite of being based on sampling (not frequent) rather than on cumulative measurements is accurate. It is also precise. This turns out from the ring test results, which will be published elsewhere in full details. Here we want to remark the following. In order to compare the different results it is helpful to normalize the biodegradation values of each laboratory to the corresponding value of cellulose (considered as reference material with its biodegradation fixed equal to 100%). It turns out that the normalised results of the present study are very close to the ringtest normalized averages: Biopol degradation is 104.1% of cellulose (ringtest average 106.7%) and paper is 89.1% (average 91.3%).

The apparatus installed and experimented in this trial has shown satisfactory features: it is relatively cheap; even if not automatized, it is not laborious and time consuming, at least in comparison with other respirometers (i.e. Sturm test). The CO_2 concentration is determined by an IR gas monitor originally designed to be used as a gas-alarm in cellars and breweries. It turned out to be fast (it takes few minutes to give the final value), precise (as verified by using different calibration gas mixtures) and stable (the drift after many weeks is negligible). The following protocol improvements were adopted. To obtain a regular flow rate at outlet and to facilitate air flow reading, expanded clay balls (used in horticulture and floriculture) were added to the composting mixture to increase porosity of the substrate. Water evaporation was controlled by weighing each vessel periodically and restoring the initial weight by adding water. This to avoid the risk of plugging the rotameters with condensed water by using saturated air as prescribed by the method. The system of CO_2 removal from air stream was nearly depleted at the end of the experiment. The real need of using CO_2 free air is questionable. In future experiments we intend to measure at every sampling the CO_2 concentration entering the vessels (I), the CO_2 concentration at the outlet (O), and correcting the reading: 'net' CO_2 concentration = O − I we will take in consideration only the part of CO_2 which has been produced by degradation. This is applicable only if O>>I.

The CEN method is very interesting because it offers the possibility to know the biodegradation of a material – as CO_2 evolution, namely measuring a clear, unambiguous parameter – when subjected to environmental and microbiological conditions which reproduce the typical composting conditions. A frequent objection is that in real composting the test material is mixed with fresh waste rather than with a mature, stabilized compost. On the other hand, a fresh solid waste would be unsuitable for this test because it would produce a too high CO_2 background which would mask the CO_2 evolved by test material. A study to understand the influence of the composting substrate on the biodegradation behaviour of solid materials indicated that the test methods based on the use of mature compost can possibly underestimate the biodegradation occurring in fresh waste, i.e. in real composting plants, and are to be considered as conservative test methods (manuscript in preparation).

Effects of the Spreading into the Soil of Olive Mill Wastes on the Physico-chemical Properties of the Humic acids. I. Interaction of natural and Synthetic Humic Acids with Caffeic Acid.

[1]Deiana S., [2]Gessa C., [1]Manunza B., [1]Pistidda C., [3]Rausa R. and [1]Solinas V.

Introduction

Organic wastes from different sources determine the variation of several soil physico-chemical parameters such as pH, heavy metal content, redox potential, complexing activity and degree of humification, what may affect the soil fertility in a positive or negative way. The organic fraction of low molecular weight is involved in ionic mobilization and humification processes, its presence is particularly important in soils with a low level of organic matter and micronutrients.

As a part of a wider project, whose objective is the evaluation of the modifications that olive mill wastes induce on the structural and physico-chemical properties of humic acids, we studied the interaction between caffeic acid (CAF), a relevant constituent of these biomasses, and natural and syntetic humic acid (NHA and RHA, respectively). In this paper we report on both the interaction between NHA and CAF and the redox properties of the NHA-CAF compound.

Materials and Methods

The NHA were extracted from an Andosol following the procedure of the International Humic Substances Society (HISS). The chemical and physical characteristics of the RHA, obtained via coal oxidation, and of the NHA as well as the analytical procedure for the Iron(II) and Iron(III) determination are reported in previous papers (Rausa et al., 1994 ; Deiana et al., 1995). CAF concentration was measured with an HPLC DIONEX equipped with a HC-ODS/PHA column supplied by Perkin Elmer. The IR spectra were performed on KBr pellets by using a Nicolet 205 spectrometer.

1 Di.S.A.A.B.A., Sezione Chimica del Suolo, Facoltà di Agraria, Viale Italia 39, 07100 Sassari, Italy.
2 Istituto di Chimica Agraria, Facoltà di Agraria, Viale Berti-Pichat 10, 40127 Bologna, Italy.
3 Eniricerche, via Maritano 26, 20097 S. Donato Milanese, Italy

Results and discussion

The interaction of CAF by NHA was studied by examining the effect of NHA-CAF ratios (keeping constant the amount of the organic material), solution pH and reaction time. In Figure1 the kinetic curves of absorption of CAF by NHA at pH 2.0, 3.0, 4.0, 5.0 and 6.0 are shown. The plots indicate that at pH 2.0 the absorption of CAF is strongly inhibited, while it is favoured in the 3.0–6.0 pH range. The absorption kinetics at pH 6.0 is slower probably due to the increase of the carboxylate groups concentration of both the NHA and CAF. The concentration of such charged groups is greater than that of the same systems at lower pH values. In these conditions both the humic acids and CAF behave as hydrophilic, negatively charged molecules, thus increasing repulsive forces between them and hampering the formation of aggregates.

Figure 2 reports the amount of CAF absorbed at various initial CAF concentrations at pH 4.0. These plots shown that the maximum of CAF absorbed is about 1.73 mmoles g-1.

The CAF interacts very strongly with the organic matrix: in fact, an increase in the H+ concentration (pH 2.0) does not cause the release of CAF absorbed. The ocurrence of a strong interaction between CAF and NHA is also shown by the reaction between NHA–CAF and iron(III) when compared to that of NHA and NHA–Fe(III) with iron(III) and CAF, respectively. The NHA–CAF and Fe(III) reaction gives a yield in Fe(II) of about 50% after 100 hours of reaction. In contrast, the reduction of Fe(III) by NHA and of the Fe(III) complexed by NHA operated by CAF is complete and occurs quickly. This indicates that the phenolic –OH groups of both NHA and CAF, which are responsible for the reduction of Fe(III) (Rausa et al.,1994; Deiana et al., 1992; 1995), are involved in the NHA–CAF interaction.

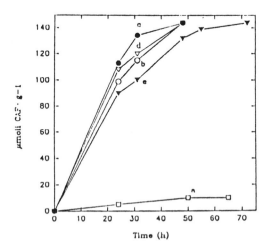

Figure 1. CAF absorption (mmoles g–1), at different pH values, as a function of time in NHA–CAF systems. Starting conditions: 25 mg NHA; 0.01 M NaClO4 at pH 2.0 (a); pH 3.0 (b); pH 4.0 (c); pH 5.0 (d) and pH 6.0 (e). The initial CAF concentration was 72 mM. Reaction volume 50 mL.

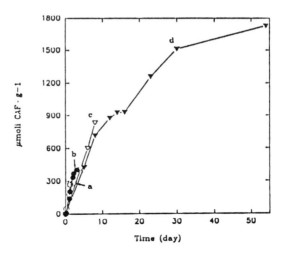

Figure 2 CAF absorption (mmoles g–1), at different CAF concentrations, as a function of time. Starting conditions: 25 mg NHA; 0.01M NaClO4; pH 4.0. The initial CAF concentrations (mM) were 144 (a), 216 (b), 432 (c), 864 (d). Reaction volume 50 mL.

In order to assess the role of the phenolic groups on the interaction between NHA and CAF we carried out the same tests on RHA, which were chosen because most of their functional groups are constituted by carboxylic and phenolic groups (Rausa et al.,1994). Results obtained by using these compounds confirmed that phenolic groups are involved in the formation of the NHA-CAF adducts.

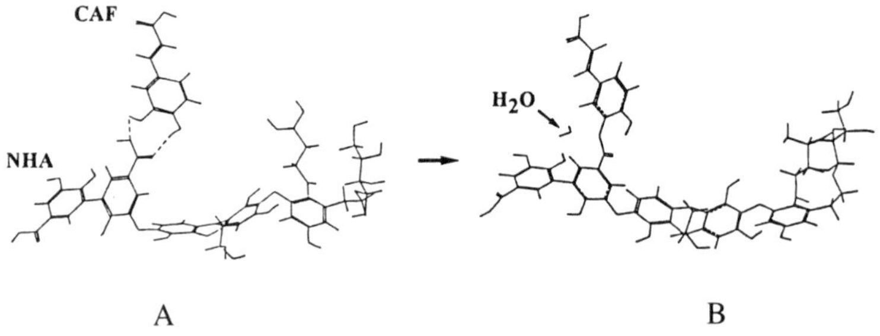

Figure 3 Proposed structure of the NHA-CAF adduct as obtained by QMC.

The IR spectrum of the NHA is similar to that of the NHA-CAF compound, suggesting that a strict interaction occurs between the CAF molecule and the humic substrate.

Preliminary quantum mechanics calculations (QMC) (Fig. 3) show that both the carboxylic and phenolic groups of the humic substances and CAF can be active in this reaction. In particular two adducts can be proposed: the hydrogen bonded adduct A which evolves to the B adduct with the elimination of a water molecule and a energy lowering of about 50 Kcal mole–1. Further studies at this regard are in progress.

Conclusions

This study, which constitutes a first step in the survey of more complex systems, indicates that humic acids can be used to lower the concentration of phenolic compounds, which can induce toxic effects on the soil microorganisms and plants. However, it is important to underline that both HA and CAF, once that the adduct forms, lose some of their reduction properties.
Acknowledgement
Financial support from CNR (Rome) and MURST (40%).

References

Deiana S., Gessa C., Manunza B., Rausa R andSolinas V. 1995. Iron(III) reduction by natural humic acids: a potentiometric and spectroscopic study. E.J. of Soil Science, 46, 000–000.
Deiana S., Gessa C., Manunza B., Marchetti M. and Usai M. 1992. Mechanism and stoichiometry of the redox reaction between iron(III) and caffeic acid. Plant and Soil,145,287–294.
Rausa R., Deiana S and Gessa C. 1994. Reduction of iron(III) by synthetic humic acids obtained via coal oxidation. In: Humic Substances in the Global Environment and Implications on Human Healt (eds N. Senesi & T.M. Miano), pp.1177–1182. Elsevier, Amsterdam.

The use of Dairy Manure Compost for Maize Production and its Effects on Soil Nutrients, Maize Maturity and Maize Nutrition

V C DE TOLEDO, H C LEE, T A WATT and J M LOPEZ-REAL – Sustainable Agriculture Research Group, Wye College, University of London, Wye, Kent TN25 5AH, UK

Introduction

Maize (Zea mays L.) is widely distributed in tropical and temperate regions, being an important cereal used for both human and animal consumption. Inorganic fertilizers can be too costly and out of reach for small producers, being unable to sustain soil fertility and also posing environmental risks of nitrate pollution of ground water table (Maynard 1994).

A production system using renewable resources within the farm was studied, recycling crop and animal residues into organic fertilizer, and aiming at reducing dependence on external inputs. Dairy manure and cereal straw were used to produce compost, and its effects on soil nutritional status, maize nutrition and crop production were studied. Further to results obtained in a previous experiment (de Toledo et al 1994), the possible effect of compost in delaying maize maturity was also examined.

Materials and Methods

A field experiment with forage maize was carried out at Wye College Farm, UK, on Orchard Field (calcareous silt loam soil, pH 8.0). The experiment, drilled on 6th May 1994, compared inorganic fertilizer (as Nitram, 34.5%N) and dairy manure compost at two rates of application. Compost rates were equivalent to Nitram in terms of nitrogen availability (Brinton 1985).

The trial had Nitram at 50 and 100 kg N.ha–1, applied as topdressing in one application only, one week after drilling. Compost (25 and 50 t.ha–1) was uniformly spread on the soil surface and incorporated before drilling. An untreated control was included. Herbicide (atrazine, as Gesaprim, 3 l.ha–1) was sprayed in all plots for weed control, on the day of crop emergence. A split-plot design with four replicates was used, to allow for a comparison of all the treatments with two maize varieties, LG2080 and Aziz – an earlier maturing variety.

Plots were 3 m wide and 16 m long, and had four rows of maize, with sampling from the two inner rows only. Soil samples were taken in June, August and September for nutrient analyses (total N, NO_3–N, NH_4–N, P, K, Ca, Mg, dry matter and organic matter). Maize samples were taken at three stages of crop development for biomass, leaf area measurement and nutrient analyses (results not presented). Final maize yield and nutrient uptake were determined at harvest (5th October 1994). All data were analysed using analysis of variance.

Results

No difference in soil nutrients was found between varieties LG2080 and Aziz. Results of the mean of the two varieties are shown for soil NO3–N (Table 1) and soil K (Table 2). Higher soil NO3–N content ($p < 0.01$) was found under Nitram application, but this was not associated with better maize yields. Plots treated with compost had higher soil K levels ($p < 0.001$), and higher soil total N, P and Mg contents (data not presented). Only traces of NH4–N were found in soil under all treatments. Soil Ca (1500 mg.kg–1 soil) and soil organic matter (4.5%) were high, varying little within treatments at all samplings. At early stages, soil dry matter was higher ($p < 0.05$) in Nitram treated plots, thus indicating a higher soil water content under compost application. Due to heterogeneity of variance, statistical analyses of soil nutrients were made using log transformed data.

Table 1 Soil NO3–N levels (mg.kg–1 soil) at three stages (mean of two varieties)

treatments	June		August		September	
(kgN.ha–1)	mean	log	mean	log	mean	log
control 0	13.4	1.1	5.8	0.8	1.7	0.4
Nitram 50	27.1	1.4	13.5	1.1	4.5	0.6
Nitram 100	35.5	1.5	19.7	1.2	9.4	1.0
compost 50	17.5	1.2	7.3	0.8	4.1	0.6
compost 100	19.7	1.3	8.2	0.9	4.3	0.7
s.e.d.		0.06		0.16		0.12

Table 2 Soil K levels (mg.kg–1 soil) at three stages (mean of two varieties)

treatments	June		August		September	
(kgN.ha–1)	mean	log	mean	log	mean	log
control 0	184.4	2.2	128.9	2.1	191.6	2.3
Nitram 50	209.0	2.3	104.8	1.9	199.8	2.3
Nitram 100	139.2	2.1	70.4	1.7	195.7	2.3
compost 50	291.2	2.4	163.3	2.2	302.6	2.5
compost 100	476.0	2.6	214.9	2.3	434.2	2.6
s.e.d.		0.12		0.21		0.07

Nitrogen uptake by maize was higher under Nitram ($p < 0.001$) at the high level of application (100 kg N.ha–1), but did not differ from compost at the lower level. However, maize had higher accumulation of potassium ($p < 0.001$) in compost treated plots at both levels of application (Table 3). Variety Aziz showed poorer responses in compost treated plots than did LG2080, particularly at the higher level of application, but performed well in plots treated with Nitram.

Table 3 Maize N and K uptake at harvest in above ground plant parts (kg.ha–1)

treatments (kgN.ha–1)	N uptake variety Aziz	variety LG2080	K uptake variety Aziz	variety LG2080
control 0	112.8	137.9	93.0	118.2
Nitram 50	142.4	126.4	108.3	104.1
Nitram 100	157.0	150.5	102.8	102.4
compost 50	135.1	133.4	118.3	123.5
compost 100	126.2	141.8	117.8	135.1
s.e.d.	10.49	10.49	10.96	10.96

Maize fresh yield was higher ($p < 0.01$) in compost treated plots at both levels of application, and higher in variety LG2080 than in Aziz in all treatments ($p < 0.05$), but no significant difference was found in maize dry yield (Table 4). Variety Aziz had a higher dry matter content (average 38%) than variety LG2080 (average 33%) at harvest ($p < 0.01$), in all treatments.

Table 4. Maize fresh and dry weight yields at harvest (t.ha–1)

treatments (kgN.ha–1)	fresh weight variety Aziz	variety LG2080	dry weight variety Aziz	variety LG2080
control 0	29.5	37.7	11.7	13.0
Nitram 50	34.6	36.3	13.4	12.0
Nitram 100	33.5	39.4	13.6	12.9
compost 50	36.3	40.7	13.4	13.2
compost 100	34.9	41.7	12.2	13.1
s.e.d.	1.70	1.70	0.78	0.78

Discussion

Nitram treated plots have a higher potential for nitrate pollution of ground water table than compost ones, as the excess of NO3–N in soil tends to leach (Maynard 1994).

The high levels of organic matter in the soil, and the sensitivity of maize to low soil temperature could have masked the effects of treatments. There might be better responses in areas warmer than UK, and/or with lower soil organic matter content.

The use of an early variety other than Aziz might show a correlation between higher fresh yield and higher dry yield with compost application. Higher fresh yield under compost could be explained by the higher soil water content in compost treated plots at early stages, or by the higher accumulation of potassium in maize, as plants well supplied with K show increased uptake of water and reduced water loss (Mengel and Kirkby 1987).

Acknowledgements

This work was carried out with financial support from CAPES (Brazil).

References

Brinton Jr W F. 1985. Nitrogen response of maize to fresh and composted manure. Biological Agriculture and Horticulture 3: 55–64.

Maynard A A. 1994. Sustained vegetable production for three years using composted animal manures. Compost Science & Utilization 2 (1): 88–96.

Mengel K and Kirkby E A. 1987. Principles of Plant Nutrition. 4th edition. International Potash Institute, Bern, pp 436–438.

de Toledo V C, Lee H C and Watt T A. 1994. Sustainable maize production – the effects of compost on soil nitrogen content and crop and weed biomass. Proceedings 3rd European Society for Agronomy Congress. Abano–Padova, pp 680–681.

Sewage Sludge Composting: Study of Nitrogen Mineralization Using Electroultrafiltration

DÍAZ-BURGOS M.A., DÍEZ J.A. and POLO A. – C.C. Medioambientales (CSIC). Madrid. Spain.

Introduction

Mediterranean areas are specially affected by a soil organic matter decrease due to the drop of agricultural productivity and the deterioration of soil (desertization process, salinization, forest fire, etc.). On the same time, this waste may pollute the pedological environment because of its high contents of heavy metals and plant-damaging organic compounds.

The use of composted waste is an alternative to overcome problems in agricultural use of this material. Composting can reduce the contents of harmful compounds. However, generalizations about the process and the composts produced from different wastes cannot be made, therefore data on each composting process are required.

The composting of sewage sludge has been extensively studied (LOBO et al., 1987; GARCIA-IZQUIERDO et al., 1987; KATAYAMA et al., 1986), nevertheless, the criteria for evaluating the process and the final composts remain subjects for discussion (HAPADA et al., 1980; SAVIOZZI et al., 1986; ZUCCONI et al., 1981; DIAZ-BURGOS and POLO, 1992).

This paper shows an approach to these problems using the electroultrafiltration (EUF) technique. It was used for monitoring transformations taking place in the nitrogenous fractions of two anaerobic sewage sludges mixed with two different additives.

Material and methods

Two anaerobic sewage sludges from a waste water treatment plant of Madrid city (urban: U and industrial: I origin) were mixed with chips (C) from tree pruning or vine shoots (V) in a 2/1 weight ratio (sludge/additive) and composted for 14 weeks. The selection of the sewage sludge was based on the different heavy metal content which may influence the microbial populations mainly responsible for the composting process. The additive was selected in order to reuse the agricultural and urban wastes after the treatment.

Composting was performed by keeping the mixtures under controlled humidity and temperature conditions (28 °C and at water holding capacity) in a thermostated chamber. Sampling was carried out after turning over the mixtures, thus ensuring homogenization and preventing local anaerobic processes, after 1, 2, 4, 6, 8, 10, 12 and 14 weeks of incubation.

Variations of nitrogen compounds were evaluated using the electroultrafiltration technique for monitoring transformations taking place. Different fractions can be distinguished: total N (EUF-N), organic N (EUF-Norg), NO_3^- (EUF-NO_3) and $NH4^+$ (EUF-NH_4). The method allowed to evaluate the microbial activity and to determine the availability of nitrogen through the ratios of EUF-Norg/EUF-NO_3 and EUF-NO_3/EUF-N respectively. The conditions for the electroultrafiltration and the determinations nitrogen fractions in the extracts were reported in a previous paper by DIEZ (1988). The characterization of wastes and compost was carried out by DIAZ-BURGOS (1990).

Results

The analysis of EUF-N evolution curves (figures 1 and 2) enables to differentiate four phases, which present the specific characteristics in function of the sewage sludge type and its additive.

In general, a first phase was characterized by an increase of total nitrogen and nitrate and a decrease of ammonia. This phase implicated a strong oxidation process and a low proteolitic action. The sludge type conditioned the nitrogen availability (table 1) which increased in the case of the urban sewage sludge. For the industrial sewage sludge mixed with vine shoot, this phase was preceded by a reorganization period in which all these fractions slightly diminished. The microbial activity was affected by the additive type: the chips increased the activity while the vine shoots decreased it (Table 1). The later additive had an inhibitor effect (LOBO, 1987) in the first composting stages.

The second phase was characterized by a low change in the nitrogen levels; the EUF-NH4 fraction disappeared and the availability of nitrogen decreased. A higher microbial activity occurred (Table 1) caused by the microbial flora being replaced, leading to appreciable nitrogen immobilization. In the industrial sewage sludge a decrease in the total nitrogen content was observed. This second phase was retarded 6–7 weeks in the industrial sludge with vine shoot compost. This may be due to a greater presence of heavy metals added and to the initial toxicity of vine shoots compost on the different microbial populations.

This phase predisposes the behaviour of the third phase, which was characterized by an acceleration of the mineralization process. After the microbial population has developed, they act on the short chains of peptones and aminoacids, quickly turning the nitrogenous radicals into nitrates, which contributes to increase the EUF-NO3, according to previous experiences (DIAZ-BURGOS et al., 1993; GARCIA et al., 1992). In the urban sewage sludge further effects observed were, the increase of N-availability and the nitrate.

Table 1 Indices of microbial activity and Ntrogeri availability during sewage sludge composting

SAMPLE[1]	Micr. Act.[2]	N-availab.[3]	SAMPLE[1]	Micr.Act.[2]	N-availab.[3]
UC1	2,96	0,06	UV1	5,51	0,12
UC2	2,71	0,06	UV2	0,52	0,44
UC4	1,09	0,27	UV4	1,12	0,43
UC6	0,54	0,40	UV6	0,59	0,54
UC8	0,42	0,66	UV8	1,36	0,41
UC10	0,40	0,68	UV10	0,27	0,77
UC12	1,37	0,19	UV12	0,15	0,86
UC14	0,29	0,76	UV14	0,18	0,82
IC1	1,09	0,52	UC1	0,51	0,33
1C2	0,91	0,47	UC2	0,46	0,33
1C4	1,00	0,50	UC4	—	—
IC6	0,39	0,28	UC6	0,28	0,22
IC8	0,57	0,36	UC8	0,56	0,36
1C10	0,53	0,35	UC10	0,48	0,32
1C12	0,53	0,35	UC12	0,44	0,30
1C14	0,84	0,45	UC14	0,36	0,26

(1) The number refers to composting time in weeks. (2) Microbial activity: EUF-Norg/EUF-NO3.
(3) Nitrogen availability: EUF-N03/EUF-Nt.

In the last phase (IV) all EUF parameters dropped which was characteristic for the exhaustion of the energy substrate, preventing the nitrifying bacteria from continuing to exercise their function. This phase was not observed in all cases. The crossing of nitrate and ammonium curves was used as a criterion of maturity for the compost (DIAZ-BURGOS et al., 1991; Van de KERKHOVE, 1990). The time needed showed the different behaviour of the compost studied. Only the sewage sludge with chips showed an NH_4^+–NO_3^- crossing point after 6 weeks, which did not occur in the other composts where toxicity affected the EUF-NH4 production from the beginning of composting.

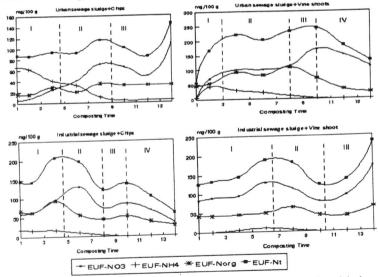

Figure 1 Changes in the EUF Nitrogen fraction with time during composting of municipal sewage sludge

References

DIAZ-BURGOS MA y POLO A (1991). Evolución de las sustancias húmicas durante el compostaje de lodos residuales. Suelo y Planta 3: 57–70.

DIAZ-BURGOS, M.A., CECCANTI, B. and POLO, A. (1993) Monitoring biochemical activity during sewage sludge composting. Biol Fertil Soil, 16: 145–150.

DIAZ-BURGOS, M.A., CECCANTI, B. and POLO, A. (1991). Changes in the inorganic nitrogen fraction during sewage sludge composting. In: Treatment and use of sewage sludge and liquid agricultural wastes. E. L'Hermite Ed. Elsev. Appl. Sci. Athenes.

DIEZ, J.A. (1988). Revisión del método de determinación automatizada de nitrógeno UV oxidable en extracto de suelo. En. Edafol. Agrobiol. 47, 1029–1039.

GARCIA-IZQUERDO CJ, COSTA F y HERDANDEZ MT (1987). Evolución de parámetros químicos durante el compostaje. VII Congreso Nacional de Química. Química Agrícola y Alimentaria-2. Volumen II. Sevilla. pp 103–110.

GARCIA C, HERNANDEZ MT, COSTA F, CECCANTI B and CIARDI C (1992). Changes in ATP content, enzyme activity and inorganic nitrogen species during composting of organic wastes. Can J Soil Sci 72: 243–253.

HARADAY., INOKO, A. (1980). Relationship between cation exchange capacity and degree of maturity refuse compost. Soil Sci. Plant. Nutr., 26, 127–134.

KATAYAMA A, KER KC, HIRAI M, SHODA M and KUBOTA H (1986). Stabilization process of sewage sludge compost in soil. In: Bertoldi, Ferranti, L'Hermite and Zucconi (eds.). Compost: Production, quality and use. Elsev Appl Sci Publish. Udine, pp 341–350.

LOBO MC, DIAZ-BURGOS MA y POLO A (1987). Evolución del grado de madurez de diferentes compost mediante el estudio electroforético de la fracción hidrosoluble. VII Congreso Nacional de Química. Química Agrícola y Alimentaria-2. Vol II. Sevilla. pp 111–118.

LOBO MC (1988). The effect of compost from vine shoots on the growth of barley. Biological Wastes 25: 281–290.

SAVIOZZI, A., RIFFALDI, R. and LEVI-MINZI, R. (1986) Compost maturity by water extract analyses. In: Compost: production, quality and use. Eds. Bertoldi, Ferranti, L'Hermite and Zucconi. Elsev. Appl. Sci. Publish. Udine, 359–367.

Van De KERKHOVE, J.M. (1990) Evolution de la maturité de trois déchets urbains en cours de compostage. Thésè d'Etat. INPL. Nancy. France.

ZUCCONI, F., FORTE, S., MONACO, A. and De BERTOLDI, M. (1981) Biological evaluation of compost maturity. Biocycle, 22, 27–29.

Co-Composting Process of (Sugarbeet) Vinasse and Grape Marc

DÍAZ, M.J.*, MADEJÓN, E.**, LÓPEZ, R.*, RON VAZ M.D.* and CABRERA, F.*

Beet vinasse, a high density liquid waste from the sugar industry, and grape marc, a primary by-product of wine production, could be recycled as fertilizers due to their high OM and nutrient contents. The direct incorporation of grape marc into agricultural land, a common practice, has become a serious problem because degradation products inhibiting root growth are released (Inbar *et al.*, 1991). Furthermore, the direct aplication of vinasse has also several shortcomings because of its high salinity (EC 250–300 dS m^{-1}), low P content (P$_2$O$_5$ 0.012 %) and liquid dense character (1.3 g cm^{-3}). An alternative to overcome these disadvantages is the co-composting of vinasse with grape marc.

A compost was obtained from a mixture of vinasse, V (2.5% N, 0.06% P$_2$O$_5$, 3.6% K$_2$O, 27% OM) with grape marc, GM (1.4% N, 0.63% P$_2$O$_5$, 1.16% K$_2$O, 71.6% OM). Sugarbeet factory lime (F) containig 50% CaCO$_3$ was added to the mixture to overcome the vinasse acidity (pH 4.7). The proportion of each component in the mixture was 17% of V, 82% of GM and 1% of F, where GM and F are expressed on a dry matter basis.

The co-composting process was carried out in a static pile, under cover, with forced aeration following the Rutgers method with a total mass of ca. 15 metric tons. The pile (3m length x 5 m width x 1.5 m height) was centered over three perforated pipes installed in grooves in the concrete floor. The air required for the process was provided by a blower at a rate of 0.87 kg O$_2$ h^{-1}.

During the co-composting process, samples were taken periodically at two depths (0–30 and 40–100 cm) in three zones of the pile. Temperature was measured in four zones at the same depths. The following parameters were determined throughout the whole process: moisture content, pH$_{H_2O}$ (1:5 w:v), OM, EC$_{H_2O}$ (1:5 w:v), total–N, NH$_4$–N, NO$_3$–N and CEC. Macro and micronutrients were measured after 1, 80 and 190 days of composting. Elemental analysis (C, N, and H) of the mixture was carried out before and after composting.

The temperature profiles at two depths in the compost pile are shown in Figure 1. The temperature rose rapidly reaching an averaged value of 53°C at day 8. The thermophilic phase was completed after 54 days of composting. Similar temperature profiles have been described elsewhere (Pérez García and Iglesias Jiménez,

* Instituto de Recursos Naturales y Agrobiología de Sevilla (CSIC). Apdo. 1052, 41080 Sevilla, Spain.
* * EBRO AGRICOLAS S.A. Apdo. 9, 41300 S. José de la Rinconada, Sevilla, Spain.

1984), where the thermophilic phases were too long due to the large size of the piles. The moisture content was maintained during the thermophilic period at around 50%, this being considered the optimum. After 54 days, the mesophilic stage started lasting one month. The compost was then left to mature for the following four months. No water was added throughout the mesophilic and maturing stages. The final moisture content of the compost was 30% being under the maximum water content (40%) permitted by EEC regulation.

The pH increased from 7.1 to a maximun of 9.5 after 25 days of composting. Afterwards, the pH decreased to 8.3, remaining approximately constant during the maturity period. This pH value has been considered as standard for compost stabilization (Poincelot, 1974). The losses of organic matter and nitrogen were calculated by a mass balance following Haug (1980). The OM decreased by ca. 24%. The OM content of the final product was 50%. The N loss accounted for 0.22 moles NH_3 kg^{-1} compost, which represents 15% of the total-N, this loss being similar to that reported by De Bertoldi et al.(1982).

The NO_3–N and NH_4–N evolutions in the deep samples through the composting process are shown in Figure 2. A high NO_3–N content of the mixture at the beginning of composting was observed. This could be attributed to an initial N mineralization of the susbtrates during their storage because of their high moisture contents in addition to the aeration produced in the conveyance. The NO_3–N content decreased from 1066 to 311 mg NO_3–N kg^{-1} in the thermophilic period and then, increased up to 697 mg NO_3–N kg^{-1} in the mesophilic stage. The NH_4–N content increased rapidly from 25 to 958 mg NH_4–N kg^{-1} after 10 days of composting. Afterwards, NH_4–N content tends to decrease reaching the original values after 180 days of composting.

The evolutions of C/N and CEC are given in Table I. Generally, composts having a C/N<20 are considered mature. However, a C/N<20 cannot be used as an indicator of maturity when a N–rich waste, is added to the compost mixture (Mathur et al., 1993). Nevertheless, the evolution of C/N and CEC, among others, have been used as an index to control the maturity of the final products (Harada et al., 1981). The CEC values increased steadily to reach a constant value of ca. 142 $cmol_c$ kg^{-1} in 4 months. Furthermore, a highly significant linear relationship between CEC and C/N was found: ln(CEC) =12.18 –2.93 ln (C/N), r^2=0.90 (P<0.001). Similar findings have been reported by Inbar et al. (1991) during grape marc composting. This relationship indicated that the CEC values could be used for estimating the degree of maturity of this compost.

The N (2.1%), P_2O_5 (0.70%) and K_2O (1.3%) contents of the compost were well above the minimum nutritional specifications described by Zucconi and De Bertoldi (1987). Sodium, Ca, Mg and micronutrient contents for the compost (data not shown) were within the range of similar products (Chen et al., 1988). Lixiviation of the macro and micronutrients throughout the composting process was too small to be considered important.

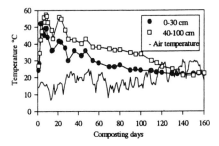

Figure 1 Temperature profile at two depths

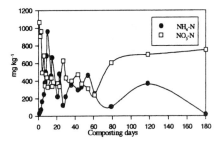

Figure 2 Evolution of NO3–N and NH4–N during composting

The compost showed a high EC value (11.5 dS m^{-1}) because of the high salinity of the vinasse. Nevertheless, this compost could be added to soils at low rates to avoid detrimental effect of its salinity on plant growth.

The co-composting process of beet vinasse and grape marc followed the classic pattern described in the literature. The compost maturity was achieved in four months. The final product, despite of its high salinity, has a high agriculture value because of its OM, N and K contents.

Tabla 1 Evolution of the maturity parameters

	C/N	C.E.C.* cmolc kg–1
Day		
1	15.4	66
32	14.6	74
52	13.1	83
79	12.5	124
120	11.5	142
190	11.9	146

*Ash-free material basis

Referencs

De Bertoldi, M., Vallini, G. and Pera, A. (1982). Comparison of three windrow compost systems. Biocycle, March/April, 45–50.

Chen, Y., Inbar, Y. and Hadar, Y. (1988). Composted agricultural wastes as potting media for ornamental plants. Soil Sci. 145, 298–303.

Haug, R.T. (1980). Compost engineering: principles and practice. Ann Arbor Science Publishers Inc. Lancaster. pp. 655.

Inbar, Y., Chen, Y. and Hadar, Y. (1991). Carbon-13 CPMAS NMR FTIR spectroscopic analysis of organic matter transformations during composting of solid wastes from wineries. Soil Sci. 152, 272–282.

Harada, Y., Inoko, A., Tadaki, M. and Izawa, T. (1981). Maturing process of city refuse compost during piling. Soil Sci. Plant Nutr. 27, 357–364.

Mathur, S.P., Owen, G., Dinel, H. and Schnitzer, M. (1993). Determination of compost biomaturity. I. Literature review. Biol. Agric. Hortic. 10, 65–85.

Pérez García, V. and Iglesias Jiménez, E. (1984). Compost a partir de R.S.U. de la Isla de Tenerife.

Características fisico-químicas. Proc. II. Congreso Nacional sobre recuperación Residuos-Tecnologías. Soria. pp. 973–995.

Poincelot, R.P. (1974). A scientific examination of the principles and practice of composting. Compost Sci. Summer, 24–31.

Zucconi, F. and De Bertoldi, M. (1987). Compost specifications for the production and characterization of compost from municipal waste. In M. De Bertoldi, M. Ferranti, P. L'Hermite and F. Zucconi. (ed.) 'Compost: production, quality and use'. Elsevier Applied Science. London, pp. 30–50.

Chemical Characterization of Three Compost of (Sugarbeet) Vinasse with other Agroindustrial Residues

DÍAZ, M.J.*, MADEJÓN, E.**, LÓPEZ, R.*, RON VAZ, M.D.* and CABRERA, F.*

Increasing amounts of liquid and solid wastes are produced by the food and agricultural industries in Andalusia. However, the recycling of these residues is not always billed and their elimination, is, at times, an environmental problem. Beet vinasse, a high density liquid waste from the sugar industry, contains high levels of OM (35%), N (3%) and K (3%), which make the vinasse a potential fertilizer. However, the direct application of concentrated vinasse on agricultural land may lead to economical and environmental problems because of high salinity (EC 250–300 dS m^{-1}), low P content (P2O5 0.012%) and its liquid dense character (1.3g cm^{-3}). The co-composting of vinasse with other agricultural residues could be used to overcome these disadvantages by producing a compost easily handled with higher P content and lower salinity.

Three composts were obtained from mixtures of vinasse (V) (2.5% N, 0.06% P_2O_5, 3.6% K_2O, 27% OM) with each of the three following agroindustrial residues: grape marc (1.4% N, 0.63% P_2O_5, 1.16% K_2O, 71.6% OM) (compost OC), olive pressed cake (1.03% N, 0.06% P_2O_5, 0.9% K_2O, 71.6% OM) (compost CC), and cotton gin trash (1.45% N, 0.35% P_2O_5, 2.53% K_2O, 67.8% OM) (compost GC). Sugarbeet factory lime (F) containing 50% $CaCO_3$ was added to the mixtures to increase pH. This was carried out to overcome the vinasse acidity (pH 4.7). The mixtures containing olive pressed cake and cotton gin trash were complemented with leonardite (L) (0.3% N, 0.04% P_2O_5, 0.16% K_2O, 48.7% OM), a low maturity lignite containing 25% (w/w) of humic acids. The proportion of V, GM, OC, CC, F and L for each mixture was as follows: Pile 1: GM (82%) 1 V (17%) 1 F (1%); Pile 2: OC (76%) 1 v (17%) 1 F (1%) 1 L (6%); Pile 3: CC (47%) 1 V (49%) 1 F (1%) 1 L (3%), where GM, OC, CC, F and L are expressed on a dry matter basis.

Co-composting was carried out in static piles, under cover, with forced aeration and in controlled conditions during three months. During the process, the piles were watered regularly to maintain moisture contents to 50% for pile 1,40% for pile 2 and 40% for pile 3. After this stage, the composts were left to mature during

*Instituto de Recursos Naturales y Agrobiología de Sevilla (CSIC). Apdo. 1052, 1080 Sevilla, Spain.
**EBRO AGRICOLAS S.A. Apdo. 9, 41300 S. José de la Rinconada, Sevilla, Spain.

the following four months. Samples were taken periodically at two depths (0–30 and 40–100 cm) in three zones of the pile. The following parameters were determined throughout the whole process: moisture content, pH_{H2O} (1:5 w:v), OM, EC_{H2O} (1:5 w:v), total–N, NH_4–N, NO_3–N, P_2O_5, K_2O, CEC, Na, Ca, Mg, Fe, Cu, Mn and Zn. The main chemical characteristics of the mixtures before and after composting are shown in Table I. After composting, the lipidic fraction composition of the composts was also determined to study the presence of anthropogenic compounds.

The maturity of the composts was completed after 90 days where the C/N and CEC reached a constant value. The evolution of these parameters, among others, have been used as an index to control the maturity of the final products (Harada *et al.*, 1981). For composts GC and CC, the C/N and CEC values after 210 days of composting were within the values of mature composts (Mathur et a/., 1993). This was not the case for compost OC which showed a high C/N ratio and a low CEC value. However, these two parameters did not show any change from day 90 onwards and therefore this compost could be considered mature (Harada et a/., 1981).

Table 1. Main chemical characteristics of the mixtures before and after composting. Pile 1=compost GC, Pile 2= compost OC and Pile 3= compost CC.

	Moist (%)	pH	OM (%)	C/N	N %	NH_4–N mg kg–1	NO_3–N mg kg–1	K_2O (%)	P_2O_5 (%)	CEC* cmol$_c$ kg⁻¹	EC dS m–1
Pile 1											
0 days	52	7.1	54	15	1.9	33.0	1066	1.5	0.69	66	11.8
210 days	35	8.3	50	12	2.1	17.0	749	1.3	0.70	146	11.5
Pile 2											
0 days	41	6.8	85	41.6	1.2	34.6	1104	0.8	0.07	53	12.1
210 days	32	8.5	70	34.0	1.0	90.0	174	0.9	0.13	79	12.7
Pile 3											
0 days	42	7.3	65	12	3.2	200	3039	1.9	0.20	72	13.9
210 days	24	8.1	51	8	2.6	534	683	2.1	0.28	137	29.3

*Ash-free material basis

The three composts were under the maximum water content (40%) permitted by EEC regulation. The pH values were stabilized above 8, being these values slightly higher than those reported as optimum for compost stabilization (Nogales and Gallardo-Lara, 1984). However, similar pH values have been reported for composts derived from similar agroindustrial residues (García-Izquierdo et a/., 1987).

For the three composts, the OM, total-N, NH_4–N, NO_3–N and K_2O contents were well above the minimum nutritional specifications described by Zucconi and De Bertoldi (1987). Similar results for other agroindustrial composted products have been reported elsewhere (Baca-García et a/., 1987). For pile 3, the largest pile, the ammonium accumulation at the end of the process was probably due to a lack of oxygen since the aeration system used was not powerful enough to provide higher oxygen concentration. Furthermore, the ammonium concentration in pile 3 was higher in the samples taken in the deepest zones (data not shown).

For the composts OC (Pile 2) and CC (Pile 3), the P2O5 contents were below the minimum required (0.5%) for agricultural use (Zucconi and De Bertoldi, 1987). Nevertheless, the nutritional status and yield of corn on soil fertilized with these composts have been compared with those for soils treated with a mineral fertilizer (Madejon et a/., 1994). These authors did not find any significant differences in the P contents in leaves for all treatments, where the amount of P2O5 added to the soils with the mineral fertilizer was ten- and five-fold higher than those added with the composts OC and CC, respectively.

The three composts showed a high EC value because of the high salinity of the vinasse. However, these composts could be added to soils at low rates to avoid detrimental effects for plant growth.

Sodium, Ca, Mg and micronutrients contents for the three composts (data not shown) were within the range of similar products (Chen et a/., 1988). Lixiviation of the macro and micronutrients throughout the composting process was too small to be considered important.

Analysis of the lipidic fraction for the three composts revealed the absence of anthropogenic compounds which could have been derived from pesticides added to the crops, and therefore they may have been present in the agroindustrial residues used for composting.

From the chemical characterization of these composts, it could be concluded that the co-composting of vinasse with grape marc (compost GC, pile 1) and with cotton gin trash (compost CC, pile 3) resulted in final products with a high value from an agricultural standpoint (rich in OM and mineral nutrients, low C/N ratio and high CEC values). However the compost OC, mixture of vinasse with olive pressed cake, showed a high C/N ratio and a low CEC value to be considered a suitable organic fertilizer. The addition of these composts as organic fertilizers has been tested in field experiments. Soils fertilized with composts GC and CC gave a better quality crop and higher yields than those treated with mineral fertilizers alone (Madejón et al., 1994).

References

Baca García, M.T., Esteban Velasco, E. and Sánchez Raya, A.J. (1987). Características fisico-químicas de diferentes composts obtenidos a partir de sub-productos agroindustriales. Proc. VII Congreso Nacional de Química (Química Agricola y Alimentaria II), Sevilla. Vol 2, 87–94.

Chen, Y., Inbar, Y. and Hadar, Y. (1988). Composted agricultural wastes as potting media for ornamental plants. Soil Sci. **145**, 298–303.

García Izquierdo, C., Costa Yagüe, F. and Hernandez Fernández, M.T. (1987). Evolución de parámetros químicos durante el proceso de compostaje. Proc. VII Congreso Nacional de Química (Química Agricola y Alimentaria II), Sevilla. Vol 2, 103–109.

Harada, Y., Inoko, A., Tadaki, M. and Izawa, T. (1981). Maturing process of city refuse compost during piling. Soil Sci. Plant. Nutr. **27**, 357–364.

Madejón, E., Díaz, M., López, R. Murillo, J.M. and Cabrera, F. (1995). Corn fertilization with three (sugarbeet) vinasse composts. Fresenius Environ. Bull. 4(4) (in press).

Mathur, S.P., Owen, G., Dinel, H. and Schnitzer, M. (1993). Determination of Compost Biomaturity. I. Literature Review. Biol. Agric. Hortic. **10**, 65–85.

Nogales, R., and Gallardo-Lara, F. (1984). Criterios para la determinación del grado de madurez de los composts de basura urbana. Proc. II Congreso Nacional sobre recuperación de Residuos-Tecnologías, Soria. pp. 941–951.

Zucconi, F. and De Bertoldi, M. (1987). Compost specifications for the production and characterization of compost from municipal waste. In M. De Bertoldi, M. Fetranti, P. L'Hermite and F. Zucconi. (ed.) 'Compost: production, quality and use'. Elsevier Applied Science. London, pp. 30–50.

Monitoring of Organic Matter During Composting

M. DOMFIZEL, N. VALENTIN, C. MASSIANI, P. LLOPART – Laboratoire de Chimie et Environnement, Université de Provence, 3 place Victor Hugo, 13331 Marseille Cedex 3 (France).

Introduction

The composting of town refuse is an interesting tool for waste management: compost is a major source of organic matter and fits completely into the framework of the recycling of organic waste. However, in order to be able to spread compost on different soils and to meet the various agricultural requirements, it is necessary to know perfectly the evolution of organic matter and the characteristics of the end-product. The composting process can be divided into two major stages: a rapid degradation of the easily degradable insoluble components into water-soluble compounds, and subsequently the metabolization by microorganisms of the latter and of the less degradable compounds. Many tests and criteria have been experimented for the long term monitoring of organic matter during the composting process as a whole whereas a less attention has been paid to monitoring merely the first stage of the process. Butyric, propionic and acetic acids are water-soluble intermediate compounds formed as organic matter are broken down. A better follow-up of the first stage of the process is necessary to avoid the presence of these phytotoxic compounds (DeVleeschauwer et al., 1981; Cocucci et al., 1989) in the end-product. Whereas volatile fatty acid (VFA) concentration is a parameter which has been widely used to monitor the anaerobic digestion (Azinari di San Marzano et al., 1981; Henson et al., 1986, Mata-Alvarez, 1990) of organic waste, it has rarely been used for the evaluation of the composting process evaluation (Saviozzi et al., 1992). In this work, it is proposed to use the VFA as the means to monitor organic matter during the first stage of composting while, for the characterization of the maturation level, a new index of maturity, based on rapid spectrophotometric measurements, has been proposed in previous works (Prudent et al., 1994, 1995).

Material and methods

The organic fraction of domestic waste underwent accelerated composting (air injection) in a laboratory scale pilot. Classic parameters such as pH, temperature, moisture and total volatile solids, were measured. The volatile fatty acid concentrations of water extracts were determined by gaseous chromatography. The chromatograph was equipped with a capillary column (WCOT) and a flame ionization detector.

Results and discussion

The initial material includes 66% water and its content in total volatile solids represents 70% of the dry weight (d.w.).

In the first experiment (E1) the beginning of the biological process takes place almost immediately. This is attested by a rapid decrease of the pH to 4.5, linked to an increase in temperature (Fig. 1 and 2, curves E1). After the acidic phase, pH increases and then, after 7 days, stabilizes at a (Fig. 1 and 2, curves E1). After the acidic phase, pH increases and then, after 7 days, stabilizes at a value of approximately 8.5. After about 10 days, the temperature reaches its maximum and levels (thermophilic phase, T = 65 °C). These phases correspond to the hydrolysis of the easily biodegradable compounds and to the maximum rate of volatile acid production (Fig. 3). The concentrations of acetic acid are greater than those of propionic and butyric acids.

Between days 10 and 15 a drop is observed in the temperature curve. This drop is generally attributed to a depletion of easily decomposable material and is an index of waste stabilization. However, observation of the volatile fatty acid curves (Fig. 3) shows that during the same period acetic, propionic and – more intensively – butyric acids are accumulated. A slight decrease of pH is also noticeable. These results are linked to the decrease, during the same period, of the water content (Fig.2) to a critical value (less than 50%). With the re-establishment of favourable conditions the temperature increases and the concentrations of volatile fatty acids again fall below detection limits. When stabilization is achieved, about 25 days after the beginning of the experiment, volatile fatty acids are undetectable and temperature returns to ambient level. As observed in anaerobic digestion, the transient accumulation of volatile fatty acids is a response to stress and to change in environmental conditions.

Another experiment (E2) is conducted with a high water content, around 75%. In these conditions pH remains acidic for a longer period and after a slight increase, temperature levels at 40 °C (Fig. 1 and 2, curves E2). During the same time, butyric acid accumulates while acetic acid concentrations decrease to very low values indicating that the biodegradation process is blocked (Fig. 4). Such a high water concentration is incompatible with the structure of the material. It is probable that water fills the interstitial spaces and the circulation of air is limited

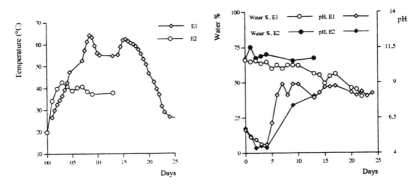

Figure 1 Evolution of temperature versus time.   **Figure 2** Evolution of PH and water percentage versus time.

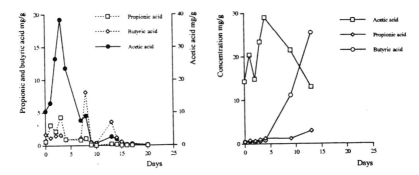

Figure 3 Evolution of VFA concentration versus   **Figure 4** Evolution of VFA concentlation versus
              time (E1).   time (E2).

by its diffusion in the aqueous phase which is a slow process. Local anaerobic conditions appear which slow the breakdown of intermediate degradation compounds down and favour their accumulation. However the persistence of unfavourable conditions can lead to harmful concentrations of volatile fatty acids, what is observed in this experiment.

Conclusion

Usually volatile fatty acid concentrations are measured to evaluate the phytotoxicity of the end-product. The results obtained in this work show that, as in anaerobic digestion, accumulation of these compounds, whether transient or not, is a response to an environmental stress. It is a more reliable parameter than those generally used and is well adapted to the monitoring of organic matter during the mesophilic and thermophilic phases of accelerated composting. It allows a rapid intervention for the re-establishment of favourable conditions.

References

Asinari di San Marzano C. M., Binot R., Bol T., Fripiat J. L., Hutschemakers J., Melchior J. L., Perez I., Naveau H., Nyns E. I. (1981). Volatile fatty acids, an important state parameter for the control of the reliability and the productivities of methane anaerobic digestions. Biomass, I, 47–59.

DeVleeschauwer D., Verdonck O., Van Assche P. (1981). Phytotoxicity of refuse compost. Biocycle, 22, 44–46.

Henson J.M., Bordeaux F.M., Richard C.J., Smith P.H. (1986). Quantitative influences of butyrate and propionate on thermophilic production of methane from biomass. Applied environmental microbiology, 51(2), 288–292.

Mata-Alvarez J., Cecchi F., Pavan P., Labres P. (1990). The performances of digesters treating the organic fraction of municipal solid wastes differently sorted. Biological wastes, 33, 181–199.

Prudent P., Massiani C., Thomas O. (1994). Fast determination of a maturation index of compost and soil organic matter. The Pittsburgh conference, Pittcon '94

Prudent P., Domeizel M., Massiani C., Thomas O. (1995). Gel chromatography separation and U.V. spectroscopic characterization of humic-like substances in urban composts. The Science of the Total Environment, in press.

Saviozzi A., Levi-Minzi R., Riffaldi R, Benetti A. (1992). Evaluating compost garbage. Biocycle, 33, 72–75.

Effects of Bulking Agents in Composting of Pig Slurries

J. Domínguez, C. Elvira, L. Sampedro, M. García and S. Mato Dpto. Recursos Naturais e Medio Ambiente. Universidade de Vigo. Apt. 874. E-36200. Vigo, Spain

The large amounts of slurry wastes produced in intensive farms cause serious disposal and pollution problems. Composting of slurries from animal waste offers an answer to odour nuisance and water pollution associated with their management. The aim of this research was to carry out a preliminar laboratory study trying to transform slurries into compost so they can be used as horticultural substrates, organic soil conditioners or fertilizers.

Experimental

The experiments were performed in 1 l. volume vessels (dewar glass) thermically isolated (fig. 1). The vessel lid was equipped with an air inlet and outlet, and a hole for inserting 1 J type thermocouple and there is a false perforated floor to foster the air difussion at the botton. Compressed air was supplied to the system at a constant pressure in ventilation cycles of 75 sec. every 15 min. Air was scrubbed of CO_2 by means a 200 ml. NaOH 1N trap.

Temperature was monitored via thermocouples type J placed at the compost matrix center and registered continously in a hibrid register SEKONIC SD-100 M.

Carbon dioxide evolution was measured as an indicator of microbial respiration (Anderson, 1982) using 200 ml. NaOH 1N.

The selfheating efficiency of pig slurry mixed with different agricultural and forest wastes (straw, pine bark, pine needles, fern and oak leaves) at two different ratios (2:1 and 5:1 w:w) has been tested. These materials behave as bulking agents reducing the slurry moisture content, supplying C for the C/N ratio improvement and avoiding N losses.

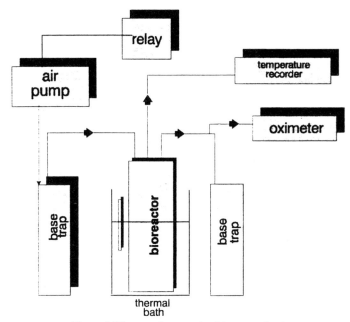

Figure 1 Diagram of composting laboratory simulator

Table 1 Some chemical characteristics of the pig slurry

Moisture (%)	81,45
Total Solids (%)	18,55
Ashes (%)	71,68
pH (H2O)	8,7
Conductivity (mS.cm-1)	4,4
Total Nitrogen (g.kg–1)	14,81
NH$_4$ (g.kg–1)	39,29

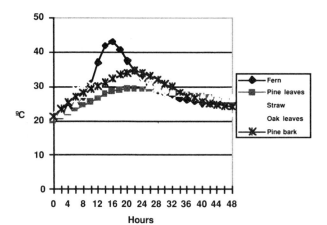

Figure 2 Temperature evolution of pig slurry mixed with different agricultural and forest wastes
(2:1 w:w)

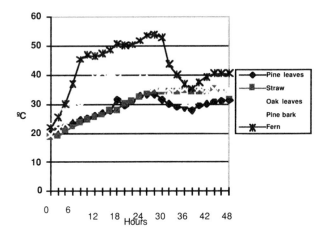

Figure 3 Temperature evolution of pig slurry mixed with different agricultural and forest wastes (5:1 w:w)

Table 2 Initial and final chemical characteristics of pig slurry mixed with different agricultural and forest wastes (2:1 w:w)

2:1	MOISTURE %		pH 1:10		NH4 g.kg–1		Cond. mS.cm–1		CO$_2$ %	NH$_4$ %
	initial	final	initial	final	initial	final	initial	final	final	final
pine needles	54,96	54,85	8,40	8,50	8,47	7,21	0,93	0,66	3,57	3,57
oak leaves	65,4	58,30	8,30	7,73	9,10	7,93	0,90	0,53	4,92	4,92
pine bark	60,72	62,82	8,40	8,21	9,34	11,02	0,88	0,70	3,43	3,43
fern	57,14	58,21	7,90	8,50	7,90	10,61	1,90	1,45	9,42	9,42
straw	57,02	58,49	8,20	8,60	7,33	7,00	1,62	1,56	5,41	5,41

Table 3 Initial and final chemical characteristics of pig slurry mixed with different agricultural and forest wastes (5:1 w:w)

5:1	MOISTURE %		pH 1:10		NH4 g.kg–1		Cond. mS.cm–1		CO$_2$ %	NH$_4$ %
	initial	final	initial	final	initial	final	initial	final	final	final
pine needles	71,11	70,21	9	8,85	8,47	7,21	0,93	0,66	3,57	3,57
oak leaves	68,69	68,24	8,78	8,66	9,10	7,93	0,90	0,53	4,92	4,92
pine bark	70,42	70,90	8,9	9,01	9,34	11,02	0,88	0,70	3,43	3,43
fern	68,87	71,99	8,9	9,15	7,90	10,61	1,90	1,45	9,42	9,42
straw	69,55	71,20	69,55	71,20	7,33	7,00	1,62	1,56	5,41	5,41

In spite of in 2:1(w:w) mixtures the moisture contents are the most appropriate for composting processes, the fermentative activity does not reach the thermophilic phase in none mixture, probably due to the thermic exigencies of the microreactor.

Nevertheless, increasing the amount of slurry (mixtures 5:1 (w:w)), in spite of that the moisture contents exceed the 70%, the thermophilic phase was reached in the microreactor in three mixtures: fern, straw, and oak leaves.

This let us to propose that for composting of pig slurry it is possible save large amounts of bulking material, because composting works in spite of the moisture surpasses the limits advised by Haug (1985). The different bulking materials improve the composting process in function of their major or less facility to be degraded, and then, the best results have been obtained in the mixtures with fern, straw and oak leaves. This preliminar study offers an important information for batch scale experiments in which we are involved actually.

References

Anderson, J.P.E. (1982) Soil respiration. In Methods of Soil Analysis. Part 2 (R.H. Miller&D.R. Keeney, eds.). Madison, Wisconsin, USA: American Society of Agronomy, Inc., pp. 830–871.
Haug, R.T. (1980) Compost Engineering. Principles and Practice. Ann. Arbor Science Publishers Inc. 655 pp.

Sewage Sludge Composting with Thermpopostage ™ Process (The Platform of Arenthon *Haute Savoie, France*)

(MICHEL DURILLON*, CHRISTINE FONTAINE** , HUGUES BAZAN** , CLAUDE PREVOT***)

Landfillings restrictions and more stringent regulations on product spread on the fields oblige sludges and wastes producers to improve practices and quality.

In that context, composting process appears to be a good way for sludge and waste disposal.

The aim of that paper is not the description of the biological process, applied for a long time for waste treatment: existing composting places designed with windrow system are in operation on large area with or without forced aeration .Depending on waste characteristics and parameter control, trouble like bad smells, polluted leakages could occur during active and aging phase of the process;

Few years ago, appeared new composting concepts based on the total process control .

THERMOPOSTAGE ™ is one of these new concepts, under 3 years operation for sludge treatment, on the Arenthon waste water plant, near Annecy (Haute-Savoie, France).

THERMOPOSTAGE ™ process is based on the use of mobile reactors, making the system modular and easily adapted to the amount of sludges or wastes to be processed : this approach guarantees flexible operation and a clean site.

Description of the composting unit

In a previous workshop, the dewatered sludge, issued from a belt press, is mixed with sawdust, pine barks or recycled compost, providing an optimal porosity necessary for this forced and controled aeration process.

Mixing operation is the most important one on a composting unit : sludge to carbonaceous elements ratio is adapted to sludge dryness and rheological characteristic. In Arenthon place, the volumic ratio sludge/sawdust/recycled compost or

* DEGREMONT EXPLOITATION , 24 avenue René Cassin BP 9131 69263 LYON FRANCE Tel : 78 64 88 38

** DEGREMONT , Sludges and Wastes Division, 183 avenue du 18 Juin 1940 92500 RUEIL MAL-MAISON FRANCE Tel : 1 46 25 60 00

*** DEGREMONT , R§D Division, 87 Chemin de Ronde 78290 CROISSY / Seine FRANCE Tel : 1 34 80 48 08

barks is on average 1/1/1. Doubled shaft with paddles mixing systems are amongst the best devices to obtain suitable porosity and avoid any anaerobic fermentation in the reactor.

Then, the mix is loaded in the biological reactors which are transported to the aeration site by a lorry.These 30 m^3 reactors, are equipped with an air distribution system at the bottom, included a leachate collector .Each reactor is well closed to confine any odorous compounds .

During 3 weeks, sludge mixed with co-elements is aerated under succion; air flow-rate is regulated to maintain an inside temperature level nearly by 70°. In aging and drying period, reactors are continuously blown during one or two weeks. Centralized succion and aeration pipes, equipped with automatic valves, are connected on each reactor; duration and frequency of succion or blowing period can be adapted on an hourly basis .

Compost is then screened – if barks elements have been used – and stored for further aging or sold for agricultural reuse and as soil remediation product.

Mixing, loading, and screening operations are managed by an automate so the presence of operators on the composting plant is limited.

Design parameters

The waste water plant of Arenthon received a 35000 population equivalent effluent with a high food industry ratio.

The composting site of Arenthon located on the same plant, is now designed to treat 2500 m^3/ year of dewatered anaerobic digested sludges with a dryness comprised between 23 to 26 % .

For this capacity, 15 composting biological reactors are needed ; This modular process will permit to double the capacity just by adding 15 others reactors .

2500 m3 of dewatered sludges will produce from 3500 to 4000 m3 of compost with a dryness between 40 to 50%, odour – free, stable and hygienized ; this compost reaches the french regulations on heavy metals for agricultural reuse.

New developments

THERMOPOSTAGE ™ process is well adapted for others organic wastes treatments such as greenwastes and sorting household wastes. Additional water can automatically be spread on mixture inside reactors to sufficient humidity level for optimal micro-organisms development. Each reactor is easily connected to a biological or chemical odour treatment if necessary. Tests are carrying on to optimize mixing operation and ratios between wastes and co-elements ; thus we propose now a well experienced technology for sludge and organic wastes stabilization and ecological disposal.

Biodegradation of 'Thiram' in Composting Process

A. FANTONI*, P. MURARO*, C. PICCO**,
G. ZORZI***

Foreword

Composting is proposed as a solution capable of meeting the twofold requirement of giving a correct solution to the disposing of waste biomasses and waste of different origins, pursuing recovery and valorization standards,while producing plenty of cheap organic fertilizer.

The composting treatement of agrofood residues is an interesting means of recovering to agronomic ends such waste whose landfill disposal is becoming more and more expensive and whose improper management can cause enviromental pollution.

In the recent past years the main chemical firms dealing with cereal seeds for agricolture have had to undertake heavy expenses to solve the problem of disposing of left-over materials.

Within this context 'Eco-Pol Spa' has worked out a test programme in order to check whether the chemically treated matter is fit for composting, judging both the specific degradation of the fungicidal employed and the agronomic quality of the resulting product (compost).

Materials and Methods

The test programme has employed maize seeds ('Zea mays') furnished by 'Pioneer Hi-Bred Italia Spa'. They have been treated with THIRAM (tetramethylthiuramdisulphide) dose: 15 gr/ql expressed on 48% a.p. : dithiocarbammate generally used in the treatement of seeds, bulbs and tubers to protect them against Tilletia caries and Ustilago maydis by performing a contact action on the treated surface (Engst – Schnaak 1974).

The degradation system chosen by 'Eco-Pol Spa' is composed by an accellerated biooxidation phase (20 days) in horizontal reactor equipped with an overhead

* Eco – Pol Spa, Bagnolo Mella (Bs)
** Centro Ricerche Chimiche, Montichiari (Bs)
*** Istituto Agrario S.Michele all'Adige , (Tn)

– travelling crane fitted with two conveyors. The mass front is 80 mt long and 10 mt wide. Its maximum height is about 1,8 mt. Odours, that may be produced, are properly treated with an ozone emission control system. After the thermophile phase, the biomass is arranged in windrows having triangular cross section (4,4 mt wide x 1,9 mt heigh) on outdoors covered curing areas for a period of about 50 days. A turning machine equipped with two horizontal rotors keeps the masses revolving thus granting constant layer – turning and proper gas – turnover in every point.

The programme settled on the composition of the following mixture.

- 60% maize seeds ('Zea mays') treated with THIRAM;
- 20% bilogical sludge
- 20% lignocellulosic residues ⟶ pre-fermented mixture

Wooden material has been properly shredded to 15–20 cm pieces.

Temperature, moisture, ph, a.p. concentration and phytotoxicity have been checked by monitoring the whole process.

The biooxidation treatement included a set of mechanical turnings planned as follows: during the thermophile phase every 3 days till about the 20th and then every 5–7–10–12–14–15 days until the end of the process.

Sampling consisted in the taking of 10 sub-samples from the centre of the mass and along the diagonal.

Analytical determination of dithiocarbammate in maize seeds

The method that has been employed is provided for in M.U. n.118, vol. I°.

It bases on the hydrolysis of dithiocarbammate in an acid environment: it develops carbon sulfide that is fixed with Cullen's reagent (copper acetoacetate and diethanolamine) after sodium carbonate and benzene wash.

The quantitative determination is carried out by means of spectrophotometrical measurement at 435 nm of the resulting coloured compound.

The calibration standards have been predetermined adding to untreated seeds known quantities of Thiram which have subsequently been submitted to the same treatement of the samples.

The conversion factor, employed to transform the result in ppm of Thiram , is 1,57.

Results

The check of data (fig. 1) underlines interesting results as to Thiram biodegradetion in the composting process. The temperature profile recorded during the process shows the thermophile phase followed by a mesophile one one marked by a specific microbic microflora.

After about 20 days from the beginning of the process the concentration of Thiram decreases passing from 28,8 ppm to about 3 ppm; then it decreases slowlier according to the temperature.

From table n.1 can be inferred an increase in the concentration of the active principle in the first days of the process, increase wich was certainly due to the sampling of a heterogeneous mixture (seeds,biologic sludge,lignocellulosic residues). The presence, in the mizture, of a fungicidal such as Thiram did not affect the progress of the process, ruled by different microbic groups and finally solved in a more than average lapse of time (about 75 days).

The matter has undergone phitotoxicity tests to check the agronomic suitability and the results are completely favourable. The analytical characteristics of the compost obtained from treated material underline the full stability of the organic component, low content of pollulants, balanced C/N ratio, low salinity.

Conclusions

The compete analysis of the obtained data allows the following remarks:

- maize seeds from Thiram fungicidal treatement can be profitably composted, thus removing them from the amount of waste meant for landfill incinerator;
- Thiram biodegradetion occurs only during the thermophile phase, in the first days of the treatement; the above mentioned matter is thus allowed to be considered fit for composting as any other agrofood waste;
- the quality of the compost obtained from the biooxidative transformation of the test matter is considered good both for itys agronomic properties and for its degree of environmental security:

THIRAM

TETRAMETHYLTHIURAMDISULPHIDE

Bibliography

Regione Piemonte 'Metodo di analisi dei compost' – Collana Ambiente – 1992
P.Muraro – G. Zorzi 'Gestione tecnica della trasformazione di rifiuti urbani selezionati ed agro -industriali in compost di qualità' – Simposio 'Ingegneria della trasformazione del compost' – Bari 1994.

Figura 1   BIODEGRADATION THIRAM

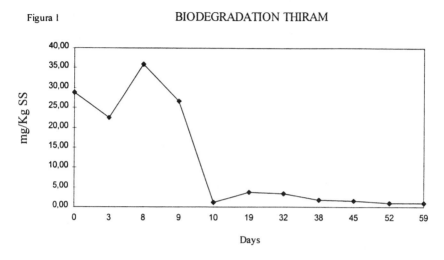

Figura 2   TEMPERATURE VARIATIONS DURING THE PERIOD

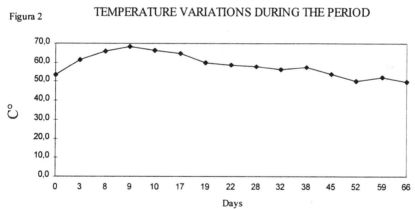

ANALYTICAL CHARACTERISTICS OF COMPOSTCOMPARED TO OTHER ORGANIC
CONDITIONERS

	COMPOST	MANURE	PEAT
MOISTURE (%)	42,7	72,3	61,6
pH	6,4	8,38	5,8
EC (µS/cm)	500	2424	630
ORGANIC CARBON (%)	26,48	36,7	41,9
ORGANIC MATTER (%)	45,65	63,2	72,2
C/N	13,1	18,8	44,1
N (%)	2,02	1,93	0,95
P2O5 (%)	1,7	1,73	0,24
K2O (%)	1,4	1,78	0,18
Cu (mg/Kg)	13	54	19
Pb (mg/Kg)	9,9	32	9
Col (mg/Kg)	< 1	< 1	< 1
Zn (mg/Kg)	45	245	47
Ni (mg/Kg)	4,8	10	4
Cr (mg/Kg)	6,6	30	42
Germination index (%)	85	63	88

(ALL VALUES ARE EXPRESSED ON A DRY MATTER BASIS WITH EXCEPTION OF
MOISTURE – PH AND EC)

First Experiments of Compost Suppressiveness to Some Phytopathogens

FERRARA A.M., AVATANEO M., NAPPI P. – Istituto per le Piante da Legno e l'Ambiente I.P.L.A. Torino

Compost can be used in floriculture and in horticulture within integrated and biological disease control programs (1–2–4). It is in fact well known that vegetables and ornamental plants are often damaged by pathogenous fungi causing damping off and root rot.

Thanks to studies on waste recovery and recycling which I.P.L.A. started long ago, some composts were examined in order to verify if compost utilization in agriculture and floriculture could have a useful repressive function.

Two lines of research were followed:

1. cultivation trials in pot, under temperature and moisture controlled conditions, in order to evaluate compost repressive action against: Phytium ultimum on cucumber and beet; Rhizoctonia solani on bean and basil; Fusarium oxisporum f sp. basilicum on basil.
2. compost microbiological evaluation by isolating microorganisms and testing their antagonistic capacity by means of in vitro and in vivo (pot) experiments.

Materials and methods

In pot experiments: the following composts were tested:
Compost A = from municipal sewage sludges and poplar bark
Compost B = from industrial sludges and poplar bark
Compost C = from poplar bark
Compost D = from the organic fraction of MSW.

P. ultimum and R. solani inocula were obtained by growing their mycelium on a sterile medium made of corn and hemp seeds; to breed F. oxysporum f sp. basilicum, conidia-in- casein hydrolizate suspensions were prepared, which were then dispersed in talc and let to dry. Composts were mixed to a soil conducive to pathogens in percentages of 25% and 50%; 100% of the same soil was used as control. Test plant sowing was carried out soon after pot inoculation. The trials were organised on the basis of 2 tests for each compost, and each test was repeated three times. Disease severity was checked by noting down number of emerged plants, and by later subdividing them into healthy and dead plants.

The second line of research aimed at isolating some compost microorganisms such as *Trichoderma* spp. strains by successive dilutions (up to l0210) and using selective substrates. The suppressive activity of *Trichoderma* spp. strains (1761, 1769, 1780), isolated respectively from composts A, B and C, was tested in vivo by means of pot trials. For this purpose a sterile substrate conducive to pathogens was infested using *Trichoderma* spp. mycelium, and P. ultimum inoculum. Cucumber plantlets, used as host plants, were seeded soon after infesting the substrate with the pathogen and the antagonist.

Results

Table 1 shows the results of plant disease suppression bioassays carried out in pots for some phytopahogens. In trials on cucumber grown on a soil-compost mix infested by P. u*lt*imum, percentages of emerged plantlets are not very dissimilar in the test pots and in the control. In the case of green beet with P. *ultimu*m, the number of emerged plantlets was greater on compost-soil mixes (especially composts A and B), than in the control. Post-emergence damages were also reduced with the exception of C25.

All tested composts proved to repress damages due to F. *oxysporum* f.spp. basilicum on basil before and after plant emergence. On basil again a reduction of damages due to R. *solani* was also observed with composts A25, B50, C25 and D50 before plant emergence; with composts B50, C25 and D50 after plant emergence. On bean symptoms of R. *solani* disease were limited in all trials. Suppressiveness was stronger after plant emergence especially with compost C25. P. ultimum damages to *Impatiens* were restrained. Suppressiveness was rather intense with all composts at all percentages with the exception of D25.

As for the second line of research see table 2. One can notice that the use of Tric*h*oderma spp. strains isolated from composts A(1769), B(1780) and C(1761) limits damages due to P. u*lt*imum on cucumber before emergence. The most active strain is n° 1761.

Conclusions

The tested composts generally showed a good disease control in all trials with pathogen inoculation. The best results could be obtained with the most mature composts, it is thought in fact that during the maturation process substances such as antibiotics are produced inhibiting phytopathogen development (3).

All composts with the exception of D proved suppressive to F. *oxysporu*m on basil. Trials *in vivo* using P. u*lt*imum were satisfying both for green beet, for cucumber and Im*pati*ens. Damages due to R solani could be limited in trials where basil and bean had been seeded especially with composts B and C.

Encouraging results were attained in *in vivo* trials as to the antagonistic action of

Table 1 Compost suppressiveness to some phytopathogens

COMPOST DOSES	Cucumber		Phytium ultimum Green beet		Impatiens		Fusarium oxysporum Basil		Rhizoctonia solani Basil		Bean	
	emerged plants %	healthy plants/ emerged plants %	emerged plants %	healthy plants/ emerged plants %	emerged plants %	healthy plants/ emerged plants %	emerged plants %	healthy plants/ emerged plants %	emerged plants %	healthy plants/ emerged plants %	emerged plants %	healthy plants/ emerged plants %
A 25	77 a	52 abc	11 abcd	9 abcd	78 e	76 e	83 de	82 f	59 cde	46 abc	73 a	40 ab
A 50	100 d	93 c	14 de	12 cd	70 de	66 cd	81 cde	75 def	44 abcd	39 ab	77 a	76 de
B 25	93 bcd	72 bc	12 cd	8 abc	70 de	66 cd	79 cde	77 e	37 a	33 a	77 a	78 de
B 50	97 cd	97 c	11 abcd	7 abc	78 e	74 e	74 bcde	60 cde	52 abcd	44 abc	77 a	44 ab
C 25	87 bcd	85 bc	7 abc	4 a	68 cde	64 cd	75 bcde	72 def	62 de	60 cd	67 a	98 e
C 50	87 bcd	85 bc	6 a	5 ab	68 cde	64 cd	71 bcde	57 cd	38 ab	34 a	83 a	52 bcd
D 25	83 bcd	80 c	10 abcd	7 ab	22 a	10 a	65 abc	58 cd	39 ab	35 ab	77 a	64 ab
D 50	87 bcd	73 bc	7 ab	5 ab	74 e	7-2 e	67 abcd	52 bc	56 abcd	53 abcd	67 a	32 a
Control	80 abcd	4 a	7 abc	4 a	48 bcd	24 ab	55 a	27 a	42 abc	36 ab	63 a	26 a

Means followed by the same letters are not significantly different (p , 0,05).

some *Trichoderma* spp. strains isolated from composts. Strain n° 1761 suppressiveness to plant disease further confirmed the results of in vitro tests.

The research was carried out with the cooperation of DI.VA.P.R.A. – Phytopathology Institute of the Faculty of Agronomy of Turin University – and with the financial contribution of Piedmont Region Govemement.

Table 2 Suppressiveness of *Trichodermia* spp. strains n° 1780, 1769 and 1761 to P. ultimum in pot trials with cucumber

Strains	Emerged plants %	Healthy plants/Emerged plants %
1780	63 b	57 bc
1769	67 b	37 ab
1761	70 b	70 cd
Control	37 a	13 a

Literature

Garibaldi A. and M.L. Gullino, 1989 – La lotta biologica contro i funghi fitopatogeni. Inf. fitopatol. 10: 9–17.

Hoitink H.A.J. and P.C. Fahy, 1986 – Basis for the control of soilborne plant pathogens with compost. Ann. Rev. Phytopathol. 24: 93–114.

Hoitink H.A.J. and M.E. Grebus, 1994 – Status of biological control of plant diseases with composts. Compost Science & Utilization, Spring 6–12.

Ozores-Hampton M., H. Bryan and R. McMilan 1994 – Suppressing disease in field crop. Biocycle, July 60–61.

Reduction in Phytotoxicity of Olive Mill Waste by Incubation with Compost and its Influence on the Soil-plant System

1Franco, I.; 1Baca, MT; 1Hurtarte, M.; 2 Leita, L.; 2De Nobili, M. and 1Gallardo-Lara, F.

Key words: Alpechín, organic matter, phytotoxicity, compost.

Summary

Olive mill waste waste (alpechín) is a liquid agroindustrial residue produced during the extraction of olive oil. Disposal of waste water is a problem because of its phytotoxicity and difficulty of handling. We tried to eliminate the phytotoxicity of alpechín by absorbing it on composted organic matter and posterior incubation. When this waste water was absorbed on compost, a great part of its phytotoxicity was chemically neutralized, and also had positive effects on plant growth. These effects increased with time of incubation of the alpechín/compost mixtures.

Introduction

Disposal of agricultural wastes in order to improve the use of existing resources and to reduce environmental risk, has become a critical concern in many areas of the world (Janer del Valle, 1980). Composting satisfies the health and aesthetic aspects of waste disposal by destroing pathogens. In addition, the product is agriculturally beneficial as a soil conditioner and fertilizer (Baca et al., 1992)..

In Spain, where about 30% of world´s olive oil is produced . Its liquid residue: alpechín (olive mill waste) represents one of the most important polluting residues of the mediterranean countries. However, it is rich in mineral nutrients and could be a perfect candidate to be used as a fertilizer (Gallardo-Lara & Pérez, 1990). The addition of this waste water directly to soil has harmful effect on plant development by the presence of polyphenolic substances, which have also a high biotoxicity (Martínez Nieto et al., 1993).

Soil biotic activity is the driving force in the degradation and conversion of

1 Estación Experimental del Zaidín, CSIC. P.O. Box 419, E-18080 Granada, Spain.
2 Ist. Prod. Veg. Tec. Agr.; Fac. di Agraria, Univers. Udine, I-33100 Udine, Italy.

exogenous materials, transformations of organic matter and evolution of soil structure. This activity plays a primary function in nutrient cycling and supports plant life (Dick, 1992). The aim of this study was to try to neutralize the biotoxicity of alpechín by absorbing with a composted organic matter followed by subsequent incubation periods with and without parallel plant growth.

Materials and Methods

The soil (S) in our experiment was a silty loamy plough layer; with an organic matter content of 3.5%. The organic materials used were: Alpechín (A) with 2% of dry matter, compost of cotton waste (C) with 72% of volatil solids. The treatments were: Soil (C), soil-compost-alpechín (SCA), and soil-alpechín (SA). The proportions of mixtures utilized were: 80:15 (v:v) soil-alpechín with or without 15:5 (v:v) composted cotton waste. Data were obtained at the begining of the experiment and after 12 days (equilibration) and after 50 days of incubation with and without parallel plant growth; the incubation with parallel plant growth was performed in 100 ml plastic pots where seeds of Lactuca sativa L. were planted. The experimental design was: 1 plant/pot and 5 repetitions by treatment. The assay was performed in a greenhouse under controlled conditions of light, temperature and humidity.

Total extractable organic carbon (TEC) and the humification index were determined following the methodology of Sequi et al., 1986. Biomass content was determined by the F-E method proposed by Brookes et al., (1985), and carbon and nitrogen contents following the methods proposed by Vance et al., (1987) and Joergensen and Brookes, (1992), respectively.

Results and Discussion

Plant development was practically blocked in the treatment incubated with alpechín (Table I). The presence of this waste water produced a strong inhibition of plant growth. Nevertheless, in the SCA treatment, plants grew better than the controls, probably because this substrate not only neutralized the alpechín phytotoxicity but also acted as a fertilizer.

Table 1 Dry weight of shoots of plants grown for 50 days.

	Control	SA	SCA
Dry weight	0,428	0,031	1,206

Total organic carbon (TEC) extracted in alkaline solution remained constant throughout the experiment in the control and in SCA (Fig. 1A). The adition of the C+A mixture increased the TEC considerably but this increase remained constant as index of stability in the new equilibrium created. Most of the alpechín was probably neutralized chemically from the beginning and the biotoxic activity reduced as can be see by the biomass content in fig. 1B. This biomass was probably reduced initially but fastly recovered at the end of the process. The treatment where the alpechín was incorporated to the soil presented a initial high TEC, which decreased after 12 days considerably, probably by different ways: volatilization, recombination with SOM or biological activity. The soil biomass was strongly reduced initially when the waste water was added and slowly recovered up to the values found in the soil during the studied period.

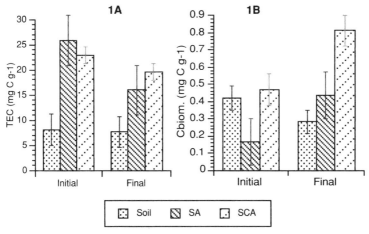

Fig. 1A Total organic carbon (TEC) extracted in alkaline solution at the beginning and the final phase in the three treatments. Bars indicate standart error.
Fig. 1B Biomass carbon content determined by the F-E method at the beginning and the final phase in the three tratments. Bars indicate standart error.

At the end of the incubation period, changes observed in TEC in the treatments, with soil (S) and soil suplemented of alpechín (SA), were independent from the fact that the incubation was carried out with or without plants (fig. 2A). The SCA treatment did not change its TEC when the incubation was developed without plant growth from the beginning (equilibration value) but the presence of plants caused significative decrease of the TEC content. Parallely, biomass did not change in any treatment after the incubation without plant (fig. 2B). Nevertheless, the presence of the plant roots resulted in an increase of biomass of three times in the soil (S) and in the SCA treatments. Root exudates sastified nutrient requirements for biomass development increasing its content in control treatment and its activity in SCA treatment to degrade available organic matter as can be see in the TEC decrease found in SCA (fig. 2A).

Humification index

The index indicated the state of equilibrium of the soil organic matter when adding into soil. In S and SCA treatments, this index did not change appreciably during the experiment indicating that the original organic matter in the soil and that additioned as C+A were in adequate state of transformation. Nevertheless, in the SA treatment, this index changed appreciably through the equilibration and posterior incubation period passing from values like 1.6 to values near to 0.2, lower of that founded in soil or SCA treatments indicating that the organic matter of this treatment was not yet stabilized (Table II). The decrease in the humification index at the end of the experiment in SA treatment, when plants where grown in the mixture, is the consequence of the high polyphenolic content of alpechín and of its phitotoxicity, which did not allow the increase in HI consequent to plant growth. An intermediate effect was observed in SCA as could be reasonably expected.

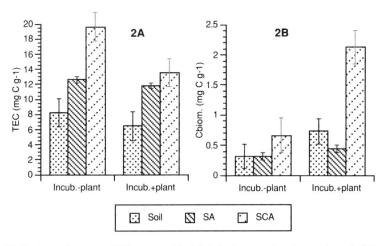

Fig. 2A Total organic carbon (TEC) extracted in 0.1M alkaline pyrophosphate at the end of the incubation period of 50 days with and without plant growth. Bars indicate standart error.
Fig. 2B Biomass carbon content determined by F-E method at the end of the incubation period of 50 days with and without plant growth. Bars indicate standart error.

Table II. Humification index (HI) in the treatments with and whitout plant growth.

Treatments	Control	SA	SCA
Initial	0.5	1.6	0.8
Final	0.5	0.6	0.7
HI (without plant)	0.5	0.6	0.6
HI (with plant)	0.58	0.25	0.54

Conclusions

The toxicity of alpechín was neutralized when this waste was added to soil mixed with composted organic matter.

Plant growth was better on soil amended with compost plus alpechín showing that this substrate can act as an organic fertilizer.

Soil microbial biomass was negatively influenced when alpechín was added directly to soil.

Acknowledgements

The authors express their appreciation to the C.S.I.C. which provided funding for the present study. This work was supported by the project N0 NAT 90–0823

References

Baca, M.T.; Fornasier, F. and de Nobili, M. 1992. J. Ferment.Bioeng. 74(3): 179–184.

Brookes, P.C.; Landman, A; Pruden, G and Jenkinson, D.S. (1985). Soil Biol. Biochem. 17, 837–842.

Dick, R.P. 1992. Agriculture, Ecosystems and Environment, 40, 25–36.

Gallardo-Lara, F. & Pérez, J.D. (1990). Environ. Sci. Health, 25 (3), 379–394.

Janer del Valle, L. (1980). Grasas y Aceites, 31 (4), 273–279.

Joergensen, R.G. and Brookes, P.C. (1992). Soil Biol. Biochem. 22.

Martínez Nieto, L.; Garrido Hoyos, S.E.; Camacho Rubio, F.; García Pareja, M.P. y Ramos Cormenzana, A. 1993. Biores. Technol.43, 215–219.

Sequi, de Nobili, M.; Leita, L. and Cercignani, G. (1986). Agrochimica Vol XXX 1–2.

Vance, E.D.; Brookes, P.C. and Jenkinson, D.S. (1987). Soil Biol. Biochem. 19, 703–707.

An Assessment of the Agronomic Value of Co-composted MSW and Sewage Sludge

MICK FULLER[1], ROB PARKINSON[1], SAM JURY[1&2], GEORGE VANTARAKIS[1] and ANDY GROENHOF[2]

Introduction

Modern crop production in Europe does not have sufficient farm animal waste by-products to maintain crop yields at current levels and as a consequence relies heavily on the use of cheap easily available mineral fertilizers. A deterioration in soil organic matter and soil structure has been a consequence of intensive crop production and is especially evident with intensive cereal production. In the UK the area of cereal production has reduced slightly in the last 2 years due to CAP Reform measures and the introduction of set-aside policy but there has been a marked increase in the area of arable fodder crops grown particularly forage maize. Traditionally forage maize is grown in the UK without the use of much fertilizer but with the use of large quantities of farm animal waste particularly cow slurry. However, the expansion of the area of forage maize now exceeds cow slurry availability.

Agriculture offers one of the largest disposal routes for biodegradable human waste streams provided that attention is paid to toxicity factors in the waste streams. Properly composted materials offer an alternative fertilizer source to crop growers and a material that has the capability to improve soil structure and stability with repeated usage.

This paper reports the results of two years of field experimentation using forage maize to assess the value of co-composted Municipal Solid Waste (MSW) and Sewage Sludge.

[1]. Seale-Hayne Faculty of Agriculture, Food and Land Use, University of Plymouth, Newton Abbot, Devon, TQ12 6NQ, UK.
[2]. Ecological Sciences Ltd, Wolfson Laboratory, Higher Hoopern Lane, Exeter, Devon, EX14 8YU, UK.

Materials and Methods

Compost

The compost used was supplied by Ecological Sciences Ltd and was produced by composting a 50:50 mixture (by weight) of non-separated MSW and sewage sludge (Groenhof *et al* 1995). The compost was manufactured at Chelston Meadow COWS project site in Plymouth, Devon in Autumn/Spring 1992/3. The material destined for the second year of trials was stored on a concrete pad without cover until use in 1994.

Analysis of the compost revealed sub-toxic levels of heavy metals and high levels of available nutrients especially nitrates and potash (See Table 2 Parkinson *et al* 1995)

Field Trials

The effect of applying compost to field trials of forage maize was assessed. The trial was repeated in both 1993 and 1994 using fresh land for the 1994 trial. In addition the residual effects of the 1993 treatments were assessed by growing maize on the same plots during 1994 but without additional applications of compost or fertilizer (this trial is designated as 1994R).

A randomised block field trial design was used with a total of 16 plots per trial divided into 4 replicate blocks each with 4 treatments. The experimental treatments were a factorial combination of fertilizer and compost (Table 1)

Plots measured 12m x 3.75m and comprised 5 rows of forage maize (variety Cyrano) plants (row width 0.75m). Seed population sown was 12 per square metre. Plots were surrounded by guard rows of discard maize. All plots were harvested by hand with sub-samples taken for dry matter determination and dry matter partitioning analysis. Non-destructive records of crop growth and development were made throughout the growing seasons.

Table 1 Experimental treatments

		Compost application	
		None	50t/ha
Fertilizer Application	None	Control	Compost only
	$N:P_2O_5:K_2O$ 125:63:63 kg/ha l	Fertilizer only	Compost + Fertilizer

Results

Total crop yields and grain yields are given for each trial in Table 2. There were differences in yield between the two years of the investigation with higher yields in 1993 compared to 1994. Differential responses to the treatments were obtained in the two trial years with compost applications performing better in 1993 than in 1994 but over the two years the mean effect of compost applications was equal to or slightly better than the use of mineral fertilizer and when applied in conjunction with fertilizer consistently improved crop yield by a factor of 30% compared to fertilizer alone. This improvement was 25% with respect to grain yield.

Other significant results obtained were the raising of establishment percentage in 1993 with the application of compost which was correlated with higher soil water content. Faster development of the crop was recorded where compost was applied in 1993 as well as higher leaf area index and leaf chlorophyll content. In the 1994 trial these effects were also recorded with the highest yielding treatments Compost + Fertilizer and Fertilizer only.

Results of nutrient movement in the soils under the various treatments is given elsewhere (Parkinson et al, 1995).

Table 2 Dry matter yields of maize for 1993 and 1994 grown with and without compost and fertilizer and residual effects (R) in 1994. Within a trial differences which are smaller than the LSD (Least Significant Difference) are not significant.

Total Crop Yields t/ha	Control	Fertilizer only	Compost only	Compost +Fert.	LSD 0.05
1993	8.81	9.12	12.83	12.65	1.40
1994	5.58	9.06	6.39	10.88	1.24
Mean	7.20	9.09	9.61	11.77	
1994(R)	7.59	7.60	7.97	7.71	1.01
Grain Yields t/ha	Control	Fertilizer only	Compost only	Compost +Fert.	LSD 0.05
1993	3.73	3.85	5.50	5.18	0.71
1994	2.75	4.51	3.15	5.24	0.73
Mean	3.24	4.18	4.33	5.21	
1994(R)	3.80	3.78	3.99	3.74	0.91

Discussion

There were significant improvements in yield when compost was used in 1993 and the yield improvements were associated with improvements in crop establishment and crop growth with taller plants, with larger leaves and a higher chlorophyll concentration when compared to control plants and fertilizer only plants. This was indicative of a nutrient effect on the crop particularly a nitrogen effect. In 1993 the compost applied at 50 t/ha was capable of supplying all of the nutrient needs of the crop since when compost and fertilizer were applied together there was not a significant further increase in yield. The same increases in yield due to compost application were not evident in 1994 with the crop yields significantly below the yields when fertilizer alone was used. This was also attributed to a nutrient effect (negative) since compost treated plants showed smaller plant size, smaller leaf area and lower chlorophyll contents. When the results of the two years are meaned a positive effect of the compost comparable to using fertilizer was recorded and when compost and fertilizer were applied together then a substantial increase in crop and grain yield over fertilizer alone of 25 to 30% was obtained.

These results indicate that the nutrient content of the compost and in particular the nitrate content, was lowered substantially during the storage of the compost from 1993 to 1994. It is evident that if compost is to be stored outside over winter then there can be a very large reduction in its fertilizer value for agriculture. This has serious consequences when composting companies are faced with large quantities of compost to store from one year to another.

The assessment of the residual effects of the compost between years showed that a nutrient effect was non-existent in the second year after application. It appeared that there was no sustained breakdown of compost to release nutrients in the second year after application. There was a significant site effect with the 1994R control yield being significantly higher than the 1994 control yield.

It is concluded that compost derived from co-composted MSW and sewage sludge does have a beneficial effect on the yield of forage maize but that the benefit can be variable. In order that such composts can be marketed to agriculture in the UK then emphasis must be made of their nutritive value in the first year of application. Furthermore, their main use would be as a fertilizer replacement and as such an accurate estimate of their N:P:K value must be made so that farmers can adjust application rates according to the needs of their crops.

Acknowledgements.

The University of Plymouth and Ecological Sciences Ltd gratefully acknowledge the financial support of the DTI EUREKA scheme for support with the 1994 trials.

References

Groenhof, A.C., Dale, B., Tucker,G.C., Heeley, E.C. and Young, T.F.E. (1995). A feasibility study on the co-composting of MSW and Sewage Sludge. Proceedings of   the   International Symposium The Science of Composting, Bologna, Italy, 30 May – 2   June 1995.

Parkinson, R.J., Fuller, M.P., Jury,S. and Groenhof, A. (1995). An evaluation of soil   n u t r i e n t status following the application of co-composted MSW and sewage sludge and   greenwaste to maize. Proceedings of the International Symposium The Science of Composting,   Bologna, Italy, 30 May – 2 June 1995.

New Bulking Agents for Composting Sewage Sludge (*pteridium* sp. and *ulex* sp.), a Laboratory Scale Evaluation

M. GARCÍA, D. OTERO and S. MATO – Dpto. Recusos Naturales y Medio Ambiente Facultad de Ciencias. Universidad de Vigo Apdo.874 36200 VIGO (Pontevedra) SPAIN

Introduction

In Galicia (NW of Spain) there are wide extensions of forest destined to wood exploitation. *Pteridium* sp. and *Ulex* sp. are two common scrub species and they became into forestry wastes when cleaning works are made to minimise burning risks.

The implementation of EU rules regarding sewage treatment is increasing the amount of sewage sludge for disposal in the region. Composting could be a cheap solution for co-disposal of sewage sludge and forestry waste. For testing the self-heating capacity of sludge mixed with forest residues composting experiments were performed using a reactor vessel with forced ventilation.

Material and Methods

Composting reactor is a cylindrical vessel of 15 litres of volume, made of PVC and termo-isolated with several layers of cork sheet and closed up by a polypropylene lid airtight with a neoprene gasket.. The vessel is located into a box isolated with expanded polystyrene

Through the lid of the vessel four termocoupled measured and continuously recorded the temperature of the composting mass. CO_2 and NH_3 evolved were collected into gas traps and the final gas emission was chemically determined. After gas traps, oxygen percent is recorded by a paramagnetic oxymeter (Fig. 3).

Air is supply by a compressor through the base of the composting vessel. Air supply is controlled by a timer until the mass reach 55ºC. The program of the timer is 75 seconds ventilation on and 15 minutes ventilation off. When the system reaches 55 ºC air is supply by temperature feed-back control.

Air dried forest wastes were mixed with sewage sludge to get an initial moisture content of 60% wet weight and the proportions in dry weight of each component were: Ulex/sewage sludge = 3 : 1 and Pteridium/sewage sludge = 5.5 :1.

When composting finished the composting mass was separated in three layers, air dried, finely ground and analysed. The measured parameters were % Moisture (Mois), pH, Conductivity (C_{ond}), Total (N_{ke}), Ammoniacal (NH_4) and Oxidized Nitrogen (N_{ox}), %Carbon (C), Water-soluble (C_{ws}) and Extractable Carbon (C_{ex}), Humic (C_{ha}), Fulvic (C_{fa}) and No humic fractions (C_{nh}), Humification index by Sequi et al. (IH), K, P, Fe, Mn, Zn, Cd, Pb, Cu, Ni.

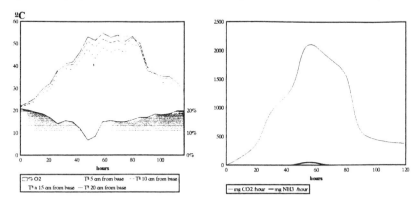

Figure 1 Evolution of temperature, % Oxygen and Production of CO2 and Ammonia in Pteridium-sewage sludge experience.

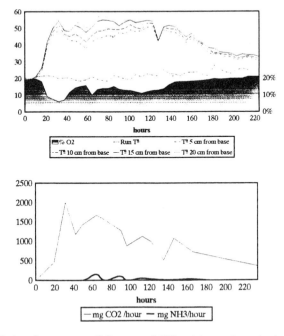

Figure 2 Evolution of temperature, % Oxygen and CO2 and Ammonia production in Ulex-sewage sludge experience.

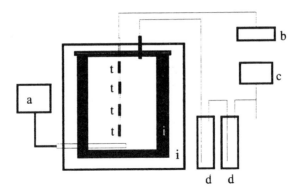

Figure 3 Diagram of laboratory composter a) air supply. b) temperature register. c) oxygen register d) gas traps. i) isolation t) termocouples.

Results

The evolution of Temperature, CO_2, NH_3, and Oxygen are reported in the figures 1 and 2 and the change of chemical parameters along composting in the tables 1–3.

Both mixtures showed similar temperature profiles. Three phases typical of composting with forced ventilation are clear. For heating phase (0–40 hours) the temperature increase to termophilic values, CO_2 evolved is correlated with temperature ascent. Oxygen into the mass reach values below 10 % at the end of heating phase when temperature is close to 50°C. Low oxygen values point that air supply by the timer is not enough for meeting biological demand.

Heating phase is follow for a period with temperature controlled by feed-back (40–120 hours for Ulex mixture and 40–80 hours for Pteridum mixture). Temperature profile is constant and a clear stratification, with lower part colder than upper, is stablished. Under temperature control, oxygen values always remain above 10% guaranteeing aerobic conditions and oxygen in excess of biological demand.

Table 1

Final moisture of the Pteridium-sewage sludge mixture	Final moisture of the Ulex-sewage sludge mixture
Upper layer : 74.9 %	Upper layer : 71.0 %
Center layer : 58.8 %	Center layer : 50.8 %
Base layer : 56.2 %	Base layer : 46.4 %

Cooling phase begin when biological activity decrease due to substrate depletion or other environmental factor became critical. Along cooling phase CO_2 and temperature gradually turn to surrounding values.

Duration of temperature control phase is longer in Ulex mixture than Pteridium because Ulex mixture contains more sewage sludge and so more time is needed for biological conversion.

Moisture stratification is clear for both mixtures (Table 1). Ventilation and temperature stratification lead to a graduation in moisture content with higher values in the upper layer. This effect is widely reported in the literature as typical of forced ventilation systems.

Emissions of Greenhouse and Enviromental Relevant Gases by the Decomposition of Organic Waste from Households

Dr. ANDREAS GRONAUER, Dipl.-Ing. MARKUS HELM, Dipl.-Ing. SILVIA SCHATTNER-SCHMIDT – Institut für Landtechnik, Technische Universität München-Weihenstephan; Dr. BETTINA HELLMAN – GSF Neuherberg

Introduction

Composting is an aerob process. Aerob microorganismens need at least an oxygen-concentration of 3 Vol.–% to take part on the degradation of organic matter. If the oxygen-concentration is getting too low, anaerobic organismens become dominant. They produce oder and greenhouse gases. To avoid this effect all instruments must be used to increae the O_2–concentration in the plant.

Factors influencing the O_2–concentration in the plant

The concentration of oxygen in the air of the plant is depending on the microbial aktivity (O_2-user) and the gas-exchange between the atmosphere and the plant. There are four possibilities to influence the O_2-concentration in the plant:

- mixing the input-materials to provide the structure of the plant; by this way a better gas exchange can take place
- limiting the size of the plant to render a diffusion of oxygen into the center of the plant
- turning the plant
- active aeration

You can find quiet a lot of functional systems for an active aeration, but the financial costs are to high for small composting yards.

There are cheaper possibilities to ensure the oxygen concentration in the plant, for example by optimising the handling of the process.

Size of the plant

Triangular plants have a good supposition to ensure a sufficiend O2-level in all parts of the plant. Between 3 and 21 Vol.–% O2 of the O2-concentration is corresponding to the CO2-concentration in a linear model. By measuring the CO2-concentration all dates can be transformated into Vol.–% CO2. The triangular plant was three weeks old and built up with 50 % organic wastes from the households and 50 % structurial material. Only in the center of the plant the CO2-concentration is nearly 20 %, so that anaerobic processes can not be excluded. By plants of this form there is obviously a special circulation: air streaming into the plant from the side, being warmed up in the plant and exhausting at the top of the plant. This circulation leeds to a continious gas exchange between the plant and the atmosphere.

Turning and structure-material

Gas exchange is the result of turning the plant. During the turning oxygen is streaming into the plant. As we could see in other experiences, turning the plant is necessary for mixing the organic material to make new surfaces for microorganismens. On the other side it does not ensure a continious oxygen concentration for the aerobic microorganisms. A sufficient oxygen concentration is only possible by limiting the size of the plant, by enough structure material in the mixture or by active radiation.

Odor emissions by composting

Carbondioxid- or oxygen concentration in the air of the plant give information about the situation of the aerobic microorganisms. One main problem by composting organic wastes in open compost-yards in Germany are odor emissions. To evaluate the emission-potential of odor, the concentration of H2S and Total Organic Carbon in the plant has been measured. H2S is a odor compound. Between the TOC-concentration and olfactorial measurements there is a correlation. With triangular plants and a structure material part of 60 Vol.-%, the turning intervall is influencing the formation of H2S and TOC in the plant.

Olfactorial measurements of the odor-concentration in the exhausting air confirm these results: With short turning intervalls (e.g. one day) the process is starting qiuckly and also the peak of the odor-concentration is reached very early. It stays on a low level and is falling soon. With the turning intervall of 3 days, the increasing period is starting later, and the peak is a little bit higher. With the turning intervall of seven days the increase of the oder-concentration in the exhausting air starts later and the peak is lower. But there is a higher concentration of odor compounds in the exhausting air for a period of almost 3 weeks.

Another factor influencing the bilding of odor compounds is the part of structure material in the plant. Plants formed with organic waste from households only, develope a high emission potential of H2S and TOC. As other experiments showed, only a part of 20 Vol.–% structurial material is reducing the formation of H2S and TOC very effectivly. With usual mixing rates for open triangular plants between 40 and 60 Vol.–% structur material, the formation of H2S and TOC is very low which means low emission rates by this plants.

Emissions of greenhouse gases by composting

Odor is a great problem for composting-yards which haven't got enough distance to their neighbourhood, but it has got less enviromental relevance. Greenhouse gases such as CO2, CH4 and N2O do not disturb anybody, but they might have an global importance. To estimate the importance of the emission of greenhouse gases by composting organic wastes, we measured the emission rates from the plants and during the turning of the plant.

Methods

The emission rates of the surface of the plant were measured with the closed-chamber-method. To get the emission during the turning of the plant, a turning machine has been modified: On the top of the machine a radiator was installed, which could exhaust the air streaming of the plant during the turning. The radiator was as strong, that it was possible to ensure, that no gas could escape at the front- or the backside of the turning machine. In the tower on the other side of the radiator it was possible to measure the air-flow of the exhausting air. At this place an aliquote of the exhausting air was separated for analysing the gas-concentrations in a mobile laboratory.

Results

The example of N2O shows, that the emissions of the greenhouse gases are depending on the part of structure material in the plant. With growing parts of structure material the N2O emissions got less. The part of N2O that was emitting during turning was low compared with the part emitting of the surface of the plant. With CO2 and CH4 the same results were reached. The importance of the emission of greenhouse gases is shown by the following example: If all organic waste in Germany would be treated by composting, the emission of CH4 would rise for 0.5% of the actual emission from agriculture.

The Compost Information Kit – Business Tools for Success

JAMES R. HOLLYER – Agricultural Economist, Department of Agricultural and Resource Economics, University of Hawaii at Manoa, Honolulu, Hawaii 96822 USA. RODNEY TYLER – Recycling Consultant, Browning Ferris Industries, Oberlin, Ohio 44074 USA

Introduction

Composting is Mother Nature's original form of recycling. Simply, composting is the controlled biological decomposition and conversion of solid and liquid organic materials into a humus-like product. Managed composting, while somewhat limited in the U.S. since the green revolution, is a popular and successful animal waste and plant residue management option world-wide. It is more preferable to landfilling, incineration and roadside dumping. The benefits of the act of composting include a reduction in volume of compost inputs, destruction of pathogens, reduction in weed seeds, the degradation of many potentially harmful organic compounds, the binding-up of high-nitrogen materials such as manures, and the production of a useful soil amendment.

While most composting of organic materials is done involuntarily via unattended degradation (e.g. grasscycling or on the forest floor), the organized composting industry in the U.S. is growing. Stueteville and Goldstein (BioCycle, 1994) estimate a growth rate of these semi-commercial operations at about 35 percent per year since 1988. This translates into nearly 3,000 yard waste composting facilities in 1992, up from about 800 in 1988, and about 320 biosolids composting projects by the end of 1993. This is to say nothing of the many new on-farm composting operations which are springing up across the Nation everyday.

To the person who is involved with the regulation, production or sales of compost every day this new growth is not surprising, yet a proactive approach to compost use and sales education will do much to stem the tide of any compost surpluses. A collection of sales tools, *The Compost Information Kit*, has been developed to help aid compost marketers in their quest for profitable sales and manageable levels of product.

The Compost Information Kit

In a first of its kind study for the Composting Council (formerly known as the Solid Waste Composting Council), Potential U.S. Applications for Compost, gave a general estimate of the potential volume of compost that could be used in the U.S. – over a billion cubic yards per year (1992). Although many in the business question that figure, due to the rather low buyer knowledge in the U.S. about the value of compost and the ability to charge for such a relatively unknown product, what most people will agree on is that the market could be very, very large. Thus there is the potential on the one hand to produce more compost than a particular local economy could handle and yet on the other hand all that compost could potentially be easily absorbed if consumers were informed about compost's value. The solution then would be to provide learning aids to help compost producers effectively sell their high quality product.

The Composting Council has developed a Compost Information Kit to help inform sellers and buyers of the benefits of compost. The kit includes a slide rule to estimate compost use rates; a compost-use brochure for the bulk buyer; a compost-use brochure for the homeowner; a wall chart which explains the marketing of compost, and finally a compost business financing guide.

To illustrate the nature of the Kit, below are selected, miniaturized 'stages' of the wall chart. The idea of the chart is to give compost producers an educational tool which can be readily accessed and used to inform staff and clients. The entire Kit is this user-friendly.

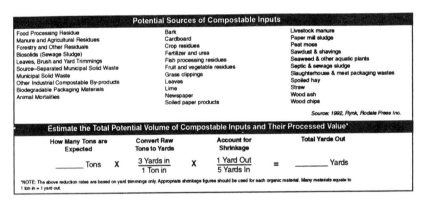

Stage 1 Identify Compost Feedstock

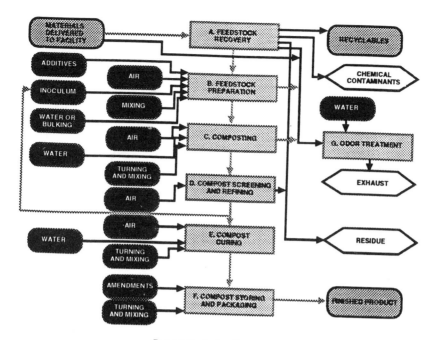

Stage 3 Producing Compost

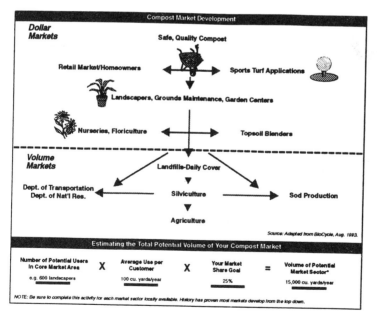

Stage 5 Available Markets

Information and Activities to Consider	
Educational Information • Scientific evidence of quality/performance • Instructions for use • Economic or other benefits • Join Composting Council	**Research & Development** • New formulations and technologies • Continuous finished product testing • Plant responses in various soils
Testimonials • Where has it been used? • Who has used it? • How have they used it? • What are the results?	**Demonstrations** • City parks • Community gardens • Road strips • Schools and campuses • Major theme parks • Open houses • Compost demonstration site sponsorship
Pricing Strategy • Available substitutes • Penetration pricing • Market-share pricing • Long-term considerations	**Adapt Project Specifications to Allow for Compost Use** • Work with landscape architects • Work with building specification departments • Work with DOT/DNR agencies
Advertising • Magazines • Trade Shows • Direct Mail • Associations • Newspapers • Conferences • Radio and TV • Word of mouth	**Product Refinement & Market Targeting** • Niche products for new markets • Superior service • Equipment rental for applications • Technical support
Promotions • Specials • Meeting sponsorships • Campaigns • Community projects • Giveaways • Talks at schools and civic • Contests groups	**Sales Skills** • What's in it for your customer? • Understanding product features and benefits • Customer needs analysis

Stage 9 Marketing & Education

The Compost Information Kit can be purchased from The Composting Council, 114 South Pitt Street, Alexandria, VA 22314 USA. Phone (703) 739–2401 / Fax (703)739–2407. Funding for the Kit's development was provided by the University of Hawaii via USDA–CSRS Grant 91–COOP–6159.

The Development of Low-Input, On-Farm Composting of High C:N Ratio Residues

D. B. CHURCHILL, W. R. HORWATH, and
L. F. ELLIOTT

Abstract

The field straw windrow composting study showed that a minimum of two turns and natural rainfall was required to reduce straw residue volume by 80% or greater in 16 to 21 wk. The decomposition of the high C:N ratio straw residue is contrary to established composting methodology where a C:N ratio of 30:1 or less is thought to be required to compost grass straw. The microbial biomass required more C and less N to function in the thermophilic treatment compared to the LT treatment. The microbial biomass N requirement was less than 4% of total straw residue N in both treatments. The low N requirement of the decomposer biomass indicates that the form or compartmentalization of N is more indicative of substrate quality than the concept of combined substrate C:N ratio. The increased requirement of the thermophilic biomass for C resulted in an increase in the decomposition of the lignin fraction compared to the LT treatment. The increased lignin fraction decomposition was detected through changes in the element composition of this fraction. For these reasons, the composting of grass straw residue is feasible without initially lowering the C:N ratio. Composting has value both as an avenue of straw disposal and for its potential for utilization in the cropping system. Volume reductions of straw windrows in this study were as high as 88% over 32 wk making in-field composting a viable straw disposal alternative to open field burning.

Key words: low-input, composting, ryegrass, microbial biomass, crop residues, lignin

Contribution of the USDA-ARS, in cooperation with the Agricultural Experiment Station, Oregon State University, Corvallis, Oregon, 97331 (Technical Paper No. ; D. B. Churchill, W. L. Horwath, and L. F. Elliott, USDA-Agricultural Research Service, National Forage Seed Production Research Center, Corvallis, OR 97331

Introduction

Crop residues are critical contributors to processes that maintain soil quality and conserve nutrients. Recently, public concern for the environmental consequences of open field burning of grass seed straw has led to legislation that severely restricts this practice (Mackey 1991). Utilization of grass straw in the field, such as by shredding and chopping, have been investigated (Young et al, 1993), but these approaches are unsatisfactory for some grass species. Alternate straw management practices are needed to preserve grass seed yields, to control pests, and to develop sustainable cropping systems. The low-input, on-site composting of grass seed straw and other crop residues is an attractive alternative to thermal and mechanical residue removal (Churchill et al., 1995a). However, composting of organic material is thought to require a C:N ratio of 30:1 or less (Biddlestone et al., 1987; Golueke 1991; Hammouda and Adams 1987). Development of a low-input grass straw composting system would provide an avenue for straw utilization.

Materials and Methods

Field study

The threshed perennial ryegrass straw was collected into windrows approximately 2 m high by 6 m wide. Timing of windrow turns in the Willamette Valley was based on site access and on having at least 3 wk between consecutive turns. To determine the minimum number of turns required to produce a spreadable compost material, windrows were turned 0, 2, 4, or 6 times with a commercial compost turner throughout the rainy season during the winter of 1992–93. The study was repeated in the winter of 1993–94 using a tractor-mounted, front-end loader for turning. Procedures and analyses are described by Churchill et al. (1995a).
Laboratory study
    Perennial ryegrass straw was incubated at 25°C and 50°C and analyzed as reported by Horwath and Elliott (1995b).

Results and Discussion

Field study

Greatest straw windrow volume reduction occurred with two or more turns of the straw (Figure 1). The majority of the volume reduction occurred after 16–21 wk of composting. At the end of the study (after 32 wk) the final volume of the zero-turn plots was significantly greater (p<0.05) than plots turned two, four or six times. The C:N ratios of the straw decreased from 57:1 to 16:1 by March 11 which was about 28 wk after initiation (Table 1). However, appreciable rainfall did not occur until late October, 10 wk after the windrows were formed. These results suggest that growers composting this type of straw can achieve near-maximum volume reduction with as few as two turns. During this study, it was noted that when the straw was turned, straw windrow temperatures increased rapidly. Adequate moisture must also be present. During the winter of 1993–94, the compost was turned with a tractor-mounted, front-end loader and the same results were obtained (D. B. Churchill, unpublished results).

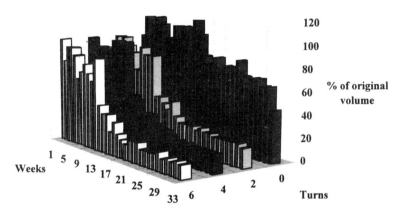

Figure 1 Percent of original volume remaining straw windrows with 0 to 6 turns (Churchill et al., 1995).

Table 1 Long straw turning dates showing accumulated rain fall, plot number, date sample was taken and C:N ratio.

DATE TURNED	PLOT No.	ACCUMULATED RAINFALL (mm)	SAMPLE DATE	C:N
10/29/92	2, 4, 6	22.10	8/18/92	57/1
12/9/92	6	250.70	12/8/92	49/1
1/13/93	*2, 4, 6 (combined)	375.16	1/26/93	22/1
3/11/93	4, 6	511.05	2/23/93	16/1
4/20/93	6	684.78	3/30/93	15/1
5/17/93	4, 6	759.97	4/27/93	16/1
6/15/93		887.48	5/25/93	16/1

* Combined three windrows into one for each of the turned treatments

During the composting process, internal temperatures of the straw windrows pile increased. Temperatures of 55–60°C for a period of a few minutes to a few days are considered sufficient to kill most types of seeds and disease propagules (Biddlestone et al., 1987; Golueke 1991). Sustained temperatures of 50°C or more were attained in the straw windrows turned a minimum of 4 times. More recent studies have shown that weed seeds and disease propagules were destroyed during the composting process (Churchill et al., 1995b, In press).

Laboratory study

Microbial biomass C was highest (22 mg of C g–1 straw) after 3 d of incubation in both the LT and HT treatments (HT treatment maintained at 25°C for first 5 d) (Figure 2a). Microbial C in the LT (low temperature) treatment remained relatively constant through 20 d and then declined to 8 mg of C g–1 straw by 45 d (Figure 2a, Horwath and Elliott 1995a). Microbial biomass C production lagged after the 5–day incubation at 25°C, then peaked and gradually decreased to 4 mg of C g–1 straw after the temperature was increased to 50°C in the HT (high temperature, Figure 2a).

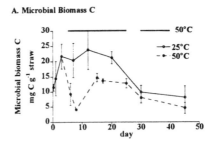

A. Microbial Biomass C

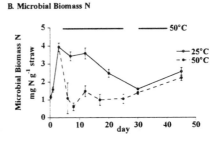

B. Microbial Biomass N

Figure 2 Microbial biomass C (A.) and N (B.) in the LT and HT treatments during a 45 d incubation. Lines indicate the duration of the 50°C HT treatment. Standard deviation of the mean shown as line bars, n=3 (Horwath and Elliott 1995a).

Microbial biomass N increased to 4.0 mg of N g–1 straw after 3 d in both the LT and HT treatments (Figure 2b). In the LT treatment, microbial biomass N remained unchanged during the first 12 d, declined to 1.7 mg of N g–1 straw by 30 d and increased to 2.8 mg of N g–1 straw at the end of the incubation. In the HT treatment, microbial biomass N declined to less than 1 mg of N g–1 straw after the temperature was increased to 50°C and then gradually increased to 2.1 mg of N g–1 straw at the end of the incubation. The increase in microbial N may indicate an accumulation of microbial by-products and humic substances since microbial C declined constantly in both treatments (Hammouda and Adams 1987). Carbon mineralization occurred rapidly in the LT and HT treatments (Figure 3a). The amount of C mineralized was 185 g and 210 g of C kg–1 straw in the LT and HT treatments, respectively. Initially, the straw contained 400 g of C kg–1 (Horwath and Elliott 1995b). The majority of C mineralization from both of the temperature treatments (LT and HT) occurred by 20 d. The similarity in straw C mineralization in the LT and HT treatments was explained by relating the C mineralization activity to microbial biomass C (Figure 3b). The respiratory quotient (total C mineralized/microbial biomass C) for the HT treatment was approximately twice that of the LT treatment. In the LT treatment, the increased C mineralization was associated with respiratory activity not an increase in the size of the microbial biomass. The thermophiles appeared to require less biomass C and N than mesophiles to decompose approximately twice as much C per unit of microbial biomass.

Nitrogen in the lignin fraction increased 12% in the LT treatment and 16% in the HT treatment (Table 2). The loss of lignin H was similar to C in both treatments. The mass of O remained similar to undecomposed straw lignin O in the HT treatment and increased to 127% in the LT treatment. The constant or increased level of O and loss of C and H indicated that 94% of the lignin fraction was oxidized or altered during the decomposition process. Reviews of degradative reactions during the decomposition of lignin have indicated that O content increases through the oxidative splitting of side chains and oxidative ring cleavage to form carboxylic acid groups (Kirk 1971; Flaig et al., 1975; Chang et al., 1980; Crawford 1981; Kogel-Knabner 1993).

Table 2 The concentration of elements in the initial lignin fraction and the percent remaining for the LT and HT treatments. Standard deviation of the mean shown in parentheses.

Treatment	C	H	O	N
Day 0		g kg–1		
265.8	640.2 (51.9)	(44.5) 12.1	81.9 (1.4)	(6.2)
		% remaining of original		
LT	75.0 (0.4)	75.1 (0.9)	126.5 (5.5)	111.6 (9.2)
HT	61.3 (0.3)	60.0 (0.8)	98.2 (3.7)	116.1 (6.7)

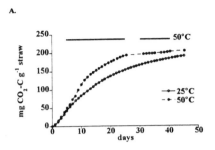

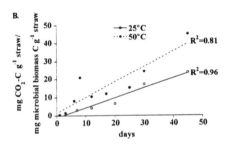

Figure 3 C mineralization (A.) and respiratory quotient (B.) is shown for the LT and HT treatments during a 45 d incubation. Lines indicate the duration of the 50°C HT treatment. Standard deviation of the mean shown as line bars are not shown when smaller than symbols, n=3 (Horwath and Elliott 1995a).

Acknowledgment

Partial support for these studies was provided by the Oregon Department of Agriculture and the Oregon Agricultural Experiment Station. We thank Amanda Warren, Thomas Edgar, Douglas Bilsland, and Hudson S. Minshew for the technical assistance.

References

Biddlestone, A.J., K.R. Gray, and C.A. Day. 1987. Composting and straw decomposition. In: Environmental Biotechnology. C.F. Forster and D.A. Wase (eds.). John Wiley and Sons, New York, pp. 135–175.

Chang, H., C. Chen and T.K. Kirk. 1980. The chemistry of lignin degradation by white-rot fungi. In: Lignin Biodegradation: Microbiology, Chemistry and Applications. T.K. Kirk, T. Higuchi, and H. Chang (eds.). CRC Press, Boca Raton, Florida, pp. 215–230.

Churchill, D.B., S.C. Alderman, G.W. Mueller-Warrant, L.F. Elliott and D.M. Bilsland. 1995b. Survival of Weed Seed and Seed Disease Propagules in Composted Grass Seed Straw, ASAE, St. Joseph, MI. (In Press).

Churchill, D.B., D.M. Bilsland and L.F. Elliott. 1995a. Methods for composting grass Seed Straw Residue. Applied Engineering In Agriculture 11: (In Press).

Crawford, R.L. 1981. Lignin biodegradation and transformation. John Wiley and Sons, New York.

Flaig, W., H. Beutelsacher and E. Rietz. 1975. Chemical composition and physical Properties of humic substances. In: Soil Components: Volume 1, Organic Components. J.E. Gieseking (eds.). Springer-Verlag, New York, p. 534.

Golueke, C.G. 1991. Principles of composting. In: The Biocycle Guide to the Art and Science of Composting, The JG Press, Inc., Emmaus, Pennsylvania.

Hammouda, G.H.H. and W.A. Adams. 1987. The decomposition, humification and fate of nitrogen during composting of some plant residues. In: Compost: Production, Quality and Use. M.D. Bertoldi, M.P. Ferranti, P. L'Hermite, and F. Zucconi (eds.). Elsevier Applied Science, New York, pp. 245–253.

Horwath, W.R. and L.F. Elliott. 1995a. Microbial C and N dynamics during mesophilic and thermophilic incubations of ryegrass. Biology and Fertility of Soils 18: (In Press).

Horwath, W.R. and L.F. Elliott. 1995b. Ryegrass straw component decomposition during mesophilic and thermophilic incubations. Biology and Fertility of Soils 18: (In Press).

Kirk, T.K. 1971. Effects of microorganisms on lignin. Annual Review of Phytopathology 9:185–210.

Kögel–Knabner, I. 1993. Biodegradation and humification processes in forest soils. In: Soil Biochemistry. J.M. Bollag and G. Stotzky (eds.). Marcel Dekker, Inc., New York,   p   p 101–135.

Mackey, J.E. 1991. Opportunities in grass straw utilization. Prepared for the Oregon Economic Development and the Oregon Department of Agriculture. Prepared by CH2M-Hill in conjunction with Oregon State University.

Young, W.C. III, T.B. Silberstein and D.O. Chilcote. 1993. Evaluation of equipment used by Willamette Valley grass seed growers as a substitute for open-field burning. Final report prepared for the Oregon Dept. of Agriculture's alternatives to open-field burning research program.

Decentralized Compost-Management: Case-Study of a District of 77.000 Inhabitants – the 'Kulmbach Model'

Axel Kolb Dipl.-Ing.(FH) Eichner & Kolb Kompost GmbH, Project Development Group Dörnhof 1, 95326 Kulmbach, Germany

Summary

The following case study evaluates the development of a system of separate collection and composting of 'greenwaste'(from garden– and park–areas) and 'biowaste'(from the separate collection of municipal solid waste) in an Administrational District around the city of Kulmbach in Northern Bavaria/Germany. Greenwaste is brought by the citizens to decentralized, farm-based compost-sites. Biowaste is collected by the municipality in a separate 'bio-bin' and brought to biowaste-compost-places, which are connected to greenwaste-sites.

Seven years after start 17.500 tons per year, or 170 kg per citizen are recycled on 14 compost-places and two biowaste-sites. Costs per input-ton are low, citizens accept the sites well, the system is modularly adaptable to changing requirements and can be directed by one central office.

Introduction to the Case Study

The following case study resumes the past 7 years of a system of separate collection and recycling of organic municipal waste fractions, which was developed in a region in Northern Bavaria/Germany, an 'administrational district' (called 'Landkreis') around the city of Kulmbach. The district has 77.000 inhabitants with a city center of 30.000 and rural communities of 47.000 citizens.

The District Administration ('Landratsamt') is legally responsible for all the municipal waste management, recycling policy, contracting and finance for this purpose.

In 1987 the District Administration decided to stop incinerating *greenwaste* (organic waste from garden, park and landscaping areas, like grass and branches).

Instead of this a compost-site was installed on farm ground.

After short, the site was so well received by the public that successively 14 compost-sites were put into place. The district administration contracted farmers to build and run the sites on a long-term contract basis. The locations had to be well accessible from public roads and spread over the district, so that each citizen can reach one within a ten-minutes drive by car.

The quantity of recycled greenwaste increased from 500 tons in 1988 to 16.200 tons in 1994. This represents ca. 156 kg of Greenwaste from private households per citizen in 1994.

In 1991 a *pilot-project* for the separate collection and recycling of biowaste was started (mainly kitchen waste, mainly of botanical origin, food left-overs in small quantities, seperately collected in a 'bio-bin'). The aim was to introduce, study and improve:

- the introduction-campaign for the citizens,
- the administrational measurements (regulations, fees)
- lquestions of hygiene and odours of the 'bio-bin'
- the logistic costs
- the recycling/composting process.

In 1993 an introduction campaign of the 'bio-bin' for all the district citizens was run and two biowaste-compost-places were installed - next to existing greenwaste compost-sites.

In 1994, the first year of the 'bio-bin', 1377 tons of Biowaste were collected, respectively 59 kg per each citizen connected to this system. (Many citizens chose to do 'garden-composting' as alternative to the bio-bin.)

Basic Comparison: Centralized or Decentralized Composting Systems

Centralized System

One or more complete districts are linked to one central processing-/compost-plant. This is dimensioned for input of up to 100.000 tons/year. These plants are fully enclosed and automated. The radius of the connected area is up to 50 km.

Decentralized System

One District splits its area and installs small compost-sites, directly accessible for citizens (max 10 minutes drive by car) and thus provides a short and direct way for organic materials to be recycled close to agricultural areas.

Table 1: Basic Overview: Decentral and Central Composting Systems

	Decentral	Central
Greenwaste		
Input Material/ Collection	'Bring-System', citizens bring greenwaste directly to site, no input-transportation	citizens deliver to 'Recycling yard' Transport to site
Processing	open piles low costs	plant should be enclosed
Final Product/ Distribution	directly to fields or: local marketing as soil conditioner	transportation to farms extra or: regional marketing system
Biowaste		
Input Material/ Collection	Municipal Pick-Up-System in separate garbage-bins	Municipal Pick-Up-System in separate garbage-bins
Processing	open winrows, if necessary under roof intensive winrowing with specialized machinery	automated and fully enclosed plant/system high investment cost and energy consumption per ton
Final Product/ Distribution	to fields or: local marketing as soil conditioner	transportation to farms extra or: regional marketing system

Chosen Solution for the Kulmbach District: Decentral

14 Compost-Sites for Greenwaste

These consist of areas for cars, fresh material, shredded materials and for further composting-stages and storage. Piles are turned by payloaders.

Final-products: fresh or ripened compost, sometimes screened and sold.

The sites in Kulmbach for greenwaste are open, fenceless, without collection of fees. Fees for the citizens are included in the annual garbage fee. Input comes very clean.

2 Biowaste-compost-places for biowaste (one of them also for Industrial Biowastes)

These consist of areas for the intensive rotting-process, storage areas open and under roof. Intensive winrowing on small winrows using a winrow-turner. Process-Management according to O2-consumption, frequent measurements and analyses during composting to optimize the rotting-process and to have early information on features of use. Final material: fresh, ripened, screened and classified compost-types, depending on use. According to Bundesimissionsschutzgesetz (Federal Law for the Control of Gaseous Emissions) and 'Technische Anleitung Siedlungsabfall' (Technical Instruction for Waste Management) these plants with less than 0,75 tons/h input-capacity (ca. 6300 tons/year), are regarded as 'small' and thus, only the requirements of building-regulations have to be met. This accelerates the application-process for building-permissions considerably.

Management

Initially, during the pilot-phase, single contracts were made between the administration and the greenwaste-sites. Later – for the biowaste-compost-places – one central 'Compost-Ring-Company' was founded to have one single partner for the administration.

Greenwaste

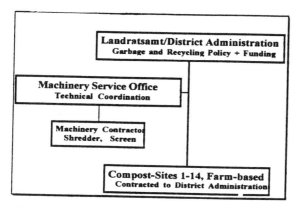

Figure 1 Management of Greenwaste-Composting in Kulmbach District

The District Administration contracts farmers, who build the site on agricultural ground, well accessible from public roads. Contracts include: investment, maintenance, all services of running the place, composting and spreading into fields. The 'Machinery Service Office' does all technical and contractual coordination.

Biowaste

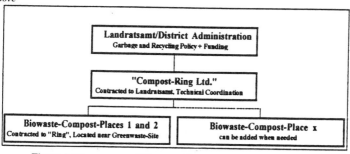

Figure 2 Management of Biowaste-Composting in Kulmbach District

The District Administration contracts the 'Compost-Ring-Company' (Limited Company), which in turn contracts the biowaste-sites. These sites are run as independent companies owned and run by farmers owning already a greenwaste compost-place. According to the quantities of biowaste collected, more of these 'modules' can be installed. Payment by tonnage input.

Quantitative Development of Composted Input between 1987 and 1994

Fig. 3 illustrates the total-input of organic waste-materials into all decentral compost-places per year. These figures demonstrate the immense potential of garbage-diversion. Especially the large portion of greenwaste is relatively easy to collect and process, the quantities of biowaste from the bio-bin are considerably smaller than in alternative centralized systems.

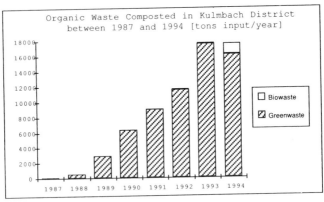

Figure 3 Development of Organic Waste Composted between 1987 and 1994
Source: Landratsamt Kulmbach, 1995

Conclusions

This case study of an Administrational District in Northern Bavaria/Germany – a region of 77.000 inhabitants shows, that a decentralized compost-site-system can:

– collect, process and re-use a majority of municipal organic waste materials
– achieve a high acceptance by the citizens – ('our site'-effect, few 'disturbing materials')
– supply a low cost-base
– be very reliably run by farmers
– be easily installed and managed by one central-office
– be adapted modularly to the changing needs of environment – and waste-policies
–strongly contribute to environmental protection: save landfill space, reduce climatically dangerous methane-release from dumps.

Microbial Succession in a Technical Composting Process

T. KUBOCZ and C.E. GRÜNEKLEE – Herhof Umwelttechnik GmbH, Riemannstr.1, 35606 Solms, Germany

Introduction

During the process of composting, organic material is turned into the stable final product compost. It is performed by microorganisms causing the degradation and conversion of the organic material. Therefore, the main target of processing is to create optimal living conditions for the microorganisms.

The assignment of this study was to observe the succession of some groups of microorganisms from the delivery of separately collected biowaste to finished compost. Investigations were carried out in June and August 1994 at the composting plant in Beilstein (Hessen). The composting system used is the Herhof-Rottebox.

Herhof technique

The Herhof-Rottebox is a fully encapsulated system. Approximate 60 cubic meters of biowaste can be decomposed under controlled conditions. Air is directed into the composting matter through holes in the floor. For heating the material up, circulation air is used. Normaly fresh air with a higher oxygen content than circulation air is pressed through the filled box. With the expelled air carbon dioxyde, some organic compounds like organic acids, which tend to have a penetrate smell, and humidity leave the box. The organic compounds are condensated out in a heat exchanger and are removed together with the air humidity. The final stage of the air treatment is a biofilter.

Experimentel Disgine

124 tons of separately collected biowaste were mixed with 28tons of green matter and shredded to a defined particle size of approximate 150mm. This material

was filled in four Herhof-Boxes and the decomposition process was started. During 6 days the organic matter had a temperature of about 45°C. The material was then heated up to 60°C for 3 days. This is the hygienization phase. After cooling down, the fresh compost was taken out of the box.

For a second passage, the fresh compost was crushed in a hammer mill. This is essential to create new surface area for the microorganisms. Water was also added for a optimal moisture content of the compost. This preparation is called „dynamic step'. In the second passage, the process temperature was 45°C during 9 days. After these 9 days of decomposition, the dynamic step was repeated. In the third passage, the temperature was 40°C. After 7 days, the further decomposition was stopped, resulting in a finished compost (according to the standards of the Federal Quality Commision RAL-Gütesiegel).

During the decomposition time of 26 days compost samples were taken as follows:

- at the end of the first decomposition passage (after 10 days).
- at the end of the second decomposition passage (after 19 days).
- at the end of the third decomposition passage (after 26 days).

Samples were also taken from the shreddered input material, which was mixed with green matter.

The samples were analysed qualitatively and quantitatively for the following procaryontic and eucaryontic microorganisms.

- aerobic bacteria (medium: Plate Count Agar)
- anaerobic bacteria (medium: Columbia 5% SB)
- aerobic sporogenous bacteria (medium: Columbia 5% SB)
- anaerobic sporogenous bacteria (medium: Columbia 5% SB)
- Enterobacteriaceae (medium: VRBD-Agar)
- mold (medium: YGC-Agar)
- yeast (medium: YGC-Agar)

The following compost parameters were also analysed (according to the instructions of the Federal Quality Commission RAL-Gütesiegel).

- pH-value
- water content
- decomposition degree
- loss on ignition

Results

Aerobic and anaerobic bacteria

The highest quantity of aerobic and anaerobic bacteria were measured in the input material of the first box-passage. After the first passage, both values of total count

were decreased. The reason for the decreased number of bacteria is the hygienization phase at the end of the box-passage. It is unknown, if the amount of bacteria increases before the hygienization phase begins.

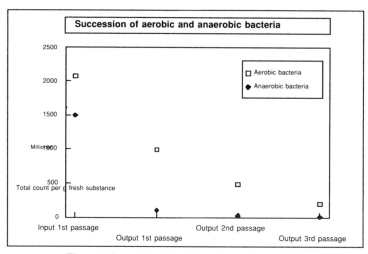

Figure 1 Succession of aerobic and anaerobic bacteria.

At the end of the next two box-passages (including dynamic step) the amount of
bacteria were decreased continuously. This was happened without a further hygienization step. A clear result is a considerable drop in the total number of anaerobic bacteria. Based on the aeration, the anaerobic species are selected out. The
ratio between aerobic and anaerobic bacteria was increased drastically. At the
beginning of decomposition the ratio was nearly 1. This means, that the number of
anaerobic and aerobic bacteria was nearly equal. At the end of the decomposition,
we measured more than thirty times more aerobic bacteria than anaerobic bacteria.
The effectivety of the hygienization step can be seen from the number of
Enterobacteriaceaes. The amount of this group of microorganisms, which are able
to live under aerobic and also under anaerobic conditions (facultatively aerobic),
decreased after the hygienization step to a 10th of the input number. At the end of
the third decomposition passage, the amount of Enterobacteriaceae was only 1%
of the amount of the input material.

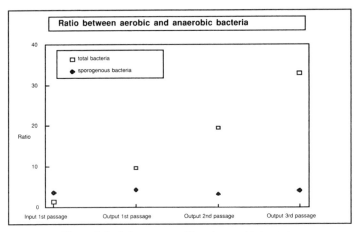

Figure 2 Ratio between aerobic and anaerobic bacteria.

Yeast and mold

Yeast and especially mold, grow much slower than bacteria, but both can grow under conditions which are unfavourable for most bacteria. In our investigations, the number of yeasts were the highest in the input material. After the first box-passage, only 5% of this amount could be detected. At the end of the second step, yeasts are quasi eliminated.

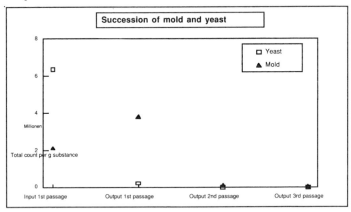

Figure 3 Succession of mold and yeast.

In our investigations, the number of yeasts were the highest in the input material. After the first box-passage, only 5% of this amount could be detected. At the end of the second step, yeasts are quasi eliminated.

In comparison with all measured microorganisms, the highest contents of mold were found in the output material from the first box-passage. That means that mold not only survived the hygienization step (probablly as spore), but multiplied during the first 10 days. This could be seen by opening the door of the box: a white carpet of mold covered the compost. After the second box-passage, mold was minimized. Aerobic and anaerobic sporogenous bacteria

In order to avoid death under unfavourable living conditions, some bacterias can form spores, which germinate and grow under better living conditions. In our investigations, neither the aerobic sporogenous bacteria, nor the anaerobic sporogenous bacteria have been effected through the different living conditions inside the box.

Living conditions inside the box

The changes in the amount of the examined microorganisms groups can be explained with the changing environment inside the box. Not only the temperature is decisive for the succession. Physical and chemical parameters are important, but also the relationships between the microorganisms.

pH-value

At the beginning of the decomposition, the pH-value is about 5,5. During the collection of organic material in households, the organic waste is often deposited under anaerobic conditions. This is the reason for a relatively high concentration of organic acids in the input material with a corresponding low pH-value. The anaerobic conditions promote the anaerobic organisms while the low ph-value promotes the yeasts. During the decomposition, the pH rises to a basic value. Together with the aeration, the higher pH stimulate the aerobic bacteria.

Water content

Indeed, not the water content in the organic material influences the microorganisms, but the water activity (the water activity aw is a size for the availability of water). Distilled water has the aw = 1). For microorganisms, a high water content is essential. Most bacteria particularly need a high aw value. The majority of yeasts and also mold are somewhat less sensitive to a lowering of aw. In order to promote the growth of microorganisms, water was added during the dynamic step.

Decomposition activity

The main target of decomposition is the decomposition of the so-called easily degradable organic matter. As easier a matter is degradable, as easier this matter can be used as nutrition for microorganisms. During decomposition, organic material is decomposed to carbon dioxyde, water, and energy. This signifies a decreasing amount of feed for the microorganisms.

Résumé

The succession of some groups of microorganisms were investigated during a 3-stage-box-passge (System Herhof).

– Along with modified living conditions, the population density of this microorganism groups has also changed.
– Population density of all microorganisms were the highest in the input material, except the amount of mold.
– Because of the aeration, the aerobic bacteria were promoted, while the growth of anaerobic bacteria were slowed down.
– The low pH-value at the beginning of the decomposition supported the yeast growing.
– The hygienization phase killed most Enterobacteriaceae and yeasts.
– At the end of the box decomposition, the compost was a finished compost according to the Federal Quality Commision (RAL-Gütesiegel).

Assessment of Compost Maturity

MICHAEL KÜHNER Chem.- Biol. ANDREAS SIHLER DIPL. BIOL – University of Stuttgart Institute for Sanitary Engineering, Water Quality and Solid Waste Management

Abstract

For composting firms, compost producers and compost users it is very important to measure the compost maturity, cause of process control, process optimation, comparison of the process efficiency of different composting systems and application effects of the compost itself. Founded on experiments in aerated and non-aerated composting systems new aspects for the assessment of stabilization during biowaste composting have been achieved.

It could be shown that the parameters respiratory activity and self heating rate are influenced by organic acids. This dependency seems to be the most important reason for the failures made in the measurement for compost maturity.

Based on these results the suitability of the parameters to characterize the real rotting process can be shown. Also suggestions for improvement are given to minimize the failures and to increase the transferability and the quality of the results.

Introduction

During biowaste composting it is very important to have standardized parameters with reliable methods to assess the stabilisation during biowaste composting. The degree of compost maturity can help

- to optimize and to control the composting process (Minimization of composting time, required area and working costs),
- to guarantee the effectiveness of the system,
- to assess the process efficiency of different composting systems,
- to ensure the quality and the application effects of the product (fresh compost or mature compost)

Based upon the research of Jourdan (JOURDAN (1988)) the LAGA established the self-heating test and the respiration activity test for the assessment of compost maiturity (LAGA M10, 1985). The maximum heat (Tmax) produced in a 1,5 l Dewar-vessel (self-heating test) and the oxygen consumption in four days (respiration activity AT4) classify the compost in five degrees of maturity (I – V) called „Rottegrad (RG)' [fresh compost (RG II–III) and mature compost (RG IV–V)].

The suitability of these parameters to characterize the real composting process must be questioned because several experiments showed that there are difficulties in correlating the two parameters with other methods for the assessment of stabilisation during biowaste composting. It happened that during an aerated composting process the degree of maturity (RG) suddenly decreased from V to II when the compost was no longer areated. SIHLER und BIDLINGMAIER (1992) showed that the degree of maturity could vary between II and V although the degree of decomposition was higher than 50–60%. Futhermore in a parallel interlaboratory test of the „Bundesgütegemeinschaft Kompost e. V.' (BIDLING-MAIER, W., MAILE, A:, 1994) the degree of composting determined by the self-heating test was emphasized as an particularly critical parameter. One third of the laboratories definitely overestimated the degree of maturity.

One reason for this could be that these two methods were not usuable for biowaste compost because they were evaluated on garbage and sewage sludge composts.

The following results give suggestions to minimize the failures and to increase the transferability and the quality of the assessment of compost maturity.

Methods

Different methods for the assessment of compost maturity of biowaste compost have been investigated in areated and non-aerated composting systems. The main parameters tested were the respiration activity, self-heating rate, BOD_5 and COD of the water extract, degree of decomposition, C/N–, NH_4–, NO_2– and NO_3– ratio. To get the optimum humidity in the compost sample for the biological tests the so-called „Faustprobe (fist-check)' from the BGK (1994) was used.

Results

Influence from very easy degradable, water soluble substances on the parameterss self-heating rate and respiratory activity.

Beyond the „classical' parameters the BOD5 and the COD from the water extract of the compost sample were investigated. It was proved that the appearance of these water soluble substances have a very big influence on the self-heating rate and respiratory activity and so the characterization of the real rotting process was not possible. To minimize this influence, a modified parameter, the corrected respiration activity AT4korr was introduced.

$$AT_{4korr} = AT_4 - BOD_5$$

Fig. 1 and Fig. 2 shows the composting process by different parameters in a non-areated static pile composting.

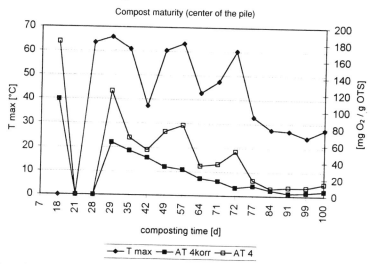

Figure.1 Assessment of compost maturity samples taken on the center of an non-areated static pile

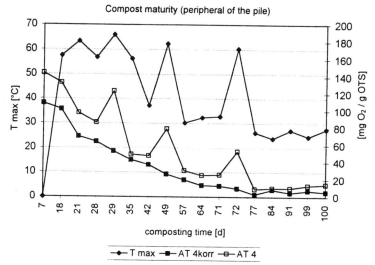

Figure 2 Assessment of Compost maturity. Samples taken from the periphery of a non-areated static pile

It could be shown that

- the self-heating test and the respiration activity were extremely influenced by this BOD5- fraction so that the real degree of compost maturity could not be classified.
- high production of heat in the self-heating test after 5 weeks of composting depends on the existence of a great BOD5- fraction. Water soluble substances are necessary for the heat production.
- The place where the sample is taken from (anaerobic zone in the center of the pile or aerobic zone on the periphery of the pile) and the time when the sample is taken [i.e. before or after a mechanical composting process (e.g. sieving, shreddering, turning over)] have an influence on the amount of the BOD5 and as a consequence also on the degree of maturity (e.g. before shreddering RG IV, after sheddrering RG I).
- only the modified respiration activity [AT4korr] characterizes the real rottting process.

Characterisation of the BOD5 - fraction

Parallel the examination of the compost eluate on DEV H21, modified from Kapp (1984), characterized the amount of organic acids in this fraction. The theoretical oxygen demand of this organic acids (acetic-, propionic-, butric- and valerianic acid) was nearly 90% of the BOD5 of this fraction. It could be said that mainly these organic acids caused the high oxygen consumption in the respirometric test and the heat in the self-heating test.

The concentration of these organic acids vary extremely during the composting process in dependence of the availability of oxygen for the microorganisms. The organic acids could be built or degradated in-between hours by the microorganisms. As a consequence samples with a long lasting sample transport under aerobic conditions will be overestimated in the degree of maturity (versus RG V) (because of strong microbilogical degredation of these organic acids) compared to the same sample tested immediatly. On the other hand it is also possible that a good areated sample during a long transport under anaerobic conditions will be underestimated in the degree of maturity (versus RG I) because of the increase of organic acids being formed by the hydrolysis of organic compounds by microorganisms.

Special features in aerated systems

The strong dependency of the tests on organic acids mentioned above could also be found in aerated systems. But in contrast to non-aerated systems there is the possibility of stripping these water soluable organic components out of the pile by air. This leads to an overestimated compost maturity (versus RG V). Experiments which tested different post composting conditions after an areated pre composting on the assessment of compost maturity showed that without aeration the degree of maturity RG decreased (versus RG I) in different times.

First results pointed out: in the same period of time the degree of maturity decreased (versus RG I) the more the higher the water content and the temperature and the lower the availability of oxygen in the post composting was. In addition to this it should be mentioned that the shorter the time of pre composting is the more decreases the degree of maturity.

Table 1 shows the change of the respiration activity under different conditions during post composting.

Thereby it is very interesting that these conditions have not only an influence on the existence of organic acids (increase of the BOD5), they also could increase the AT4korr. To explain this phenomenon further scientific research will be carried out.

Table 1 Simulation of an non aerated post compostion following an aerated compostion

| composting time [d] | original sample | | | sample after 7 days 20°C non-aerated s | | | sample after 7 days 40°C non-aerated | | |
| | [mgO$_2$/gOTS] | | | [mgO$_2$/gOTS] | | | [mgO$_2$/gOTS] | | |
	AT_4	BOD_5	AT_{4korr}	AT_4	BOD_5	AT_{4korr}	AT_4	BOD_5	AT_{4korr}
17	63,8	25,8	38,0	93,0	26,9	66,1	116,2	49,0	67,2

Problem with the method (LAGA M10)

Further tests which are not presented in this paper show that

- the adjustment of water content of the compost sample according to LAGA M10 (35 % for the self-heating test and 50% for the respiration test) is not applicable. Better results to reach the optimum humidity have been achieved by using the so-called 'Faustprobe (fist-check)'.
- the degree of maturity determined by the respiration activity is lower (versus RGI) than determined by the self-heating test. Therefore a new definition of the scale seems to be necessary.

Summary and new prospects

The examination of the maturing process during biowaste composting in non-areated and aerated systems showed that the parameters self-heating test and respiration activity can be extremely influenced by organic acids. This can lead to an incorrect assessment oft compost maturity dependent of external factors like time of sample taking (influences of operations during the process), non-homogenity of the pile, sample transport and -preparation and composting technique.

This influence can be minimized by more exactly studying the respiration activity and by using the modified parameter AT4korr. As the organic acids are decomposed under aerobic conditions inbetween 48 hours, the respiration activity should be measured after that period of time. The extention of the test time (or higher microbial activity because of higher test temperature) can increase the amount of organic registered (2%–10%) in the test, which leads to a better characterization oft compost maturity.

Furthermore buffer systems /mineral media can be used for obtaining comparable conditions.

Nevertheless other methods, using material specific isolation and analytics, should be taken into consideration to get relevant and valuable information about the degradability of the material and for the assessment of compost maturity.

Taking the uncertainities mentioned above into consideration the quality of the method of assessing process efficiency and compost quality, based upon the self-heating test and respiration activity, has to be questioned.

References

BGK, 1994 Methodenhandbuch zur Analyse von Kompost. Bundesgütegemeinschaft e.V. (Edit.), Köln
JOURDAN, B., 1988 Zur Kennzeichnung des Rottegrades von Müll- und Klärschlammkomposten. Stuttgarter Berichte zur Abfallwirtschaft, 30, Erich Schmidt Verlag, Bielefeld
KAPP, H., 1984 Schlammfaulung mit hohem Feststoffgehalt Stuttgarter Berichte zur Siedlungswasserwirtschaft, Band 86, Erich Schmidt Verlag, Bielefeld
LAGA M10, 1985 Qualitätskriterien und Anwendungsempfehlungen für Kompost aus Müll und Müll/Klärschlamm. Merkblatt 10 der Länderarbeitsgemeinschaft Abfall. Müllhandbuch, Erich Schmidt Verlag, Berlin
SIHLER, A., BIDLINGMAIER, W., 1992 Korrelation zwischen Abbaugrad und Rottegrad unveröffentlichter Bericht, Universität Stuttgart, Institut für Siedlungswasserbau, Wassergüte- und Abfallwirtschaft, Abteilung Siedlungsabfall

Accelerated Composting in Tunnels

Charlotta LindbergGicom Composting Systems B.V
Plein 11–13, 8256 AZ Biddinghuizen, The Netherlands

Gicom B.V. is developing and designing facilities for composting in tunnels. The composting process is fast, the system is closed, and the process control is optimized. The Gicom tunnel composting system is able to transform a variety of organic waste streams into compost, such as:

- biowaste (source separated domestic waste)
- municipal solid waste
- sewage sludge
- industrial sludges
- manure
- anaerobic digestion process residues
- contaminated soil

Since 1991, some 20 full-scale facilities for composting of biowaste or sludge, designed by Gicom, has been built or are under projection. They are located in the Netherlands, Germany, Spain, Austria and the United States. The treatment capacities range from 8.000 tons/year to 100.000 tons/year.

The Gicom tunnel composting system is based on accelerated batch composting in a closed box: the tunnel. The tunnel is divided by a patented perforated floor into two parts; the aeration plenum and the process room. The process air is blown from the plenum through the perforated floor into the process room where the composting takes place.

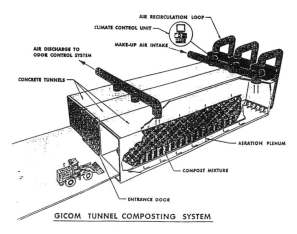

GICOM TUNNEL COMPOSTING SYSTEM

Biowaste is transformed into compost in two steps of together three weeks. The normal overall mass reduction is 60%. The content of dry matter in the compost is around 70%. When composting sewage sludge, the total loss of weight is about 85% after 14 days. The dry content of the composted sludge is about 60%.

The percolate is recirculated, and normally there is no wastewater effluent at all. Also the process air is recirculated, and is cleared from ammonia and odour by use of a scrubber and a biofilter prior to being let out. The environmental impact of a Gicom tunnel composting facility is thus minimized.

The recirculation of the process air makes it possible to obtain an axcellent process control. The climate in each tunnel is measured and regulated by a process control computer. All computer systems used are developed by Gicom. The following parameters are monitored for each tunnel:

• compost temperature

- air temperature and humidity
- oxygen/carbon dioxide concentration
- fresh air flow
- recirculation air flow
- static pressure

The results of the measurements are sent to the Gicom process computer, where they are compared with setpoint data. On basis of these comparisons, the process computer controls the climatisation unit. The climatisation unit, which is also used in the mushroom growing industry, is used at more than 180 plants all over the world.

Each Gicom composting facility is individually designed, to suit the needs of the client, including waste stream quantity and quality, the compost quality required and the economical situation.

The advantages with the Gicom tunnel composting system can be summarized as follows:

- excellent process control
- decreased detention time to produce compost
- flexibility towards input material and end products
- sophisticated odour control
- no devices in the process room - no corrosion
- working area separated from process area
- automatic process handling
- no waste water effluent
- fully enclosed active processing
- re-use of process energy
- world wide experience with full scale plants

Composting of Organic Garden and Kitchen Waste in Open-air Windrows: Influence of Turning Frequency on the Development of Aspergillus Fumigatus

JOHANNA LOTT FISCHER PIERRE-FRANÇOIS LYON, TRELLO BEFFA, and MICHEL ARAGNO – Microbiology Laboratory, University of Neuchâtel, rue Emile-Argand 11,2007 Neuchâtel, Switzerland

Introduction

Composting at industrial scale (installations treating . 100 tons per year) is increasingly used in Switzerland for the stabilization of organic waste and the recycling of humigenic materials. In 1992, 320.000 tons of garden and kitchen waste were treated in 150 composting sites, 80% thereof treating more than 1.000 t/a (1). Even if mature composts are generally of satisfactory quality, the process itself could often be considerably improved. Our research team is currently involved in a three-year study that aims at optimizing the thermogenic phase of composting processes, in order to ensure a good thermo-hygienization (elimination of allergenic and pathogenic micro-organisms, e.g. the thermotolerant mold Aspergillus fumigatus) and to improve the number and diversity of the thermophilic micro-organisms. This will lead to a rapid degradation of the organic material. By this, phytotoxicity hazards can also be avoided.

If the composting process is not correctly managed, it can induce the proliferation and dispersion of potentially pathogenic and/or allergenic, thermotolerant or thermophilic molds and bacteria, among which the most widespread is the mold Aspergillus fumigatus. Spore numbers exceeding 106/m3 air were measured at composting sites, especially during the turning of the windrows (2). This represents a bio-hazard for the personnel working at the composting site, but also for people living in the close vicinity of it. Aspergillus fumigatus is not only an opportunistic pathogen which can cause infections (aspergilloma, infectious aspergillosis) in immuno-depressed people, but also a powerful allergen, provoking immunoallergic diseases like allergic broncho-pulmonary aspergillosis and allergic alveolitis. Immunoallergic responses of individuals depend on their genetic disposition, on the frequency of exposure and on the number of inhaled spores(3).

For our study, we are working together with several industries, representing some of the most frequently used composting systems in Switzerland:

– About 2/3 of the compost in our country is produced in classic, open-air windrow systems. The windrows are mostly turned with specialized windrow

turning machines, the size of the trapezoid heaps is thus given by the dimensions of the turning machine (normally 3–4 m wide, and 1.5 m high). If necessary, water can be added to the compost during the turning.
- 1/10 of the compost is produced in boxes, which are either roofed or are inside a closed hall. An aeration system, controlled by the temperature measured inside the heaps, provides air to the compost to ensure sufficient oxygenation. Several automatic systems are used to turn the heaps, water can be added to the compost. Box composting is mostly used in larger composting plants.
- At several composting sites anaerobic fermentation of mainly unstructured, moisture-rich kitchen waste is done prior to composting. The biogas produced is used to turnish energy (electricity and heat) to the composting plant.

At the different composting sites of our industrial partners, we investigated the occurrence, the growth conditions and the dispersion of molds and bacteria during the whole length of the composting process, from the fresh substrate to the final compost. The results demonstrated a hygienization effect or, on the contrary, a colonization potential in function of the nature and composition of the initial substrate, the composting system and management, the stratification of the composting mass, the physico-chemical parameters (temperature, oxygenation, moisture, pH) and the degree of maturity.

In this paper, we will show the results of an experiment conducted at a classic open-air windrow composting plant, as this system is most currently used in Switzerland. The aim of this experiment was to monitor, during 8 weeks, the bacterial biodiversity and the thermohygienization towards Aspergillus fumigatus.

Material and Methods

At the composting site of Grenchen (SO), two windrows were put up, both 3.3 m wide, 1.2 m high and 25 m long (= 50 m³), consisting of 70% green waste, 10% kitchen waste, and 20% shredded wood. The initial C/N ratio was 25–28.

Windrow A was turned 5 x/week (= intensive management), windrow B was turned 1 x/week moderately extensive management) with a specialized windrow turning machine. Samples were taken daily for the first two weeks and then every second to third day, from the surface of the windrows (lateral at − 20 cm) and from the center (lateral at − 60 cm).

Temperature evolution was monitored on-line.

Colony forming units (cf) of Aspergillus fumigatus in compost were enumerated on Malt Extract Agar with antibiotics (Streptomycin and Novobiocin), after incubation at 40 °C for 48h. Results were expressed as cfu/g dry weight of compost.

Results and discussion

Figures 1A (intensive management) and 1B (extensive management) show the temperature evolution in the center and at the surface of the compost.

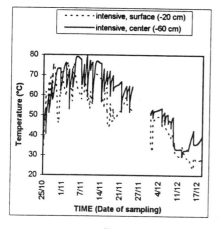

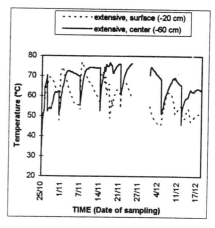

Fig. 1A   Fig. 1B

In the intensively treated compost, temperatures rose faster, more homogeneously, and reached higher values than in the extensively treated one. There, surface temperatures exceeded only momentarily (after the turnings) 70 °C, and fell for the rest of the time below 60 °C. Peak temperatures were reached after 2 weeks of composting for windrow A, and after 4–5 weeks for windrow B.

Figures 2A (intensive management) and 2B (extensive management) show the concentration of A.*spergillus fu*miga*t*us in the center and at the surface of the compost.

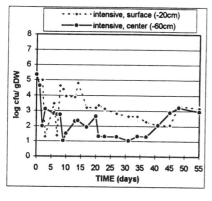

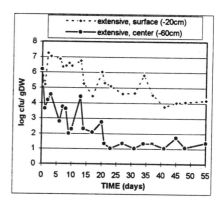

Fig. 2A   Fig. 2B

Concentration of Asperg*illus fumigatus* were low (between 10 and 100 cfu/gDW of compost) at the center of both windrows, after 2 weeks of composting. However, at the surface, A*spergill*us fumigatus counts were > 10.000 cfu/gDW during the whole composting process for the extensively treated windrow.

Measurements of the concentration of Asperg*illus fumigatus* spores and mycelium dispersed in the air 2 m behind the turning machine showed also clearly lower values when the intensively treated windrow was turned (results not shown).

Conclusions

– Daily turnings of compost effected a faster temperature rise in the windrow compared to turnings carried out only once a week. Also, degradation of organic matter, as shown by the temperature decline after three weeks of composting, was favored by an intensive treatment.
– Due to the frequent mixing of the composting material in the intensively treated windrow, Aspergillus fumigatus counts were low at the center as well as at the surface of the windrow. Whereas for the extensively treated windrow, a strong proliferation of Aspergillus fumigatus was observed in the colder surface layers.
– The lower Aspergillus fumigatus counts in the intensively turned windrow were reflected in the lower concentrations of mold propagates measured in the air during the turning of this windrow.
– Health risks for the personnel working on the composting sites can be diminished by an intensive treatment of the compost.

Acknowledgments

This research was funded by the Swiss National Science Foundation (grant 5002-038921) and by Alfred Muller AG, Baar and BRV SA, Bôle; Bühler AG, Uzwil; Compag AG, Gossau (SG) and Zweckverband Kompostier-anlage Kreuzlingen-Tägerwilen; Vollenweider Reisen und Transporte AG, Grenchen.

Literature cited

Schleiss K. and Chardonnens M. (1994): Stand und Entwicklung der Kompostierung in der Schweiz, 1993. Umweltmaterialien Nr. 21, Abfälle. Ed. by Bundesamt für Umwelt, Wald und Landschaft, Bern (CH)

Beffa T. et al. (1994): Anwesenheit, Verteilung und medizinische Aspekte von Schimmelpilzen (im besonderen von Aspergillus fumigatus) in verschiedenen Kompostsystemen der Schweiz. In: Gesundheitsrisiken bei der Entsorgung Kommunaler Abfälle. Ed. by Stalder K. and Verkoyen C, Verlag Die Werkstatt, Göttingen (D)

Staib F. (1991) Zunehmende Inzidenz tiefer Mykosen- Bundesgesundhbl. 5, 212–216

Sugarbeet Fertilization with Three (Sugarbeet) Vinasse Composts

MADEJÓN, E.*, DÍAZ, M.J.**, LÓPEZ, R.**, MURILLO, J.M.** and CABRERA, F.**

The recycling of the organic wastes from different industries could satisfy the increasing demand for organic materials in agriculture and horticulture.

Beet molasses are used as raw material for production of alcohol by distillation. For each litre of alcohol, nearly fifteen litres of a dark brown effluent known as vinasse are generated. The high salt content of the vinasse produced in the south of Spain limits its use for animal feeding. Therefore the use of this waste as fertilizer is being studied at present (López et al., 1993).Vinasse has three major problems for direct application as fertilizer: (i) high salt content (EC 250–300 dS m21), (ii) low P content (P_2O_5 0.012%) and (iii) its liquid dense character (1.3g cm23). These problems may be overcome through the co-composting of vinasse with agricultural solid wastes, thus obtaining a compost which can be used as fertilizer.

In this paper, the effect of deep fertilization with three vinasse composts as an alternative to traditional mineral fertilizer on sugarbeet is considered. Nutritional status, yield and quality of sugarbeet cultivated in a sandy loam soil fertilized with three vinasse composts and a mineral fertilizer were compared.

Three mixtures of vinasse and agricultural solid wastes were co-composted in static piles with forced aeration during four months. The initial proportion of solid wastes and vinasse were: Compost G: grape marc (82%) + sugarbeet factory lime (1%) + vinasse (17%); Compost O: olive pressed cake (76%) + sugarbeet factory lime (1%) + leonardite (6%) + vinasse (17%); Compost C: cotton gin trash (47%) + sugarbeet factory lime (1%) + leonardite (3%) + vinasse (49%). The chemical analysis of the three composts is shown in Table 1.

Some relevant characteristics of the soil at two different depths (20 and 40 cm) are given in Table 2. Field experiments were carried out in duplicated plots of 10 × 15 m, in which five treatments were tested. Each plot were subdivided in four subplots from where plants and roots samples were taken.The following doses for treatments were applied: TG 14,000 kg ha^{-1} of G; TO 22,000 kg ha^{-1} of O, TC 15,000 kg ha^{-1} of C; TF 600 kg ha^{-1} of a 9–18–27 N–P–K mineral fertilizer. Treatments TO and TC were complemented with 158 and 122 kg ha^{-1} of P_2O5 as superphosphate, respectively. A treatment, TB, without fertilization was used as control. All treatments, except TB, received two top dressings of urea (46% N),

EBRO AGRICOLAS S.A. Apdo. 9, 41300 S. José de la Rinconada, Sevilla, Spain. Instituto de Recursos Naturales y Agrobiología de Sedlla (CSIC). Apdo. 1052, 41080 Sevilla, Spain.

equivalent to 2×90 kg ha^{-1}. Sugarbeet c.v. Taurus was the test variety used for the experiment. Plant material was collected at 33 and 164 days after sowing. Mineral elements in leaves were analyzed acording to Jones et al., (1990). The data were analyzed by ANOVA and the differences between treatments were compared by Tukey's test.

Table 1 Chemical composition of the compost (Oven-dry basis)

COMPOST		G	O	C
Moisture	%	31	25	18
N-Kjeldahl	%	2.10	1.00	2.60
Plo5	%	0.70	0.13	0.28
K20	%	1.30	0.90	2.10
0M	%	50	70	51
Ha	%	1.70	1.30	2.40
Ca	%	2.80	1.90	1.30
Mg	%	0.30	0.20	0.40
C/N	%	12	34	8.2

The nutrients contents in leaves at 33 days after the sowing did not differ significantly among treatments (data not shown). The N, P, K, Ca, Mg and Ha contents in leaves at 164 days after the sowing are shown in Table 3. The nutrient contents in composts and mineral fertilizer treatments were higher than for TB treatment. For the compost treatments, the sodium contents were similar to that of the mineral treatment, despite the high Ha contents of the composts. Nutrient contents were within the usual ranges reported for similar climate conditions (Cantos, 1988).

Table 2 Analytical characteristics of the soil

		DEPTH(cm)	
PARAMETERS		0–20	20–40
Sand	(%)	79.4	81.3
Silt	(%)	10.6	9.7
Clay	(%)	10	9
pH (H$_2$O)		8.1	8.2
CaCO$_3$	(%)	8.8	7.2
OM	(%)	0.8	
Kjeldahi–N	mg kg^{-1}	676	675
Available–P	mg kg^{-1}	16	
Available–K	mg kg^{-1}	175	205

Sugarbeet root yields for all treatments are shown in Figure 1. There was an apparent treatment effect on sugarbeet root yield. Plots fertilized with either of the composts or the mineral fertilizer gave significantly higher yields than plots without fertilization. For treatments TG, TC and TF, sugarbeet root yields were three-fold higher than for treatment TB, while the sugarbeet root for treatments TO was only two-fold higher than for treatment TB. No significant differences on sugarbeet root yield were observed between each of the three compost treatments and the mineral fertilizer treatment. The production of sugar followed the same pattern, where the highest sugar production was obtained for treatment TG (Figure 1).

Parameters indicating the quality of sugarbeet juice for sugar production were determined following the standard methods of the British Sugar Company (Table 4).

Table 3 Nutritional content in the ear leaf at 164 days afrer the sowing

TREATMENTS	N %	P %	K %	Ha %	Ca %	Mg %
TB	3.93 a	0.33 a	3.30 a	2.81 a	1.11 a	0.90 a
TG	4.83 c	0.41 c	4.34 c	3.68 b	1.28 a	1.41 c
TO	4.60 bc	0.37 ab	4.53 ab	4.37 c	1.13 a	1.08 b
TC	4.88 c	0.40 bc	3.94 bc	4.32 c	1.03 a	1.33 c
TF	4.53 bc	0.38 bc	4.09 bc	4.08 c	1.11 a	1.18 b

Values following by the same letter in the sane column do not differ significantly (P < 0.05).

Table 4 Sugarbeet quality

TREATMENTS	Red. sugar* %	Sugar %	Na meq/100g	K meq/100g	α-aminoacid meq/100g
TB	0.17 a	15.9 a	2.67 a	5.71 a	0.55 a
TG	0.15 a	15.8 a	3.07 a	6.41 bc	1.45 ab
TO	0.18 a	16.3 a	2.71 a	5.79 ab	1.30 ab
TC	0.16 a	16.6 a	3.01 a	7.01 c	2.01 b
TF	0.14 a	16.6 a	3.50 a	6.60c	2.06 b

Values following by the sane letter in the sane column do not differ significantly (P < 0.05).
* Reducing sugar.

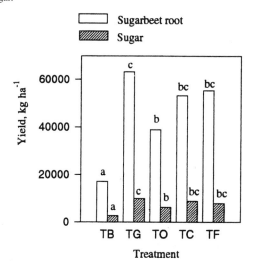

Figure 1 Yield of sugarbeet root and sugar for the different treatments. Means of data columns for each yield with same letters are not significantly different (P , 0.05).

There were not significant differences among treatment on the percentage of reducing sugar, sugar and Ha contents in beet root. For treatments TG, TC and TF, the K contents were significantly higher than for TB.

The negative influence that high N fertilization may produce on the technological sugarbeet quality (Draycott, et al., 1977) was not apparent since the a-aminoacid contents for composts and mineral fertilizer treatments were lower than the normal value (6.4 meq/100g) for sugarbeet under similar climate conditions (Cantos, 1988).

Results highlighted the use of compost as an alternative of traditional mineral fertilizer. Compost of vinasse and agroindustrial wastes had not detrimental effects on sugarbeet (yield, nutritional status and quality). Best results were observed for vinasse-grape marc compost (G).

References

Cantos, M. (1988). Calidad tecnológica de la remolacha azucarera de siembra otoñal en la zona sur de Andalucía Occidental. Ph. D. University of Cordoba, Spain.

Draycott, A.P., Durrant, M.J. and Last, P.J. (1974). Effect of fertilizers on sugar beet quality. Int. Sugar J. 76, 355–358.

Jones, J.B.Jr., Eck, H.V. y Voss, R. (1990). Plant analysis as an aid in fertilizing sugarbeet. In Westermen, R.L. (ed.) 'Soil testing and plant analysis'. Chap. 16 Madison, SSSA.

López, R., Cabrera, F. y Murillo, J.M. (1993). Effect of beet vinasse on radish seeding emergence and fresh weight production. Acta Horticulture 335, 115–119.

Waste Collection Utility Development

MAGAGNI A. and TOSETTI A.

Introduction

The CEC Directive 271/91 requires more stringent nitrogen and phosphorous standards in the wastewater discharged from treatment works. The reduction of the heavy euthropication problems in several European areas of the Mediterranean Sea (high Adriatic Sea) is a prioritary task, and the action is oriented at reducing the MSW (Municipal Solid Waste) addressed to land fills and /or incineration by increasing the performances of wastewater treatment plants.

This aspect, together with the necessity to find a solution to the urban solid waste disposal service,has led to the recovery of not very recent already-made projects that can propose new developments.

The solid waste collection utility (collection, transport and treatment phases) is organized in a different way from the urban liquid waste that is treated through the city sewer system.

In the F.U.S. 20 project the organic waste is treated through the sewer system, after trituration effected under the domestic wash-basin. Then wastewater are carried to a depuration district plant, where water and organic fraction are separated. In fact, the action planned by F.U.S. 20 project must be preceeded by transfering the source sorted municipal soild waste from the traditional road pick up system to the sewer line introducing food waste disposer in a quarter of Padova, and educate the families to use it properly.

Chronicle and feasibility development

In the '60 in the U.S.A. there was already the food waste disposer (particular tests were carried out in Los Angeles and in New York more recently).

In California it was given up for problems of water supplying. In fact great delivery of water was required to obtain the necessary fluency to the disposer operation, because waste is made for about 60/70% of dry elements.

The F.U.S. 20 project concerns only organic waste both in the domestic treatment and in its utilization, requiring a low water quantity.

Today it is a topical subject as far as waste collection is concerned, to consider both the selective collection of humid waste and dry waste.

In the same way it is proposed the water 'carrier' and tritured waste but only for the humid elements.

Project aims

This research programme is focused on the possibility to realize an effective integration between the anaerobic solid waste treatment process and the biological wastewater tretaments processes.

The aim of this project can be divided into two options.

1) The first option is mainly based on the optimisation of the BNR through anaerobic fermentation of the organic fraction of municipal solid waste. The objectives are to increase the concentration of volatile fatty acid in the influent, approaching the activation of the high-rate phosphorous release kinetic (doubling the speed of this section of the process) and to check the effectiveness of the internal carbon source production. in the project it will be made use of an existing small wastewater plant (3.000 E.I. Equivalent Inhabitants) where the incoming flowrates are wastewaters of an experimental civil area, in which alla houses have food waste disposer to discharge the organic fraction of municipal solid wastes into the sewage. The proposal id to experiment the wastewater plant into two different configurations either utilising readily biodegradable carbon or slightly biodegradable carbon produced by the food waste disposer source sorted organic fraction of municipal solid wastes. This fraction could be obtained following two strategies:
a) using a balancing tank, located at the head of the plant, in which the fermentation processes will be performed in no striclty anaerobic conditions;
b) adopting an anaerobic fermenter, in which the primary sludgewhich contains high percentages of organics coming from the shredding of organic fraction of municipal solid waste, will be fermented in a mesophilic range of temperature.

The second option is mainly devoted to the energy balance of the wastewater plant and regards the possibility of co-digestion between the source sorted organic fraction of municipal solid waste coming from the families through the comminutors in the sewers and sewage sludge. This choice can be adopted by treating the mixed sludge (the shredded organic fraction of municipal solid waste and the sludge normally produced in the wastewater facility) using the anaerobic process.

Expected results

The results expected from these actions are related to the improvement of the performances of these biotechnologic techniques applied to the environmental protection. the two options previously described could give an imposrtant verification on a semi-real scale. In particular, these are the expected results:

- information about the sewers network behaviour when the source sorted organic fraction of municipal solid waste is added to the normal flowrate. Design guidelines commonly used by american engineers assume a BOD load of 100 gr. per person per day where municpal solid waste are used and BODS load of 80 gr. where municipal solid waste are not used.
- the wastewater inlet reachable concentration, in terms of increase of readily biodegradable carob (and especially specific, in terms of acetate and volatile fatty acids concentration) obtainable though the addition of source sorted organic fraction of municipal solid waste to the normal flowrate using this approach;
- an effective evaluation of the improvement of permrmances in a conventional configuration wastewater treatment plant snd in a biological nutrient removal configuration: more specifically to verify that it has been reached the high-rate kinetic about phosphorous release and acetate uptake described by Wentzel (1988);
- the feasibility of an energy recovery which allows an important reduction of management costs of the integrated process adopting the approach of the co-digestion of source sorted organic fraction of municipal solid waste and sewage sludge.
- to elucidate mechanisms governing both the hydrolysis of particulate substrates and the biological excess of phosphorous removal.
- the production of fuel gas from the supercritical wet gasification facility with a max of 20% organic fraction in the liquid phase, and in particular :
- production of a very clean gas easŷ to use in any kind of internal combustion engine without ignition trouble;
- possibility to feed the reactor with a organic fraction containing a few parts of papaer and plastic;
- short reaction time (5:10 min.) with the possibility to treat waste materials in small size reactors close to the production of waste.

Suitability of Composted Household Waste of Helsinki Metropolitan Area for Agriculture

MÄKELÄ-KURTTO, R.[1], SIPPOLA, J.[1], HÄNNINEN, K.[2] and PAAVILAINEN, J.[3]

Introduction

The number of inhabitants in Helsinki metropolitan area is 850000. They produce annually over 300000 tons of municipal waste, of which about one third is of rapidly decomposable biological origin. In April 1993, collection of source separated household waste and municipal biowaste was started in a large scale in the area by the the Helsinki Metropolitan Area Council (HMAC). This study is a part of the larger project conducted by the HMAC and the Technical Research Centre of Finland (VTT) which will find out ways to recycle the source separated biowaste. The aim of the present study was to clarify the suitability of the biowaste for agriculture with the aid of physico-chemical analyses and pot experiments which were carried out by the Agricultural Research Centre of Finland (MTT).

Material and methods

Source separated household waste was composted with wood chips in a volume ratio of 2 : 1 in windrows (2.5 m × 50 m × 40.0 m) in a sanitary landfill of Ämmässuo. Since August 1993, the ratio was 1 : 1. At first, a turning cycle was three weeks and since December 1993, two weeks.

During the present study, 12 composts were studied. Ten samples from different-aged, 2, 4, 6, 8, 9, 10, 12 and 18 month old, composts were analysed for general characteristics and for total and soluble macro- and microelements. Furthermore, two samples, one from a drum compost and another from a sewage sludge compost, were studied in a similar way. During the growing seasons of 1993 and 1994, experiments with barley (*Hordeum vulgare* L.) were carried out in Kick-Brauman pots in outdoor conditions supplied with watering. Compost fertilization respecting the application rates of 24 and 96 tons ha^{-1} DM was studied in

1 Agricultural Research Centre of Finland, Institute of Soils and Environment, FIN-31600 Jokioinen, Finland
2 Technical Research Centre of Finland, Energy, Koivurannantie 1, FIN-40100 Jyv∑skyl∑, Finland
3 Helsinki Metropolitan Area Council, Opastinsilta 6 A, FIN-00520 Helsinki, Finland

finesand (6 liter of soil/pot) with or without N supplement. Comparisons were made to NPK-fertilization and no fertilization.

Results

Pot experiments indicated that household waste composted for 2 or 4 months was unsuitable for agriculture due to inhibiting effects on barley growth. These composts considered as unmature had a high NH_4–N/NO_3–N -ratio and a pH less than 7. Household waste composted for 6, 8, 9, 10, 12 and 18 months had not growth inhibiting effects on barley and seemed to be suitable for cereal cultivation. When the soil received only compost fertilization at application rates of 24 or 96 tons DM ha^{-1}, green mass yields of barley were of the same low level as without any fertilization. When the compost fertilization was supplemented with N, grain and straw yields of barley reached or even exceeded those obtained with NPK-fertilization. In addition to this, compost fertilization supplemented with N had positive effects on the quality of the yields increasing the concentrations of P, Cu and Zn in grains. In general, compost fertilization increased soil humus content and fertility and decreased soil acidity.

Fertilizing value of the composts seems to be based on P and especially on K, but to a lesser extent on N. Mean concentrations of total N, P and K in the eight household composts suitable for agriculture were 17, 5 and 19 g kg^{-1} DW, respectively. Only 2.5% of N, 35% of P and 55% of K was in a soluble form, on the average. In addition to this, these composts contained Ca 23.1, Mg 3.2, Na 2.0, Al 20.4, Fe 8.4 g and Mn 211 mg kg^{-1} DW and B 2.6 mg l^{-1} of air dried material. Mean pH(H_2O) of the composts was 7.5, humus content 41% and C/N -ratio 14. Concentrations of heavy metals of all the 12 composts studied were very low: Hg 0.171, Cd 0.57, As 7.8, Ni 7.3, Pb 40, Cu 81 and Zn 182 mg kg^{-1} DW, on the average. The concentrations were clearly lower than the limit values set for soil improvers by the Finnish government in 1994: Hg 2.0, Cd 3.0, As 50, Ni 100, Pb 150, Cu 600 and Zn 1500 mg kg^{-1} FW. In general, solubility of most nutrients and heavy metals in composts decreased, when the composting time proceeded.

Conclusions

Household waste composted in windrows for 6 months or more seem to be mature and suitable for agriculture. These composts also meet the quality requirements set by the authorities for soil improvers. The composts can be considered as PK-fertilizers rather than as NPK-fertilizers. In addition to this, they are good liming and soil improving agents. Research results obtained in the present study give support to recycling of household waste in agriculture which, in turn, will proceed sustainable development in the society.

Reference

Hänninen, K. & Mäkelä-Kurtto, R. 1995. Erilliskerätyn biojätteen aumakompostointi ja kompostin käytt÷kelpoisuus (Abstract: Windrow Composting and Use of Source Separated Biowaste). Final report. VTT, Energy, PL 1603, FIN-40101 Jyväskylä and MTT, Institute of Soils and Environment, FIN-31600 Jokioinen. 58 p.

Effects of Compost on Soil Biological Fertility and Maize (*zea mays*) Production

S MARINARI, L BADALUCCO, S GREGO – Dipartimento di Agrobiologia e Agrochimica, Università della Tuscia, via S. Camillo de Lellis 01100 Viterbo Italy VC TOLEDO – Wye college Univ of London, Kent TN255AH UK

The use of compost for sustainable agriculture can suitably furnish the nutrients to plants and improve the soil fertility in terms of biological and chemical status. Effects of compost have been evaluated during the maize growth season as changes in the forage yield, the soil microbial biomass content, the soil metabolic quotient and the availability of nitrogen. Soil was analysed in correspondence of four different maize growth stages (1st sampling before compost and inorganic N application, 2nd sampling during stem elongation, 3th during flowering and 4th during milk maturity of maize).

Methods

Forage maize (var.LG2080) has been grown with different treatments:1) composted manure (50 t ha^{-1}); 2) composted manure (100 t ha^{-1}); 3) nitram (100 kg N ha^{-1}); 4) and nitram (200 kg N ha^{-1}); 5) control (no fertilizer). Soil samples have been taken at 5–10 cm of depth (6 subsamples for each block). The CO_2 evolution of soil, needed in order to estimate the metabolic quotient, has been determinated by trapping in 1N NaOH, according to Badalucco et al (1992). The fumigation-extraction method was used to estimate the soil microbial biomass content (Jenkinson 1988) in maize trials. Experimental blocks were replicated 3 times.

Results

Table 1 Carbon-Biomass at different maize growth stages.
Standard deviation never exceeded 13% of the mean value; ds= dry soil.

Treatments	1st sampling μg C g–1 ds	2nd sampling μg C g–1 ds	3rd sampling μg C g–1 ds	4th sampling μg C g–1 ds
Control	305.22	286.28	245.31	273.30
Compost 50	305.22	325.34	296.14	299.32
Compost 100	305.22	360.00	266.27	355.00
Nitram 100	305.22	272.27	361.92	246.43
Nitram 200	305.22	336.74	324.75	331.11

Table 2 Metabolic-Quotient of soil microbial biomass at different maize growth stages
Standard deviation never exceeded 15% of the mean value; ds= dry soil

Treatments	1st sampling $\mu g CO_2 h–1 g^{-1}$	2nd sampling $\mu g CO_2 h^{-1}/Bc$	3rd sampling $\mu g CO_2 h^{-1}/Bc$	4th sampling $\mu g CO_2 h^{-1}/Bc$
Control	0.006421	0.006636	0.006382	0.007317
Compost 50	0.006421	0.008893	0.007068	0.009410
Compost 100	0.006421	0.014586	0.006422	0.007593
Nitram 100	0.006421	0.006904	0.007140	0.008210
Nitram 200	0.006421	0.006751	0.007092	0.007704

Table 3 Yield of maize
Standard deviation never exceeded 18% of the mean value.

	Control	Compost 50	Compost 100	Nitram 100	Nitram 200
Forage yield dry weight $txha^{-1}$	4.10	3.16	5.70	5.13	6.50

Conclusion

After compost application, during stem elongation of maize, both soil microbial biomass (Table 1) and its activity increased (Table 2), whereas in soil treated with inorganic fertilizer these two parameters were not sgnificantly different from control (no fertilizer application). At flowering stage of maize the metabolic quotient was not significantly different between treatments presumably because the plant uptake induced a considerable competition with soil microbial biomass. As a general trend all treatments showed an increase of microbial biomass at the end of maize cycle, probably due to the new substrate for microorganisms came from rhizodeposition and plant tissue senescence. The forage maize production was higher with inorganic fertilizer application (Nitram 200 kg ha^{-1}), the improvement of soil biological activity after compost application had not repercussion on plant productivity in terms of forage maize yield (dry weight).

References

Badalucco L, Grego S, Dell'Orco S, Nannipieri P (1992 b) Effect of liming on some chemical, bio-chemical and microbiological properties of acid soils under spruce (Picea abies L.). Biol Fertil Soils 14:76–83.

Jenkinson DS 1988 In JR Wilson Ed, 'Advances in Nitrogen Cycling in Agricultural Ecosystems'. CAB International, Wallingford.

The Effect of Municipal Waste Compost on the Development and Viability of Ascarid Eggs.

H. J. MEEKINGS, E. I. STENTIFORD and D. L LEE.

Abstract

Ascariasis effects approximately one quarter of the world's population and is particularly prevalent in developing countries. As improvements in sanitation in most of these countries becomes more widespread, composting of human excreta and sewage sludge has become a popular and cost effective method for the disposal of these products, as well as offering a valuable source of organic fertiliser. If, however, the composting process is not properly controlled there may be a potential risk from pathogen and helminth infection to workers handling the compost and consumers of foodstuffs grown on land treated by the compost.

The work carried out has examined the effect of compost on the development and the viability of ascarid embryos within their protective eggshell, in an attempt to establish any mechanisms which may be involved in the destruction of these eggs. *Ascaridia galli* eggs were used as a model for *Ascaris lumbricoides* eggs. These were suspended for a period of two to three weeks in compost and micro-organism free compost extract, as well as in distilled water controls. In the first set of experiments, approximately four week old municipal waste compost was used, while in the following set of experiments approximately one week old compost was used. The results showed that both the one week old and the four week old compost extract retarded the development of the *Ascaridia galli* eggs suspended in them. This effect was seen to be more pronounced in the one week old compost extracts. No such delay in the development of the *A. galli* eggs was observed in the micro-organism free compost media. In spite of the observed delay in the development of the *A. galli* eggs suspended in the compost media, the viability of the eggs was unaffected. The observed delay in development was thought to have resulted from some form of microbial and/or fungal activity which would be present within the compost suspensions but not the micro-organisms free environments.

Introduction

Ascariasis is believed to effect one quarter of the worlds population causing 1 million cases of morbidity and 20 000 cases of mortality per year (Crompton *et al.*, 1985). The disease is caused by the presence within the intestine of a human host of an adult nematode worm *Ascaris lumbricoides*. If two or more adult *A. lumbricoides* worms are present inside an intestine, male and female, eggs are produced and passed in the faeces of the human host, in a single-cell state (plate 1). This disease may then be transmitted through the ingestion an egg once it has developed to its viable state (plate 2). Ascariasis is most prevalent in areas of poor sanitation, primarily in the developing world. With improvements in sanitation in these countries, composting of excreta, sewage sludge and municipal waste has become popular being seen as an effective method for their treatment (Mara & Cairncross, 1989). If the composting of these materials is carried out successfully, the final product offers a valuable, pathogen free, organic fertiliser. However, if the process is not properly controlled, then there may be a risk from pathogens, including Ascaris. Due to the resistant nature of Ascaris eggs, the presence or absence of the eggs in a compost can be used as an indicator of the pathogenic nature of the compost (Feachem, *et al.*, 1981). Present literature indicates that Ascaris eggs are destroyed by exposure to temperatures in excess of 50 (C for a period of greater than 1 day, this should easily be attained during theromphilic composting (Feachem, *et al.*, 1983). However, no mechanism other than heat has been thoroughly investigated as having any effect on Ascaris eggs during the composting process. A better understanding of how composting effects Ascaris eggs may lead to a re-examination of the temperature ceilings presently used, which may in turn have implications in operation time and costs.

The work carried out for this paper has began to examine weather factors other than temperature may be contributing towards the degeneration of Ascaris eggs during composting.

Methods

The effect of four week old and of one week old municipal waste composts on the development and viability of *Ascaridia galli* eggs were examined. *A. galli* eggs (eggs of the chicken ascarid) were used as a model for *Ascaris lumbricoides*. For each 'compost' sample under test, 5 g. of compost was suspended into 50 ml. of distilled water to which several thousand *A. galli* eggs were added. The effect of 'compost' and 'filtered compost' samples were compared to a control of distilled water. Each 'filtered compost' sample of 5 g. of compost added to 150 ml. of distilled water was filtered through a 0.45 (m. membrane before the *A. galli* eggs were added. To the 150 ml. volumes of distilled water used as controls several hundred *A. galli* eggs were added. All the samples (three of each type) were placed in a shaking incubator at 30 (C and aerated constantly (see fig. 1). This set-up was used in an attempt to simulate actual composting conditions without water or air becoming limiting. From each sample sub-samples were taken on a regular basis. These were examined to assess the stage of development of the *A. galli* eggs from the sub-sample, before being incubated for 28 days in 0.1N H_2SO_4 at 28 (2 (C under non-oxygen limiting conditions, after which the sub-samples were re-examined and the viability determined.

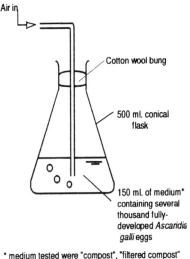

Air in

Cotton wool bung

500 ml. conical flask

150 ml. of medium* containing several thousand fully-developed *Ascaridia galli* eggs

* medium tested were "compost", "filtered compost" and a distilled water control

Figure 1 Example of apparatus used for one sample

Results and discussion

The 4 week old compost

On initial examination of the *A. galli* eggs from the 'compost' sub-samples the process of cell-division and development from single-cell zygotes to fully-developed larvae was observed. However, this rate of development when compared with that of the *A. galli* eggs from the 'filtered compost' and the distilled water control sub-samples, was retarded. By the end of the 14 day sampling period, not all the eggs examined from the 'compost' sub-samples had reached full-development. When compared with the distilled water control and 'filtered compost' samples (fig. 2a), significantly fewer fully-developed eggs were found from the 'compost' environment than from the distilled water control and 'filtered compost' environments, which were found to be similar to one another with respects to the percentages of fully-developed eggs observed (p 0.05). However, no significant differences were found between the increasing percentages of degenerate eggs found from the three environments (fig. 2b).

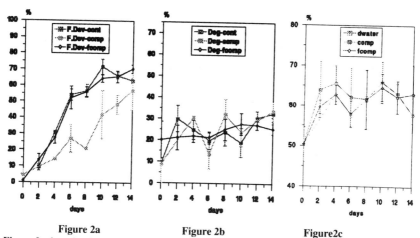

Figure 2a   **Figure 2b**   **Figure2c**

Figure 2a, b and c. Showing the percentages of fully-developed (a), degenerate (b) and viable (c). A galli eggs observed in the sub-samples from the 4 week old municipal waste compost experiment.

Following the 28 days incubation period, the percentages of fully-developed, viable A. galli eggs found from the control, 'filtered compost' and 'compost' environments were all found to be constant (p 0.05), with mean percentages of 63, 60 and 60% and were shown to be statistically similar to one another (p 0.05) (fig. 2c).

The 1 week old compost.

The development of the A. galli eggs from the 1 week old 'compost' was retarded when compared with the eggs sub-sampled from the distilled water control and 'filtered compost' environments. Over the 18 day sampling period the increasing percentage of fully-developed eggs was significantly smaller from the 'compost' sub-samples than from the 'filtered compost' and control, which were found to be significantly similar to one another (p 0.06) (fig. 3a). The percentages of degenerate eggs significantly increased over the sampling period in all the environments (p 0.05). Most degenerate eggs were observed in the 'compost' environment, while fewer were observed from the 'filtered compost' and control (fig. 3b).

After the 28 day incubation period, no significant differences were found between the percentages of fully-developed, viable eggs from the three environments (p 0.05)(fig. 3c). The percentages of fully-developed eggs found in all environments were found to be statistically constant with mean percentages of 35, 49 and 36 % from the distilled water control, 'filtered compost' and 'compost' environments, respectively.

Comparison between 1 week old and 4 week old composts.

A comparison of the trends in percentages of fully-developed and degenerate A.

galli eggs from the 1 week old and the 4 week old composts was undertaken. Allowances were made for the fact that the experiments were carried out on different occasions and had different controls. Comparing the 'compost' environments prior to incubation, a larger percentage of fully-developed *A. galli* eggs and fewer degenerate eggs were found in the sub-samples taken from the 4 week old compost. After incubation, a greater percentage of eggs from the 4 week old compost were found to be fully-developed and viable.

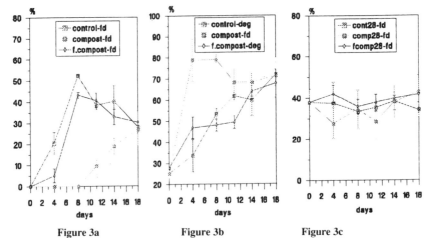

| Figure 3a | Figure 3b | Figure 3c |

Figure 3a, b and c. Showing the percentages of fully-developed (a), degenerate (b) and viable (c) A galli eggs observed in the sub-samples from the 1 week old municipal waste compost experiment.

The retarded development of the *A. galli* eggs only under the compost conditions during both experiments, was thought to be a result of the actions of the micro-organisms present in the compost. This was believed to be the case as the development of the eggs suspended in the 'filtered compost' were unaffected by their environment. The 'compost' and 'filtered compost' differed primarily from each other by the presence and absence of micro-organisms.

From the results, it appeared that something within the 'compost' environment and not within either the 'filtered compost' or distilled water control was responsible for the retarded development of *A. galli* eggs.

Once the eggs had been removed from the 'compost' environment and placed in favourable conditions for incubation (in 0.1 N H_2SO_4 for 28 days at 28 °C under non-oxygen limiting conditions), the delay in development was overcome and the observed viability was similar to those eggs expose to the control and 'filtered compost' environments. The most probable explanation for these observations is that it was the result of microbial action from within the compost environment. It was not believed that the chemical composition of the compost played a role in the delayed development. If this had been the case, one might expect the development of the eggs exposed to the 'filtered compost' conditions to be similar to those from the 'compost' conditions and not that of the distilled water control, as was observed. The maturity of the compost also was seen to have an effect on the rate of retardation. The more mature the compost, the less the retarding effect it had. This may be explained by the greater microbial action that would be expected during the earlier stages of composting. Specific oxygen uptake rates (method as described by Robinson, 1991, with minor adjustments) were determined for the two composts. The 1 week old compost showed a greater uptake rate than the 4 week old compost, indicating a greater amount of microbial action taking place in the less mature compost.

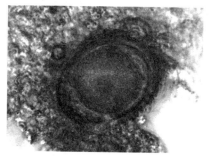 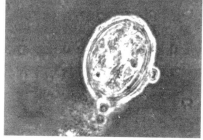

Plate 1 single-cell, undeveloped ascarid zygote   **Plate 2** Fully -developed, potentially viable ascarid ova

Conclusions

Compost was shown to have the effect of retarding the development of ascarid eggs for a short period of time. This delay in development of the eggs was greater when the eggs were exposed to more immature compost. Once the eggs were removed from the 'compost' environment and placed into more favourable incubation conditions, the delay in development was overcome and the eggs developed to viable, potentially infective eggs.

References

1. Crompton D. W. T., Neshem, M. C., Pawlowski, Z. S. *Ascariasis and its prevention and control.* 1989, Taylor Francis Ltd. London.
2. Feachem, R. G., Bradley, D. J., Garelick, H., Mara, D. D. *Sanitation and Disease - Health aspects of excreta and wastewater management.* , 1981, The World Bank.
3. Feachem, R. G., Bradley, D. J., Garelick, H, Mara, D. D. *Evaluation of composting process performance.* from *Proceedings of the International Conference on composting of solid wastas and slurries.* 1983, The University of Leeds.
4. Mara, D. D. & Cairncross, S. *Guidelines for the safe use of wastewater and excreta in agriculture and aquaculture.* 1989, WHO.

Fate of Lawn Care Pesticides During the Composting of Yard Trimmings

FREDERICK C. MICHEL Jr., DAN GRAEBER, LARRY J. FOMEY, C. ADINARAYANA REDDY – NSF-Center for Microbial Ecology, Michigan State University, East Lansing, MI 48824 phone: 517-355-6499, fax 353-1926; e-mail: 21394fcm@msu.edu

There is a growing emphasis on the composting of yard trimmings because of governmental bans on land-filling and incineration of leaves, grass and brush in many parts of the U.S. A primary concern with the composting of yard trimmings is the lack of knowledge about the fate of commonly used lawn care pesticides during composting. The purpose of this report is to summarize our recent findings on the fate of the most widely used lawn care pesticides 2,4–D (2,4–dichlorophenoxy acetic acid), Diazinon (O,O–diethyl O–[2–isopropyl–4–methyl–6–pyrimidinyl] phosphorothioate) and Pendimethalin (N–1–[ethylpropyl] 3,4–dimethyl 2,6–dinitrobenzamine) during the composting of yard trimmings.

The extent of 2,4–D, Diazinon and Pendimethalin mineralization, incorporation into humic matter, volatilization, and sorption during the composting of yard trimmings was determined using a laboratory scale compost reactor system described previously by Michel et al. (1995). Yard trimmings (2:1 leaves:grass, w/w) were amended with ^{14}C–ring-labeled pesticides at levels usually seen soon after application on grass (10–20 mg/kg dry weight) and water was added to 60% moisture content. The reactor system was programmed to rise from 25 °C to 60 °C over an eight day period and remain at this temperature through day 50. The amount of $[^{14}C]$ remaining in the composts was determined by oxidizing samples and trapping the evolved $^{14}CO_2$. The extent of pesticide volatilization was determined by passing the exhaust gas through two polyurethane foam traps which were extracted with organic solvents and counted as described by Michel et al., (1995). The distribution of the unmineralized pesticide in various fractions of the compost matrix was determined by sequential extraction of the compost as described by McCall et al. (1981). The organic solvents used were: ether for 2,4–D, chloroform for Diazinon, and hexane for Pendimethalin. The amount of $[^{14}C]$ in the humin fractions was determined by oxidizing the extracted compost and trapping and counting the evolved $^{14}CO_2$. The molecular weight of the humic transformation products in the NaOH extracts was determined by gel permeation chromatography as described by Michel et al., (1995). To determine the potential for pesticide residue leaching from compost, 10g of the [^{14}C] pesticide amended composts were pressed with a soil press until no more liquid was expressed and the [^{14}C] present in the expressed liquid was determined.

Results showed that during composting, 46% of the organic matter (OM) present in the yard trimmings was lost as CO_2, and the compost was stable, with an oxygen uptake rate of 0.09 (mg O_2/g OM/h), and was well humified (humification index = 0.39).

The 2,4–D mineralization was 48% after 50 days of composting. Most of the remaining 2,4–D was transformed into high molecular weight humic compounds or was unextractable (Table 1). In autoclaved control composters, the distribution of 2,4–D remained relatively unchanged as compared to the initial distribution indicating that the mineralization and humification processes were microbially mediated.

Diazinon was not readily mineralized and underwent transformation to products which were water soluble and potentially leachable (Table 1). Thinlayer chromatographic analysis of the water soluble extracts revealed that most of the added Diazinon was converted to the metabolite isopropyl-methyl-hydroxy-pyrimidine (IMHP) by day 10. Two other metabolites, the appearance of which coincided with an increase in the rate of mineralization of Diazinon in the compost, were also observed. About 20% of the ^{14}C from the ^{14}C Diazinon was present in the NaOH extractable fraction (Table 1). As with 2,4–D, the ^{14}C in this fraction had a molecular weight much higher than the parent compound.

Pendimethalin was not readily mineralized and a large fraction (65%) was transformed into unextractable residues during composting (Table 1).

Table 1 Distribution of [^{14}C] 2,4-D, Diazinon and Pendimethalin in various compost fractions

| | Pesticide [a,b] | | | | | |
| | 2,4-D | | Diazinon | | Pendimethalin | |
Fraction	Initial	Final[c]	Initial	Final[c]	Initial	Final[c]
^{14}C-CO_2	0	47.7 ± 6.7	0	10.9 ± 2.9	0	13.3 ± 4.9
Volatiles	0	0.4 ± 0.2	0	0.18 ± 0.1	0	2.5 ± 2.9
Organic[d]	92.4 ± 2.6	1.3 ± 0.5	82.9 ± 6.7	0.8 ± 0.2	66 ± 8.7	8.2 ± 5.5
Water	3.4 ± 2.2	2.5 ± 0.5	6.7 ± 0.4	36.3 ± 6.6	1.0 ± 0.3	3.9 ± 0.6
NaOH[e]	3.7 ± 0.7	29.1 ± 2.4	4.9 ± 4.4	19.8 ± 3.8	0.3 ± 0.5	6.6 ± 1.0
Unextractable	0.5 ± 0.1	19.5 ± 4.6	5.5 ± 0.6	31.9 ± 1.6	32.6 ± 9.4	65.4 ± 3.6
Recovery[f]	100 ± 4	97 ± 10	91 ± 4	90 ± 11	90 ± 2	96 ± 18

a-Values represent % distribution of added ^{14}C ± standard deviations for triplicate composters divided by recovery.

b-The initial amount of the pesticides added was 11 mg/kg of 2,4-D, 10 mg/kg of Diazinon and 10 mg/kg of Pendimethalin on a wet weight basis.

c-Final composts after 50–54 days of composting in laboratory scale compost reactor.

d-Organic solvents used were ether (2,4-D), chloroform (Diazinon) and hexane (Pendimethalin).

e-0.1 M NaOH + 0.1 M Na Pyrophosphate.

f-% Recovery equals the amount of radioactivity in the fractions divided by the initial amount of ^{14}C added.

Leachates from large-scale yard waste composting facilities are of concern since these may contain pesticide and other residues which could contaminate ground water. Although much of the added radioactivity from the ^{14}C pesticides remained in the final compost, much of the parent compound has either been transformed or mineralized or both and less than 2% of the original Diazinon, 1% of the 2,4–D, and 5% of the Pendimethalin was extractable using organic solvents. None of the pesticides was volatilized to any significant extent during composting, even at 60 °C. Most of the 2,4–D was bound to the water insoluble component of the compost matrix and was not readily leachable. Diazinon metabolites on the other hand became more water soluble and in the final compost, more than 30% of the applied ^{14}C was found in the expressed compost liquid. On the other hand, negligible amounts of the Pendimethalin were present in the expressed compost liquid after day 1.

In conclusion, the commonly used lawn care pesticides 2,4–D, Diazinon and Pendimethalin have very different fates during yard waste composting. 2,4–D is rapidly mineralized and transformed into high molecular weight compounds and very little of the added 2,4–D appears to be leachable during composting. Diazinon, on the other hand, is not readily mineralized but appears to undergo rapid transformation to water soluble metabolites during composting. Pendimethalin is converted primarily into unextractable residues. Our results also indicate that active mineralization of 2,4–D and biotransformation of Diazinon and Pendimethalin occurs at thermophilic composting temperaturers of 55–60 °C.

References

Michel Jr., F.C., Reddy C.A. and Forney L.J., (1993). Yard waste composting: Studies using different mixes of leaves and grass in a laboratory scale system. Compost Science and Utilization, **1**, 85–96.

Michel Jr., F.C., Reddy, C.A. and Forney L.J., (1995). Microbial degradation and humification of 2,4–Dichlorophenoxy acetic acid during the composting of yard trimmings. Applied and Environmental Microbiology, in press.

McCall, P.J., Vrona, S.A. and Kelley, S.S., (1981). Fate of uniformly carbon–14 ring labeled 2,4,5.trichlorophenoxyacetic acid and 2,4.dichlorophenoxyacetic acid. Journal of Agricultural and Food Chemistry, **29**, 100–107.

Characterization and Composting of Source-Separated Food Store Organics

FREDERICK C. MICHEL, Jr.,1 SUSAN DREW#,
LARRY J. FORNEY,1 C. ADINARAYANA REDDY1,*
– NSF-Center for Microbial Ecology, Michigan State
University, East Lansing, MI 48824

Composting of food store organic wastes is an attractive alternative to incineration, or disposal of these wastes in land-fills. In this study, the total wastes as well as the organic wastes generated by individual departments in a typical U.S. food store were quantified and characterized and the organic fraction of the wastes was composted with leaves to determine the qualities of the finished compost.

The volumes of total wastes generated at a representative food store which currently recycles corrugated cardboard, aluminum, plastic pallet wrap, wood pallets, polystyrene, meat scraps, plastic grocery bags and out-dated and damaged items was determined. The total amount of wastes generated was 1.90 ± 0.32 m^3/d which represented an average of 0.35 ± 0.03 m^3/employee/week and with an average bulk density of 79 kg/m^3. Paper bags and boxes used by customers to return cans and bottles, and waste generated at check-out stands (bottle/front end) constituted the greatest portion of waste (0.53 m^3/d) by volume. However, this waste had a low bulk density (13 kg/m^3) and accounted for only 6% of the total mass of waste generated. The produce, bakery, and delicatessen departments, respectively, accounted for 32%, 26%, and 21% of the total store waste by weight, and had high bulk densities of 241, 105, and 116 kg/m^3.

The wastes collected from each department were sorted and weighed to determine the relative amount of organic and non-compostable material produced in each department. The results showed that organics account for an average of 81% of the total wastes generated. In general, the compostable fraction was more dense than the non-compostable fraction (which was largely plastic containers and plastic film) such that only 64% of the total waste on a volume basis was compostable.

The organic fractions were reduced to a particle size of between 3 to 30 mm using a garden shredder and/or hammer mill. The moisture contents varied between 6.6% (bottle return/front end section) and 83% (produce section) and the organic matter contents were high ($> 93\%$). Organic wastes from the delicatessen, dairy and bakery departments had C:N ratios of 21, 21 and 17, respectively. The

#Resource Recycling Systems, Inc., 416 Longshore Drive, Ann Arbor, MI 48105
phone: 517–355–6499, fax 353–1926; e–mail: 21394fcm@msu.edu

floral and produce department wastes had C:N ratios of 31 while the bottle/front end department organics had a C:N ratio typical of paper wastes (C:N = 52 1).

The organic waste fractions were mixed in relative proportion to the total volume of these wastes produced by the food store chain as a whole and the moisture content was adjusted by adding dry leaves and water to give an initial moisture content of 55%. Three feed stocks were composted; the store organics alone, and 1:1 (Mix 1) and 1:2 (Mix 2) mixtures of store organics and leaves. Composting was done in a controlled laboratory scale compost reactor system which simulates the temperature (60 °C) and aeration environment found during windrow composting of yard trimmings (Michel *et al.* 1993).

The food store organics alone made a poor compost feed stock in that less than 1% of the organic matter was converted to CO_2 during the first 10 days of composting. A strong fatty acid odor developed and the oxygen uptake rate, an indicator of microbial activity, was relatively low (0.9 mg O_2/g OM/hr) as compared to a value of 2.3 mg O_2/g 0M/hr for fresh leaves and between 1.5 and 1.9 mg O_2/g OM/hr for Mixes 1 and 2, respectively. The low initial pH (4.8) of the store organics did not change during the first ten days of composting. The slow composting rate for the store organics may have been due to this low pH which would be expected to inhibit microbial growth. Because of the poor composting properties, the composting of the food store organics alone was stopped atter 10 days.

At the end of 54 days of composting, 60% of the organic matter originally present in Mix 1 and 47% of that present in Mix 2 was lost. The rate of organic matter loss from Mix 2 was relatively rapid even after 54 days of composting, while the loss of organic matter from Mix 1 slowed considerably after 40 days. The pH of the mixes climbed to pH 8.1 within 10 days and remained at that level through the end of the composting run.

Table 1 Characteristics of composts produced from two mixes of food store organics and leaves

Characteristic	Mix 1 (1:1)[a,c]	Mix 2 (1:2)[a,c]
Organic matter content (%)	53	54
Carbon (%)	27.8	28.8
Nitrogen (%)	2.3	1.8
C:N ratio	12	16
pH	8.15	8.20
Nitrogen loss (%)	25	17
Total phosphorus (%)	0.35	0.24
Total potassium (%)	0.91	0.50
Soluble salts (mMho)	4.75	2.34
Sodium[b] (ppm)	510	210
Chloride[b] (ppm)	2042	986
Stability (mg O_2/g OM/h)	0.09	0.19
Humification Index	0.19	0.22

a-Values are averages for replicate composters.
b-Values are averages for duplicate experimental runs determined by the water saturation method (Wade and Krauskopf 1983).
c-Mix 1 contained leaves:stores organics 1:1 and Mix 2 contained leaves: stores organics 2:1 on a dry weight basis.

The characteristics of the final composts produced from Mixes 1 and 2 (see Table 1) indicated that both Mixes had undergone extensive changes during the 54 day composting period. The GM ratio of the final composts was 12 for Mix 1 and 16 for Mix 2, down from the initial values of 32 and 36, respectively. At the end of composting, approximately 5 mg of nitrogen was lost per gram (dry weight) of Mix 1 but only 1.5 mg of nitrogen was lost per gram of Mix 2.

An analysis of the total amount of various mineral nutrients present in the finished composts from the two mixes showed that Mix 1, which contained 50% store organics, had roughly twice as much sodium, potassium and phosphorus as Mix 2, which contained 33% store organics (Table 1). Water saturation extracts (Warnke et al., 1983) of the finished composts from Mixes 1 and 2 both had low levels of nitrate and acceptable levels of phosphorus (Warnke and Krauskop1, 1983). The levels of magnesium, calcium and potassium, which compete for similar uptake sites in plants, were relatively unbalanced. Unacceptably high levels of available potassium, sodium, and Chloride were observed in both finished composts (Table 1).

The composts were stable and well humified after 54 days of composting. Mix 1 had an oxygen uptake rate of 0.09 while compost derived from Mix 2 had an 02 uptake rate of 0.19. Both Mix 1 and Mix 2 had humification index (Sequi et al., 1986) values of less than 0.4 indicating that both samples were well humified.

In conclusion, a large proportion (81% by weight) of food store wastes are compostable organics. These organics have a high bulk density, a high moisture content, a low C:N ratio and a low pH which make them difficult to compost without amendment. Amendment of store organics with leaves at a 1:1 and 1:2 ratio led to rapid rates of organic matter conversion, and the production of a stable and well humified compost end-product. The final composts produced were rich in mineral nutrients; however, the relative amounts of these nutrients were somewhat unbalanced. High salt levels were associated with higher levels of food store organics in the composting mix.

References

Michel Jr., F.C., Reddy, C.A. and Forney, L.J. (1993). Yard waste composting: Studies using different mixes of leaves and grass in a laboratory scale system. Compost Science and Utilization, 1, 85–96.

Sequi, P., M.DeNobili, M., Leita, L. and Cercignani, G. (1986). A new index of humification. Agrochimica, 30, 175–179.

Warncke, D.D. and Krauskopf, D.M. (1983). Greenhouse growth media: testing and nutrient guidelines, extension bulletin E-1736, MSU Cooperative Extension Service.

Heavy Metals Determination in MSW Merceological Classes and Derived Compost

MORSELLI LUCIANO, DONATI ALFREDO,
FORMIGONI DANIELE – Dipartimento di Chimica
Industriale e Dei Materiali Università degli Studi di
Bologna – Viale Risorgimento 4 – Bologna (Italy)

Introduction

Heavy metals (HM) are both hazard to human health and risk to the envinronment, the presence in composts has been estabilished as the main problem involved in the their use. The HM that are considered in italian legislation are As, Cd, Cr, Cu, Hg, Ni, Pb, Zn and limits are fixed for them in the compost and in soil (Ref. 1).

This paper deal with the HM content in Municipal Solid Waste(MSW), the relative distribution in the different merceological classes and the influence of each class on the compost characteristic. This is a focal point to select the method to reduce HM content (possible methods are for example: discarding screening fraction, preselection of waste, collection of refuse in separate bins, treatment technology to produce compost poor in HM) (Ref. 2, 3) .

In this paper were considered samples of wastes collected in three areas, characterized by a different human activities, in the Province of Mantova, North Italy. Generally is very difficult to realize a correct balance Input/Output of HM in a wastes treatment plants due to the heterogenity of the original material, different humidity percentage, different organic content, presence of hazardous wastes. This explains the difficulty in obtaining statistically meantingful samples on which carry out the pollutants trace analysis (Ref. 4).

To obtain a correct balance is necessary to use a correct methodology, including the following step: knowledge of the characteristics of refuse; principal productive activities in the area; diffusion of waste's separate collection ; merceological analysis of material in input to the composting plant; analytical analysis for determination of content in HM in each MSW merceological class ; evaluation of HM presence in the final product (compost); knowledge of technological condition and working process of plant

Experimental

In the province of Mantova sixtynine towns are involved in the ' CIME ' consortium, and in this area were collected the samples of waste analyzed refered to three town: city of Mantova, main city of the province (about 150.000 inahabitants), S.Benedetto Po, (characterized by a high penetration of pubblic separated collection of refuse), and Revere, (where the agriculture is the prevailing activity). The composition of waste was determined as indicated by official methods (Ref. 5), and the analysis were conducted as reported in bibliography (Ref. 6). The merceological classes composition is reported in Tab.1.

Table 1 MSW merceological composition (1992)

Merceological Class	Province of Mantova %	Italy %
Umidity	39	35
Textiles– Wood	11	9
Wastev size 0 –20 mm	18	10
Paper – Cartboard	23	28
Plastic – Rubber	13	15
Metals	5	5
Glass	3	5
Organic substances	26	28

Materials as batteries, glasses, papers, pharmaceutical products are collected separately in the three town considered and in S. Benedetto Po the pubblic collection is extended to some patrts of wood and textile also.

The contents of HM in the different merceological classes is reported in Tab.2 . After the treatment of refuse in the preselection plant (DANECO), the output material used to prepare the compost was composed respectively by 55% of waste size from 0 to 20 mm, 41% by organic substances and 3% by other classes of material (wood, paper, cartboard). During the composting process, the mass is subject to a relevant contraction of volume and weight (3:1) that caused a concentration of contaminants in the final product.

The teoric HM values estimated on the basis of HM content in the different merceological classes , the analytical values obtained and the law limit values(Ref. 1) are reported in fig.1. . No significative differences are found between teoric and mean analytical values for copper, nickel and lead, while for zinc the difference is significative. The maximum analytical values registred for Cadmium, Chromium and Nickel are near to teoric values. Arsenic and Mercury are present in compost in very low quantity (Ref.3)

Table 2 Heavy metals in MSW merceological classes (mg/kg)

Merc Cl.	Area	As	Cd	Cr	Cu	Hg	Ni	Pb	Sb	Zn
MSW	S. Ben	1,1	5,1	239,4	709,6	2,8	102,6	86	1,8	709
	Revere	4,8	0,7	41,8	37,5	12,0	46,9	62	0,3	38
	Mantova	1,4	0,1	4,2	37,9	11,7	2,1	118	n.d.	38
	Mean	2,4	2,0	85,1	261,7	8,8	50,5	89	0,7	262
waste	S. Ben.	1,6	15,0	59,9	184,0	3,6	258,6	155	1,3	184
size 0–20	Revere	1,8	1,2	24,6	34,7	4,8	20,6	56	0,2	35
	Mantova	4,4	0,5	1,5	82,0	25,5	2,9	106	0,1	82
	Mean	2,6	5,5	28,7	100,0	11,3	94,0	106	0,5	100
textiles	S. Ben.	0,9	3,8	41,3	27,9	n.r.	18,2	123	2,4	28
	Revere	n.d.	1,1	302,9	42,3	7,8	14,8	45	n.d.	42
	Mantova	1,4	1,4	n.d.	30,0	8,0	2,7	50	2,8	30
	Mean	0,7	2,1	114,7	33,4	5,3	11,9	73	1,7	33
wood	S. Ben.	n.d.	3,7	n.r.	22,7	3,3	31,2	15	0,6	23
	Revere	n.d.	0,3	4,5	17,2	5,7	n.d.	60	n.r.	17
	Mantova	n.d.	0,1	1,2	10,9	3,9	n.d.	75	n.d.	11
	Mean	n.d.	1,4	1,9	17,0	4,3	10,4	50	0,2	17
paper	S. Ben.	2,8	6,2	n.d.	150,7	n.r.	19,6	51	0,2	151
	Revere	1,7	0,7	8,6	19,7	6,5	4,9	98	0,2	20
	Mantova	n.d.	n.d.	n.d.	16,9	9,9	n.d.	96	n.d.	17
	Mean	1,5	2,3	2,9	62,4	5,5	8,2	82	0,1	62
cartb.	S. Ben.	0,8	2,8	23,1	25,0	n.d.	25,7	36	1,7	25
	Revere	0,7	0,6	n.d.	29,7	2,1	n.d.	n.d.	0,6	30
	Mantova	n.d.	n.d.	0,9	66,4	5,2	7,8	45	n.d.	66
	Mean	0,5	11,1	8,0	40,4	2,4	11,2	39	0,8	40
film pl.	S. Ben.	2,7	5,7	n.d.	117,7	5,7	12,7	278	3,4	118
	Revere	0,2	0,8	47,8	31,2	6,9	29,7	216	3,9	31
	Mantova	2,5	n.d.	45,0	18,0	3,4	n.d.	806	n.d.	18
	Mean	1,8	2,2	31,0	55,7	5,3	14,1	433	2,4	56
hard pl.	S. Ben.	n.d.	4,6	n.r.	28,6	2,8	179,7	91,	4,9	29
	Revere	n.d.	n.d.	18,5	32,9	1,8	n.d.	34	3,8	33
	Mantova	n.d.	0,7	n.d.	18,1	n.d.	n.d.	35	5,7	18
	Mean	n.d.	1,8	6,2	26,5	1,5	59,9	53	4,8	27
org. sub.	S. Ben.	0,2	5,0	26,6	44,6	8,9	22,0	53	n.r.	45
	Revere	22,4	0,9	19,3	31,7	2,2	12,2	31	n.r.	32
	Mantova	n.r.	n.r.	5,3	29,0	10,3	2,3	33	0,14	30
	Mean	7,5	2,0	17,1	35,4	7,2	11,9	39	0,05	36

Legend n.d. = not detectable

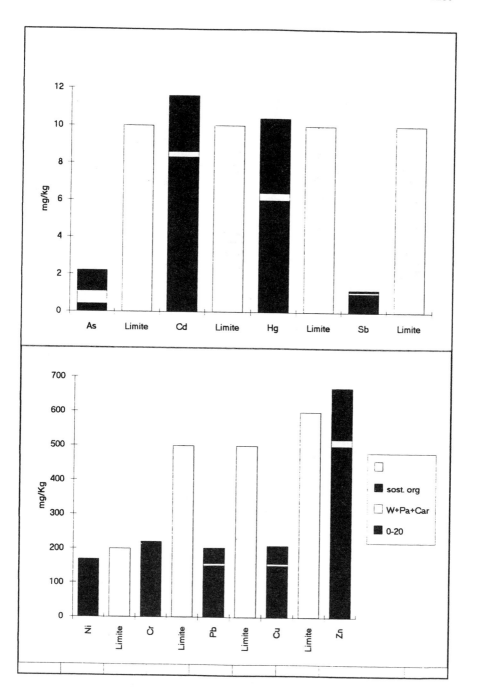

Figure 1 HM teoric values estimated/law limits

Conclusion

The nature of the heavy metals in municipal refuse influence the amount of HM in compost; fine fractions of refuse (<10 mm) are responsable for most of the heavy metal burden. In addition, batteries (mercury, cadmium and zinc), leather (chromium), paint (chromium, cadmium, lead, cadmium), plastics (cadmium) and paper (lead) have fallen under suspicion. Other studies(Ref.3) find that 'the native organic fraction' >40 mm, purified manually from remaining paper, plastics, glass, visible metal pieces and other nonorganics was characterized by high concentrations of Pb, Zn,Hg, Cu and Cd Further information concerning the nature of the heavy metals in the dense fraction was obtained by microscopical examination: metallic particles and partcles of metallic oxides were found in the dense fraction of <8 mm, 8–40mm and >40 mm organic material as well as in the corrisponding fraction of paper and plastics.

Potential lead souces are: pieces of lead-foil (e.g. from caps of wine bottles or from polaroid films), silver tinsel, bullets of air-guns, solder. copper was mainly found as wire and as particles of brass (small cog-weels, snap fasteners, zip fasteners, refills of balls pens). Zinc was found as foil-pieces and as brass.It is important to note, that small particles as a pieces of copper wire was found in the < 8mm material as well as in the >40 mm material.

High amount of HM in compost seem to be the main problem related to inorganic pollutants. Fine fraction of refuse (<20mm) are responsable for most of the HM burden (Ref. 8).

In order to reduce the heavy metal content in compost the possible measures are :

– separation of magnetic sensitive material
– separation of sieve fractions e.g. rejection of the <10 mm-material
– processing of matured compost (e.g. stoner or air table)
– pubblic colletions of domestic hazardous wastes
– another method of separating heavy metals from the remaining waste is to take advantage of their high density using air classification.

In addition there are some attempts to separately collect the non-polluted organic fraction (wastes from kitchen and garden).

Compost from municipal refuse poor in heavy metals can be obtained by different attempts. the most promising are separate collection of biomass and sophisticated technology for waste treatment using the phisical properties of metals (such as their high density). Intensive public heavy metal collection (e.g. by small plastic bags) is useful in addition to 'traditional' composting plants.

This paper represent an approach to understand the sources and fate of HM in composting process and to prevent cantaminations between correct input materials, composting techniques, working process of plants to refer law limits.

Biblography

Ref 1 : DPR 915/82, italian law 915/82 and successive Decree 27/7/84

Ref. 2 Morselli Inquinamento sett. 1993

Ref. 3 Krauss, Blessing, Korrherr ; in Compost : Production, Quality and Use –
Ed. Elsevier Applied

Ref. 4 Morselli L., Zappoli S., Milliterno S. 1993 ; Toxicological and Environmental Chemistriy vol.
37, pp 139–145

Ref 5 CNR 1980, PEE – LB3 1980

Ref. 6 IRSA CNR 1984 QUADERNO 64 FDR

Ref. 7 Morselli, Zappoli, Tirabassi Chemosphere Vol. 24 No. 12, 1992

Ref. 8 Lechner . International Congress Energy and material recovery from wastes.

Ref. 9 Genevini, Zaccheo . International Congress Energy and material recovery from wastes. Perugia
1988

Giardia Die Off in Anaerobically Digested Wastewater Sludge During Composting

N R MORT. H T HOFSTEDE and R A GIBBS – Institute for Environmental Science, Murdoch University, Perth, Western Australia 6150.

Introduction

In the Perth metropolitan area, sludge from primary and secondary treatment is anaerobically digested in a two phase, mesophilic process, before being dewatered by belt press or centrifugation. Currently, all of the sludge produced is either stored on site, or composted by contractors and sold as a soil amendment. The composting processes and the quality of the marketed product are currently not subject to regulation. However, sludge use guidelines are expected to be released, which will specify a range of requirements for sludge based products intended for land application.

Previous research by Gibbs et a*l*. (1994) and Hu (1994) has suggested that current sludge composting practices by at least one commercial operator in Perth are inadequate for pathogen destruction and result in a product which contains levels of salmonellae which are above the levels set in US EPA, Queensland and New South Wales sludge guidelines for sludge products suitable for unrestricted marketing (Beavers, 1993; Ross et a*l*., 1991; Sieger, Hermarn, 1993).

Research was undertaken to assess and optimize compost substrates containing anaerobically digested sewage sludge in terms of maximising pathogen die-off. Substrates were composted in small scale reactors, prior to selecting promising substrates for larger scale studies. The concentrations of a number of micro-organisms were monitored, including salmonellae and feacal coliforms. This paper focuses on Giardia die-off during anaerobic sludge composting in the small scale reactors.

Methods

Composting. Substrates were composted in 30 litre, insulated, polyethylene bins. Humidified air was passed through the bins at a rate of $1 \ \mathrm{l.min^{-1}}$ below 55 °C. At temperatures above 55 °C, air flow was increased to 3 l.min^{-1}. The system was developed for research by Hofstede (1994).

Six mixtures were selected in order to achieve initial C:N ratios (TOC:TKN) of greater than 20, moisture contents of between 50 and 60% and an approximately 3:1 ratio of bulking agent to sludge by volume. Substrate 5 contained unammended sludge to act as a control. A mixture of red mud (a bauxite refining residue) and 10% gypsum was added to substrates 6 and 8 to assess its effect on pathogen die-off.

Table 1 Compost mixtures used for pathogen die-off experiments

Material (kg)	1	2	3	4	5	6	7	8
Sludge	7.55	6.61	6.78	4.94	10	10	5.53	5.53
Skimmings	–	–	0.68	–	–	–	–	–
Brewery Waste	–	0.66	–	–	–	–	–	–
Grass Clippings	–	–	–	0.49	–	–	2.6	2.6
Sawdust	–	–	–	–	–	–	1.3	1.3
Shredded Tree Waste	3.3	3.3	3.3	3.3	–	–	2.6	2.6
RMG	–	–	–	–	–	1.11	1.05	
GM ratio	23.0	31.98	34.74	39.32	7.48	7.48	19.51	19.51
Moisture Content	60.27	59.44	58.22	53.92	79.3	73.22	59.01	64.19

Sampling. Samples were made up of approximately 20 × 25g sub-samples, taken randomly throughout the top two thirds of the compost mixture. Samples were taken with sterile spatulas or forceps and stored in sterile 2 litre plastic beakers. Samples were homogenized by adding 500 ml of sterile PBS (phosphate buffered saline) and mixing with a long sterile spatula or a sterilized hand held kitchen blender. Samples were analyzed for total and volatile solids (APHA et al., 1989) and pH (1:10 dilution).

Microbial analysis. Giardia, feacal coliform and feacal streptococci concentrations in the homogenate were determined using the method described by Gibbs et al. (1993).

Results and Discussion

The temperature, pH and solid loss of each compost substrate is sumarised in table 2. Temperature profiles of compost mixtures were below those specified in the US EPA sludge use guidelines for processes for the further reduction of pathogens, despite efforts to optimize starting conditions. This appeared to be due to a lack of insulation in the small scale bins. A 100m^3 enclosed forced aeration static windrow of substrate 2 maintained temperatures through out of greater than 55 °C for five days (unpublished data).

Samples for microbial analysis were taken at the start of composting, after the completion of the thermophilic phase, and one week after the end of the thermophilic phase. Giardia concentrations in mixtures 1–8 are shown in figure 1. The variation in initial concentrations is a result of dilution effects and variations in cyst recovery associated with the method used (60–99%, unpublished data).

Table 2 Summary of compost performance of substrates 1–8 in small scale reactors

Substrate	Temp. (Max.)	Days Temp. > 55 °C	Days Temp. > 45 °C	Min. pH	Max. pH	Total Solid Loss (kg)
1	46.1	0	3.12	6.55	8.40	0.08
2	52.9	0	1.38	6.65	8.67	0.48
3	57.6	0.33	2.00	678	9.02	0.38
4	52.3	0	4.17	7.35	8.46	0.23
5	31.1	0	0	7.10	8.31	0.38
6	26.7	0	0	7.02	8.15	0.68
7	52.7	0	1.62	6.64	8.06	0.38
8	59.0	0.79	1.67	6.5	8.18	0.68

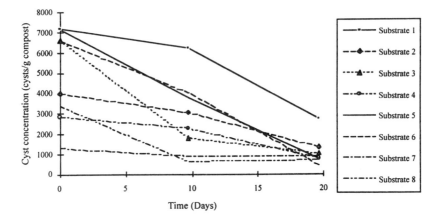

Figure 1 Giardia concentrations in compost substrates over time

For runs 1–8, final concentrations of Giardia cysts remained high, ranging from 4.8×10^2 cysts g^{-1} to 2.76×10 cysts g^{-1}. According to Gibbs et al. (1993), an individual ingesting 100 viable cysts has a 25% chance of contracting Giardiasis. At the highest level observed in the final product, 0.04g of compost contained 100 cysts. Unfortunately, the viability of the isolated cysts could not be determined due to the lack of a suitable method, although only intact cysts were counted.

No significant correlation between the concentration of Giardia cysts at the end of the composting process and the concentration of indicator organisms (feacal coliforms and feacal streptococcus) was observed (Mort 1995).

No significant differences between individual treatments (including a sludge only control) and die-off rates, log reduction times or final cyst concentrations were identified. High temperatures and moisture loss did not appear to promote the reduction of Giardia concentrations in the compost. Cyst concentrations and log cyst concentrations for each sample were pooled and compared to a range of factors such as moisture content, volatile solids, pH, feacal coliform and feacal streptococci concentrations. Again, no significant associations were identified. No effect on Giardia reduction from the addition of RMG was observed. However, differences between initial and final concentrations across all bins were significant (Mort, 1995).

Using the same methods as were used in this study, Gibbs et al., (1994) examined Giardia die-off in stored sludge over one year. In the first four weeks, a rapid decrease in Giardia concentration from an average of 4000 cysts per gram of sludge to 1000 cysts per gram was observed. Interestingly, after 60 weeks, Giardia concentrations remained around 1000 cysts per gram.

Conclusions

Based on this short term study, composting as a means of reducing Giardia cyst concentrations does not appear to be any more effective than simply storing the sludge over a similar period of time. Greater cyst destruction may occur in larger, more insulated systems which allow high temperatures to be maintained over longer periods.

Although Gibbs and Ho (1993) found that one of the greatest risks associated with sludge re-use was Giardia infection, the validity of monitoring Giardia cysts in a compost system has not been established. A means of testing the viability of Giardia cysts in sludge samples needs to be developed before the significance of cysts in garden products can be assessed, particularly given some reports that the viability of seeded cysts under mesophillic anaerobic digester conditions is reduced by 99.9% (Van Praagh et al., 1993; Gavaghan et al., 1993).

References

American Public Health Association, American Water Works Association and Water Pollution Control Federation. 1989. *Standard Methods for the Examination of Waste and Wastewater* (17th ed.). Edited by M. Franson. APHA, Washington.

Beavers, P.D. (1993). Guidelines for the use of biosolids – The Queensland scene. *Water*, December 1993., 23–26.

Gavaghan, P.D, Sykora, J.L., Jakubowski, W., Sorber, C.A., Sninsky, A.M., Lichte, M.D. and Keleti, G. (1993). Inactivation of *Giardia* by anaerobic digestion of sludge. Water Science and Technology, 27 (3–4), pp.111–114.

Gibbs, R.A., Ho, G.E., (1993). Health risks from pathogens in untreated wastewater sludge – Implications for Australian sludge management guidelines. *Water*, February, 1993. pp 17–22.

Gibbs, R.A., Hu., C., Ho, G., Unkovich, 1., Phillips, P. (1994). Die-off of human pathogens in stored wastewater sludge and sludge applied to land. UWRAA Research Project. 55–51 (91/58).

Hofstede, H.T. (1994). *Use Of Bauxite Refining Residue To Reduce The Mobility Of Heavy Metals In Municipal Waste Compost* - PhD Thedis. Murdoch University (Perth).

Hu, C. (1994). *Development Of A Presence/Absence Test For Salmonellae In Wastewater Sludge.* Honors Thesis (Environmental Science, Murdoch University).

Mort, N.R. (1995). *Disinfection of Anaerobically Digested Sewage Sludge Through Composting.* Honors Thesis (Environmental Science, Murdoch University).

Sieger, R.B., Hermann, G.J. (1993). Land requirements of the new sludge rules. *Water Engineering and Management*, August 1993, 30–32, 35.

Ross, A.D., Lawrie, R.A., Whatmuff M.S., Keneally, J.P., Awad, A.S. (1991). *Guidelines For The Use Of Sewage Sludge On Agricultural Land.* NSW Agriculture.

Van Praagh, A.D., Gavaghan, P.D., Sykora, J.L. (1993). *Giardia muris* cyst inactivation in anaerobically digested sludge. *Water Science Technology*, 27(3/4), pp. 105–109.

Effect of Sweet Sorghum Bagasse Compost on Sweet Sorghum Productivity in Pots.

NEGRO, M.J., CARRASCO, J.E., SAEZ, F., CIRIA, P., and SOLANO, M.L – Instituto de Energias Renovables. CIEMAT. Avda. Complutense, 22. 28040 MADRID. SPAIN

Abstract

The main purpose of this work is to study the effect of sweet sorghum bagasse compost application on biomass production of sweet sorghum (*Sorghum bicolor* (L.) Moench cv Dale). Biomass production was studied in terms of dry matter and stalk sugar content, at the end of the crop cycle. The addition of compost to soil produced significant increases on sorghum productivity with regard to the control. Best results have been achieved with the composts obtained from a sweet sorghum bagasse and pig manure mixture (30t/ha). In this case, the aerial dry biomass increments with regard to the mineral fertilizer treatment was 37%, where the irrigation was 2/3 of the available water.

Introduction

In the late years, sweet sorghum is being considered as a potential alternative crop for energy and industry (1) mainly because it can yield high biomass productivities as well as fermentable sugars.

Sugar extraction from sorghum stalks provides a cellulosic residue - bagasse- as byproduct. This residue represents about a 40% of the plant fresh weight and its elimination and recycling through composting has a great interest.

Different kinds of composts were obtained by using the bagasse of sweet sorghum as the main substrate when mixed with other types of residues. In general, all the products obtained had physical, physico-chemical and chemical characteristics suitable for being used in agriculture. However, it should be taken into account that physico-chemical characteristics of sorghum compost only provides an approximate information about its capacity as a fertilizer, because it is only possible to obtain a more accurate idea of the compost efficacy through the vegetal response. According to this, this work has been focused to the study of the effect that these composts could produce on a sorghum crop.

Materials and methods

In order to study the effects of compost addition to soil on sorghum growth and productivity. *Sorghum bicolor* vr. Dale was grown in 32 l capacity pots in open land. A poor soil from an agronomical point of view was used (pH 7, 0.6% organic matter, 0.05% total nitrogen and 6.5 and 25 ppm assimilable P and K respectively, texture: 68% sand, 26% silt and 6% clay). Two different composts were used. One of them was obtained from sweet sorghum bagasse and sewage-sludge mixture (SBSS) (75% organic matte, 39% carbon, 3% total nitrogen, 2% P_2O_5, 0.7 K_2O, pH (1:5 w/v water extract) 5.3, electrical conductivity (1:5 w/v water extract) 2.98 mmhos/cm) and the other one from bagasse and pig manure mixtures (SBPM) (75% organic matter, 37.6% carbon, 3% total nitrogen, 3.7% P_2O_5, 1% K_2O, pH 6.0, 3.4 mmhos/cm electrical conductivity).

Composting was performed in turning piles. The doses of compost from sorghum bagasse and pig manure added to the soil were 15 and 30 t/ha whereas in the case of the compost from sorghum bagasse and sewage sludge only the 15 t/ha dose was applied. In all cases, pots were supplemented, when it was necessary, with N, P, K mineral fertilization up to complete the following equivalent doses: N 120 Kg/ha, P2O5 90 Kg/ha and K2O 90 Kg/ha, considering that the annual availability of nitrogen, phosphorous and potassium from compost was taken to be equal to one third of the total N, K, and P contents. In order to study the influence of composts addition on the soil water retention, two different irrigation regimes were utilized (1/3 ($I_{1/3}$)and 2/3 ($I_{2/3}$) available water). Fifteen pots were tested for each treatment.

Unamended soils was used as a control. The crop was stablished on May 26th 1994 at CEDER in Lubia (Soria, North-central Spain Region) the harvesting was made on October 11th 1994.

The effect of compost was studied concerning to the following parameters: a) biomass production, results were expressed in Kg dry matter/m^2 and b) stalk sugar content. Sugar content were determined by HPLC. Results were expressed in percentage on stalks dry weight basis (glucose+saccharose+fructose).

All data were subjected to an analysis of variance (ANOVA).

Results

Compost obtained out sorghum bagasse and pig manure mixture, at 30t/ha dose, yielded highest productions either in total or aerial dry biomass in both irrigation rates, even yields higher than those reached with inorganic fertilizer were obtained (table I). When 15t/ha of this compost was added the yield obtained was similar to those reached with the inorganic fertilizer.

Table 1 Roots dry weight (RDW), aerial dry weight (ADW), total dry weight (TDW) production and total sugar content (TSC) percentage.

Irrigation rates		CONTROL	INORGANIC FERTILIZER	SBPM 15t/ha	SBPM 30t/ha	SBSS 15t/h
I2/3	RDW kg/m2	0.7c	1.4b	1.6ab	1.4ab	1.1c
	ADW kg/m2	1.4c	3.4b	3.5b	4.5a	2.7b
	TDW kg/m2	2.2d	5.1b	5.1b	6.2a	3.8c
	TSC (%)	17.2a	21.3a	20.3a	22.8a	21.0a
I1/3	RDW kg/m2	1.0c	1.9a	1.6b	1.5b	1.1c
	ADW kg/m2	1.4b	3.5a	3.6a	4.4a	3.8a
	TDW kg/m2	2.5c	5.6a	5.4a	5.8a	3.9b
	TSC (%)	17.3a	20.6a	20.4a	19.6a	20.0a

Figures followed by same letter in the same line are not significant differents ($p<0.05$).

These results are similar to those reported by other authors (2).

Lowest yields were obtained when compost from sorghum bagasse and sewage sludge mixture at a 15 t/ha dosage was applicated, that could be due to the compost itself or to a lowest organic matter mineralization rate, for this reason it could be necessary further studies in next years, in order to study the residual effect of the organic matter in the soil.

No significant differents were found between the two irrigation rates, except as far as roots are concerned when mineral fertilizer was used. Higher yields were obtained with the lowest irrigation rate.

Figure 1 shows the effect of the type of fertilization on the aerial dry biomass yield respect to the control (non amendment). When compost SBPM was utilized at a loading rate of 30t/ha the aerial biomass production is three times higher than the control, which resulting in 37% increasing yields compared to chemical fertilizer addition.

Concerning to stalk sugars production no significant differences between treatments and irrigation regimens, were observed. These results are similar to those obtained by Dercas and cols (3).

Sugar concentartion ranged about 20%. Their production is considered far below the results cited in literature for this cultivar. This could be due to the low temperatures registred by September-mids (average temperatures lower than 10ºC) what could have been resulting in that the cycle were not finished and then, the metabolism for surgar production had not been produced. Alternatively, this result could be related to an insufficient potassium fertilization.

To conclude, and according to the results obtained it can be said that compost utilization as soil amendment could reduce the amount of commercial nitrogen fertilizer applied, since the yields obtained with the assayed composts were similar or even higher to those obtained when inorganic fertilizer was used.

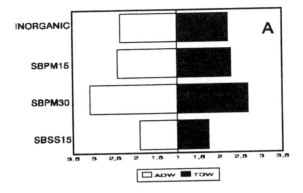

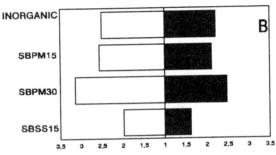

Figure 1 Relative values compared to the control(treatment biomass productivity/control biomass productivity)

References

1- Fernandez, J."Sweet sorghum: an alternative crop for Spain". In Sweet Sorghum, Ed. G. Grassi and G.Gosse, CEC DG XII, Brussels, 1990, 57-66.
2- Shiralipour, A.; McConnell, D.B. and Smith, W.H. (1992). Use and benefits of MSW compost: a review and an assessment". *Biomass and Bioenergy*, vol 2, n 3-4:267-279.
3- Dercas, P.; Panoutsou, S.; Dalianis, D. and Sooter, A. "Sweet sorghum (*Sorghum bicolor* (L.) Moench) response to four irrigation and two nitrogen fertilization rates". 8th European Conference on Biomass for Energy, Environment, Agriculture and Industry, Vienna 1994.

Acknowledgements

The authors thank A. González for HPLC analysis, A. Navarro and P.Soriano for their technical assistance and the CICYT (AGF93-1407-CE) and UE (AIR1-CT92-0041) for the financial support.

The Compodan Composting Process Features Rapid Process Set-off, Uniform Process Conditions and a Built-in Biofilter.

LARS KROGSGAARD NIELSEN

The composting plants for source separated household waste in Aarhus, Denmark, and Creusot-Montceau, France, are designed for the newly developed and patented Compodan process. The plant in Aarhus was put into operation in March 1995. The Compodan process features rapid process set-off, uniform process conditions and a built-in biofilter. In the reactor, the waste admitted on a daily basis is decomposed in 35 days.

The reactor is divided into at least 4 no-separated zones according to the extent of decomposition. The waste is introduced into zone 1 and compost is removed from zone 4, after being transported through the zones.

For the Compodan process a special aeration system design is applied. The process air from zones 2 and 3 is sucked out and recirculated to provide uniform process conditions in the compost.

Part of the air recirculated from zone 2 is blown through zone 1 from the bottom to heat the incoming waste so that the process sets off rapidly.

All exhaust air is sucked out through zone 4 and emitted into the atmosphere through a stack. The exhaust air is replaced by fresh air from the surroundings. Zone 4, which contains the most stable compost, thereby functions as a biofilter incorporated into the reactor where the filter material is continually renewed.

The air supply to zones 2 and 3 is adjusted according to a temperature set-point which ensures optimum decomposition. By sucking the process air down through the material an average air temperature is measured for each zone.

The Compodan process is developed on the basis of previous test results with one-way air injection (temperature set-points) and a dynamic model that simulates the composting processes in general. The dynamic model calculates the mass balance of decomposition of volatile solids, evaporation of water and amount of air injected to keep a steady temperature. The energy balance is calculated of energy generated from the decomposition, energy loss to heat up the air and waste, to evaporate the water and for conduction.

Compodan-process

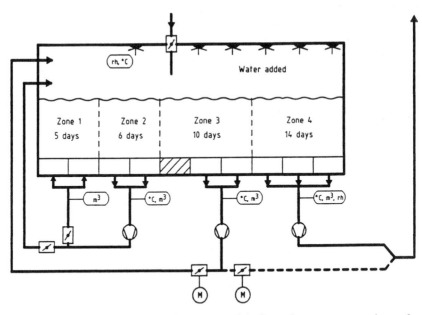

Table 1 Results of a static model for the air parameters of the Compodan process per metric ton of waste.

		zone 1	zone 2	zone 3	zone 4	zone 5
Retention time	days	5	6	10	14	–
VS decomposition	% of VS–start	10	19	14	5	–
Air flow	m3/h	3.5	10.5	10.5	9.5	–
Air flow	m3/(h*dry ton)	9.0	26.9	26.9	24.4	–
Energy to air	kJ/h	393	1124	828	296	–
Energy to air	kJ/kg	111	104	76.1	29.9	–
Temperature, air	_C	60	55	55	55	47.6
Relative humidity, air	%	99	98	90	76	100
Water, air	kg/kg	0.152	0.112	0.101	0.0836	0.0792
Enthalpy, air	kJ/kg	457	346	319	273	244

A static model for the air parameters in the Compodan process has been developed for the 4 zones and the air chamber over the compost (zone 5). Results of a calculation with the model based on data from Aarhus are shown in table 1.

Air flow through zones 2, 3 and 4 is calculated to be 10.5, 10.5 and 9.5 m³/(h*ton added waste) of which 3.5 m³/(h*ton added waste) is led to zone 1. In zone 5, air temperature is 47.6°C and relative humidity exceeds 100%.

The calculation is based on the assumption that the waste received contains 40% DS, of which 65% VS, and based on a retention time of 5, 6, 10 and 14 days, respectively, a corresponding VS degradation of 10, 19, 14 and 5%, respectively, and a relative humidity of 99, 98, 90 and 76, respectively in the 4 zones. The assumptions are shown in bold in table 1.

Until now no empirical data from the plant in Aarhus are available about the relative humidity in the process air. It seems reasonable that the value is dependent on the water content in the compost. Therefore a mathematic equation describing the relative humidity as a function of water content in the compost is used. Other variables such as the structure of the compost, the air velocity through the compost and the temperature may influence as well.

Based on the assumption that the compost contains 65% DS, the model calculates the water balance (per ton of waste); 610 kg H_2O in waste, 85 kg H_2O is generated during the respiration (0.7 kg H_2O/kg VS decomposed), 106 kg H_2O is added to fit the energy balance, 657 kg H_2O evaporates and 144 kg H_2O is found in the compost.

The energy balance shows that a total of 2385 MJ/ton of added waste is produced and 179 MJ is introduced with the air. 2266 MJ is refound in the exhaust air as the major part of the energy produced is consumed for evaporation of water. 167 MJ is used for heating of the waste and only 131 MJ is lost to conduction as the reactor is presumed to be insulated.

It is planned to combine the static and the dynamic model to achieve a dynamic model for the Compodan process. The model must incorporate empirical data about water evaporation to predict the relative humidity in the process air.

Demonstration of the Influence of Mg Vermiculite on the Activity of Cellulosic Agents and Diazotrophs During Composting of Lignocellulosic Residues

NUNTAGIJ A., KAEMMERER M., BIDEGAIN R and BRUN G – Laboratoire Ingenierie Agronomique, ENSAT-INP,145 Av. de Muret, 31076 Toulouse, France

Introduction

Adsorption and desorption properties, of a magnesian vermiculite (VMg), have been used to follow the dynamic of the NH_4^+ ion ,which has been produced during the 'cellulotytic-diazotroph' association appeared during the aerobic fermentation of a mixture of wheat straw and poplar sawdust.This fermentation has been regularly followed for three months in a mini digestor, and temperature, ventilation, pH, humidity, and C/N ratio of macro-elements were controlled. The same is done for the associative activity of cellulolytics and diazotrophes (NUNTAGIJ, *et* al., 1989). However, if fermentation is not carried out under optimal conditions, there is a loss of ammoniacal nitrogen, and that often occurs during large scale composting of organic residues (MARTIN, 1991). Clay properties can be used to control this loss. Four types of vermiculites are chosen to do so: a natural vermiculite (VMg), an NH_4^+ saturated vermiculite (NH_4^+), a calcium and ammonium-saturated vermiculite, only in exchangeable position $(VCaNH_4^+)$ and a vermiculite having exchangeable oligoelements (VMgCaCuCoMnZn).

Material and methods

- The aerobic fermentation is done in a horizontal mini digestor (2,4 litres)
- The lignocellulosic substrate is a mixture of wheat straw sieved to reach 2 mm and poplar sawdust grinded to reach 5 mm (3:1 w/w), to which has been added supplementary nutrients to facilitate the microbiological activity;
- The innoculum (1,5g) comes from a mixture in a process of being composted;–
    The magnesian vermiculite has a cation exchange capacity of 120 meq/100g and a period of 14.4Å (ANDRE,1972). The granulometry of the tinsels is located between 200 and 500μm thus, lateral outer surfaces are negligible rel-

ative to interfoliar inner surfaces. Moreover, the size of these clay particules allows them to be better recovered in the straw-sandust mixture.
- CO_2: the quantity of CO_2 discharged is determined by weighing absorption tubes filled with Soda lime (with Self-indicating).
- Decomposition rate: it is obtained by the ash rate
- Nitrogen: the nitrogen rate is expressed in relation to the initial dry matter

Results and discussion

The CO_2 productions, the variations of the organic matter decomposition rate and the nitrogen content are represented in figure 1. It has been observed that:

- in the presence of ammonium-'saturated' vermiculite ($VNH_4 1$) after the first peak of CO_2 production, identical to other fermentations one (9.5g CO_2Mg MS/j) with or without vermiculite (reference), due to glucose and the more easily biodegradable molecules consumption (De LASSUS, 1986), the microbiological activity remains very weak (between 2 and 3g/Kg of MS/j) between the 10th and the 120th day the final organic matter decomposition rate reaches 20% but no nitrogen fixation is observed.‰
- in the presence of magnesian vermiculite, calcium-ammonium vermiculite and vermiculite having oligoelements, global activity is close to the reference one the decomposition rate is near 50% the molecular nitrogen fixation began about the 25th day (VMg) or the 30th day ($VCNH_4 1$, VMgCaCoCuMnZn). The total quantities of fixed nitrogen reach 3.6‰ for VMgCaCoCuMnZn, 3.7‰ for VMg and 4.2‰ for VCaNH41, which have to be compared to the 3.5‰ of the reference. Total outputs of nitrogen fixation are 8.8 mg N/g of consumed substrate (C.S) for VMgCaCoCuMnZn, 9.2 mg N/g. C.S for VMg and 10.0 mg N/g. C.S for $VCaNH_4 1$, to 7.4 mg N/g C.S for the reference.

The NH_4^+ interfoliar content varies with time, according to the type of vermiculite (Tab. 1); with VNH_4^+, the content decreases from 70.7 to 53.3 meq/100g in 120 days, i.e. a 17.4 meq/100g V loss with $VCaNH_4^+$, it decreases from 46.9 to 54.5 meq/100g, i.e. a 8 meq/100g V gain; with VMg and VMgCaCoCuMnZn, it increases from to 0 to 22.2 and 19.6 meq/100 g V, i.e. a 2.5 increase during the same period of time.

All these results (total microbial activity, N2 fixation and interfoliar content) show that there is a relation between microbial activities and interfoliar content:

- in the presence of an interfoliar NH_4^+-saturated vermiculite, microorganisms use in priority interfoliar NH_4^+ nitrogen and diazotrophs do not fix N_2.
- in the presence of a $VCaNH_4^+$, VMg or VMgCaCoCuMnZn vermiculite, the absence or a lack of organic or mineral nitrogen, after a week of activity, favour the N_2 fixation. This fixation is all the more efficient since the nitrogen which is liberated, during the metabolism, as NH_4^+, is put into the interfoliar spaces.

This situation is temporary (ANDERSON et al, 1985) (and here again proved using VNH_4^+) like all exchangeable (Mc CALLA, 1939) or non exchangeable (Van BRAGG et al, 1980) mineral elements.

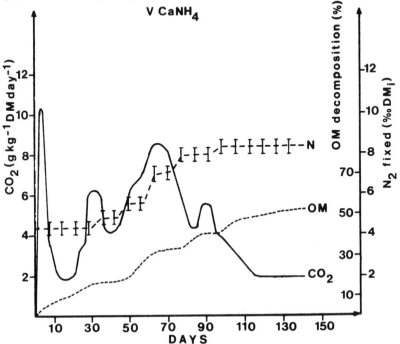

Fig. 1 CO_2 production; organic matter (OM) decomposition; N2 fixed, in a 1:3 w/w straw:sandust mixture with $VCaNH_4^+$

Table 1 The NH_4^+ interfoliar content with the time and the type of vermiculite

VERMICULITE	NH_4^+ interfoliar (meq/100g V)					
	0j	20j	30j	60j	90j	120j
V	–	–	–	–	–	–
VNH_4^+	70.7		58.4	55.3	54.3	53.3
$VCaNH_4^+$	46.9	48.0	51.6	54.5	54.5	
VMg	0	ε	10.0	12.5	16.1	22.2
VMgCaCoCuZn	0	ε	9.3	11.5	14.3	19.6

Conclusion

The presence of vermiculite was therefore shown to improve diazotrophic micro-biological activity. This was partly due to the release of certain interfoliar mineral elements (Ca^{++}, mg^{++} and oligoelements) and partly to the storage of NH_4^+, thereby permitting distribution of the nitrogen supply to plants over time and attenuating the risks of pollution due to poor control of chemical nitrogen fertilization.

Bbibliography

Anderson H.A. and Vaughan D., 1985. Soil nitrogen: its extraction, distribution and dynamics. In 'Soil organic matter and biological activity' ed. D. Vaughan and R.E. Malcolm. Martinus Nijhoff.

Andre I; 1972. Contribution à l'étude des mécanismes d'échange de cations dans les vermiculites tri-octaédriques. Thèse n° 503, Université Paul Sabatier, Toulouse.

De Lassus C., 1986. Fixation d'azote moléculaire par la sciure au cours de sa biodégradation. Thèse Docteur-Ingénieur, Institut National Polytechnique, Toulouse.

Mc Calla T.M., 1939. The adsorbed ions of colloidal clay as a factor in nitrogen fixation by Azotobacter. Soil Sci., 281–286.

Nuntagij A., De Lassus C., Sayag D. and Andre L., 1989. Aerobic nitrogen fixation during the biodegradation of lignocellulosic wastes. Biological Wastes, 29,43–61.

Van Bragg H.I., Fisher V. and Riga A., 1980. Fate of fertilizer nitrogen applied to winter wheat as Na NO3 and (NH4)2SO4 studied in microplots through a four-course rotation. 2, Fixed ammonium turnover and nitrogen reversion. Soil Sci., 130, 100–105.

A Comparison Between Chemical and Biological Index for Measuring Compost Quality

D.OTERO, M.GARCÍA and S. MATO Dpto. Recursos Naturales y Medio Ambiente Facultad de Ciencias. Universidad de Vigo Apdo. 874, 36200 Vigo (Pontevedra) SPAIN

Introduction

Although technologies for making compost are well developed, index for assessing the quality of product is a difficult issue. There are several parameters in the literature and the subject have been reviewed several times. Some problems related with topic of compost quality or maturity are the wide diversity of substrates which can be composted, the wide number of index proposed and the lack of results of applying the index to compost from different origins and nature. In our point of view the main difficult is the lack of a patron.

The poster presents a comparison between chemical and biological methods reported in the literature when are applying to compost from different origins and nature.

Material and Methods

Four types of samples were analysed:

a) Fresh biodegradable fraction of Municipal Solid Waste (**1a**) and mixed Municipal Solid Waste before composting (**2a**)
b) Compost from mixed Municipal Solid Waste with <2 mm grade and 90 day of composting (**3c,3d,3e**), <20mm grade and 15 days of composting sample (**2b**). Sample (1b) have a <20 mm grade and 15 days of composting. Samples coming from pilot plant experiments applying static and turning composting.
c) Sewage sludge mixed with forestry waste after 7 days of composting (**3f**).
d) Commercial samples of vermicompost made from animal manure (**2c,3a,3b**).

Chemical determinations were: C/N, Oxidable Carbon, Ammoniacal Nitrogen, Extractable and Water soluble Carbon, Fulvic and Humic fractions and ratios. The biological test was the germination index developed by Zucconi *et al.* (1981) for *Lepidium sativum*. Determinations were made with water soluble extract diluted to 30%.

Table 1 Chemical characteristics of samples.

	1a	1b	2a	2b	2c	3a	3b	3c	3d	3e	3f
Cox %dw	39	25	24	33	23	23	17	14	14	8	35
N %dw	2.4	2.2	1.0	0.9	1.4	1.9	1.3	1.7	1.2	0.9	1.1
N-NH4 mg/Kg	400	902	276	70	<10	<10	108	74	70	60	669
N-NOx mg/Kg	38	154	30	<10	59	<10	1498	45	20	273	32
pH	6.1	7.1	6.7	7.3	6.4	5.8	7.2	8.0	7.8	7.8	6.0
Cond. mS/cm	11.2	10	3.8	4.0	2.6	4.4	3.0	2.8	1.4	2.4	0.5
C/N	16	11	24	36	16	12	12	8	11	9	32

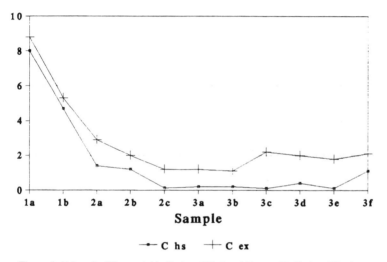

Figure 1 Values for Water-soluble Carbon (C hs) and Extractable Carbon (C ex)

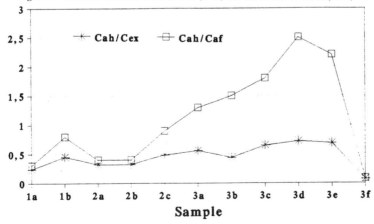

Figure 2 Values for Humic Carbon/Fulvic Carbon ratio Cah/Caf and Humic Carbon/Extractable Carbon/Extractable ratio Cah/Cex

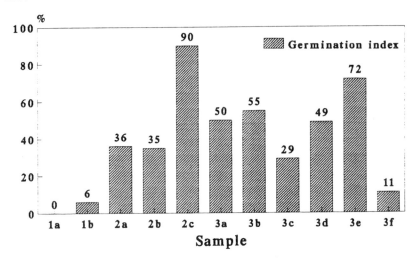

Figure 3 Values for Germination index

Results

Figure 1 and 2 show sample values for some maturity index reported in the literature and Figure 3 show values for Zucconi *et al.* germination index. Samples can be divided into three classes. Class 1 samples are in the first stages of composting and show high values for water-soluble, extractable Carbon, and low values for Humic Carbon/Extractable Carbon and Fulvic Carbon/Humic Carbon ratio. Class 3 samples show low values of water-soluble, extractable Carbon and Humic Carbon/Fulvic Carbon ratio. Samples of Class 2 have halfway values between other class. Class 2 and 3 samples are compost and is difficult to choose a value of chemical index for classifying Class 2 and 3 as mature because chemical index change stepwise. Often in the literature, time of composting is choosing as pointer of maturity, and chemical values at the end of composting as reference values. However it can lead to misunderstanding; samples 3c and 3e have identical composting time with different values for Humic Carbon/Extractable Carbon ratio.

When germination index is applied for classified samples, results are not so clear. Some samples can be classified as mature with chemical index but not if germination index is employed. Samples 3d and 3e have quite similar values for organic matter composition however different seed response. Identical response can be reported for samples 2c and 3a. For sample 3f, 1a and 1b high values of Ammonical nitrogen, salt content can be explain the differences between chemical and biological behaviour.

Reference

ZUCCONI, F., PERA, A., FORTE, M. & DE BERTOLDI, M. (1981). Evaluating toxicity of inmature compost. Biocycle. March/April: 54–57

Sewage Sludge – Soil Conditioner and Nutrient Source: I. Phosphorus availability and its uptake by ryegrass (*Lolium perenne L.*) grown in a pot experiment

*ERASMUS OTABBONG, STEFAN ATTERWALL, JAN PERSSON and ENOK HAAK – Department of Soil Sciences, Swedish University of Agricultural Sciences, Box 7014, S–750 07 Uppsala, Sweden

Abstract

The plant availability of municipal sewage sludge-borne phosphorus (MSS–P) and its effects on various soil P forms were investigated in a pot experiment using single superphosphate P (SSP–P) as the standard P source. Unlimed (control) and limed (pH 6.64, 7.19) samples of a silt loam soil [$pH(H_2O)$ = 5.72] were mixed with 134 mg P kg^{-1} dry soil in the form of SSP or MSS, labelled with 32P tracer and gown with ryegrass (Lolium perenne L.). After harvesting the rye-grass plants at the earing stage, the soils were sampled and extracted sequentially with resin in the Cl^- form followed by 0.5 M NaHCO3 (pH 8.5), 0.1 M NaOH and 1.0 M HCl solutions for P analysis. The sum of resin- and NaHC03–P levels as well as NaOH– and HCl–P levels were designated as the labile and nonlabile P pools, respectively. Relative MSS–P uptake established by 32P tracer averaged 8% (range 7.4–8.9), whereas the corresponding average MSS–P uptake established by the difference method was 7.5% (range 3.8–14.5), with lower values recorded in the limed soils. Liming and applications of SSP and MSS significantly (P = 0.05) increased soil levels of resin– and inorganic NaHCO3–P. Liming decreased organic NaOH–P levels, whereas organic NaHCO3–P and HCl–P levels did not significantly (P = 0.05) differ from that in the control treatment. MSS–treatments were, on average, less effective than the corresponding SSP–treatments at elevating the labile P pool sizes, whereas they were more effective than the latter at increasing nonlabile P pool sizes.

*5 Corresponding author.

Introduction

Household and industrial wastes are being produced at rates that exceed society's present capacity to safely dispose of them. Municipal sewage sludge (MSS) can be incorporated into agricultural land to improve soil physical properties and nutrient supplies, provided that it does not contain unacceptable amounts of toxic elements, such as heavy metals (Vigerust & Selmer–Olsen, 1985). Our objective was to assess the value of MSS as a source of phosphorus (P) for plants.

Materials and methods

Soil samples were taken from the 0–20 cm layer of a silt loam soil: pH in water 5.72, contents of clay, silt and sand = 33.4, 56.9 and 1.0% respectively. (Crops respond to P application on this soil.) Single superphosphate (SSP), the standard P source, or MSS stabilized with FeCl3 and containing 3.34% total P and 27.7% total organic C, the test P source, was applied to four replicates each of 1 kg unlimed and limed soils at a rate giving 134 mg P. In addition, 400 mg N and 400 mg K kg^{-1} dry soil pot^{-1} were applied to each replicate. Treatments were as follows: unlimed, limed (pH 6.64), limed (7.19), unlimed + SSP, unlimed + MSS, limed (pH 6.64) + SSP, limed (pH 7.19) + SSP, limed (pH 6.64) + MSS and limed (pH 7.19) + MSS. Carrier-free (31P) solution (Haak, 1993) had been added to all pots five days before ryegrass (Lolium perenne L.) seeds were sown. After harvesting the ryegrass plants at the earing stage, the soils were sampled and extracted sequentially with resin in the Cl$^-$ form followed by 0.5 M NaHCO3 (pH 8.5), 0.1 M NaOH and 1.0 M HCl solutions for P analysis. Analytical details are described elsewhere (Otabbong & Persson, 1994). The sum of resin– and NaHCO3–P levels as well as NaOH– and HCl–P levels were designated as labile and nonlabile P pools, respectively. The relative effect of the adding of MSS–P on labile P pool was calculated as follows: [(labile P level in MSS–treatment)/(labile P level in corresponding SSP treatment]* 100. Similar calculations were conducted for nonlabile P pool. The relative MSS–P uptake was computed using the 32P technique and difference method (D–method).

Table 1 Comparision of relative MSS-P uptake by plants established using the ^{32}P-technique and the D-method, with SSP-P used as the standard P source

pH(H$_2$O)	^{32}P-technique	D-method
4.98	7.7	14.1
6.64	8.9	3.8
7.19	7.4	4.5
Mean	8.0	7.5
LSD$_{(0.05)}$	0.6	2.9

Table 2 Effects of pH(H$_2$O), SSP- and MSS-P on soil levels (mg kg^1 dry soil) of resin- and NaHCO$_3$-extractable P fractions (labile P-pool = ΣP-L)

pH	P-addition	Resin-P	NaHCO$_3$-P		ΣP-L	Relative value
			Inorganic	Organic		
4.98	Nil	41	75	185	301	100
6.64	Nil	55	62	171	288	96
7.19	Nil	56	76	144	276	92
4.98	SSP	93	119	197	409	136
6.64	SSP	124	105	166	395	131
7.19	SSP	128	106	167	401	133
4.98	MSS	71	121	191	383	127
6.64	MSS	75	98	167	340	113
7.19	MSS	66	110	162	338	112
CV%		5	5	5	–	–
LSD$_{(0.05)}$		11	9	15	–	–

Table 3 Effects of pH(H$_2$O), SSP- and MSS-P on soil levels (mg kg^1 dry soil) of HCl- and NaOH-extractable P fractions (nonlabile P pool = ΣP-Nl)

pH	P-addition	HCl-P	NaOH-P		Σ(P-Nl)	Relative value
			Inorganic	Organic		
4.98	Nil	138	216	345	699	100
6.64	Nil	148	204	272	624	89
7.19	Nil	147	191	209	547	78
4.98	SSP	135	260	317	712	102
6.64	SSP	179	248	279	707	101
7.19	SSP	150	219	213	582	83
4.98	MSS	132	300	339	771	110
6.64	MSS	145	313	277	735	105
7.19	MSS	146	292	219	657	94
CV%		5	3	7	–	–
LSD$_{(0.05)}$		22	41	39	–	–

Table 4 Relative effects (%) of MSS-P on labile and nonlabile P pools

P pool	pH(H$_2$O) value			Average
	4.98	6.64	7.19	
Labile	94	86	84	88
Nonlabile	108	104	113	108

Results and discussion

Up take of MSS–P

The relative magnitude of MSS–P uptake by the plants established by the 32P technique was fairly similar (7.4–8.9%) regardless of the soil pH (Table 1), with the highest value recorded at a pH of 6.5. These results are in agreement with those published elsewhere (Kelling et al., 1977). Our results suggest that to obtain a level of P uptake similar to that reached in soil receiving a given amount of SSP–P, about ten times as MSS–P would be needed. However, the results sharply contrast with the ones established by the D–method, which ranged from 3.8 to 14.1 and indicated that MSS–P uptake decreased in the limed soils. This indicates that MSS–P solubility decreases in response to increasing pH relative to SSP–P.

Liming and applications of SSP and MSS markedly increased soil levels of resin- and inorganic NaHCO3–P (Table 2). Organic NaHCO3– and HCl–P levels did not significantly (P = 0.05) change relative to the control treatment, whereas organic NaOH–P levels decreased (Table 3). The latter P fraction was probably mineralized and subsequently entered the labile P pool.

On average, MSS–treatments were less effective than the corresponding SSP–treatments at elevating the labile P pool sizes, whereas they surpassed the latter at elevating the nonlabile P pool sizes. This pattern of effects was expected, however, and reminds that of phosphate rock and fused phosphate which solubilize slowly in soils (Otabbong & Persson, 1992, 1994). The P in the standard fertilizer, SSP, is in the form of $CaHPO4$ which is soluble in water, whereas P in the test fertilizer, MSS, was predominantly in the form of $FePO4$ which is soluble in alkaline solutions.

Concluding Statement

Based on the results presented in this paper, it is concluded that MSS–P solubilizes slowly and can, therefore, be used for corrective P–fertilization.

Acknowledgement

We are grateful to Dr Hans Lönsjö for constructive comments on the manuscript.

References

Haak, E. 1985. *Aspekter på användning av spårämnes teknik i mark växtforskning.* (Aspects on tracer use in soil-plant research.) In Swedish. Depart. Soil Sci. Box 7014, S–750 07 Uppsala.

Kelling, H.A., Peterson, A. E., Walsh, L.M., Ryan, J.A. & Keeny, D.R. 1977. Effect on crop yield and uptake of N and P. *J Environ. Qual.* **6**, 339–345.

Otabbong, E. & Persson, 1992. Relative agronomic merit of fused calcium phosphate. II. Dry matter production and P yields of rye grass (*Lolium perenne* L.) and barley (*Hordeum vulgare* L.) in pot experiment. *Fert. Res.* **32**, 269–277.

Otabbong. E. and Persson, J. 1994. Relative agronomic merit of fused calcium phosphate. III. Forms of phosphorus in soils repeatedly cropped in pot experiment. *Acta Agric. Scand. Sect. B. Soil and Plant Sci.* **44**, 2–11.

Vigerust, E. & Selmer-Olsen, 1985. Basis for metal limits relevant to sludge utilization. *Series B 4/85. Inst. for Jordkultur, Norges Landbrukshogskole*, 1432 Ås-NLH.

Composting of Fresh and Pond-Stored Olive-Mill Wastewater by the Rutgers Sytem

C. PARADES[a], J. CEGARRA[a], M.A. SÉNCHEZ MONEDERO[a], E. GALLI[a], F. FIORELLI[b].

Introduction.

In the Mediterranean areas a great stream of liquid waste, roughly estimated in about 10 million tons, is yearly produced by olive mill industry in a short rainy season (November–February). In spite of the existing laws, olive mill wastewaters (O.M.W.) are often disposed of in the environment or collected in lagoons, with a consequent pollution linked to odours, insect proliferation and sludge production. To solve the problem, both clearing and recycling have been proposed (Fiestas Ros de Ursinos & Borja Padilla, 1992). At present, recycling rather than cleaning seems to be the most suitable solution. From the standpoint of the waste recycling for agronomical uses, the direct disposal onto soil has been taken into consideration (Andrich et al., 1992; Riffaldi et al., 1993).

The aim of this study is to verify the efficiency of the composting process for transforming both fresh and stored O.M.W. into compost to supply onto soil as organic fertilizers.

Materials and methods.

Two different mixtures:
   MSO: 53.0% maize straw + 47.0% sewage sludge + fresh O.M.W. 1.765 l/Kg.
   MOs: 11.1% maize straw + 88.9% stored O.M.W.
were composted in two trapezoidal piles of 1–1.5 m high with a 2 × 3 m base in a pilot plant based on the Rutgers static pile system (Finstein et al., 1985). The air was blown from the base of pile, the timer was set for 30 s. ventilation every 15 minutes and ceiling temperature for continuous air blowing was 55 °C. After the biooxidative phase of composting, The air-blowing was then stopped to allow the compost to mature over a period of two months. The piles were sampled weekly

[a]Department of Soil and Water Conservation and Organic Waste Management, Centro de Edafología y Biología Aplicada del Segura, CSIC. P.O. Box 4195, 30080 Murcia, Spain.
[b]Istituto di Biochimica ed Ecofisiologia Vegetali, CNR. ADR Roma, Via Salaria Km 29.300, 00016 Monterotondo Scalo (Roma), Italy.

till the end of biooxidative phase and, once again, after the maturation period. Moisture content was assayed by drying at 105 °C, organic matter (OM) by loss-on ignition at 430 °C for 24 h., total nitrogen (Nt) and organic carbon (Co) by automatic microanalysis, cation exchange capacity (CEC) by BaCl2 at pH 8.1. NH_4-N was extracted with 2 M KCl (1:20 w/v) and determined by a colorimetric method based on Berthelot's reaction. Humic-like substances were isolated by treating compost samples with 0.1 M NaOH, later separating humic from fulvic acids by acid precipitation and centrifugation. Both extractable organic carbon and fulvic acid-like carbon (C_{fa}) were analyzed by automatic microanalysis. Humic acid-like carbon (C_{ha}) was then calculated by difference. Nitrogen fixation was determined by gas chromatography according to Cacciari et al., (1989) on compost samples (1 g d.w.) incubated for 72 h. at 30 °C in NaCl 0.9 % (25 cc). Losses of OM and Nt were calculated from the initial and final ash contents according to the equation of Viel et al., (1987). Phytotoxicity was determined according to Zucconi et al., (1981). Phenols are extracted according to Balice et al., (1985) and detected by a ^1H NMR Bruker AMX 600 operating at 600.13 MHZ.

Results and discussion.

The biooxidative phase of the two piles lasted different time, 9 weeks for MSO and 13 weeks for MOs. As a consequence, a lower mineralization occurred in MSO in comparison to MOs (Fig. 1). A great OM loss was recorded for MSO during the first four weeks of the composting process probably depended on the abundance of indigenous micro-organisms present in the sewage sludge and their fast growth, which provoke vigorous attack of the labile organic compounds (Iglesias & Pérez, 1992). Later, the OM mineralization continued in both piles, the effect was more pronounced in the MOs mixture.

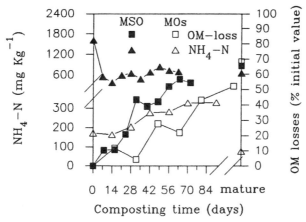

Fig. 1 Losses of organic matter and changes in amonium nitrogen during composting.

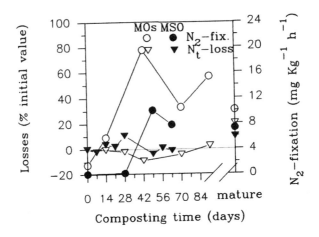

Fig. 2 Losses and fixation of nitrogen during composting.

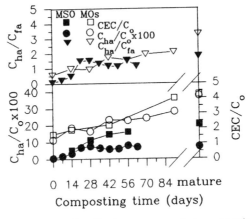

Fig. 3 Evolution of humification parameters during romposting.

As a result of the N mineralization, the evolution of the NH4–N content strongly depended on the OM losses in both piles (Fig. 1). An important reduction of the NH4–N in MSO was observed during the first week coinciding with a nearly insignificant loss of Nt (Fig. 2), which may be related to N organization by microbial immobilization of the NH4–N (Bernal et al.,1993; Mahimairaja et al., 1994). Because N is mainly lost as NH3 by volatilization during composting (Bishop & Godfrey, 1983), losses of Nt were lower in MOs than in MSO in agreement with the lower concentration of NH4–N in the former compost. N fixation occurred in both piles (Fig. 2), detecting increases of Nt when temperature was lower than 40 °C (from day 42 to 56 in MSO and from 28 to 63 in MOs).

Biological fixation of N was also higher in MOs than in MSO in good agreement with the lower NH4–N content in the former pile during the process. According to De Bertoldi et al., (1983), temperatures higher than 40 °C and high amounts of NH4–N inhibit biological fixation of N.

As shown in Fig. 3, an increase of the Ch_a/C_{fa} ration was observed in both mixtures as composting progressed, MOs exhibiting a higher value of the ratio than MSO after maturity. The above findings could be explained by the progressive formation of polycondensated substances derived from the lignin fraction of raw refuse materials and metabolic compounds from the microbial biomass, and/or a progressive degradation of substances present in the fulvic acid-like fraction (Iglesias & Pérez,1992). An increase was also detected in the $Ch_a/C_o \times 100$ ratio during process as well as in the CEC/C_o ratio, both changes indicating rising humification of composts with progress of composting. As reported by Lax et al., (1986), rise of the latter ratio should be related to the generation of new carboxilic and hydroxy-phenolic groups by decomposition and oxidation of the OM, processes which certainly occur during composting.

As reported in Table 1, phytotoxicity in MSO ceased in the 4th week while lasted till the 13th week in the other mixture, values of the germination index (G.I.) being clearly higher than 50 in both mature composts, which means an appropriate degree of maturity as reported by Zucconi et al.,(1981).

Table 1 Changes in G.I. of the mixtures during composting

Days	0	14	28	42	63	91	mature
MSO	19.6	–	105.4	–	83.6	–	91.3
MOs	0.0	4.9	–	41.0	–	66.2	69.8

Phenols disappeared during the composting process, as showed by ^{1}H NMR spectra. The phenolic compounds present at the beginning of the composting processes, made evident by the peaks in the 7.5–6.5 ppm region, were not any more detectable in the spectra performed at the end of the biooxidative phase in both piles (data not shown).

References

Andrich, G., Balzini, S., Zinnai, A., Silvestri, S., Fiorentini, R. (1992). Agr. Med., **122**, 97–100.

Balice, V., Carrieri, C., Liberti, L., Passino, R., Santori, M. (1965). Ing. Sanitaria, **2**, 69–73.

Bernal, M.P., López-Real, J.M., Scott, K.M. (1993). Bioresource Technology, **43**, 35–39.

Bishop, P.L. & Godfrey, C. (1983). BioCycle, August, 34–39.

Cacciari, I., Lippi, D., Ippoliti, S., Pietrosanti, T., Pietrosanti, W. (1989). Arch. Microbiol., **152**, 111–114.

De Bertoldi, M., Vallini, G., Pera, A. (1983). Waste Management and Research, **1**, 157–176.

Fiestas Ros de Ursinos, J.A. & Borja Padilla, R. (1992). Grasas y Aceites, **43**, 101–106.

Finstein, M.S., Miller, F.C., MacGregor, S.T., Psaranos, K.M. (1985). EPA Project Summary (EPA/600/S2–85/059) U.S. EPA, Washington.

Iglesias, E. & Pérez, V. (1992). Resources Conservation and Recycling, **6**, 243–257.

Lax, A., Roig, A., Costa, F. (1986). Plant and Soil, **94**: 349–355.

Mahimairaja, S., Bolan, N.S., Hedley, M.J., MacGregor, A.N. (1994). Bioresource Technology, **47**, 265–273.

Riffaldi, R., Levi–Minzi, R., Saviozzi, A., Vanni, A., Scagnozzi, A. (1993). Water, Air and Soil Pollution, **69**, 257–264.

Viel, M., Sayag, D., Peyre, A., André, L. (1987). Biol. Wastes, **20**, 167–185.

Zucconi, F.; Pera, A.; Forte, M.; De Bertoldi, M. (1981). BioCycle, **22**(2), 54–57.

Behaviour of Biodegradable Mater-BI ZI01U Plastic Layers in a Composting Pilot plant.

SERGIO PICCININI[1], LORELLA ROSSI[1], FRANCESCO DEGLI INNOCENTI[2], MAURIZIO TOSIN[2], CATIA BASTIOLI[2]

Introduction

The European Parliament and the Council of the European Union have recently finalized a Directive on Packaging and Packaging Waste. This foresees that, no later than five years from the date by which the Directive must be implemented in national law, between 50–65% by weight of the packaging waste should be recovered from the waste stream through recycling, reuse, or incineration with energy recovery. Recycling will cover the 25–45% of the totality of packaging materials contained in packaging waste. Composting is considered as a form of recycling of biodegradable packagings and it will have an important role in the rational management of waste. The definition of the criteria by which a material can be considered as compostable and recycled by composting is, therefore, a topical issue. International organisms such as the ISR/ASTM, CEN, ORCA have been constituted with the aim to define the compostability criteria. The fundamental requirements are: biodegradability, absence of negative effects on the quality of the compost, absence of negative effect on composting process. The first two conditions are determined at laboratory scale following the mineralisation of the test material and analysing the final compost quality. The laboratory data and the effect on the process should be verified in a composting plant.

The Mater-Bi ZI01U of NOVAMONT is a multipurpose biodegradable thermoplastic product and its biodegradation has been well studied at laboratory level. Biodegradation measured by the ASTM D 5338–92 test was 105 + 7 % while cellulose (reference) was 89.4 + 6.8% (J.Boelens, 1992). Terrestrial toxicity tests and physical-chemical characterization of the compost showed the absence of negative effects due to Mater-Bi degradation (work in progress). Biodegradation determined with a test described in the 12/7/1990 Italian Decree was higher than food-contact paper, in compliance with the prescriptions of the decree (Molinari, 1993). An analysis of Mater-Bi ZI01U has evidenced the absence of heavy metals (Bottazzini, 1994).

Aim of this field trial was to prove that the Mater-Bi ZI01U, added to the organic waste at a realistic concentration, neither damages the composting pro-

1. Centro Ricerche Produzioni Animali (C.R.P.A.), via Crispi 3, 42100 Reggio Emilia, Italy
2. NOVAMONT S.p.A., via Fauser 8, 28100 Novara, Italy

duction nor affects the final compost quality (chemical, physical, and phytotoxi-
cological parameters) and that degrades at a rate compatible with the composting
process.

Materials and methods

The composting plant located in Limidi di Soliera, Modena Italy, belonging to the
AMIU (Municipal Waste Treatment Department) of Modena , is a 3 meters wide
and 60 meters long trench, inside a greenhouse, and endowed with a turning
machine moving on a track. The composting mix was formed by: source separated organic waste (SSOW) daily collected at the fruit and vegetable market of
Modena; food scraps from restaurants and canteens; shredded trimmings of plane-
trees. Rectangular sheets of Mater-Bi ZI01U (100–150 cm^2; thickness 450 μm)
were introduced in the composting mass. Two composting experiments were run,
in sequence, in this work: SSOW 3 (April-September 1994) and SSOW 4 added
with Mater-Bi at a concentration of 0.12% (June-November 1994). The composting processes lasted about 4 months. During the first two months, corresponding to
the thermophilic phase, the composting mixtures were left in the horizontal trench
reactor and mixed frequently (2–3 times weekly). Subsequently, in the maturation
phase, the mixtures were removed under a shelter and turned 2–3 times in total
with a caterpillar.

Results and discussion

The composting process named SSOW 4 was not impaired by the ZI01U sheets
added to the organic waste. Both composting masses underwent to a strong aerobic degradation process. The analytical data concerning the two composting runs
(pH, total solids, volatile solids, total Kjeldahl-nitrogen, ammonia nitrogen, total
phosphorus, heavy metals, total organic carbon) did not differ significantly, showing that the Mater-Bi addition did not affect either the composting process or the
composition of the final product. All the values, but the pH, were within not only
the limits prescribed by the Italian low in force, but also within the new more
restrictive limits outlined in a draft bill. The composts obtained in the two trials
were both of high quality as shown by the humification parameters (Extractable
Organic Carbon, Humic acids, Fulvic Acids). The germination index of SSOW 4
compost was even higher than the control (SSOW 3) indicating the absence of
phytotoxic compounds affecting the *Lepidum sativum* germination.

   The degradation of Mater-Bi sheets was monitored very frequently by inspection of the composting mass. Apparently the degradation happened during the
thermophilic phase. At the end of the process a manual search was carried out
screening 3 aliquots of compost (about 20 Kg each) to find Mater-Bi residues with
a thin mesh (6 mm). The identification of the oversize fraction was rather difficult

because the residues were of tiny dimensions, dark and dirty of compost. The residues, tentatively recognized as Mater-Bi, were pooled, weighed, and treated with CH_2Cl_2, a solvent typically used to extract the synthetic part of ZI01U (poly-epsilon-caprolacton). No matter was extracted. Then, the residues were treated with pancreatin to digest residual starch. A 55.08% weight loss was detected. The final residues had a fibrous, vegetable appearance. These findings would exclude that the oversize fraction detected in sampling is composed by Mater-Bi residues. However, even assuming that the residues are Mater-Bi we can estimate a final degradation of 98.9%. In fact residue recovered was 3.787 g out of 57.6 Kg of compost screened, in total. Applying this ratio to the total mass of final compost (5100 Kg) we can estimate a total of 335 grams of residues present at the end of the process, representing the 1.1% of the initial amount. The Mater-Bi degradation level is extremely high but is based on a visual, qualitative approach. Without the mineralization data collected at laboratory scale it would have been impossible to discern whether this disappearance was just due to solubilisation or disintegration processes rather than to a real biodegradation. On the other hand, degradation data produced at laboratory scale are always susceptible of criticisms if not substantiated by field results. A systematic study to find a correlation between the laboratory scale results and the fate in a real composting plant would be of great advantage in this field because it would simplify the testing activities. The present work is a piece of information towards this objective. The findings of this work have been obtained using an initial Mater-Bi concentration of 0.12% (w/w). This amount, low in terms of weight, it is rather high by a visual point of view. Therefore, it was possible to easily follow the degradation. In any case, we consider that this amount is a realistic estimate of the concentration of the degradable plastics in biowaste feedstocks in a near future. This assumption is based also on the fact that household waste, source of degradable plastics, is usually mixed with other feedstocks devoid of plastics, such as sludge, wood residues, leaves etc. decreasing the final plastics concentration. Higher amounts will have to be tested if degradable plastics concentration factually found in composting should prove to be much higher than the 0.12 %, checked in this work.

References

J. Boleans. 1992. Final Report: Aerobic Biodegradation under Controlled Composting Conditions of Test Substance ZI01U.
Organic Waste System, Gent, B.
N. Bottazzini. 1994. Determinazione dei metalli pesanti in campioni di vostro interesse. Bollettino analitico Enichem - Istituto Guido Donegani, Novara, I.
G.P. Molinari. 1993. Saggio di biodegradabilità aerobica secondo D.M. del 07/12/1990: Biodegradazione del polimero ZI01U. Relazione finale. Università Cattolica del Sacro Cuore – Piacenza, I.

Acknowledgements

This work was partly financed by Regione Emilia-Romagna through the co-ordination of A.Biotec, Forlì (I). We wish to thank Dr. Marco Versari for his valuable managing efforts. Many thanks also to Dr. Bigliardi and AMIU of Modena for their interest in compostable plastics and providing the composting facilities.

The Emilia-Romagna Experiment in Animal Manure Composting

SERGIO PICCININI[1], LORELLA ROSSI[1], GIUSEPPE BONAZZI[1], GIULIO DALL'ORSO[2]

Introduction

Animal manure, with the exception of poultry, generally has a low fertilizer content. Consequently spreading cost is higher than chemical fertilizers. The organic matter content of animal manure is not good enough to raise its fertilizer value; in fact the low degree of humification helps little in raising the soil organic matter content. Composting is a good way of improving the fertilizer value of animal manure. This paper presents the results of a series of experiments performed by the C.R.P.A. and supported by funds from the Emilia-Romagna Region, Agricultural Department.

Materials and Methods

Table 1 shows the composition of the animal manures submitted to composting tests.

With regard to pig slurry we performed composting tests of the solid fraction resulting from centrifugal separation of raw slurry (referred to as SFC from now on); as bulking agents we used straw and wood chips. With regard to dairy cattle manure we performed composting tests of the solid fraction resulting from a press-screw separator and of the straw bedding used in bedded areas with and without addition of slurry coming from concrete surfaces (paddock, feed areas...).

We also carried out a composting test of the dewatered sludge (DS) (with a belt filter press) coming from a purification plant (anaerobic + aerobic reactors) treating pig and beef cattle slurry; as bulking agents we used wood chips. With regard to poultry manure we performed composting tests both with manure from laying hens (with or without in-house drying) and with litter from broilers.

We have begun and are still involved in composting tests of biomass beds (wood

[1]Research Centre for Animal Production (C.R.P.A.), Reggio Emilia (Italy)
[2]Agricultural Department, Emilia-Romagna Region, Bologna (Italy)

Table 1 Composition of the animal manures submitted to composting tests

		SFC (n = 68)		Cattle solid fraction (n = 21)		Cattle straw bedding (n = 7)		DS (n = 1)	Laying hens dried manure (n = 52)		Laying hens manure (n = 30)	
		x	s	x	s	x	s	x	x	s	x	s
TS	(%)	25.6	3.5	23.1	3.5	22.9	1.8	24.4	42.7	13.3	21.8	2.0
VS	(% TS)	71.2	14.6	86.5	5.2	75.6	3.6	54.3	65.8	6.7	74.8	1.8
TKN	(% TS)	3.7	0.78	1.59	0.43	2.93	0.33	4.15	5.6	1.05	5.4	1.1
NH4-N	(% TKN)	34.8	11.0	19.2	9.1	33.8	7.7	29.4	20.4	14.7	34.3	9.1
P	(% TS)	3.42	0.82	0.44	0.35	0.8	0.16	3.74	2.19	0.46	2.1	0.2
K	(% TS)	0.66	0.19	0.89	0.54	4.39	0.37	0.6	2.8	0.7	2	0.4
TOC	(% TS)	40.8	4.9	40.9	4.6	37.5	1.5	31.3	20.6 (*)		–	–
C/N		11	6	26	11	13	5	8	4		–	–
Cu	(mg/kg TS)	288	144	23	17	–	–	587	78 (*)		–	–
Zn	(mg/kg TS)	1235	543	153	75	–	–	1464	215		–	–

x = average; s = standard deviation
SFC = solid fraction from centrifuging of pig slurry.
DS = dewatered sludge with a belt filter press from a purification plant treating pig and beef cattle.   (*) = 1 sample

shaving and wood chips from municipal yard waste) with the addition of pig slurry.

The majority of the trials were carried out in an experimental composting plant consisting of a horizontal pit reactor, complete with greenhouse cover, turning machine equipment on wheels and aeration system. The pit is 3m wide and 60m long, maximum height of loaded material is 1m. The trials with dairy cattle manure were carried out at three dairy cattle farms on uncovered concrete surfaces. The windrows were turned with a bucket loader or with tractor-drawn turners. The trials with laying hens slurry and with poultry litter were done at two composting farm plants

Results and Discussion

All the composts show a good content of fertilizer elements and a better degree of humification than may be found in composts without animal manure. The heavy metal content is generally low, only with SFC and DS we found a higher copper and zinc content; in the future this content should be reduced as substitute products (probiotics and organic acids) are beginning to be used in animal feed as well as ways of administering the food which improve the pigs ability to adsorb Cu and Zn.

Values of pH and salinity are often higher than those of municipal sludge compost; this can be a problem for the utilization as peat substitute in growing media.

Animal manure composting trials showed that, with these materials, owing to their rapidly degradable organic matter content, a rapid start in the composting process with a fast rise in temperature is easily obtained.

The SFC contains about 25–30% of the total nitrogen, mostly in organic form, present in the raw pig slurry. During composting a part of the organic nitrogen becomes ammonia which is lost in part into the atmosphere on account of the high temperatures created within the heap. The amount of nitrogen volatilized as ammonia may even reach up to 35% of the initial nitrogen present. High ammonia losses reduce the agronomical value of the compost, contribute to the pollution of the atmosphere and affect acid rain production.

Nitrogen losses during composting may be reduced by adding zeolites, on account of their high cation exchange capacity and their remarkable ammonium ion selectivity, which can reduce the amount of free and therefore volatilizable ammonia. Two trials were carried out: in the first we added zeolite (55% phillipsite and 10% chabasite) at a rate of 20% (in weight) of the SFC and in the second at a rate of 10%. We obtained a substantial reduction of nitrogen losses in the heaps with added zeolite. In the first trial nitrogen losses decrease from 35.8% to 5.3% and in the second from 29.8% to 19.7%. The second test is more interesting from an economical point of view in that whilst using a coarser, more varied and therefore cheaper zeolite, ammonia emissions were significantly reduced.

The first results of the composting tests of biomass beds (wood shaving and

Table 2 Average composition of the composts obtained

		SFC + wood chips	SFC + straw	Cattle solid faction	DS + wood chips	laying hens dried manure + straw	Cattle straw bedding	Cattle straw bedding + slurry
pH		7.55	8.28	8.28	7.32	9.76	–	–
TS	(% TQ)	49.89	59.99	39.5	45.33	71	63.89	56.3
VS	(% ST)	71.07	59.15	75.0	52.65	43	36.55	33.11
TKN	(% ST)	2.96	3.68	3.13	2.55	2.18	1.59	1.75
NH4-N	(% HTK)	20.95	22.55	0.94	1.47	6.14	5.74	6.74
C/N		10	8	13	8	13	15	14
P	(% ST)	2.53	3.40	0.70	3.55	2.24	0.57	0.62
K	(% ST)	0.68	1.68	1.20	0.98	4.38	3.66	3.52
Cu	(mg/kg ST)	143	166	41	543	93	–	–
Zn	(mg/kg ST)	744	972	468	1170	502	–	–
Pb	(mg/kg ST)	4	2	–	8	3.6	–	–
N	(mg/kg ST)	27	18	–	20	9	–	–
Cd	(mg/kg ST)	0.26	1	–	0.10	0.06	–	–
Cr	(mg/kg ST)	64	67	–	15	17	–	–
Hg	(mg/kg ST)	0.09	0.25	–	0.31	0.18	–	–
G.I.	(%)	90	79	–	90	78	–	–
C.E.C.	(mS/cm)	1.7	2.6	0.6	1.6	5.7	19.8	20.0
TOC	(% ST)	30.1	31.1	39.7	21.6	23.0	–	–
TEC	(% ST)	13.3	16.13	24.93	13.95	11.4	–	–
HA + FA	(% ST)	10.2	11.71	18.85	9.22	8.18	–	–
NH	(% ST)	3.1	4.42	6.08	4.73	3.22	–	–
HI		0.31	0.38	0.32	0.51	0.39	–	–
DH	(%)	76.6	72.6	75.61	66.09	71.8	–	–
HR	(%)	33.8	37.65	47.5	42.65	35.6	–	–

G.I. = germination index

wood chips) with the addition of pig slurry, show that it is difficult to treat one square metre of bed with more than the daily slurry production of one pig. In order for it to gain interest from an economic point of view such a solution needs improvements.

With regard to dairy cattle slurry the composting tests of the solid fraction resulting from a press-screw separator show an improvement in the agronomic value of this solid fraction; in these trials the nitrogen losses, as ammonia, are low, about 6–19% of the total input.

Composting tests of the straw bedding, with addition of slurry coming from concrete surfaces, show that straw bedding can absorb the slurry produced and that the compost has good agronomic value. In this case the nitrogen losses are about 15–25% of the total input.

In the farm plant treating laying hens slurry (TS = 20%) mixed with chopped straw (bought at a high price) at the ratio 12:1 (w:w), satisfactory dehydration of the product at the end of the process was difficult to obtain during wintertime. In the case of laying hens slurry mixed with poultry litter (ratio 1.2:1, w:w) or in-house dried layers manure mixed with straw (ratio 4.8:1, w:w), when it was possible an adequate mixing ratio with cellulosic materials, the humidity and porosity of the starting matrix allowed an optimum composting process.

The major unsolved problem with poultry manure composting are the heavy nitrogen losses through ammonia volatilization, which have relevant environmental pollution concern, together with odour nuisance problems in the vicinity of the plants which are common.

The high temperature and pH reached in the first stage of the process, the low C/N ratio of the mixture, the high manure nitrogen content in ammoniacal form are the mean causes of these losses, which amounted, in our trials, to between 50 to 65% of initial nitrogen content, both with liquid manure and with the in-house dried type. For this reason it is advisable to build poultry manure composting plants as closed reactors with the possibilty of introducing exhaust air treatment systems.

References

G. Bonazzi, L. Valli, S. Piccinini, (1990) – Controlling ammonia emission at composting plants – Biocycle, June.

O. Martins, T. Dewes, (1992) – Loss of nitrogenous compounds during composting of animal wastes – Bioresource Technology, n. 42.

A. Ferrari, S. Piccinini, (1993) – *Prove di compostaggio delle deiezioni di bovine da latte* – L'Informatore Agrario, n. 4.

M. P. Bernal, J. M. López-Real, K. M. Scott, (1993) – *Application of natural zeolites for the reduction of ammonia emissions during the composting of organic wastes in a laboratory composting simulation* – Bioresource Technology, n. 43.

K. V. Lo, A. K. Lan, P. H. Liao, (1993) – *Composting of separated solid swine wastes* – Journal of Agricultural Engineering Research, n. 54.

S. Piccinini, (1994) – *Application of a phillipsite rich zeolitite during the composting of solid fractions of pig slurry* – Materials Engineering, Vol. 5, n. 2.

L. Valli, S. Piccinini, L. Cortellini, (1994) – *L'impianto sperimentale di Soliera (MO) per il compostaggio di residui agro-zootecnici e fanghi urbani*, *RS* – Rifiuti Solidi, Vol. VIII, n. 3, maggio-giugno.

Composting Wastewater Sludges Without the Addition of Bulking Agents

C J PULLIN, J L LAWRENCE, – Southern Water
Services, England, T DIBKE, W MAYER – SEVAR
Gmbh, Germany J TINGLE – Brackett Polcon, England

Abstract

Untreated wastewater sludges have been successfully composted without the addition of bulking agents - straw, paper, green or municipal wastes - to form a stable soil conditioner.

The advantages of this process are the minimal loss of nutrients, linked with a reduction in organic and volatile matter and moisture content. There are savings associated with construction, handling, transport and distribution costs.

This material is one of a range of products manufactured by the company for beneficial reuse in agricultural land. If deregulation of current sludge legislation occurs, it is anticipated that this product will be in demand for other markets.

Following successful trials at Horsham Wastewater Treatment Works, Sussex, Southern Water has formed a partnership with Sevar GmbH, Brackett Polcon and Tilbury Douglas Construction to build a pilot plant to 2000 tonnes of dry solids per annum of dewatered wastewater cake at the new Weatherlees Wastewater Treatment Plant on the East Kent Coast, at a cost of £3M.

Introduction

Southern Water is one of 10 water companies formed in 1989 when the industry was privatised. It provides 625 million litres of clean water and removes, treats and disposes of 1300 million litres of wastewater in an area of 4000 square miles. The current wastewater sludge production is 62000 tonnes of dry solids per annum, increasing to 134,000 tonnes by 1998 when an EU Directive stops the discharge of untreated wastewater into the sea.

The Recycling Group was formed in late 1993 to develop a strategy for the future treatment and beneficial reuse of bioproducts in the agricultural industry. The disposal of untreated sludges is deemed unacceptable due to limited agricultural outlets, odour nuisance and high transport costs.

The proposed bioproduct range will include those formed from digestion, drying or composting untreated sludges.

Southern Water had tried various forms of composting since 1978, using straw, municipal or green waste with wastewater sludge. However, the main disadvantage of all previous experiences has been the addition of bulking agents which increase civil, handling and transport costs. In addition, the use of municipal wastes gives rise to contamination from glass, plastic and metals. There is also concern over the consistency and reliability of supply of these bulking agents linked with additional transport movements and costs associated with raw materials and finished products. To overcome these major constraints, it was concluded that the only solution would be to compost wastewater sludge without the addition of bulking agents. A search by Southern Water identified Sevar GmbH as the only company offering a composting process that did not involve the addition of bulking agents. Already well known for sludge drying and composting in Germany, Sevar and their UK sister Company Brackett Polcon were commissioned for a major pilot scheme at Southern Water's Horsham Wastewater Treatment Works to demonstrate and develop their process for sludge only composting. During the Summer of 1994, this was successfully accomplished.

Tilbury Douglas, a major UK Civil Engineering and Mechanical Construction organisation, were then brought into the team as project managers and main civil contractors to help commercialise the process.

The Process

During the 1980's, Sevar GmbH developed a composting process based on research work by Prof Wolfgang Baader. Animal wastes including the litter were compacted into briquettes that were placed in composting towers or cells. The temperature within the cell rose to 70+°C and after 5–7 days the contents were cooled by force of air ventilation (using a blower). At the same time, moisture within briquettes was also lost due to evaporation. The treated briquettes were removed and then crushed, screened and graded, before distribution to various markets. When necessary lime was added prior to pelletisation to increase product range.

Following a visit to Schwege, Northern Germany, to see the process in operation, the briquetting and associated machinery was installed at Horsham Wastewater Treatment Works. The plan was to dry a quantity of cake sludge to 90-92% dry solid content, grind and back blend with other cake sludge to form a pliable mixture suitable for forming the briquettes. However, due to the fibre content of the sludge, this proved impossible.

After several alternative arrangements were tried the final preffered solution was simply to partially dry the untreated sludge cake to 50% DS in a low temperature Sevar dryer and place it directly in the compost tower. Careful control of ventilation ensured a rapid rise in temperature in excess of 70°C with little mal-odour being vented into the atmosphere. After 8–10 days, the box was emptied and the contents windrowed in the conventional manner, with careful control on moisture and temperature. This process was continued for a period of up to 12 weeks, when a stable product was obtained. Storage over a further 3–4 months has proved beneficial, with a further reduction in organic and volatile content.

This work has been replicated several times, the results always being within acceptable statistical limits.

For the commercial size plant a double size composting cell has been developed with the tower divided into two zones by means of pin rollers to form a false floor.

In the upper zone the temperature rises to around 70°C for at least two days. The raw compost is aerated, if required, at intervals. Forces aeration is not necessary due to the material preparation processes. Offgas purification is carried out via a compost filter. After a residence time of ca. 5-6 days there follows a steady transfer by turning of the pin rollers to the lower zone.

In the lower zone re-heating upto ca. 60°C occurs and the first fungal growth. This locks in the odour producing material. After a further residence period of 5–6 days the fresh compost produced is removed from the compost towers led by means of a chain scraper conveyor to the windrowing area.

This double composting arrangement further improves final compost quality.

Results

The feed and end products were analysed for solids, organic and volatile content, nutrients, potentially toxic elements and pathogens. Based on the test results, there was a reduction in organic and volatile matter from approximately 80% down to 54%. The moisture content can be controlled by the frequency of turning after the second maturation stage, and is normally less than 40%. Nutrient content – Nitrogen, Phosphate and Potash was only marginally reduced. Pathogens – E-coli, Streptococci and Salmonella were virtually eliminated following retention of the material at temperature in excess of 65°C for 5 days. Potentially toxic element concentrations increased in direct proportion to the destruction of organic and volatile matter.

The final product is light and friable, with a bulk density of approximately 0.55 at a moisture content of 35–40%.

The results were so encouraging that Southern Water have decided to proceed with a large demonstration plant of 2000 tonnes of dry solids per annum capacity to be located at Southern's new Weatherlees Hill Wastewater Treatment Works. This contract will be done by Tilbury Douglas, Sevar and Brackett Polcon and will be in full operation in October 1995. Its total value is approximately £3M.

Discussion

The composting process of wastewater sludges can be successfully accomplished without the addition of bulking materials providing the moisture content is reduced in the initial stages. Careful control of ventilation during the first stage limits odour problems and assists with the retention of nutrients, especially Nitrogen. Seeding the material placed in the composting cells with previously treated sludge rapidly enhances microbiological activity.

It was also noted that following a storm, when the partially matured product was saturated with rain water, elutiation of certain metals had taken place, especially Copper, Zinc and Molybdenum.

Uses

Pot trials are currently being undertaken by ADAS to determine suitability of the product for various crops. Subject to these results, plot and crop trials are expected to be completed for 1996. Deregulation of current sludge legislation will broaden market opportunities to include horticulture, amenity and perhaps even the domestic outlets.

References

1. Barbara A Carroll, P Caunt, G Cunliffe – Composting Sewage Sludge in J. IWEM N°7 1993.
2. Prof Dr Ing W Baader – Biological Sludge Drying, Seminar Antwerp, Royal Flemish Eng Associates, 6 Feb 1992.

The Potential Use of Composted Waste Materials for Cuttings

SRDJAN RADANOV – Faculty of Forestry, University of Belgrade, Yugoslavia.

Summary

The rooting response of three ornamental shrub species (Weigela florida, Deutzia gracilis, Kolkwitzia 'Pink Cloud') was tested in media made of pharmaceutical, village waste and farm yard manure composts.

The properties of media were analyzed to investigate factors which are of most importance for the development of roots. Special attention was given to the measuring of the physical properties of the media due to their crucial importance for root development.

Farm yard manure compost and village waste compost have proved to be suitable materials for rooting media. Pharmaceutical compost is not suitable for wider use due to its physical properties and high electrical conductivity.

Introduction

Different materials have been used to propagate cuttings. Peat is material most commonly used and forms the basis of many different propagation media. The use of peat is accompanied with some problems: the price of horticultural peat is high and the resources are limited.

The aim of this report was to investigate properties of some alternative materials based or composted waste materials and to evaluate possibilities for their practical use as rooting media for cuttings.

Materials and methods

The rooting media were made of three types of composted waste materials: pharmaceutical compost (PHC) made of the waste material remained after production of antibiotic tetracycline in chemical industry, village waste compost (VWC) made of organic waste collected in the village of Wye (England) and farm yard manure compost (FYMC) made of cattle manure from the diary farm. The composts were obtained from Wye Controlled Composting Unit (Kent, England).

All composts were leached prior to use with 3–4 times its volume of water in order to decrease high content of soluble salts which is usually accompanied with these types of materials. The composts were mixed with different proportions of grit so that nine mixes of 'compost media' were made. The standard rooting medium made of peat, perlite and grit acted as control medium.

The media used in the experiment were:

PHC + grit 1:1 vol. (PH 1);	VWC + grit 1:2 vol. (VW 3);
PHC + grit 1:1,5 vol. (PH 2);	FYMC + grit 1:1 vol. (FY 1);
PHC + grit 1:2 vol. (PH 3);	FYMC + grit 1:1,5 vol. (FY 2);
VWC + grit 1:1 vol. (VW 1);	FYMC + grit 1:2 vol. (FY 3);
VWC + grit 1:1,5 vol. (VW 2);	peat + perlite + grit 4:5:1 vol. (P/P).

The rooting response of three ornamental shrub species was tested Weigela florid, Deutzia *gracilis* and Kolkwitzia 'Pink Cloud'.

Table 1 Rooting results of cuttings

Medium	Weigela florida		Deutzia gracilis		Kolkwitzia Pink Cloude'	
	Rooting %	Mean dry weight of roots	Rooting %	Mean dry weight of roots	Rooting %	Mean dry weight of roots
PH 1	95.0	98.2 bc	87.5	51.7 c	62.5	47.9 c
PH 2	97.5	94.6 bc	100.0	50.4 c	77.5	48.3 c
PH 3	95.0	88.8 bc	97.5	65.0 abc	67.5	58.3 bc
VW 1	100.0	120.6 ab	97.5	57.7 bc	80.0	75.8 abc
VW 2	100.0	104.8 bc	100.0	77.8 a	95.0	74.0 abc
VW 3	87.5	100.0 bc	97.5	64.9 abc	92.5	78.7 abc
FY 1	92.5	95.6 bc	100.0	54.4 c	95.0	70.5 abc
FY 2	95.0	82.6 c	87.5	53.7 c	92.5	68.5 abc
FY 3	95.0	116.2 abc	97.5	64.5 abc	87.5	92.2 ab
P/P	97.5	149.7 a	100.0	73.6 ab	92.5	101.5 a

Means in the columns followed by the same letter do not differ significantly at P = 0.05 (Duncan's Multiple range test).

Special attention was given to the measuring of the physical properties of the media due to their crucial importance for the root development. In order to measure air space (AS), easily available water (EAW) and total porosity (TP), porous plate apparatus was used. Different suction pressures (10, 50, 100 cm) were applied on the samples of the media and the volume of outflow water was recorded. Different properties were calculated on the basis of measuring and explained by the methods of de Boodt and Verdonck (1972).

Table 2 Physical properties of rooting media

Medium	BD (g/cm³)	PD (g/cm³)	TP (% vol.)	AS (% vol.)	EAW (% vol.)
PH 1	1.05	2.34	55.2	11.7	8.2
PH 2	1.08	2.39	54.9	15.8	6.6
PH 3	1.20	2.44	50.9	13.4	3.3
VW 1	0.89	2.37	62.5	18.2	8.1
VW 2	0.96	2.42	60.4	22.5	3.5
VW 3	1.05	2.46	57.4	22.7	4.2
FY 1	0.95	2.22	57.3	17.5	2.7
FY 2	1.01	2.29	55.9	21.6	3.2
FY 3	1.09	2.35	53.7	22.8	2.8
P/P	0.27	1.87	85.6	18.7	15.7

BD – Bulk density; PD — Particle density; TP – Total porosity;
AS – Air space; EAW – Easily available water.

Results and discussion

The results of the rooting responses are presented in table 1. The results of media analysis are presented in table 2.

Although the highest mean dry weight of the roots for Weigela florida cuttings was recorded in control (P/P) medium (149.7 mg), high values were also recorded in village waste compost media, especially in VW 1 (120.6 mg). Both VW 1 and P/P medium have well balanced ratio of air space (AS) and easily available water (EAW).

The highest mean values of dry weight on Deutzia gracilis cuttings were in VW 2 (80.3 mg) and P/P (73.5 mg) media. Within media made of same compost the highest values of dry weight were for cuttings rooted in the media with high values of air space. For Deutzia gracilis cuttings EAW was not determining factor as for Weigela florida cuttings; they had more compact but fibrous root system which was able to uptake enough water for root development. The medium which provided the best conditions for root development (VW 2) had a very high value of air space – 22.5.

Many authors state the importance of air for root development but the opinions differ as to how much air is necessary in the media to sustain satisfactory growth. Acceptable volumes of air space within propagation media have been suggested at levels of 10–15% (Lee and Paul, 1976), 15% (Puustjarvi, 1969) and 20% (Arnold, 1973). The amount of air needed for root development depends however on species and it is difficult to give values of air space that would describe ideal medium.

From species tested in this experiment, Kolkwitzia 'Pink Cloud' and Deutzia gracilis cuttings responded better when the air space was high. That was also important for Weigela florida but for these cuttings, available water was the factor which greatly influenced the root development.

The problem for rooting process was expected to be due to the high electrical conductivity (EC) usually associated with waste materials. The electrical conductivity of VWC (4.13 mS cm^{-1}) and FYMC (5.62 mS cm^{-1}), after the composts were mixed with grit, did not have any detrimental effects to plants. The electrical conductivity of PHC (10.86 mS cm^{-1}) was much higher and since the rooting response in these media was generally inferior to the other media, it is possible that the high EC affected the root development.

Conclusions

The composts made of village waste and farm yard manure waste materials, mixed with certain proportions of grit, provide good properties for the rooting of cuttings.

Village waste compost is more suitable for making the rooting media due to the higher range of its pores which are releasing water at different pressures. Farm yard manure compost has more crumb structure and forms porous media with lower range of available water. The potential use would be in mixing these composts for making the rooting media which would combine their good properties.

Pharmaceutical compost does not provide good conditions for root development due to its physical properties and high electrical conductivity.

References

Arnold, B.R., (1973). Acta Horticulturae, **31**: 149–160.
De Boodt, M., Verdonck, O., (1972). Acta Horticulturae, **26**: 37–44.
Lee, C.I., Paul, J.L., (1976). J. Amer. Soc. Hort. Sci, **101**(5): 500–503.
Puustjarvi, V., (1969). Peat Plant News, 2: 43–53.

Inalca Experiment in Composting of Slaughter House Organic Wastes

LORELLA ROSSI[1], SERGIO PICCININI[1], CLAUDIO CIAVATTA[2], FERDINANDO CREMONINI[3], GIOVANNI SORLINI[3]

Introduction

The trials were carried out by the CRPA on behalf of INALCA SpA at one of the biggest slaughter houses in Italy situated in Castelvetro, near Modena, where an average 500 head of beef cattle are slaughtered daily. They show a desire to introduce composting technology into a reality where a demand for quality is essential for the success of the technique.

The experimentation carried out between December 1992 – August 1993 was to involve the composting of a mixture of two of the organic residues regularly produced during the slaughtering process: organic wastes from the press-screw separator, obtained from the purification plant and ruminal content removed from the workplace, pressed and collected in the same container where the press-screw wastes come together.

Materials and Methods

The experimentation was divided into two test cycles and took place during two different seasons: a winter cycle (Dec '92 – Apr '93) and a spring-summer cycle (May – Aug '93). In order to obtain starting material with the right characteristics, straw residues obtained by chopping up discarded wooden pallets and pine trimmings collected during the upkeep of local parks and gardens were added to the waste sludge mixture and ruminal content. The mixture was made according to accurate volume or weight ratios (table 1).

[1]Research Center for Animal Production (C.R.P.A.), Reggio Emilia, Italy
[2]Agricultural Chemistry Institute, University of Bologna, Italy
[3]INALCA S.p.A., Spilamberto (Modena), Italy

Table 1 Volume or weight ratios between woodchip and sludge during the winter cycle (Dec. '92 – Apr. '93) and the spring-summer cycle (May – Aug. '93).

	Trials	Woodchip/sludge (1) Weight (%)	Volume	Turning	Covered
1°	A	–	–	NO	NO
C	B	86	2 : 1	NO	NO
Y	C	41	1 : 1	YES	NO
C	D	43	1 : 1	YES	YES
L	E	24	1 : 2	YES	NO
E	F	23	1 : 2	YES	YES
2°	I	–	–	NO	NO
	L	16	1 : 2	YES	NO
C	M	18	1 : 2	YES	YES
Y	N	13	1 : 3	YES	NO
C	O	12	1 : 3	YES	YES
L	P	26	1 : 2	YES	NO
E	Q	10	–	YES	NO

(1) sludge: waste sludge and ruminal content mixture.

Thirteen triangular piles were formed on a waterproofed concrete base, one for each of the trials proposed, 6 in the first cycle and 7 in the second.

The composting cycle took four months; two were necessary for the active or thermophile stage during which time the heaps were turned once a week, and two months were needed for the stabilization-maturation of the product.

Results and Discussion

In heaps where straw residues were added the temperature trend testified to the evolving process of aerobic decomposition which was highly esothermic. The composting process did not start in the heaps where the woodchip was not added and the heaps were left unturned (T° <40°C).

The temperature trend tended to be somewhat regular when (in heap P) pine trimmings were used. Moreover, average temperatures were rather higher compared to other heaps in the second cycle, especially during the maturation stage. The only unsuitable hypothesis of all those tested, on account of the mixture and management criteria was test B which was carried out with a considerable ratio of woodchip. The objective was to obtain a mixture with a porosity which would mean that turning was no longer necessary, but in the end the degradation process turned out to be incomplete and unsatisfactory.

All the composts obtained (table 2) were the result of good agronomic quality. The macroelement content (N, P and K) is rather interesting. As far as the micro-biological aspect is concerned, salmonella tests proved negative in all the products, thereby confirming the hygienising effect of the composting. The results obtained with the germination sample (GI values higher than 70%) confirmed the absence of phytoxicity in the final products, already found in many cases at the end of the thermophile stage. On the basis of the confrontation with the qualitative limits set by the law in force about compost (D.P.R. n. 915/82) and about agricultural use of the waste sludge (D.Lgs n. 99/92) it emerges that all the composts produced are well within the parametres set by the law, even for the heavy metal content.

Further research was done on some of the composts obtained regarding the stabilising grade of the organic substance by determining some parametres of humification also fixed by national laws on fertilisers (Law of 19 October 1984, n. 748 and later indications and integrations Suppl. Ord. to the G.U. n. 29 of 4 February 1991).

The grade (DH) and humification rate (HR) highlighted excellent progress in the P heap with values grown progressively in time. However, the humification course in other heaps was more irregular, presumably because of the lack of homogeneity of the compost material (excessively rough fractions).

Determining the initial and final weight of the different heaps meant that it was possible to quantify the reduction that took place and which ended up being between 60 and 80% of the initial weight of the mixtures used in the process. More precisely, water loss averaging between 60 and 90% was observed compared to the initial content (excluding the rainwater for the uncovered heaps) and the loss of volatile solids was between 15 and 71%. The low reduction of volatile solids seen in some of the heaps during the second cycle is likely to have been caused by the microbic mineralisation activity due to the high level of dryness already reached at the end of the first stage of the process.

As far as total nitrogen is concerned, the relative mass weight highlighted a variable loss of between 43 and 65% compared to what was initially contained in the heaps. Ammonia concentrations in the vicinity of the heaps both in the static phase and during the turning of the heaps were checked weekly with the aid of a phial gas revealer (the totalising type DRAGER Polymeter). The highest value recorded, equal to 35 ppm corresponds to the maximum for limited periods of exposure (TLV-STEL) proposed by the ACGIH (American Conference of Governmental Industrial Hygienists) responsible for hygiene in the workplace and this value is recognised as a guideline by the national UUSSLL. As such this is a reassuring result for the worker responsible for managing the heaps.

On the whole the results obtained were positive and encouraging. The materials treated seem particularly suited to the composting treatment. The presence of ruminal content would appear to influence positively the biochemics of the process. The process worked by adding quantities of straw residues varying between 40 and 10% of the sludge weight thereby proving to be quite flexible. Finding out the minimum mixing limits is extremely important in that the quantities of woodchip at stake for future applications of the process are somewhat high. The good results obtained using the plant trimmings meant that a new category of residues which can substitute woodchip pallets was identified and, what is more, it is easily found and has similar disposal problems. In addition the trimmings give a higher quality product compared to the pallet woodchip.

Table 2 Characteristics of the composts.

		C	D	E	L	F	M	N	O	P	Q	D.P.R. n.915/82	DLG n.99/92
pH		6.6	7.0	6.8	6.9	7.4	7.3	7.3	7.2	7.4	7.5	6 – 8.5	–
TS	(%)	36.9	54.5	40.8	31.7	75.4	91.4	58.5	91.2	39.9	71.7	55	–
VS	(% TS)	89.4	85.5	85.7	82.4	90.2	88.4	83.8	86.3	80.6	84.9	–	–
TKN	(% TS)	2.20	1.91	1.77	2.49	2.29	2.63	2.38	2.40	3.23	2.27	> 1	> 1.5
NH4-N	(% TKN)	3.02	4.41	0.55	0.42	2.78	3.82	1.00	5.61	0.78	3.56	–	–
TOC	(% TS)	43.8	44.8	43.1	41.8	43.4	44.2	45.7	44.3	39.6	42.1	–	20
C/N		20	23	24	17	19	17	19	20	15	20	< 30	–
P	(% TS)	0.71	1.06	0.88	1.04	0.66	0.76	0.95	0.98	1.14	1.07	> 0.22	0.4
K	(% TS)	0.23	0.40	0.40	0.25	0.27	0.23	0.27	0.23	0.72	0.23	> 0.33	–
G.I.	(%)	78	83	82	100	74	108	105	55	85	85	–	–
Cu	(mg/kg TS)	26	44	75	53	30	71	29	88	71	34	600	1000
Zn	(mg/kg TS)	190	311	277	203	262	420	201	438	435	197	2500	2500
Ni	(mg/kg TS)	–	26	–	–	8	19	–	40	10	16	200	300
Pb	(mg/kg TS)	–	6	–	–	6	10	–	6	4	< 1	500	750
Cd	(mg/kg TS)	–	0.8	–	–	0.6	1.0	–	1.6	0.9	0.14	10	20
Cr	(mg/kg TS)	–	32	–	–	21	16	–	30	22	6	510 (*)	–
Hg	(mg/kg TS)	–	1.16	–	–	0.25	0.30	–	0.85	0.34	0.25	10	10
Salmonellae	(MPN/g TS)	–	0	0	0	0	0	0	0	0	0	0	1000
DH	(%)	80.7	80.4	79.6	–	68.4	68.5	–	72.9	89.2	75	–	–
HR	(%)	20.9	25.7	24.7	–	17.3	15.5	–	29.4	34.2	22.5	–	–

(*): 500 mg/kgTS for Cr (III) e 10 mg/kgTS for Cr (VI).

References

Zucconi F., De Bertoldi M., (1987) - Compost specifications for the production and characterization of compost from municipal solid wastes. Compost: Production, Quality, and Use. Ed. Elsevier, London.

Autori Vari (1992) – Il compostaggio come sistema integrato nel riciclaggio dei rifiuti. Speciale Acqua-Aria, n. 10.

Valli L., Piccinini S., Cortellini L., (1993) – Il compostaggio di residui agro-zootecnici e fanghi urbani. L'Informatore Agrario, Verona, XLIX (29).

Ciavatta C., Govi M., Pasotti L., Sequi P., (1993) – Changes in organic matter during stabilisation of compost from municipal solid wastes. Bioresurce Technology n. 43.

Ciavatta C., Govi M., (1993) – Use of insoluble polyvinylpyrrolidone and isoelectric focusing in the study of humic substances in soil and organic wastes. Journal of Chromatography n. 643.

Piccinini S., Rossi L., Ciavatta C., (1994) – Il compostaggio dei fanghi di depurazione e del contenuto ruminale in un macello bovino – RS Rifiuti Solidi n. 4.

The Influence of GFT–Compost Extracts[1] on the Motility of Juveniles of Heterodera Schachtii in Vitro

J. RYCKEBOER and J. COOSEMANS

Laboratory of Phytopathology and Plant Protection, Department of Applied Plant Sciences, Faculty of Agricultural and Applied Biological Sciences, Katholieke Universiteit Leuven, W. de Croylaan 42, 3001 Heverlee–Leuven, Belgium

Abstract

Organic material has undeniable an influence on the soil ecosystem, including the population of soil nematodes. In this experiment, the influence of GFT–compost extracts on the motility of juveniles of H. schachtii is investigated. As reference Hoogmeerturf (a white peat) extract was taken, because there is no suppressive character against nematodes described for peat.

A significant (p = 0,05) reduction of the motility of juveniles of H. schachtii was determinated from day 1 to day 4 after adding GFT-compost extract against as well the control (water) as the Hoogmeerturf extract (reference). The reduced motility couldn't be explained by the concentration of organic acids.

Introduction

This experiment, where the influence of GFT-compost extracts on the motility of juveniles of H. schachtii is investigated, is part of a greater study in which the usefulness of rest materials (wastes) for the biological control of plant pathogens is investigated (RYCKEBOER and COOSEMANS, 1994).

On the one side the composting process has an antipathogenic activity (nematodes and fungi were killed during the composting) and on the other hand compost has an antipathogenic potential (DITTMER et al., 1990; DITTMER, 1991; HOITINK and FAHY, 1986; HOITINK et al., 1976).

One of the aspects of GFT-compost would be an activation of the soil life. This results in a shifting of the present organisms, namely a stimulating of the saprophytic organisms. A measure for this could be the proportion of plant parasitic and non plant parasitic nematodes. From this we can conclude that the addition of organic material lead to a decline of the plant parasitic nematodes (STIRLING, 1991).

On the other hand there are some chemical components (allelochemicals) of which the nematicidal character is known, such as thiophenyls (for example terthienyl), cucurbitacin, alkaloids, phenols and tannins, isothiocynates, organic acids (for example asparatic acid), also nitrites and nitrates, ammonium, H2S, etc. (HASAN, 1992; RYCKEBOER and COOSEMANS, 1994; SEGERS, 1989).

MALEK and GARTNER (1975) report the suppression of some plant parasitic nematodes after the use of bark compost. The attack by rootknot nematodes Meloidogyne hapla en M. incognita in tomato was inferior on plants growing in bark compost than in peat as container substrate. The development of the population of Pratylenchus penetrans, Trichodorus christiei and Helicotylenchus spp. was inhibited in bark compost, but stimulated in peat. After adding compost to potting mixtures, the populations of Cephaloidae and Rhabditidae (useful, saprophytic nematodes) growth (HUNT et al., 1973). SEGERS (1989) made a table with a general view of organic materials which have or don't have any nematicidal activities. Spiteful, the rate of success is in some cases a subjective estimation. Although, we can write that some components of organic material have some influence on the soil life. Application of organic amendments can be a kind of biological control.

HASAN (1992) reports the use of various concentrations of water/organic solvent extracts of different parts of plants to kill nematodes under laboratory conditions. Analogous to this we conducted an experiment with GFT-compost.

HUNT et al. (1973) determinated the influence of water extracts of municipal refuses on the plant parasitic nematode Belonolaimus longicaudatus. Water extracts from saturated composted municipal refuse rendered sting nematodes, B. longicaudatus, immotile after immersion for 12 hr. Extract concentrated to 33% of its original volume rendered all of the 50 sting nematodes tested immotile in 3 hr. The fact that immotility doesn't result automatically in death is illustrated in the following text: over 80% of the nematodes exposed to the organic fraction for 6 hr and over 60% of those exposed for 24 hr regained motility within 1 hour after being placed in distilled water. However, after 144 hr in the organic fraction of compost extract, no nematodes regained motility when transferred to distilled water (HUNT et al., 1973).

Pot experiments of SZCZECH et al. (1993) showed no inhibitory effect of earthwormcompost (= vermicompost) on the development nor on diseases caused by the nematode species Heterodera schachtii en Meloidogyne hapla.

In summary we can say that not all kinds of compost are even successful in the suppression of plant pathogens, here in special nematodes. Compost type and maturity as well as the composting method have an influence on this quality (HOITINK and FAHY, 1986).

Because there were strong indications of a nematicidal character of GFT-compost in an preliminairy experiment with Steinernema carpocapsae, it was certainly meaningful to control if this approach also goes up for H. schachtii. We can work on different levels: (1) is there an influence on the hatching?; (2) is there an influence on the vitality of the cysts?; (3) is there an influence on the free living juveniles of H. schachtii? In this experiment we control if the last influence exist.

Materials and methods

In petri plates with a diameter of 5,5 cm a quantity of J2 of H. schachtii has been diluted in distilled water. Before this plates got any treatment we first counted the number of motile and non-motile nematodes. After this, extracts (extraction time: 1 day) or distilled water were conjugated with two different concentrations (proportion extract/nematode solution 1/1 or 1/2). On regular moments the vitality of the nematodes has been evaluated.

The incubation was at 20 °C. The number of replications was 5 for the GFT-compost and peat extracts, and 3 for the controls.

For the statistical processing (analysis of variance) SAS (Statistical Analysis System; the General Linear Models Procedure) was used. With the Bonferroni t-test pairwise comparisons were conducted on the 95% confidence level (NETER et al., 1990).

Because organic acids have a nematicidal character, the concentrations of some acids were measured.

Results

Table 1 (appendix I) gives the proportional number of motile (vs. living) juveniles in function of the treatment and the incubation time. In this experiment we count 26 observations for each day. The R-squares of the used GLM-model in function of the time are: on day 0: 0,00 (= initial, namely the motility on day 0 is 100 % for every treatment); on day 1: 0,642262; on day 2: 0,714219; on day 3: 0,884476; on day 4: 0,953748. In other words, with increasing time, the quality of the model rises.

The most interesting pairwise comparisons are shown in table 2 (Bonferroni t-test, appendix II), with the upper en the under limit of the confidence interval (95% or $p = 0,05$).

In the figures 1 and 2 (appendix III) the motility percentages are shown in function of the treatment (respectively 1/2 and 1/1 vol. proportion) and in function of the time. Table 3 (appendix III) shows the content of 4 organic acids as well as the EC (Electric Conductivity) and the pH.

Discussion

The tables (1 and 2) and the figures (1 and 2) show an obvious influence of GFT-compost extracts on the vitality of J2 of H. schachtii. Extracts of GFT-compost give already on day 1 a serious, significant reduction of the vitality, as well for the dilution 1/1 as 1/2 and opposite H2O (control) as the extracts of Hoogmeerturf. After day 4, Hoogmeerturf gives a significant reduction of the motility, but only for the dilution 1/1.

The results hold only for H. schachtii, and can not directly be transferred to all nematode species (SAYRE et al., 1965).

That other factors than the organic acids play an important role in the mortality of H. schachtii, is clearly illustrated in table 3. With this table we can't explain the higher mortality of the juveniles in the GFT-compost extracts, unless the pH-effect plays an important role on the ionization condition of the fatty acids.

In vitro GFT-compost has a clear nematicid character when compared with peat extracts. This was illustrated by relative higher concentrations. Now the question rises in which proportion GFT-compost in vivo has a nematicidal character. For this an in vivo experiment has been conducted.

Table 1 The mean motility (number of motile nematodes expressed as % of the total number of nematodes) of *Heterodera schachtii* in function of the treatment (ml extract/ml nematode solution) and the time, the number of observations, the standard deviation, and the minimum and the maximum motility

Day	Treatment*	N (obs.)	Mean motility	Std. dev.	Minimum	Maximum
0	H_2O-1/1	3	100,0	0,0	100,0	100,0
	Turf-1/1	5	100,0	0,0	100,0	100,0
	GFT-1/1	5	100,0	0,0	100,0	100,0
	H_2O-1/2	3	100,0	0,0	100,0	100,0
	Turf-1/2	5	100,0	0,0	100,0	100,0
	GFT-1/2	5	100,0	0,0	100,0	100,0
1	H_2O-1/1	3	100,0	0,0	100,0	100,0
	Turf-1/1	5	91,8	7,4	81,1	100,0
	GFT-1/1	5	65,5	23,8	46,2	100,0
	H_2O-1/2	3	99,5	0,8	98,6	100,0
	Turf-1/2	5	99,9	0,3	99,3	100,0
	GFT-1/2	5	71,4	11,7	58,9	84,4
2	H_2O-1/1	3	91,9	8,3	83,4	100,0
	Turf-1/1	5	77,4	12,9	62,6	97,2
	GFT-1/1	5	37,1	22,5	11,1	65,4
	H_2O-1/2	3	80,8	4,1	76,2	83,6
	Turf-1/2	5	85,4	4,6	80,8	91,5
	GFT-1/2	5	35,1	26,0	7,7	75,0
3	H_2O-1/1	3	86,7	4,9	83,4	92,4
	Turf-1/1	5	68,8	4,1	62,6	73,6
	GFT-1/1	5	14,4	9,4	2,8	28,4
	H_2O-1/2	3	78,8	4,8	73,9	83,6
	Turf-1/2	5	74,1	7,1	66,1	83,7
	GFT-1/2	5	26,8	21,6	7,7	55,6
4	H_2O-1/1	3	86,7	4,9	83,4	92,3
	Turf-1/1	5	66,5	4,5	62,6	73,6
	GFT-1/1	5	9,5	10,6	1,6	27,6
	H_2O-1/2	3	78,8	4,8	73,9	83,6
	Turf-1/2	5	71,8	6,4	66,1	78,8
	GFT-1/2	5	14,4	10,2	4,1	30,0

*H_2O = distilled water; Turf = Hoogmeerturf extract; GFT = GFT-compost extract

Table 2 'Pairwise comparisons' of the treatments in function of the time (*H. schachtii*)

Day	Comparison*	Upper limit	diff. between means	Under limit	Sign.**
0	H_2O-1/1 – GFT1/1	0,0	0,0	0,0	
	H_2O-1/1 – Turf-1/1	0,0	0,0	0,0	
	Turf-1/1 – GFT1/1	0,0	0,0	0,0	
	H_2O-1/2 – GFT1/2	0,0	0,0	0,0	
	H_2O-1/2 – Turf-1/2	0,0	0,0	0,0	
	Turf-1/2 – GFT-1/2	0,0	0,0	0,0	
1	H_2O-1/1 – GFT1/1	4,6	34,5	64,4	*
	H_2O-1/1 – Turf-1/1	-21,7	8,2	38,1	
	Turf-1/1 – GFT1/1	0,4	26,3	52,2	*
	H_2O-1/2 – GFT1/2	-1,8	28,1	58,0	
	H_2O-1/2 – Turf-1/2	-30,3	-0,3	29,6	
	Turf-1/2 – GFT-1/2	2,6	28,5	54,4	*
2	H_2O-1/1 – GFT1/1	13,9	54,8	95,7	*
	H_2O-1/1 – Turf-1/1	-26,4	14,5	55,4	
	Turf-1/1 – GFT1/1	4,9	40,3	75,7	*
	H_2O-1/2 – GFT1/2	4,8	45,7	86,6	*
	H_2O-1/2 – Turf-1/2	-45,5	-4,6	36,3	
	Turf-1/2 – GFT-1/2	14,8	50,3	85,7	*
3	H_2O-1/1 – GFT1/1	44,7	72,3	99,9	*
	H_2O-1/1 – Turf-1/1	-9,8	17,8	45,4	
	Turf-1/1 – GFT1/1	30,5	54,4	78,3	*
	H_2O-1/2 – GFT1/2	24,4	52,0	79,6	*
	H_2O-1/2 – Turf-1/2	-22,9	4,7	32,3	
	Turf-1/2 – GFT-1/2	23,4	47,3	71,2	*
4	H_2O-1/1 – GFT1/1	58,3	77,2	96,1	*
	H_2O-1/1 – Turf-1/1	1,3	20,1	39,0	*
	Turf-1/1 – GFT1/1	40,7	57,0	73,4	*
	H_2O-1/2 – GFT1/2	45,6	64,5	83,3	*
	H_2O-1/2 – Turf-1/2	-11,9	7,0	25,9	
	Turf-1/2 – GFT-1/2	41,1	57,5	73,8	*

*H_2O = distilled water; Turf = Hoogmeerturf extract; GFT = GFT-compost extract;
**Significance p = 0,05

Table 3 The pH (H_2O, 1/5 vol.), EC*, and the content of organic acids (mg/kg) of Hoogmeerturf and GFT-compost

Substrate	pH	EC	acetic acid	propionic acid	iso-butyric acid	butyric acid
GFT-compost	7,8	2,835	59,32	7,34	2,11	2,12
Hoogmeerturf	5,5	0,144	221,00	25,01	2,91	2,94

*EC(mS/cm; 1/5 vol; 25°C)

References

DITTMER U 1991 Untersuchungen zu den wirkungen des compostierungsprozesses und zum antiphy-topathogenen potential von composten gegen Sclerotinia trifoliorum Erikss., Sclerotinia sclerotiorum (Lib.) De Bary und Pseudocercosporella herpotrichoides (Fron.) Deigh. Inaugural-Dissertation. Bonn, Rheinischen Friederich-Wilhelm-Universität, 167p.

DITTMER U, BUDDE K, STINDT A, WELTZIEN H C 1990 Der Einflufl der Compostierung von Compostsubstraten und wässerigen Compostextrakten auf verschiedene Pflanzenkrankheitserreger. Gesunde pflanzen 7, 219-235.

HASAN A 1992 Allelopathy in the management of root-knot nematodes. In: Allelopathy: Basic an .applied aspects. Rizvi S J H, Rizvi V (eds.) London, Chapman & Hall, p413-443.

HOITINK H A J, FAHY P C 1986 Basis for the control of soilborne plant pathogens with composts. Ann. Rev. Phytopathol. 24, 93-114.

HOITINK H A J, HERR L J, SCHMITTHENNER A F 1976 Survival of some plant pathogens during composting of hardwood tree bark. Phytopathology 66, 1369-72.

HUNT P G, SMART G C Jr, ENO C F 1973 Sting nematode, Belonolaimus longicaudatus, immotility induced by extracts of composted municipal refuse. J. Nematol. 5(1), 60-63.

MALEK R, B GARTNER J B 1975 Hardwood bark as a soil amendement for suppression of plant-parasitic nematodes on container-grown plants. Hortical Science 10, 33-35.

NETER J, WASSERMANN W, KUTNER M H 1990 Applied Lineair Statistical Models. Irwin, United States of America, 1181p.

RYCKEBOER J, COOSEMANS J 1994 The use of rest materials (wastes) for the biological control of plantpathogens Case study: the usefulness of GFT-compost for the biocontrol of fungi and nematodes. Original title: Het gebruik van reststoffen voor de biologische controle van plantpathogenen Case study: de bruikbaarheid van GFT-compost voor de biocontrole van fungi en nematoden. Dissertation, Katholieke Universiteit Leuven, Leuven, Belgium, 188p.

SAYRER M, PATRICK Z A, THORPE H J 1965 Identification of a selective nematicidal component in extracts of plant residues decomposing in soil. Nematologica 11, 263-268.

SEGERS R 1989 Effect van organische materialen op nematoden. Case study: chitine versus Meloidogyne goeldi (Tylenchidae). Dissertation. Leuven, Katholieke Universiteit, 104p.

STIRLING G R 1991 Biological control of Plant Parasitic Nematodes. Progress, Problems and Prospects. UK, CAB International, 282p.

SZCZECH M, RONDOMANSKI W, BRZESKI M W, SMOLINSKA U, KOTOWSKI J F 1993 Suppressive Effect of a Commercial Earthworm Compost on some Root Infecting Pathogens of Cabbage and Tomato. Biological Agriculture and Horticulture 10, 47-52.

The Suppression of Penicillium Digitatum by Extracts of GFT-Compost*

J. RYCKEBOER and J. COOSEMANS

Laboratory of Phytopathology and Plant Protection, Departement of Applied Plant Sciences, Faculty of Agricultural and Applied Biological Sciences, Katholieke Universiteit Leuven, W. de Croylaan 42, B-3001 Leuven, Belgium

Abstract

There is a suppressive character described for several kinds of compost against soil borne pathogens (fungi). In this experiment there was a suppression of Penicillium digitatum in vitro after addition of extracts of GFT-compost or Hoogmeerturf (a white peat). Although there are only a few sources who reports a suppressive character of peat. The suppressive character of GFT-compostextract was a result of the high bacterial activity, while the great number of fungi (i.a. Trichoderma spp.) were responsible for the suppressive character of GFT-compost. The use of microfiltrations allowed to distinguish between the biotical and abiotical factors responsible for the suppression. The autoclavation of the Hoogmeerturfextract destroyed the suppressive character; in contrast the GFT-compostextract maintain his suppressive effect after autoclavating as a result of the allelochemicals.

Introduction

The purpose of this experiment was to check of GFT-compost extracts have a influence on the P. digitatum mycelium growth. Various composts may suppress fungal diseases (CHEN et al., 1988; HOITINK and FAHY, 1986; SZCZECH et al., 1993). STINDT and WELTZIEN (1988) found an inhibition of the conidia germination and the development of the germination tube of B. cinerea after the use of several compost extracts. Further they found a increasing suppression with increasing extraction time.

Research on suppression of Rhizoctonia damping-off has shown that microbial populations within media containing CHB (composted hardwood bark) are respon-

* GFT-compost is the result of the composting of the organic fraction of household and garden waste (biowaste)

sible for disease control (NELSON and HOITINK, 1982; NELSON and HOITINK, 1983). STINDT and WELTZIEN (1988) found that extracts of composted organic material losses their activities after sterilisation by filtration with a cellulose-acetate membrame filter with a pore size of 0,2 μm. With other words, the biocontrol is depending on the antagonistic organisms, who are removed by heating or microfiltration (0,2 μm). The same filter with a pore size of 12 μm has no influence on the activity.

But the influence of chemical inhibitors (allelochemicals) may not be underestimated. Inhibitors with fungicidal activity have been found in media amended with composted hardwood bark as well as in those amended with pine bark. The nature of the chemicals involved is being investigated (HOITINK, 1980).

A different influence on both factors by increasing temperature or temperature treatment can be expected. According to MILLIPORE (1991) would a 0,2 μm microfiltration remove the bacteria and a 1,2 μm filtration the algae, fungi and protozoa. When the compostextracts of different kinds of composts were autoclavated, the suppressive character disappears (MANDELBAUM et al., 1988; STINDT and WELTZIEN, 1988). The extraction proportion was in this experiment 1/5 (vol) and the extraction time 1 day.

HARDY and SIVASITHAMPARAM (1991) found a various effect, after adding of non-sterile and sterile extracts to cornmeal agar (CMA) or wateragar (WA) on the mycelium growth of Phytophthora nicotianae var. nicotianae, P. drechsleri and P. cinnamomi. The mycelium growth of the three studied Phytophthora-species was greater for the non-sterile nursery mix (NM)-extracts added to CMA. The opposite was found for P. cinnamomi and P. drechsleri when non-sterile composted eucalyptus bark (CEB)-extracts were added to CMA. The mycelium growth on CMA, which contain non-sterile CEB-extracts, was smaller than on CMA enriched with sterile CEB-extracts. Consequently, these authors conclude with a great evidence that a specific inhibitory effect go out from the microorganisms in the non-sterile CEB, which not occurre in the CMA. But the effect was not common, because the mycelium growth was not reduced on CMA when non-sterile NM-extracts were added.

Further, they postulate that biotical factors of CEB were responsible for the inhibition of the sporangium formation and abiotical factors the sporangium formation stimulates. But the biotical factors were dominant. It is clear that the factors present in the extracts, which are responsible for the suppression of the fungi, in connection with the kind of compost and extraction time not easily can be defined. The results of STINDT and WELTZIEN (1988) show that the action depends on the biological activity of the extract. Every kind of compost will have an other microbial composition after extraction. According to HOITINK et al (1977) is the suppressive effect of bark compost probaly due to chemical and biological rather than physical factors. Connected with that there can appear several antagonistic activities, antibiosis, concurrence or other fenomes. KUTER et al. (1983) determined a relation between species and disease suppression. Disease suppression was not associated with a single fungal taxon. Furthermore, quantitative differences in sample populations isolated from suppressive and conducive container media indicated that the lack of suppression in some media was due to factors limited development of high populations of Trichoderma or interfered with the antagonistic activity of these fungi. The maturity of the compost has an influence on the antagonistic activity (BOLLEN, personal communication).

Materials and methods

GFT-compost or Hoogmeerturf was diluted in autoclaved, distilled water with an extraction proportion of 1/5vol. Every extraction was repeated three times.

The whole was shaken for 1 day, this was followed by (micro)filtrations, respectively 1) Whatman 5; 2) 5 µm MF-Millipore membrame filter; 3) 1,2 µm MF-Millipore membrane filter; 4) Nalgene Bottle Top filter 0,2 µm. From each filtrate there was a part (1 or 5 ml) plate out with 10 ml Potato Dextrose Agar (PDA) 1/5, with a proportion filtrate/PDA: 1/5. To the control plates autoclaved, distilled water was added. A dose pump was used by 50 °C to fill the petriplates. After coagulation with the culture medium there was an inoculation with P. digitatum, therefore of P. digitatum growing on a culture medium of Water Agar were 3 mm plots made and put upside down to the PDA plates. The mycelium growth by 21 °C from this plots was followed in function of time. The number of observations was for 1 ml extract conjugation equal to 21 (3 extractions with 7 replicates for each extraction), for 5 ml extract conjugation 9 (3 extractions with 3 replicates) .

For the statistical processing (analysis of variance (ANOVA)) was the statistical computing package SAS (Statistical Analysis System) used. With the Bonferroni t-test pairwise comparisons were caried out on the 95% confidence level (p=0,05) (NETER et al. 1990). For this a confidence interval was made an upper and under limit, as well as a difference between de means, expressed in table 2. Only the most interesting pairwise comparisons were restrain.

Results

In table 1 (see appendix 1) is respectively illustrated: the mycelium growth (in cm) as mean of the three extracts of P. digitatum, the number of observations (N), the standard deviation, and the minimum and maximum growth of the fungus, after 3 and 6 days. Every treatment receive a code. The first figure in this code stands for the number of ml extract (or H2O) that was add to the plate, on the second place the substrate (G = GFT-compost; T = Hoogmeerturf), on the third place the way of sterilisation (0 = Whatman 5, 1 = 3 µm, 2 = 1,2 µm, 3 = 0,2 µm, A = autoclaved).

Table 2 (see appendix 2) gives for the pairwise comparisons of the mean P. digitatum growth for 3 and 6 days with the difference between the means, the under and upper limit of the confidence interval on day 3 and 6.

The R-square of the used GLM-model was for day 3 0,973920, for day 6 0,978644.

For the Hoogmeerturf extracts as well as the GFT-compost extracts there is a clear suppression of P. digitatum, in function of the way of sterilisation.

How smaller the pore size of the filter, how more microorganisms were filtrated out, what resulted in a smaller suppression of P. digitatum.

The biotic factors responsible for the suppression are either different for peat and GFT-compost: the suppression of the P. digitatum growth for Hoogmeerturf was a result of the growth of antagonistic fungi (i.a. Trichoderma spp.), while bacteria are responsible for the suppression in case of GFT-compost. This conclusion is not clear in the preceding tables, either photograph 1 (GFT-compost extracts) and 2 (Hoogmeerturf extracts) illustrated this clearly (see appendix 3). Note beside the difference in mycelium growth, a clear difference in the kind of the microbial cultures between both substrate extracts.

Discussion

The suppressive character of white peat was already shown or reported by different authors (HOITINK and FAHY, 1986; TAHVONEN, 1982a; TAHVONEN, 1982b; WOLFFHECHEL, 1988). Because the used Hoogmeerturf a white peat is, is the suppression as a result of the use of this peat understandable.

Although Hoogmeerturf a suppressive character shows, is the suppression of P. digitatum by GFT-compost in most cases signficant greater then the peat. On day 9, Penicillium could not colonize the whole plate when 5 ml autoclavated or 0,2 µm extracts of GFT-compost were added. All the other plates with the same extract addition (1 ml) and these with peat extract could colonize the plate well. So, we can conclude the following: (1) GFT-compost has a suppressive character as a result of biotic factors, but abiotic factors (allelochemicals) seem also to play a role, as showed by autoclavating. These allelochemicals give the compost also a kind of phytotoxicity, as shown in an other experiment. (2) For a higher concentration of GFT-compost extract (5 ml vs. 1 ml) in the medium is there also a higher concentration of colloids, salts, ..., in the medium, what resulted in a higher suppression as a result of abiotic factors. For the biotic factors plays the initial concentration of the extract a smaller role because of the great multiplication rate of the microorganisms. The different pigmentation after adding of GFT-compost extract show clearly the stress of the fungus.

According to HARDY and SIVASITHAMPARAM (1991) varied the mycelium growth with leachate, agar type and the fungi.

This experiment shows a clear suppression of P. digitatum after adding of GFT-compost extract to the medium. A specific sensitivity of several soil borne fungi is not excluded. Further research about this subject is going on and can give more clearness.

References

CHEN W, HOITINK H A J, SCHMITTHENNER A F, TUOVINEN O H 1988 The role of microbial activity in suppression of damping-off caused by Pythium ultimum. Phytopathology 78, 314–322.

HARDY G E St J and SIVASITHAMPARAM K 1991 Sporangial responses do not reflect microbial suppression of Phytophthora drechsleri in composted Eucalyptus bark mix. Soil Biol. Biochem. 23(8), 757–765.

HOITINK H A J 1980 Composted bark, a lightweight growth medium with fungicidal properties. Plant Disease 64(2), 142–147.

HOITINK H A J and FAHY P C 1986 Basis for the control of soilborne plant pathogens with composts.   Ann. Rev. Phytopathol. 24, 93–114.

HOITINK H A J, VANDOREN D M, SCHMITTHENNER A F 1977 Suppression of Phytophthora cinnamomi in a composted hardwood bark potting medium. Phytopathology 67, 561–565.

KUTER G A, NELSON E B, HOITINK H A J, MADDEN L V 1983 Fungal populations in container media amended with composted hardwood bark suppressive and conducive to Rhizoctonia damping-off. Phytopathology 73(10), 1450–1456.

MANDELBAUM R, HADAR Y, CHEN Y 1988 Composting of Agricultural Wastes for their Use as Container Media: Effect of Heat Treatments on Suppression of Phytium aphanidermatum and Microbial Activities in Substrates containing Compost. Biological Wastes 26, 261–274.

MILLIPORE 1991 Laboratory catalogue 1991–1992. Cat. No. EU 328/U. Millipore corporation U.S.A., Germany pp 5–15.

NELSON E B and HOITINK H A J 1982 Factors affecting suppression of Rhizoctonia solani in container media.Phytopathology 72, 275– .

NELSON E B and HOITINK H A J 1983 The role of microorganismes in the suppression of Rhizoctonia solani in container media amended with composted hardwood bark. Phytopathology 73, 274–278.

NETER J, WASSERMAN W, KUTNER, M H 1990 Applied Lineair Statistical Models. Irwin, United States of America, 1181p.

RYCKEBOER J and COOSEMANS J 1994 The use of rest materials (wastes) for the biological control of plant pathogens Case study: the usefulness of GFT-compost for the biocontrol of fungi and nematodes. Original title: Het gebruik van reststoffen voor de biologische controle van plantpathogenen Case study: de bruikbaarheid van GFT-compost voor de biocontrole van fungi en nematoden. Dissertation, Katholieke Universiteit Leuven, Leuven, Belgium, 188p.

STINDT A and WELTZIEN H C 1988 Der einfluss von waessrigen, mikrobiologisch aktiven extrakten von compostiertem organischen material auf Botrytis cinerea. Rijksuniversiteit Gent, Med. Fac Landbouww. 53(2a), 379–88.

TAHVONEN R 1982a The suppressiveniss of Finnish light coloured sphagnum peat. J. Sci. Agric. Soc. Finl. 54, 345–356.

TAHVONEN R 1982b Preliminary experiments into the use of Streptomyces spp. isolated from peat in the biological control of soil and seed-borne diseases in peat culture. J. Sci. Agric. Soc. Finl. 54, 357–59.

WOLFFHECHEL 1988 The suppressiveness of sphagnum peat to Pythium spp. Acta Horticulturae 221, 217–222.

Appendix 1

Table 1 The mean mycelium growth of P. digitatum, the number of observations (N), the standard deviation, and the minimum and maximum growth of the fungus, after 3 and 6 days

Treatment*	DAY NR. 3					DAY NR. 6				
	N	Mean	Std. dev.	Min.	Max.	N	Mean	Std. dev.	Min.	Max.
Nothing	10	2,43	0,09	2,3	2,6	10	5,76	0,12	5,6	6,0
1H2O-A	21	2,52	0,13	2,4	2,8	20	5,52	0,17	5,3	5,8
1G-A	21	2,37	0,08	2,2	2,5	20	5,40	0,28	4,8	5,8
1G-3	21	2,52	0,14	2,3	2,8	21	5,65	0,35	5,0	6,2
1G-2	21	0,30	0,00	0,3	0,3	21	0,30	0,00	0,3	0,3
1G-0	21	0,30	0,00	0,3	0,3	21	0,30	0,00	0,3	0,3
1T-A	21	2,51	0,08	2,4	2,6	21	5,70	0,38	4,6	6,2
1T-3	21	2,59	0,13	2,3	2,9	21	6,13	0,21	5,8	6,5
1T-2	21	0,98	0,35	0,5	1,9	19	2,29	0,86	1,0	4,0
1T-0	20	1,04	0,18	0,7	1,4	10	1,58	0,31	1,2	2,1
5H2O-A	9	2,32	0,10	2,2	2,5	9	4,88	0,12	4,7	5,1
5G-A	9	1,73	0,12	1,6	2,0	9	3,13	0,17	2,9	3,3
5G-3	9	1,73	0,12	1,6	2,0	8	3,01	0,27	2,5	3,3
5G-2	9	0,30	0,00	0,3	0,3	9	0,30	0,00	0,3	0,3
5G-0	9	0,30	0,00	0,3	0,3	9	0,30	0,00	0,3	0,3
5T-A	9	2,37	0,07	2,3	2,5	9	5,31	0,21	5,0	5,6
5T-3	9	2,51	0,08	2,4	2,6	9	5,58	0,17	5,4	5,9
5T-2	9	1,33	0,39	0,8	1,8	9	2,44	0,88	1,4	3,9
5T-0	9	0,82	0,07	0,7	0,9	9	0,82	0,07	0,7	0,9

*Every treatment receive a code. The first figure in this code stands for the number of ml extract (or H2O) that was add to the plate, on the second place the substrate (G = GFT-compost; T = Hoogmeerturf), on the third place the way of sterilisation (0 = Whatman 5, 1 = 3 µm, 2 = 1,2 µm, 3 = 0,2 µm, A = autoclavated).

Appendix 2

Table 2 Pairwise comparisons of the mean P. digitatum growth for 3 and 6 days with the difference between the means, the under and upper limit of the confidence interval on day 3 and 6.

Comparison*	DAY NR. 3 Under limit	Diff. betw. means	Upper limit	Sign.	DAY NR. 6 Under limit	Diff. betw. means	Upper limit	Sign.
1H2O-A – Nothing	–0,12	0,09	0,31		–0,74	–0,24	0,26	
1H2O-A – 1T–A	–0,16	0,01	0,19		–0,59	–0,18	0,22	
1H2O-A – 1T–3	–0,24	–0,06	0,11		–1,02	–0,61	–0,21	*
1H2O-A – 1T–2	1,37	1,55	1,72	*	2,82	3,23	3,64	*
1H2O-A – 1T–0	1,31	1,48	1,66	*	3,44	3,94	4,44	*
1H2O-A – 1G–A	–0,02	0,15	0,33		–0,29	0,12	0,53	
1H2O-A – 1G–3	–0,17	0,00	0,18		–0,54	–0,13	0,27	
1H2O-A – 1G–2	2,05	2,24	2,40	*	4,82	5,22	5,62	*
1H2O-A – 1G–0	2,05	2,24	2,40	*	4,82	5,22	5,62	*
1T–A – 1G–A	–0,04	0,14	0,31		–0,10	0,30	0,71	
1T–3 – 1G–3	–0,11	0,07	0,24		0,08	0,48	0,88	*
1T–2 – 1T–2	0,50	0,68	0,85	*	1,58	1,99	2,40	*
1T–0 – 1T–0	0,56	0,74	0,92	*	0,78	1,28	1,78	*
5H2O-A – Nothing	–0,37	–0,11	0,15		–1,48	–0,88	–0,29	*
5H2O-A – 5T–A	–0,31	–0,04	0,22		–1,04	–0,43	0,18	
5H2O-A – 5T–3	–0,46	–0,19	0,08		–1,31	–0,70	–0,09	*
5H2O-A – 5T–2	0,72	0,99	1,26	*	1,82	2,43	3,04	*
5H2O-A – 5T–0	1,23	1,50	1,77	*	3,44	4,06	4,67	*
5H2O-A – 5G–A	0,32	0,59	0,86	*	1,13	1,74	2,35	*
5H2O-A – 5G–3	0,32	0,59	0,86	*	1,24	1,87	2,49	*
5H2O-A – 5G–2	1,75	2,02	2,29	*	3,97	4,58	5,19	*
5H2O-A – 5G–0	1,75	2,02	2,29	*	3,97	4,58	5,19	*
5T–A – 5G–A	0,37	0,63	0,90	*	1,57	2,18	2,79	*
5T–3 – 5G–3	0,51	0,78	1,05	*	1,94	2,57	3,19	*
5T–2 – 5T–2	0,77	1,03	1,30	*	1,53	2,14	2,75	*
5T–0 – 5T–0	0,25	0,52	0,79	*	–0,09	0,52	1,13	

*Every treatment receive a code. The first figure in this code stands for the number of ml extract (or H2O) that was add to the plate, on the second place the substrate (G = GFT–compost; T = Hoogmeerturf), on the third place the way of sterilisation (0 = Whatman 5, 1 = 3 µm, 2 = 1,2 µm, 3 = 0,2 µm, A = autoclavated).

Appendix 3

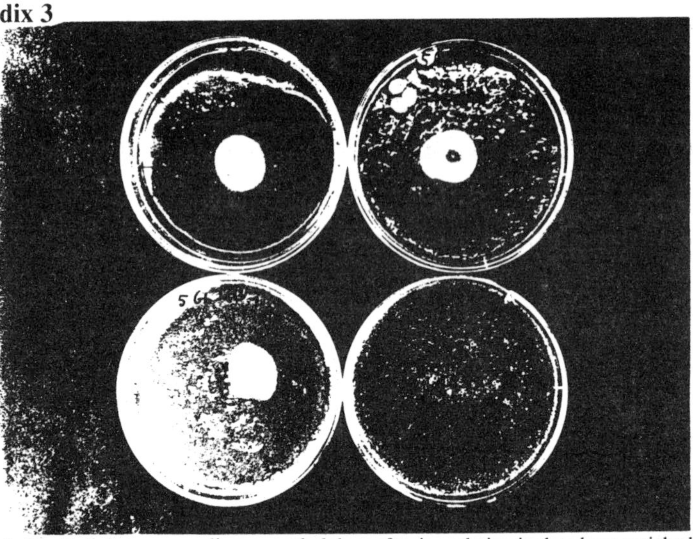

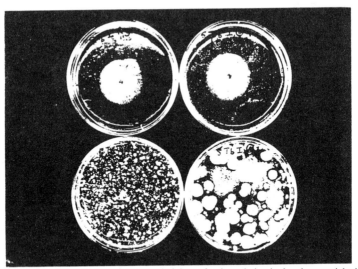

Photo 1 The P. digitatum mycelium growth 6 days after inoculation in the plates enriched with GFT–compostextract (5ml/10 ml medium) (upper left: autoclavated extract (diameter mycelium: 3,1 cm); upper right: 0,2 μm–extract (3,3 cm); under left: 1,2 μm-extract (0,3 cm; the spot visuable by this treatment is not mycelium growth, but condens); under right: Whatman 5 extract (0,3 cm); The control (autoclavated water, not shown): 4,88 cm.

Photo 2 The P. digitatum mycelium growth 6 days after inoculation in the plates enriched with Hoogmeerturf-extract (5ml/10 ml medium) (upper left: autoclavated extract (diameter mycelium: 5,1 cm); upper right: 0,2 μm-extract (5,4 cm); under left: 1,2 μm-extract (2,3 cm); under right: Whatman 5 extract (0,8 cm); The control (autoclavated water, not shown): 4,88 cm.

Cation Exchange Capacity of Manure-Straw Compost – Does Sample Preparation Modify the Results?

SAHARINEN, M.H. – Department of Biology, University of Joensuu, P.O. BOX 111, SF-80101 Finland.

Abstract

Effective cation exchange capacity (ECEC) of mature manure-straw compost was determined by saturation-exchange method. Analysis was carried out on either fresh or dried and ground composts.

Drying of the sample increased ECEC values, whereas drying accompanied with grinding did not seem to have any effect on ECEC, as compared with ECEC values analyzed from fresh compost.

This study implies that ECEC determined of samples which are dried and ground describes well the actual cation exchange properties of the mature manure-straw compost, and that fresh compost can also be used for CEC determinations.

Introduction

Cation exchange capacity of matrix, for example soil or compost, describes the quantity of negatively charged sites on this matrix and thus its ability to hold positively charged ions. Its importance as one of the variables used for determining maturity of composts is widely accepted (Mathur et al. 1993, Inbar et al. 1991, Harada and Inoko 1980).

Sample preparation modifies its visual appearance and chemical properties such as solubility of nutrients. The aim of this research was to check, if drying and grinding of mature manure-straw compost has profound effects on its effective cation exchange capacity.

Material and methods

Material was drum compost of manure-straw mixture, cured for ca. 3 months in a pile.

Pre-handled composts for cation exchange capacity analysis were: 1. field moist compost; weight field moist and CEC was determined at once, 2. dried compost; weight field moist and dried in oven at 40 C before analysing CEC, and 3. ground compost; dried in oven at 40 C and ground prior to weighing for CEC determination.

Effective cation exchange capacity was determined by saturation-exchange method (Saharinen et al. 1995).

Results and Discussion

Methods of sample preparation (drying, grinding) profoundly affected the visual appearance of the compost sample. The fresh manure-straw compost had considerably large aggregate sizes and was heterogenous in nature, and drying made its aggregate structure extremely fragile. With grounding the dried compost was thoroughly homogenized.

Drying affects various chemical properties of samples, for example solubilities of elements. Results of this study showed a slight increase in ECEC values due to drying (Table 1), which suggests that more exchange sites may have become exposed. In addition it was established that sample size (1 to 2 grams) did not have any effect on these results (Table 1).

Table 1 Effective cation exchange capacity of fresh and dried compost (drum compost after curing for ca. 3 months in a pile), determined from sample sizes of 1g and 2g

Sample	ECEC meq/100g dw	
	fresh	dried
compost (1g)	69.4 2.2	72.5 2.3
compost (2g)	69.9 1.0	72.0 2.3

Grinding is hypothesized to give overestimated ECEC-values. This was not verified in this study, as ECEC-values of dried and ground compost (67.1) did not significantly differ from those of fresh compost, as it had a considerably smaller standard deviation (0.4), due to its homogeneity, than the former.

I conclude from the results of this study that pre-handling of a compost sample does not significantly modify ECEC data, and that the data gained from ground samples reasonably well describes the cation exchange properties of the compost. Even fresh composts (as such or dried after weighing) can be used for CEC determinations, but then special concern must be taken to prevent possible corruption of the data due to accidental drying of the samples during handling and weighing procedures of the fresh samples. This study though does not exclude the possibility that drying and grounding change characteristics of the exchange sites even though the capacity of the mature manure-straw compost to hold positively charged ions is not noticeably modified.

Acknowledgements

This work was carried out as a part of the research project 'Optimization of manure composting and minimization of its environmental and hygienic disadvantages' financed by the Ministry of Environment, the Ministry of Agriculture and Forestry, The Maj and Tor Nesslings Foundation and the University of Joensuu.

References

Harada,Y. and Inoko,Y. 1980: The measurement of the cation-exchange capacity of composts for the estimation of the degree of maturity. – Soil Sci. Plant Nutr. 26: 127–134.

Inbar,Y., Chen,Y., Hadar,Y. and Hoitink, H.A.J. 1991: Approaches to determining compost maturity. In: The Staff of BioCycle (eds) 'The BioCycle guide to the art and science of composting' pp. 183–187, JG Press, Pennsylvania.

Mathur, S.P., Owen, G., Dinel, H. and Schnitzer, M. 1993: Determination of compost maturity. 1. Literary review. – Biological Agriculture and Horticulture 10: 65–85.

Saharinen, M.H., Vuorinen, A.H. and Hostikka, M. 1995: Effective cation exchange capacity of manure-straw compost of various maturity stages. (submitted)

Vermicomposting Solid Paper Pulp Mill Sludge: a Three Stages Biodegradative Process.

L. SAMPEDRO[1], C. ELVIRA[1], J. DOMÍNGUEZ[1], R. NOGALES[2] and S. MATO[1]

Introduction

The treatment of wastewater generated through paper pulp production produces large amounts of sludges; management of these residues constitutes an important environmental problem for this industry. To date, most of these sludges were land-filled or incinerated, but this only constitutes a partial solution. Incineration in the biomass boiler produces toxics and ashes and requires energy supply from these factories, and landfilling is not reliable for a long time. Thus, this waste should be recycled and exploited like an important source of organic matter.

In addition to solid paper pulp mil sludge (SPPMS) is a problematic organic refuse, it is a very steady waste, with a very slow decomposition rate. Although there are many types of SPPMS depending on manufacturing processes and waste-water treatments, they usually show high moisture content, C:N ratio, and pH which are barriers to its degradation.

Previous research showed that bioconversion is possible simply amending the SPPMS in mixture with other materials rich in nitrogen under suitable conditions. This pretreatment represents a physical, chemical and biological amelioration. In this way the C:N ratio and macronutrients levels are titled, and, at the same time, an important microorganisms population is innoculated. Using these mixtures to feed earthworms, a rapid transformation and high growth rates are obtained; then, vermicomposting could be a good method for SPPMS biogradation with a low technology system, despite its high initial moisture content.

The aim of this communication was to discuss the particular biodegradation process that SPPMS amended with pig slurry undergoes for a three months ver-micomposting.

1.– Dpto. Recursos Naturais e Medio Ambiente. Universidade de Vigo. Apto. 874 E–36200 VIGO (Po.). SPAIN.
2.– Unidad de Agroecología y Protección Vegetal. Estación Experimental Zaidín (C.S.I.C.). Prof. Albareda, 1. E-18008 GRANADA. SPAIN

Materials and Methods

Fresh SPPMS was directly obtained from the primary wastewater treatment after the filtering press of E.N.C.E. (Empresa Nacional de Celulosas), Lourizán, Po. (Spain). This factory utilizes *Eucalyptus globulus* Labill. wood to produce virgin kraft paper pulp. Table I shows the main characteristics of the sludge and slurry.

SPPMS mixed with pig slurry in a ratio 4:1 (dry weigth) has been used in a vermicomposting experiment for 90 days; 30 Kg of this initial mixture was innoculated with *Eisenia andrei* Bouché,1972 from our laboratory stock. Temperature, moisture and ashes content, oxygen, leacheate production, pH and conductivity were measured periodically; samples to analyse several parameters were taken: TOC (Total Organic Carbon); TKN (Total Kjeldhal Nitrogen); TEC (with alcaline sodium pyrophosphate); water soluble carbon (WSC); non humified (NH, labile non phenolic) and humified (humic -HA- and fulvic -FA- acids) organic fractions (Sequi *et al.*, 1986); crude fibre (CF) (AOAC, 1965) and fibrous compounds of Goering & Van Soest (1970) (neutral detergent fibre -NDF-, acid detergent fibre -ADF- and acid detergent lignin -ADL- fractions).

Results

SSPMS is a very steady waste (Table I), with a basic pH, high moisture content, and high C:N ratio and fibre content; C:N ratio droped only until 240 after one year. These difficulties to biological treatment might be ameliorated by physical, chemical and biological pretreatment. Pig slurry resulted to be a good amendment in order to fit C:N ratio to suitable values in the initial mixture.

Table I Main analytical characteristics of pig slurry and SPPMS used in the experience.

	pH	E.C.	Moisture	Ash	TOC	TKN	N–NH4	C:N	TEC	WSC	CF
SPPMS	9.3	0.32	81.1	5.4	51.5	0.20	0.01	257	2.09	0.91	70
Pig Slurry	8.7	4.40	81.5	28.3	34.1	3.92	1.48	8.7	4.31	0.66	16

(all in percentage of dry weight, except pH, and conductivity in dS.m–1)

With regard to of substrates mineralization, Table II summarizes the evolution of some organic matter parameters. TOC decreased and TKN increased reaching a final suitable ratio for agricultural purposes. C:N ratio evolution showed different slopes, firstly a rapid decrease during the first month, and then slowed down. WSC sharply increased (five times) within first 30 days, dropping to less values than initial ones. TEC increased along the experience according to the humification process. NH evolution was in agreement with WSC one. For HA and FA big increments were observed, with the highest pick in the last month for HA.

Table II Evolution of some organic matter parameters in the sustratum during the experience.

	IM	10 days	20 days	30 days	60 days	90 days
V.S.	89,9	89,2	84,7	84,4	84,8	77,2
TOC	49,6	47,9	45,6	41,86	41,6	40,7
TKN	0,8	0,98	1,18	1,9	2,17	1,72
C:N	62	49	38,6	21,6	19,2	23,6
WSC	0,77	1.457	2,39	4,28	3,83	0,5
TEC	3,37	4,14	6,01	8,22	9,01	9,29
NH	0,94	1,97	3,25	5,14	4,58	1,21
FA	0,65	0,51	0,63	0,7	1,1	1,1
HA	1,5	1,56	2,17	3,08	3,22	6,78

Days from the begining; V.S.: volatile solids, IM: initial mixture, the rest of abreviations are in the text; data are shown as percentage of dry weight

ADF–ADL (cellulose of Goering & Van Soest) is a similar technique to CF (Table III), and show quite well how the polysacharides degradation occurs, which are the fundamental component of the waste and the initial mixture. The lysis of fibrous compounds was faster within the first 30 days, since that moment it went down; final value is four times less than initial one. The experimental time resulted to be too short to allow the degradation of the lignins, this explains why the fraction identified as ADL was increasing, reaching an important value at the end, probably due only to loss of organic matter.

Table III Evolution of the fibrous compounds in the sustratum.

	IM	10 days	20 days	30 days	60 days	90 days
CF	61,05	53,9	45,3	25,8	26,6	16,35
ADF–ADL	61,76	56	40,8	21,8	20,1	19,21
ADL	9,59	11,82	14,43	17,74	18,5	27,98

Days from the begining; IM: initial mixture, the rest of abreviations are in the text; data are shown as percentage of dry weight.

Table IV shows the evolution of some usual biodegradation indexes. All these indexes are useful to assess and discuss evolution of organic wastes during its degradation and stabilization.

Table IV Evolution of some degradation indexes in the sustratums during the experience.

	IM	10 days	20 days	30 days	60 days	90 days
HI	0,43	0,95	1,15	1,35	1,08	0,15
HG	64,5	50,1	46,78	42,8	46,96	84,8
HR	4,5	4,32	6,17	9,06	10,1	19,36
FA/HA	0,43	0,32	0,29	0,23	0,31	0,16

Days from the begining; IM: initial mixture, the rest of abreviations are in the text.

Discussion

Degradation dynamic of this process is not the habitual one in vermicomposting; it is a mesophillic process, hovewer there are some great physical and chemical changes in the substratum. We could say that it is a three steps bioconversion process with two different degradation rates. The initial steady waste becomes into a very unstable material, after a fast and strong breakdown showing low exoteric activity. Then, it is slowly stabilized until to become a mature vermicompost. Humification Index (HI), a good indicator of changes trends, reflected shows these steps.

During the first phase of destabilization, the higher mineralization rate took place due to the lysis of structural polyssacharides; supplyng simple molecules (NH and WSC), and a lower C:N (from 60 to 21). After 30 days it results in a very unstable material, as the indexes and the organic matter fractions proved.

During the second phase or stabilization one, the humification of easily digerible compounds take place. FA increase during the previous phase, while HA showed a continuos increment, this produces a decrease inthe AF/AH ratio in this phase.

References

A.O.A.C. (1965). Official methods of analysis, 10th edition. *Association of Official Analytical Chemists*, Washington, D.C.

Goering H.K. and Van Soest P.J. (1970) Forage fiber analysis (Aparatus, reagents, procedures and some applications). *USDA Agriculture Handbook* 379, 1–20.

Sequi P., De Nobili M., Leita L. and Cercignani G. (1986) A new index of humification. *Agrochimica* 30, 175–179.

Evaluation of Turned and Static Pile Systems on Toxicity of Water Extracts of Biowastes

M.A. SÁNCHEZ-MONEDERO, M.P. BERNAL, A. ROIG, I. CEGARRA – Department of Soil and Water Conservation and Organic Wastes Management, Centro de Edafología y Biología Aplicada del Segura, CSIC. P.O. Box 4195. 30080–MURCIA, Spain.

Introduction

The degree of compost maturation has been widely studied in the water-soluble fraction (Chanyasak et al., 1982; Riffaldi et al., 1988; Garcia et al., 1991). Since it is there that the main biochemical transformations undergone by the microorganisms in organic matter take place. Therefore, phytotoxic substances, such as organic fatty acids of low molecular weight, ammonia and phenols and their degree of degradation during composting should be evaluated in this aqueous phase.

The aim of this work is to compare the evolution and to asses the maturity of two piles composted by the static forced ventilation system and by windrow.

Material and methods

Two composting piles were prepared with a household biowaste and sweet sorghum bagasse mixture at a volume ratio of 50:50 (95% and 5% fresh weight) with a C/N ratio of 17.7. One was composted by the Rutgers static pile system (C1) and the other by the traditional turned pile (windrow) (C$_2$). The timer in the static pile was set for 30 sec. ventilation every 15 minutes, and the ceiling temperature for continuous air blowing was kept at 55 °C. The biooxidative phase of the fermentation period lasted 11 weeks and the maturation time was a further 8 weeks in both piles. The windrow pile was turned every two days during the first week, twice a week during the second week, and once a week during the rest of the biooxidative phase.

The evolution of the composting processes was studied at the following times: 1) starting time, 2) end of the thermophilic phase (5 weeks), 3) end of the active phase (11 weeks) and 4) after maturation (19 weeks).

Water extracts were obtained with 4 gr (fresh weight) and 30 ml of water by shaking for 30 minutes at room temperature, centrifugation at 7000 rpm and then filtering through Millipore 0.45 μm filter paper (Chanyasak and Kubota, 1981). The following analyses were performed on the water extracts: water-soluble carbon and total nitrogen by elemental microanalyzer, ammonia by the colorimetric method based on Berthelot's reaction, and nitrate by ion chromatography. The organic-N was calculated by substracting the amount of inorganic-N (NH_4-N + NO3-N) from the total-N in the extract, phenols by Folin Ciocalteu's method, carbohydrates by the anthrone method, low molecular organic acids by HPLC and the germination index by the method of Zucconi et al., (1982).

Table 1 Water extract parameters as indicators of the compost process

sampling time (weeks)	Org-C (%)	carb (%)	Total-N (%)	Org-N (%)	NH_4-N (%)	NO_3-N (%)	Org-C/Org-N
C_1							
0	5.51	3.66	0.46	0.36	0.10	n.d.	15.30
5	1.71	0.36	0.35	0.24	0.11	n.d.	7.12
11	0.76	0.32	0.25	0.09	0.02	0.14	8.44
19	0.43	0.20	0.25	0.07	0.02	0.16	6.14
C_2							
0	5.51	3.66	0.46	0.36	0.10	n.d.	15.30
5	1.22	0.37	0.24	0.15	0.09	n.d.	8.13
11	0.53	0.15	0.12	0.09	0.01	0.02	5.89
19	0.28	0.10	0.13	0.05	0.01	0.07	5.60

n.d.: Not detectable.
carb: carbohydrates.

Table 2 Water extract parameters as indicators of compost phytotoxicity.

sampling time (weeks)	Ace (%)	Pro (%)	But (%)	Phenols (%)	NH_4-N (%)	G.I. (%)
C_1						
0	1.61	0.74	1.01	0.22	0.10	25.2
5	0.19	0.18	0.15	0.19	0.11	49.4
11	0.09	n.d.	n.d.	0.06	0.02	76.4
19	n.d.	n.d.	n.d.	0.04	0.02	90.4
C_2						
0	1.61	0.74	1.01	0.22	0.10	25.2
5	0.04	0.09	n.d.	0.15	0.09	92.2
11	n.d.	n.d.	n.d.	0.07	0.01	90.8
19	n.d.	n.d.	n.d.	0.03	0.01	73.7

n.d.: Not detectable.
Ace: Acetic acid; Pro: Propionic acid; But: Butyric acid.

Results and discussion

A high degradation rate was observed in the water-soluble carbon and carbohydrates (Table 1). The former had an initial value of 5.51 and, at the end of the thermophilic phase (5 weeks), the degradation was of 69.0 and 77.9 for 4 and 4,
respectively. After this time the degradation rate decreased, and the final level of
water-soluble carbon in the mature compost was 7.8 and 5.1 of the initial values
respectively. The greater degradation rate observed in C_2 could have been due to
water-soluble carbon losses through leaching, since the high temperature reached,
more than 70 °C (Fig 1), required a high amount of water for an adequate degree
of humidity to be mantained for composting.

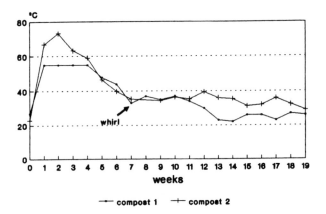

Fig 1 Temperature evolution of composts

The highest degradation rate of the soluble carbohydrates was during the first 5
weeks and by the end of the composting period the values were negligible.

The evolution of N-compounds was similar in both piles. The org-C/org-N ratio
indicated that both composts were mature after 19 weeks, although the turned pile
showed a value which may indicate that he was mature at the end of the active
phase (11 weeks), as a result of its lower water-soluble carbon content. Chanyasak
and Kubota (1981).

Table 2 shows parameters indicating the degree of phytotoxicity. Of the volatile
fatty acids, acetic acid was produced to a higher degree than propionic and butyric
acids, although all of them degraded rapidly during the composting process particularly in C_2. By week 11 only the acetic acid was detectable in C_1. Phenols,
which are also considered phytotoxic, were degraded at a similar rate in both piles.
The evolution of NH4-N showed a similar pattern in both piles. Its presence in a
compost indicates a low degree of maturity, but a very low concentration was
found in both composts after 11 weeks of composting.

The initial germination index reflects the clearly phytotoxic effect of a high
level of volatile fatty acids and NH4-N. Such phytotoxicity ceased after the ther-

mophilic phase (5 weeks) in C2, whereas in C_1 could be considered as non-phytotoxic after week 11.

The analysis of a water soluble extract has been revealed as a very useful method for studying compost phytotoxicity. However, care must be taken when the turned pile composting system is used, because of possible leaching. The water soluble carbon lost through leaching in the turned pile could be the reason for an org-C/org-N ratio close to six earlier than in the static pile, indicating its earlier maturity. Chanyasak and Kubota (1981). The mature products of both piles had no inhibitory or toxic effect upon soil fertility since the GI of both was above 50%.

Referenecs

Chanyasak, V.; Katayama, A.; Hirai, M.; Movi, S. and Kubota, H. (1983). Effects of compost maturity on growth of Komatsuma in Neubauer's pot. II. Growth inhibitory factors and assessment of degree of maturity by Org-C/org-N ratio of water extract. Soil Sci. Plant Nutr., 29, 251–259.

Chanyasak, V. and Kubota, H. (1981). Carbon/Organic nitrogen ratio in water extract as measure of composting degradation. I. Ferment. Technol. 59, 215–219.

García, C.; Hernández, T. and Costa, F. (1991). Study on water extract of sewage sludge composts. Soil Sci. Plant Nutr. 37, 399–408.

Saviozzi, A.; Riffaldi, R. and Levi-Minzi, R. (1987). Compost maturity by water extract analyses. In Compost: Production, Quality and Use. Ed. M. de Bertoldi. Elsevier Applied Science Publishers. 359–367.

Zucconi, F.; Pera, A.; Forte, M. and de Bertoldi, M. (1981). Evaluating toxicity of immature compost. BioCycle, 22, 54–57.

Maximizing Both Compost Quality and Yield
PHILIPPE SHAUNER – Procter & Gamble

It is generally accepted that even with a source-separated feedstock, the presence of nuisance materials will result in a compromise between compost quality and yield. Plants will typically produce either a low yield of high quality compost, or large amounts of lower quality compost. This poster describes how refining by selective grinding can produce both maximum yield and top-quality compost.

The project

The district of Bapaume, France, has pioneered selective kerbside collection for compostable waste: kitchen and garden waste, paper, cardboard, disposable diapers and soiled paper products are collected once a week in a green biobin. The program has been running successfully for more than four years and demonstrates several advantages associated with the collection of all paper in the biobin:

- efficiency of the collection system (household waste is split into two near-equal fractions),
- paper absorbs leachate within the biobin, reducing odours and seepage
- paper provides the structural material required during the composting process,
- the system does not rely on recycled paper markets.

Although the feedstock is based on a wide 'biowaste' definition, more than 95% of the collected material is compostable.

Full-scale experiments have shown that, after a controlled indoor window composting process, the compost meets all European heavy metals standards and is pathogen-free.

The problem

To meet the most demanding end-markets, such as home-gardening products, the compost must also be free from sharp or cutting pieces and visual impurities, especially plastic film. The problem is to remove these residual impurities without reducing compost yield.

The solution: Selective grinding

Using a hammer grinder fitted with a 20 mm-grid and a special novel aperture:

– compost is reduced to small particle size (< 3 mm),
– residual glass and stones are pulverized so are no longer sharp or visible,
– plastic pieces are little affected by the treatment so can be easily removed.

Subsequent sieving produces 3 fractions:

– 8–50 mm: plastic film fragments (2,1%)
– 3–8 mm: insufficiently degraded paper suitable for recycling within the composting process (9,1%)
– 0–3 mm: high quality compost (88,8%)

These experiments were carried out jointly by Procter & Gamble, Gondard Sovadec, and Michel Nougaret, consultant.

Fresh and Composted Pea-nut Shells Microflora

MAMADOU AMADOU SECK*; GÉRARD KILBER-TUS**

Summary

Fresh and composted pea-nut shell microfloras qualitative and quantitative compositions were studied to accelerate the pea-nut shell aerobic composting by adding 'starters' stocks. The isolated germs are well suited to the Senegalese climatic conditions and many fungi have interesting cellulolytic possibilities. This work was completed by observations using a scanning electronic microscope.

Introduction

Composting is a woody substrate microbiological transformation, accelerated by appropriate ventilation favouring the anaerobic germ catabolic activity, and by adding nitrogen and phosphates in particular enabling the product C/N to be carried to an optimal value favourable to the protein synthesis.

As cellulosys and lignolysis possibilities govern the substrate mineralisation, it is important to know the quantitative and qualitative compositions of the present microflora.

In some cases, the addition of 'Culture starters' has to be contemplated to accelerate the process (KILBERTUS, 1985).

To this end, a microbiological study and a scanning electronic microscope observation were carried out on fresh peanut shells and composted shells for 22 days, according to the technique described by SECK and KILBERTUS (1988).

MATERIAL AND METHODS

Material used

The peanut shells come from a cooking oil factory set up in Dakar. Fresh shells and composted shells were used in this study (SECK and KILBERTUS, 1988).

*Laboratoire de Biotechnologie et d'Energétique ENSUT BP 5085 Université Cheikh Anta Diop DAKAR (SENEGAL)
**Laboratoire d'Ecologie Microbienne ESSTIB BP 239 Université de NANCY I (FRANCE)

2. Microbiological analysis

Suspensions dilutions were achieved on the two categories of shells, and then sowed on:

- gelose nutritive culture medium (8 g of nutrient broth, 15 g of gelose for one liter of water) in order to seek the bacteria.
- gelose malt (15 g of malt extract, 15 g of gelose for one liter of water) to bring the fungi to the fore.

To determine the bacteria, use were made of the following works

BERGEY'S manual (1974)
KILBERTUS and SCHWARTZ (1981)
KILBERTUS *et* al. (1977)

The fungi were determined thanks to the works by :

ELLIS (1971, 1976)
BARRON (1972)
DOMSCH *et* al (1980)
REISINGER *et* al (1970)

In parallel, some sowed boxes were subjected to UV exposures with the hope that the possible effect of intense sunshine on the microorganisms will be revealed.

Observation using a scanning electronic microscope

The samples were fixed at the OsO_4 for 1h30, dehydrated in acetone increasing concentrations baths, dried at the critical point, then gold metallized.

Results

Total microflora

The dilutions suspensions carried out on the two shell types show that the total microflora of the fresh products (FS) is significantly higher than that of the composted material (CS): respectively 370.55 against 15.71×10^6 germs per gram of substrate dried at 105 °C for 24 hours (Table 1).

These differences can be linked to the substrate trophic quality: the fresh shells being likely to free a great deal more hydrosoluble products than their composted homologue, as shown by the clearly darker tint of the FS aqueous extracts.

Many experiments, with other materials, further proved that the first stages of their decomposition showed the highest number of germs (KILBERTUS 1969, 1985, BROUSSE 1983, SIFI 1984).

2. UY effects on the fresh and composted microflora

As the Sahelian grounds are subjected to strong doses of UV radiations, it has seemed interesting to check their action on the microflora present on the fresh and composted shells. To this end freshly sowed petriboxes were exposed at increasing exposure durations to the UV (fig. 1).

$\times 10^{6}$/g

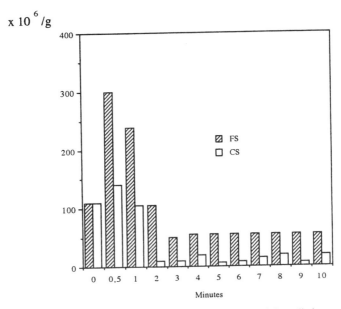

Figure 1 Total microflora evolution exposed to ulra violet radiation

Table 1 Total microflora (on nutritive culture median) $\times 10^6$ germs per gram uf shells dried at 105 °C for 24 hours

FS	46.4	113.9	440.0	415.0	581.0	601.0
CS	9.9	36.5	13.7	6.8	11.6	–

Average: FS (fresh shells)	370.55
CS (composted shells)	15.71

In the case of the fresh shells, very important stimulation of this microbial population can be observed with treatments inferior to 1 mn. Above, the germ number decreases to get stabilised between 50% and 60% of the initial population. Those percentages are still very high and suggest the presence of abundant microorganisms likely to resist to the noxious effect of electromagnetic radiation.

The results obtained with the CS, are similar to the previous ones, but at the end of the experiment, only 10 of the initial population are found. It has therefore to be admitted that the quality of the CS microflora is different from that of the FS.

Fresh and composted shell flora

Bacteria (Table 2)

The microflora is qualitatively poor: 4 species out of the FS and 5 out of the CS. The FS population is dominated by the procaryots belonging to the Bacillus king: B. megalerium and a Bacillus with deforming central spore. These 2 germs represent more than 50% of the population. The remainder includes *Microccus lute*us and Arthrobacter sp (Arthrobacter flevescens.

Table 2 Bacteria present on the FS on the CS

Fresh shells	Composted shells
Bacillus megaterium	
Bacillus sp	
Arthrobacter flavescens	*Arthrobacter flavescens*
Micrococcus luteus	*Micrococcus luteus*
	Micrococcus luteus
	(Sarcina lutea)
	Bacillus subtilis
	Bacillus sphaericus

The last two bacteria are also found on the CS where they are accompanied with Bacillus subtilis and with Bacillus sphaericus.

The Bacillus predominance (particularly on the fresh shells) has to be attributed to their endospore producing possibilities which not only protect them against the radiations but also enable them to survive when there is no water. The importance of these microorganisms has, however, to be minimized, for their possibilities of attacking the woody tissues are reduced. In wood for example, only punctuations and tracheids can be metabolized by the germs (GREAVES 1969, ELFRASJAH 1988). Thus, they probably only develop at the expense of the by products resulting from the fungus activity.

Fungi (Table 3)

The main species encountered on the FS are recorded on table 3. Neurospora sphaerica prefers the hot regions (DOMSCH *et* al. 1980). It was signalled in particular by KOBAYASI *et* al. (1968) in Ruvenzori. It is capable of decomposing the cellulosis (AGRAWAL *et* al. 1974) and producing anti-viral substance (STARRAT and ADHOSING (1967).

The presence of Rhodotorula is not surprising either as this yiest resists well to the UV. The other germs (Penicillium sp, Mucor sp) as well as the sterile white mycelium, all, except the last one, with strong sporulation are not protected by melanoid pigments. They are, however, frequently seen on fresh organic matter, at the very beginning of the organic fragment decomposition (DICKINSON and PUGH 1974).

Table 3 Fungi present on the fresh (FS) and composted (CS) shells

Fresh shells	Corn posted shells
Aspergillus blanc	*Aspergillus* blanc
Penicillium sp	*Penicillium* sp
Nigrospora sp	*Nigrospora* sp
Sterile white Mycellium	Sterile white Mycellium
Rhodotorula sp	*Rhodotorula* sp
	Trichoderma sp
	Ulocladium sp
	Sterile brown Mycelium
	Pithomyces sp
	Sphaeropside
	Fusarium sp

A part from Aspergillus, all these eucaryots were also isolated from composted shells. Furthermore, the following are also seen then:

– Ulodadium sp some species of which are met in tropical and subtropical areas (ELLIS 1976). They are able to decompose cellulosis (WHITE *et al.* 48) they also very well resist to UV (DURREL and SCHIELDS 1960).
– Trichoderma sp: the species belonging to this kind are and their growth is optimum in lighting conditions close to UV. Finally their cellulolytic aptitudes are well know (NEMEC 1969).
– Pithomycus sp: these fungi are equally stimulated by radiations close to UV (NEMEC 1969).
– As well as a melanised wall sphaeroside, a sterile brown mycellium, a green Aspergillus and finally a Fusarium.

Scanning electronic microscope study

In the case of fresh shells, the fragments surfaces are covered with many hyphes (fig. 2) or fungi spores (fig. 3). The bacteria appear rarer (fig. 4). Finally, punctuations observations do not reveal microorganism presence. This result tallies with REISINGER and KILBERTUS's theory (1980) which states that the first stages of vegetal matter biodegration are ensured by the fungi, with the procaryots only massively intervening when decomposition is advanced.

In the case of composted shells, hyphes and bacteria entanglement is often observed (fig. 5), the bacteria are locally numerous (fig. 7) and the punctuations are often colonized (fig. 6). Finally, the spiralled vessels residues are abundant, meaning cellulosis advanced decomposition. The phenomenon by procaryots has already been reported by ELFRASHAJ, in a study on epicea (1988).

In the latter case, the alteration signs show that mineralisation has already started and that the compost has reached maturity.

Fresh shells (FS)
Fig. 2 Presence of hyphes and fungi spores on the surface of fresh shells
Fig. 3 Bacteria and hyphes
Fig. 4 Fungi (probably penicillium)
Fig. 5 Non attacked punctuations

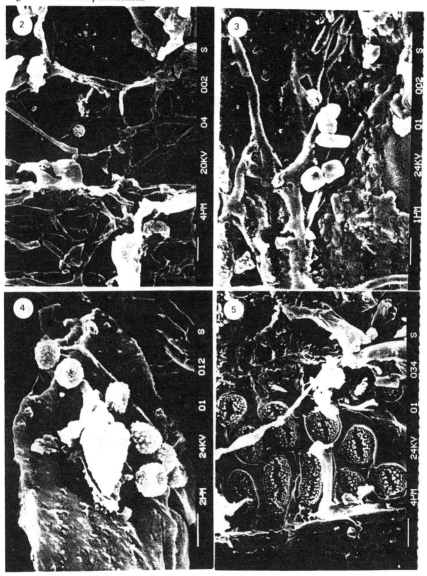

Composted shells (CS)
Fig. 6 Composted shell surface genine aspect. Presence of hyphes and bacteria
Fig. 7 Punctuations attacked by bacteria
Fig. 8 Local pile of procaryots
Fig. 9 Spiralled vessel residues

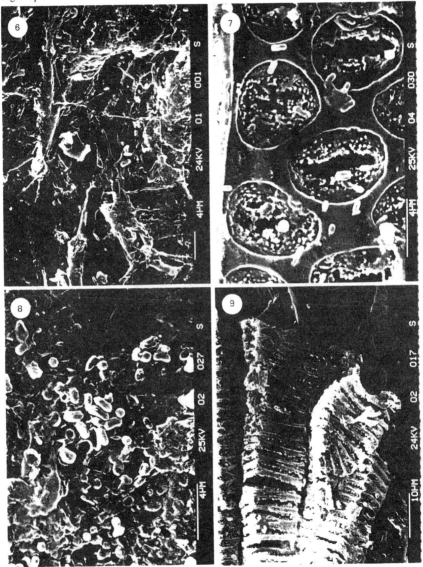

Conclusions

The total microflora evolution is according to the one obtained in the course of the composting of other woody substrates, with in particular high presence of germs during the initial phase.

From a qualitative point of view, the present bacteria either produce endospores or are chromogenous and therefore protected against solar radiation and partly against dessication.

The shell mycoflora is characterized by species famous for being resistant to UV (Rhodotorula, Ulocladium, organisms with dematiated spores (Nigropora) or with pigmented mycelium (sterile brown mycelium). Again, they are germs particularly well adapted to high sunshine conditions.

If the enzymatic possibilities of the fresh shells flora remain modest, the composted shells microflora is often specialised on the cellulosis decomposition: Trichoderma and Ulocladium in particular. Here, the phase is certainly our of advanced biodegration. As the mineralisation is active then, the microbial transformation of the substrate has gradually to put at the disposal of the plant many mineral elements necessary to its development.

Scanning electronic microscope observations confirm these results, especially the partial disappearence of the little or not ligneous tissues.

The presence of this specialized microflora which is well adapted to the local climatic conditions, allows the manufacturing of 'starters' media that will permit to put in order the manufacturing process of composts designed for Sahelian soils to be contemplated.

Bibliography

Agrawal G.W., Kulhara D., Bisen P.S. (1974) – Production of cellulolytic enzymes by Pleospora infectoria (group) and Nigrospora shaerica causing spots of pea (Pisum sativum) and bean (Dolichos lablad), Biochem., physiol. Pfl., 165, 401–405.
Barron G.L. (1968) – The genera of hyphomycetes from soil. William and Wilking, eds, Baltimore, 364 pp.
Bergey's manual of determinative bacteriology (1974) – Williams and Wilkins eds, Baltimore, 1246 pp.
Brousse J.F. (1983) – Un procédé de compostage des écorces de résineux par la voie microbienne. DEA, Université de Nancy 1, 94 pp.
Dickinson C.H., Pugh G.J.F. (1974) – Biology of plant litter decomposition. Acad. Press, London, 796 pp.
Domsch K.H., GAams W., Anderson T.H. (1980) – Compendium of soil fungi. Acad. Press. London, 859 pp.
Durrel L.W., Schields L.M. (1960) – Fungi isolated in culture from soils of the Nevada test site. Mycologia, 52, 636–641.
Efransjah (1988) – Contribution à l'étude de la préservation du bois: amélioration de l'imprégnabilité de l'épicéa (Picea sp) par un prétraitement bactérien et caractérisation du comportement du bois par l'utilisation des ultrasons. Thèse de Docteur-Ingénieur, Université de Nancy 1, 166 pp.
Greaves H. (1988) – The bacterial factor in wood decay. Wood Sci., 5, 6–16.
Kilbertus G. (1969) – Succession de champignons sur les feuilles de Brachypodium pinnatum P.B., Rev. Ecol. Biol. Sol, 5, 155–180.
Kilbertus G. (1985) – Compostage des écorces. Sapin, Epicéa, Pin, Chêne, Hêtre. Université de Nancy

1, 366 pp.

Kobayachi N., Ittaka Y., Sankawa U., Ogihara Y., Shibata S. (1968) – The crystal and molecular structure of a bromination product of (1) tetrahydrorugolosin. Tetrahedron letters, **5**, 6135–6138.

Nemec S. (1969) – Sporulation and identification of fungi isolated from root rot diceased strawberry plants. Phytopathol., **59**, 1552–1553.

Seck M.A., Kilbertus G. (1988) – Compostage des coques d'arachides. Bull. Acad. Soc. Sci., Décembre **88**, 12 pp.

Reisinger O., Kilbertus G. (1980) – Mécanismes et facteurs de biodégradation en milieu forestier (in Actualité d'écologie forestière. Sol. Flore, Faune. P. Pesson ed) Gauthier Villars, Paris, 61–86.

Reisinger O., Kilbertus G., Kiffer E. (1970) – Documents de TD de mycologie. Université de Nancy **1**, 210 pp.

Sifi B. (1984) – Compostage des écorces de chêne. DEA, Université de Nancy **1**, 82 pp.

Starrat A.N., Madhosing C. (1967) – Steril and fatty acid components of mycellium of *Fusarium oxysporum*. Can. J. Microbiol., **13**, 1351–1356.

Heat Transfer in Composting Systems

C.M.SHAW and E.I.STENTIFORD – University of Leeds, England

Abstract

In order to control the composting process, microbially generated heat must be removed from a pile. This is to prevent the temperature of the substrate, which is a good thermal insulator, from reaching levels inhibitive to the resident microbial population. A model of the heat transfer mechanisms responsible for the removal of any excess heat has been constructed. The heat transfer mechanisms involved are convection and conduction, with radiation effects being assumed negligible.

An energy balance has been constructed for the energy transfers into, within and out of a composting system. These three components of the energy balance equate to the change in energy stored within the system, which dictates the temperature within the composting substrate. Separate equations have been developed for each of these components, leading to a complete mathematical energy balance for a composting system. This balance is taken over an instant in time and is expressed as rate equations, in J/s.

Introduction

The main process control in a composting system is the self limiting interaction between heat generation by the resident microbial population and the temperature rise this causes within the pile. The heat generation occurs throughout the pile and some of the energy will be transferred to the surroundings, i.e. the surface on which the pile stands and the atmosphere. The main mechanism by which the heat generation-temperature interaction is controlled is pile ventilation. In a positive pressure aerated static pile the ventilative airflow oxygenates the pile material and also removes surplus heat to maintain a non-limiting pile temperature. This entails preventing the pile temperature rising above 60, after which the microbial population becomes inhibited, (Carlyle & Norman, 1941). Keeping the pile temperature within certain limits ensures that microbial heat generation, which results from substrate decomposition, occurs at a high rate, promoting rapid and efficient composting. In other words, after a few weeks of composting a good end product will have been obtained which, provided the appropriate temperature-time profile has been achieved, will be free of pathogenic micro-organisms.

Heat transfers

It is important to understand the way in which heat transfers occur within a composting system so that the process can continue to be refined and improved. A uniform pile temperature will promote even decomposition and drying, although temperature gradients are unavoidable. Finstein *et al* (1980), determined the heat transfer mechanisms operating within a composting system. The relative proportions of the total heat lost by each heat transfer mechanism were calculated for an aerated static pile (ASP) by measuring heat inputs and outputs to the system. They calculated that of the total heat loss from a pile 88% was due to vaporisation, 10% to dry air convection and 2% to conduction, with radiant heat transfers being assumed negligible. In such a system, where most of the heat loss is due to the evaporation of water, heat loss is closely related to pile drying. However, physical models cannot reveal the relative rates of the heat transfer mechanisms, nor which of them are taking place at different stages of the composting process. They treat a composting system as a 'black box' model, without considering what processes are actually at work inside the pile.

Heat transfer mechanisms and heat transformations within a composting system can be defined verbally. This is done by constructing a heat energy balance for a composting system as a whole, over an instant in time. Assuming that no heat is lost through radiation, the energy balance is as follows,

> heat transferred into the system by the inlet air -
>> heat transferred from the system by conduction and convection +
>>> heat generated by the microbial population =
>>>> change in heat stored within the composting substrate

Figure 1 shows diagramatically what needs to be included in a model of heat transfers of a composting system. It shows how different heat transfer mechanisms work together within the composting system. The role played by each of the heat transfer mechanisms varies throughout a composting run, depending on the configuration of the system. It is impossible, however, to depict the relative importance of each of the heat transfer mechanisms on a block diagram. This should be demonstrated by the model of the system.

Heat generation

Heat transfers into and out of a composting system are further complicated by the microbial heat generation within the substrate. In order for composting to occur, four components are necessary: organic wastes, micro-organisms, water and oxygen. The reaction occurring during the composting process is as follows:

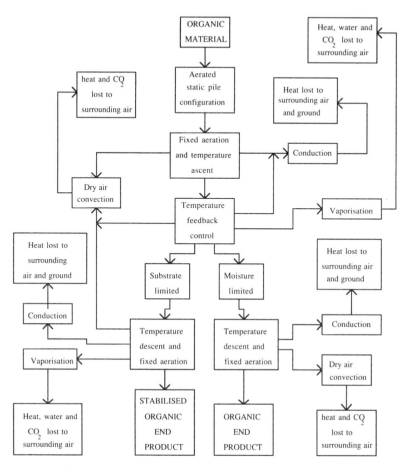

Figure 1 Block diagram showing heat transfers in an ASP system

Fresh organic waste + O_2 $\xrightarrow{\text{microbial metabolism}}$ stabilise d organic residue
$$+ CO_2 + H_2O + heat$$

(Finstein et al 1986)

It is important to know how much heat is generated within the pile by the micro-bial population. It is this heat that needs to be removed to prevent temperatures inhibitive to the resident microbial population from building up in the substrate. Some work has been done on measuring the microbial heat output of a variety of biodegradable substrates. Carlyle & Norman (1941), studied the heat outputs from straw, and Walker & Harrison (1960) and Rothbaum (1961), from wet wool. Finstein *et al* (1980), measured heat output from a sewage sludge and woodchip mixture in a field scale pile of the Rutgers ASP configuration. Their value of heat output was 0.0368 J/g(volatile solids).s at 50°C. Miller (1984), used a bench scale apparatus which measured conductive and convective heat losses. The average heat output from a sludge and woodchip mixture was 0.0144 J/g(v.s.).s when temperatures were between 50 and 60°C. Rahman (1984), used micro calorimetric techniques to measure the calorific power of sewage sludge mixtures. The average heat output of sewage sludge was 0.0134 J/g(v.s.).s at 50°C.

Discussion

In order to be able to model heat transfers within a composting system, it is necessary to have a good understanding of the composting process as a whole. The extent of microbial heat generation and factors affecting the rate of this reaction, such as temperature and moisture limitations, need to be known in order to evaluate how much heat needs to be removed from a composting system. It is also important to know which heat transfer mechanisms are dominant, and how they interact with the process configuration and process controls, such as aeration, of the composting system in question.

Once a thorough understanding of all the constituent parts of the problem has been achieved, the process of building up a model of heat transfers can begin. Mathematical equations can be constructed to fit into the verbal model of heat transfers in a composting system. These mathematical equations can then be used to simulate the temperature profile within a run of the composting system under consideration.

References

Carlyle, R.E., Norman, A.G.(1941).Microbial thermogenesis in the decomposition of plant materials, Part II.Factors involved.Journal of Bacteriology,41,699-724.

Finstein, M.S., Cirello, J., MacGregor, S.T., Miller, F.C., Psarianos, K.M.(1980).Sludge composting and utilization.Rational approach to process control. U.S. Dept. Commerce N.T.I.S., Accession no. PB 82-136243.

Finstein, M.S., Miller, F.C., Strom, P.F.(1986).Monitoring and evaluating compost process performance.Journal of the Water Pollution Control Federation, 58, 4,272-278.

Miller, F.C.(1984). Thermodynamic and matric water potential analysis in field and laboratory scale composting ecosystems. PhD dissertation, Rutgers University, UM/8424132.

Rahman, M.S. (1984). Microcalorimetric measurement of heat production and the thermophysical properties of compost. PhD dissertation, Rutgers University, UM/8520330.

Rothbaum, H.P.(1961). Heat output of thermophiles occurring on wool. Journal of Bacteriology, 81, 165–171.

Walker, I.K., Harrison, W.J.(1960).The self-heating of wet wool. New Zealand. Journal of Agricultural Research,3,6,861-895.

A Microcosm System to Determine the Gas Production of Arable Soils Amended with Different Composts

STEFANIE SIEBERT, JENS LEIFELD and INGRID KÖGEL-KNABNER – Soil Science Group, Universität of Bochum, NA 6/134, D-44780 Bochum

Introduction

The disposal of compost on arable soils is expected to be an appropriate method for restoring a large quantity of compost. The potential climatic change as a result of the atmospheric increase of CO_2 and N_2O leads to the question, if mineralisation and humification due to the disposal of compost effects CO_2 and N_2O production and release into the atmosphere.

To quantify the gas production of different composts after addition to three different soil substrates a microcosm system containing 144 columns was constructed. The subject of the paper is to describe the construction of a low price microcosm system in comparison to the automated microcosm system, e.g. the system described by HANTSCHEL (1994). This system allows a high number of replications or variations of experimental conditions, such as temperature etc., to be carried out.

Construction of the microcosm

Figure 1 shows the construction of the individual microcosm in detail. Each column of the microcosm system consists of a PVC tube with a diameter of 0,2 m and a length of 0,4 m. At a level of 0,05 m a filter plate with a void diameter of 5 μm (DURST Filtertechnik, Besingheim, Germany) is fixed with silicon paste. The sieved soils (<4 mm) at a defined water content (in this case 50% of the maximum waterholding capacity) are filled into the columns. After each 10 cm addition of soil, compaction is forced by thumping the column ten times from 20 cm height to a stable base. The upper 10 cm is a defined mixture of raw and mature compost (sieved through 10 mm sieves) with soil substrates. For comparison, the microcosm can also be filled with soil material (30 cm soil) only.

The gas fluxes can be measured in each column at regular intervalls. Therefore, they are closed by a gastight PVC-cover, containing a pasted silicon septum for sampling and a ventilator for mixing the gas phase before sampling. At the opposite site of the PVC cover a gas outlet is provided, closed by a rubber stopper, for gas exchange before accumulation of gas in a defined time interval. The volume of each microcosm is about 13 liter. The filling height of 0,3 m corresponds to the ploughed horizon of soils. However, other volumes are also possible.

Sampling

The gas sampling consists of a series of consecutive steps:

1. closing of the microcosms,
2. gas exchange in microcosms,
3. evacuation of vials and gas sampling for blank values,
4. closure microcosm for defined time interval,
5. mixing of gas phase,
6. evacuation of vials and gas sampling, and
7. opening of the columns.

The gas samples are stored in 10 ml glass vials (CS Chromatographie service, Langerwehe, Germany) before gas chromatographic measurement of CO_2 and N_2O. The vials are closed with a PTFE-coated rubber septum and have to be evacuated to a pressure of 25 mbar before sampling. This is done by a EDWARDS rotary vacuum pump. Do not evacuate the vials more than two hours before using them.

For gas exchange the columns are closed with the PVC cover, septum and rubber stopper are removed. The air in the void volume is exchanged by blowing through synthetic air (containing ca. 150 ppm CO_2) for exactly one minute. Then the cover is closed completely with septum and stopper. Immediatly, two vials are filled for blank value determination, using a bothside pointed needle, pricked first through the septum on the column and then through the vial.

Depending on the gas production rates the columns have to be kept close for different time intervals. Before sampling the gas phase is homogenised by the built-in ventilator for 20 sec, then two vials are filled. Because of the vacuum in the vial it will be filled automatically with the air of the void volume. After sampling the vials filled with gas can be stored for two weeks. The closing interval as well as the sampling depend on the gas production rate.

Measurement of CO_2 and N_2O

The amounts of CO_2 and N_2O are measured using a gas chromatograph HEWLETT PACK-ARD 5890 SERIES II (TCD, packed column, detector temperature 110YC, column temperature 50YC, carrier gas He 5.0). Calibration of the GC is made using a calibration gas mixture of 1000 ppm CO_2 and 400 ppm N_2O. The detector linearity is checked with an additional 5000 ppm CO_2 calibration gas.

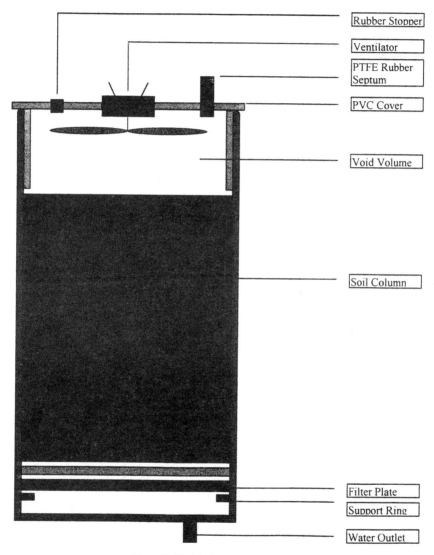

Figure 1 Model of the Microcosm

Blank values and samples are determined by injecting 250 μl using a gastight syringe (HAMILTON). Each measurement at the GC takes 15 minutes. For large sample amounts we propose to use an automatic sample injector.

Conclusions

The microcosm columns described here provide a suitable low-cost system to study the gas production of soils after addition of compost.

Because of the discontinuous gas measurement the columns can be placed in climate chambers under controlled conditions or any other suitable place, e.g. in a greenhouse.

References

HANTSCHEL, R.; H. FLESSA and F. BEESE 1994: An automated microcosm system for studying soil ecological processes. Soil Sci. Soc. Am. J. 58:401–404.

Improvement of the Composting Process

A. SILVEIRA and R. GANHO – Departamento de Ciências e Engenharia do Ambiente, Universidade Nova de Lisboa, Quinta da Torre, 2825 Monte de Caparica, Portugal

Introduction

Improvement of composting efficiency is still an important research topic. Several authors have tried to define the optimal temperature for composting (Waksman et al., 1939; Jeris and Regan, 1973; Suler and Finstein, 1977). It is often assumed that thermophilic organisms are more efficient than mesophiles. Schulze (1962) showed that it is possible to maintain a composting process continuously in the thermophilic phase. Ventilation induced through temperature feedback system (Rutgers strategy) shows a higher decomposition rate than fixed ventilation (Beltsville strategy) (MacGregor et al., 1981). Blower operating either to pull or push air into the pile (Haug, 1980), drawing air from the base or the core of the pile (Mercedes et al., 1994) and the recirculation of the air in a reactor (Miller et al, 1990) have been tried to reduce the heterogeneous conditions common to piles. The aim of the present research is to compare the conversion of dry matter under self–heating conditions (with a preset temperature limit) with composting at the optimal temperature.

Material and Methods

*Composting system.*The composting system is composed of 6 reactors immersed in a refrigerated water bath. Each reactor is a 2 liter cylindrical vessel (working inside height of 23 cm and diameter of 9 cm), made of heat resistant glass, and with a perforated plate at the bottom to distribute air supplied from a compressor. Air circulates continuously in each reactor at a constant flowrate (250 ml/min) providing a CO_2 concentration of less than 10% in outlet air when the biological activity is most intense. The outlet air is cooled to room temperature before passing into a CO_2 infrared analyser and a mass flowmeter. Thermistors are placed at different depths in the composting mass and temperatures are read every hour with a data logger. CO_2 and flowrate measurements are also recorded. A scheme of the installation is presented in figure 1.

Composting operation. The system operates under adiabatic and fixed temperature conditions. When the temperature reaches a preset value (40°C), the adiabatic experiments are conducted like the fixed temperature runs. The adiabatic runs are accomplished through a temperature controller that maintains the water temperature slightly below the temperature of the mass. In the fixed temperature trials, a thermistor placed in the centre of the mass and linked to the bath temperature controller imposes a mass temperature (40°C) by heating or cooling the water. The experiments are carried out to a CO_2 concentration in the outlet gas less than 1%. Substrate. The substrate is a mixture of whole rice hulls, finely ground rice hulls and rice flour in a fresh weight ratio of 50:34:16. This substrate was used by Hogan et al. (1989) to demonstrate a physical model of composting. The substrate is 87% volatile solid content and is well composted with 60% of moisture (adjusted with tap water). No seed is used. Previous experiments show that moisture content does not change during the composting process, pH has a typical evolution and the optimal reaction temperature for this mixture is 40°C.

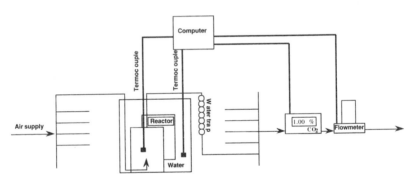

Figure 1 Scheme of the composting system

Conversion of dry matter. As CO_2 is measured continuously, its evolution rate (expressed as g/h) is used as an indicator of the microbial activity. The conversion of dry matter (X) at time t, expressed as a percentage or gram of dry matter degraded per gram of dry matter present in each run, is calculated by the equation

$$X_t = \int r_{co2}/DS_0 * Y * 100$$

where
$\int r_{co2}$ is the CO2 produced until time t (gram)
DS_0 is the initial dry matter (gram)
$Y = (DS_0 - DS_f)/\int r_{co2}$ is the gram of dry matter degraded per gram of CO2 produced
DS_f is the final dry matter (gram)

Results and Discussion

The change in temperature, CO_2 evolution rate and conversion of dry matter in the adiabatic and fixed temperature experiments are shown in figure 2.

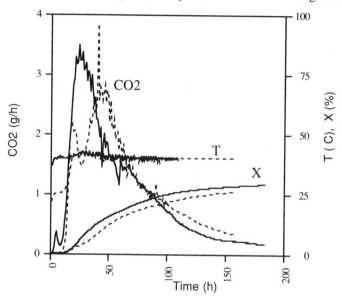

Figure 2 Composting on adiabatic (dotted line) and fixed temperature conditions (solid line). Temperature is measured at the core of the mass

The pattern of CO_2 evolution rate is similar: two peaks followed by a gradual decline in production. The lag phase is shorter in the fixed temperature than in selfheating runs. Temperature accelerates the start of the microbial activity. Schulze (1962), found that the initial phase of the composting process can be eliminated by operating continuously in the thermophilic phase. The slow temperature rise in the selfheating reactor favours the increase of the first CO_2 peak while preheating at 40°C favours the second one.

The conversion of dry matter is higher when the mass is preheated to 40°C: 25% in comparasion to 21% after 100 h of composting. Bach *et al.* (1985), working with sewage sludges at 60°C with isothermal and autothermal reactors reached the same conclusion: higher temperatures accelerate the composting process and the difference is reduced in time because the available organic matter controls the microbial growth rate in the end of the composting process. In the continuously composting system, higher conversion may be more important since fresh substrate is always available.

The temperature difference between the upper and bottom zones of the reactor is lower than 2°C in the Bach *et al.* experiments. The same difference is detected in the adiabatic temperature system but when 40°C is reached in the core of the mass, a vertical temperature gradient appears and differences of 8°C at the CO_2 peak can be reached. Despite these differences in temperature, the conversion of dry matter increases when the substrate is preheated.

In composting, the decomposition of the organic matter is a dynamic process accomplished by a succession of microorganisms. Further research has been done on the factors which affect this sucession and the role of temperature in the process.

Acknowledgements

This research has been supported by The Portuguese Ministry of The Environment and The National Institute for Scientific and Technological Research (DGQA/JNICT), Contract n° 88/91.

References

Bach, P.D.; M. Shoda; H. Kubota (1985)– Composting reaction rate of sewage sludge in an autothermal packed bed reactor. *Journal of Fermentation Technology*, 63(3), 271–278.

Jeris, J.S.; R.W. Regan (1973)– Controlling environmental parameters for optimum composting. *Compost Science*, January–February, 10–15.

Haug, R.T. (1980)– *Compost engineering. Principles and practice*. Ann Arbor Science. Michigan.

Hogan, J.A.; F.C. Miller; M.S. Finstein (1989)– Physical modelling of the composting ecosystem. *Applied and Environmental Microbiology*, 55(5), 1082–1092.

MacGregor, S.T., F.C. Miller, K.M. Psarianos, M.S. Finstein (1981)– Composting process control based on interaction between microbial heat output and temperature. *Applied and Environmental Microbiology*, 41(6), 1321–1330.

Mercedes, S.S.P.; J.T.Pereira Neto; M.A. Azevedo (1994)– Avaliação da eficiência de dois sistemas de aeração forçada na eliminação de patogénicos durante a compostagem de lixo urbano. *VI Encontro Nacional de Saneamento Básico*, APESB. Portugal.

Miller, F.C.; E.R. Harper; B.J. Macauley; A. Gulliver (1990)– Composting based on moderately thermophilic and aerobic conditions for the production of commercial mushroom growing compost. *Australian Journal of Experimental Agriculture*, 30, 287–296.

Schulze, K.L. (1962)– Continuous thermophilic composting. *Compost Science*, Spring, 22–34.

Suler, D.J. and M.S. Finstein (1977)– Effect of temperature, aeration and moisture on CO_2 formation in bench–scale, continuously thermophilic composting of solid waste. *Applied Environmental Microbiology*, 33, 345–350.

Waksman, S.A; T.C. Cordon; N. Hulpoi (1939)– Influence of temperature upon the microbiological population and decomposition processes in composts of stable manure. *Soil Science*, 47, 83–113.

Use of MATER-BI ZF03U Biodegradable Bags in Source-separated Collection and Composting of Organic Waste.

SILVIA SILVESTRI[1], GIANNI ZORZI[1], FRANCESCO DEGLI INNOCENTI[2], CATIA BASTIOLI[2]

Introduction

The 5/29/91 Italian Law by Decree foresees the source-separated collection and composting of MSW organic fraction. This implies the distribution of containers for the collection of the wet matter. The direct use of bins is not well accepted by the public because of aesthetic and hygienic problems. A specific composting bag i.e. waterproof, weight and tear resistant, easily tied after use to contain bad odours, and biodegradable is preferable. Paper bags show a poor resistance to wetting and tearing, are bulky and not easily tied. The normal polyethylene bags are not biodegradable. When torn by the bags-breakers before composting they form not sievable fragments which end up into the compost as not biodegradable inerts. Biodegradable plastic composting bags are supposed to be a solution because they combine the biodegradability of paper with the physical properties of plastics. Mater-Bi ZF03U film was specifically developed for the production of composting bags and its biodegradability well studied at laboratory scale. Biodegradation measured by the ASTM D 5338-92 test was 78% while cellulose (reference) was 85% (J.Boelens, 1992). Terrestrial toxicity tests and physical-chemical characterization of the compost showed the absence of negative effects due to Mater-Bi degradation (De Wilde and Boelens 1992). Biodegradation determined with a test described in the 12/7/1990 Italian Decree was higher than food-contact paper, in compliance with the prescriptions of the decree (Molinari, 1994).

The present study had the following main objectives: verification of the acceptance of composting bags by the users; verification of compostability; effect of Mater-Bi on the composting process and on the compost quality.

1. Istituto Agrario di S.Michele a/Adige, via E.Mach 1, S.Michele a/Adige 38010 S.Michele a/A.
2. Novamont S.p.A., via Fauser 8, 28100 Novara

Materials and methods

The collection was performed in 5 supermarkets, 3 canteens, 1 refectory. The Mater-Bi bags had the following dimensions: height: 89 cm; width: 73 cm; weight: 65 g; thickness: 30 µm. The colour: light background, red chequered. The selection and collection period lasted three weeks (from 12th to 30th July, 1993) with 3 passages a week. Questionnaire were distributed to the users. Three windrows were prepared at the composting plant (Trento). *Windrow 1* (W-1): Simulation of the action of a dilacerator device with the tearing of Mater-Bi bags filled with organics. *Windrow 2* (W-2): The control, added with composting paper bags. These bags, because of reduced dimensions and of a general distrust of users, have been introduced directly into the mass. Windrow 3 (W-3): Simulation of the action of a knife mill, cutting Mater-Bi bags into 10 x 10 cm pieces. The final composting mix was obtained by adding some extra organic matter besides MSW: sludge from a municipal depurator; poplar barks; fruit scraps; grass mowing. The mix had an apparent density of 0.6 ton/m^3, humidity around 68% and a C/N ratio near to 27. Empty bags were introduced to reach the fixed amount of 38 Kg of bioplastics, established assuming that the whole matter of each windrow (60 m^3) was delivered in 100 litres bags. In total, 585 bags were introduced into the W-1 and W-3, while 543 paper bags, corresponding to 38 Kg were introduced into W-2.

Four steel frames each containing a piece of Mater-Bi bag and a piece of paper bag, closed between two metal nets, were introduced into W-3 with the purpose to easily recover and identify the two materials during the process.

The transforming matter was turned with a composting turning machine at days: 0, 2, 5, 12, 16, 20, 29, 34, 54, 79.

Results and discussion

The rating on the bags was generally satisfactory. Some users stated to be very pleased and, in some cases, even to prefer these bags to the polyethylene ones. However, in other cases the following defects were noticed. Rupture at the seal (too thin; a more robust sealing was then adopted). Lateral tearing caused by sharpened objects (the cause of this problem was the excessive longitudinal orientation of the film. In a subsequent experience conducted at Kornenburg, Austria, with improved filming and sealing conditions the acceptability vs. paper bags was 87%.). Poor resistance at temperature around 60°C (after this complaints the bags were marked with a warding explaining the proper use and storage conditions). Dimensions: the bag did not rest to the bottom of the containers.

The temperature trend of the three windrows was similar: a short lag phase followed by a long thermophilic phase (6–8 weeks). A concomitant increase of pH, electrical conductivity, ashes content, and a phytotoxicity decrease was noticed in all the windrows as a result of the progressive mineralization and stabilisation of organic matter. The windrows with Mater-Bi (W-1 and W-3) and the windrow with paper (W-2) did not show any significant difference imputable to the different nature of the bags, even though the degradation of organic matter resulted more intense in W-1 because of an initial difference in its composition. The ripe compost, refined at 10 mm and analysed, showed valuable agronomical features, namely: absence of phytotoxic factors; neutral pH, a good amount of stabilized and sufficiently humified organic matter; high levels of nitrogen and phosphorus; optimal humidity level. An high salt levels was noticed, due to the organic fraction of MSW. The heavy metals concentration was low and in full accordance with the limits fixed by the law in force and to the standard outlined in a draft amendment of the law 748/84. At the end of process a sample of 40-50 litres of refined compost was isolated and subdivided into sub-samples and visually checked for possible Mater-Bi residues in W–1 and W–3, and for paper residues in W–2. No fragments, either Mater-Bi or paper, were found. The same survey was done on similar volumes of oversize fractions, with the following results. W–1: 23.75 g of Mater-Bi out of 45 litres of oversize fraction. W–2: 0 g of paper out of 53 litres. W–3: 12.40 g of Mater-Bi out of 40 litres. These data, if related to the total raw compost production achieved (35 m^3), represent an estimate of the residues present at the end of the process, before sieving. In W–1 the biodegradation estimate is 70.85% by weight; in W–3 the estimate is 82.87%. Therefore, film shredding before composting (W–3) improves degradation. The mincing treatment increased the surface and reduced the possibility of adhesion of foils each other. On the contrary, the whole film underwent to the action of the turning machine, which reduced it to long and curled shreds, diminishing in that way the free surface. The Mater–Bi residues conserved such a dimension to be completely removed with the oversize fraction. The materials sorted out by sieving can be re-introduced upstream in the cycle. This makes possible a further degradation of Mater-Bi. The degradation of Mater-Bi in the frames was faster than paper. After 20 days Mater-Bi degradation was already massive while paper seemed not altered at all till day 44. In any case, the metallic net prevented the mechanical action of the turning machine which seems to be important to fast biodegradation.

References

Boelens J.(1992) Aerobic Biodegradation under Controlled Composting Conditions of Test Substance ZF03U. Final Report. Organic Waste Systems. Gent, Belgium

De Wilde B., Boelens J. (1992) Compost quality tests of test substance ZF03U compost. Final Report. Organic Waste Systems. Gent, Belgium

Molinari G.P. (1994) Saggio di biodegradabilità aerobica secondo D.M. del 07/12/1990 del polimero ZF03U. Rapporto di prova. Università Cattolica del Sacro Cuore. Piacenza, Italia.

Oil-Mill Wastewater Sludge Composting

SOLANO, M.L. and NEGRO, M.J. – Instituto de Energías Renovables. CIEMAT Av. Complutense 22, 28040 MADRID. SPAIN.

Abstract

In this work, oil-mill wastewater sludge composting has been carried out.

Different lignocellulosic residues (straw, wine shoots, olive branches and olive husks) have been mixed with the sludge in order to obtain a suitable moisture and C/N ratio.

In all cases, during composting an important organic matter degradation was observed. Results obtained showed a high extent of cellulose degradation indicating a very important microbial activity on the substrate, while the lignin attack was very light.

Introduction

The disposal of olive oil-mill wastewaters represents a relevant and yet unresolved agricultural and environmental problem for olive oil producing countries. These wastewaters are characterized by a relatively high content of organic carbon that can be recycled with enormous advantages for the typically organic matter poor soils of the Mediterranean area.

Since fresh organic matter contained in this residue may result in more adverse than beneficial effects on global soil fertility and due to antimicrobial and phytotoxic effects reported for this residue (1), it could be necessary to carry out a previous transformation process.

At the present work, sludge transformation through composting has been proposed like an alternative, because composting is a reasonable, economic and safe way to obtain an organic amendment from very different kinds of organic residues. Moreover, composting allows the raw material nutrients to be kept and to become useful.

Table 1 Characterization of the residues utilized in the oil-mill wastewater sludge composting.

	Moisture (%)	C (%)	N (%)	P_2O_5 (%)	K_2O (%)	Na_2O (%)
Oil-mill wastewater sludge	68.0–97.2	52.4–64.3	1.6–2.9	0.27–1.63	2.0–1.3	0.04–0.11
Olive branches	7.0	49.3	1.4	0.18	0.72	0.01
Straw	7.7	42.5	0.4	0.11	1.4	0.02
Wine shoots	6.0	48.0	0.6	0.12	0.48	0.03
Olive husks	3.0	49.9	1.3	0.16	1.30	0.03

Materials and methods

First of all, a characterization of the main substrate – oil-mill wastewater sludge – was carried out. The oil-mill waster came from German Baena oil-mill (Córdoba, Spain) and this residue was flocculated with Z-50 (commercial product).

Different mixtures of this sludge with other lignocellulosic residues were achieved. Mixtures transformation through composting was evaluated at laboratory scale in a climatic chamber.

Lignocellulosic residues utilized for the mixtures were: wine shoots, straw, olive branches and olive husks. Main characteristics of these residues are shown in table 1.

These residues were mixed with the oil-mill wastewater sludge in a 1:1 (w/w) ratio, on a dry weight basis.

Experimental composting of the sludge obtained from the flocculation was performed in 5 l capacity PVC containers in a climatic chamber. It was used a temperature program which reproduces the temperature fluctuations recorded during the active decomposition phase, or stabilization phase of a composting pile. C/N ratio of all the mixtures was adjusted to 35. Analysis of the parameters measured during the composting process were carried out by conventional laboratory tests (2).

Results

In previous studies performed with a mixture made of oil-mill wastewater and straw or wine shoots, flocculant used showed not to inhibit the composting process (no showed data) (2).

Table 2 only shows the results corresponding to the evolution of different parameters at the initial, medium and final times determined during the composting of different mixtures assayed test. It can be seen that there is a clear decrease of carbon, as well as an increase of nitrogen which is due to a concentration effect as a result of the degradation of non nitrogenous organic matter, which originates a loss of weight and, therefore, a relative increase of nitrogen concentration.

Table 2 Evolution of main characteristics during composting process of oil-mill wastewater sludge mixed with: straw, olive branches, and olive husks

Mixture	Time month	Moisture (%)	Organic matter	Carbon (%)	Nitrogen			P_2O_5 (%)	K_2O (%)	Loss of weight	pH
					Nt ppm	NH_4 ppm	NO_3				
Straw	0	75	92	54	1.5	2392	1949	0.78	1.1	0	6.0
	6	77	83	45	3.1	2174	2579	0.94	1.7	44	8.4
	9	14	83	40	2.7	647	670	2.79	1.9	45	8.2
Olive Branches	0	66	91	52	1.6	963	1444	0.46	0.7	0	5.6
	6	76	83	45	3.3	2121	2418	0.71	1.4	49	8.4
	9	17	78	40	3.1	637	644	1.49	1.4	59	8.3
Olive husks	0	62	94	52	1.8	1399	1445	0.22	0.6	0	5.4
	5	41	92	44	2.2	580	513	0.36	1.0	21	6.4
	7	9.6	91	43	2.1	370	420	0.33	1.2	31	6.2

In all cases there was an increase of the organic matter degradation, although in olive husks and oil-mill wastewater sludge mixture the organic matter mineralization was less. pH increase up to reach alkaline values (higher than 8), except to the mixture with olive husks.

Given that the ammonium concentration got after 6 month was a high one, which could be related to the high pH values found, the product was allowed to evolve in a natural way outside a climatic chamber for a two months maturation period.

On the other hand, a decrease of the lipid fraction with regard to the original material was observed, nevertheless, in the case of hydrosoluble this decrease was not so powerful (2). Regarding the subtrates used, a clear decrease of cellulose was observed (67% and 80% degradation for mixtures of wastewater sludge and straw and olive branches respectively) the lignin fraction being the most degradation resistant fraction (25.5% and 37% degradation for mixtures of wastewater sludge and straw and olive branches respectively). Percentages of degradation both for lignin and for cellulose were lower in the case of a mixture with olive degradation both for lignin and for cellulose were lower in the case of a mixture with olive husk (20.8% and 43.7% respectively).

Final products achieved after the composting of the wastewater sludge resulting from flocculation with straw and olive branches can be considered as mature ones after 9 months from the beginning of the process, while that coming from the mixture with olive husks is mature after 7 months. In relation to the phytotoxicity tests, the high toxicity against germination showed by wastewater is progressively surpassed during composting. Germination index values higher than 90 have been got.

Cation exchange capacity values show that products obtained could be considered as mature since in all of the cases they were higher than 100 meq/100 g.

With regards to nitrogen, phosphorous, potassium and magnesium contents in final products, values over the acceptable minimum values were achieved, according to specifications recommended by the EC for mature compost.

On the other hand, total phenol contents is 82.8% lower for the mixture with straw, 73.8% lower for the mixture with olive branches and 61.5% lower in the case of a mixture with olive husks, and perhaps this lower diminishing is due to the less time of composting.

To conclude, and in spite of these results, it would be convenient to complete this study with vegetal response tests since physico-chemical characteristics of compost give only a approximate information on the fertilizing capacity thereof, and it is the plant that is going to have a final response.

Bibliography

Moreno, E.; E. Pérez, J., Ramos-Cormezana, A. and Martínez, J. (1987). Antimicrobial effect of waste water from olive oil extraction plants selecting soil bacteria after incubation with diluted waste. Microbios **51**, pp 169–174.

Negro, M.J.; Cabañas, A.; Carrasco, J. and Solano, L. Compostaje de fangos de alpechin. Proceeding III Congreso Internacional de Química de la ANQUE. Residuos Sólidos y Líquidos: su mejor destino. Tenerife 1994.

Acknowledgements

This research was supported by CICYT (PTR93-0019). The authors thank Ana Navarro for the help during some of the present experiments.

Sugarcane Filtercake Compost Influence on Tomato Emergence, Seedling Growth, and Yields.

PETER J. STOFFELLA – University of Florida, IFAS, Agricultural Research and Education Center, 2199 South Rock Road, Fort Pierce, Florida 34945-3138 USA. DONALD A. GRAETZ – University of Florida, IFAS, Soil and Water Science Dept., Box 110510, Gainesville, Florida 32611-0510 USA

Abstract

Tomato (*Lycopersicon esculentum* Mill.)('Sunny') seeds were sown in Speedling trays filled with sugarcane filtercake (a waste byproduct of sugarcane processing), plug-mix, field soil (sandy), compost:soil mixture (1:1, v:v) under greenhouse conditions. Mean days to emergence, final percent emergence, or mean days to fully expanded cotyledons were not different among treatments. In a second study, tomato 'Sunny' transplants were placed in 3.7 L pots filled with compost, field soil, or compost:soil mixture (1:1, v:v) under greenhouse conditions. After 25 days, plant grown in compost, or compost:soil mixture had heavier shoots and roots, thicker stems, and taller plants than plants grown in field soil. In a field study, `Sunny' plants were transplanted in plots without and with compost (224 mt·ha^{-1}) at 0, 76.5N-67P-140K, or 153N-134P-280K fertilizer rates. Early and total marketable yields, shoot heights, diameters, and weights were higher in plots with compost than without compost, regardless of fertilizer rates. These results suggest that plots with incorporated sugarcane filtercake compost produced higher tomato yields and larger plants than plots without compost.

Introduction

Several benefits of composts incorporated into vegetable crop production systems have been reported. Composts can be a potential alternative to polyethylene mulch (Roe, et al. 1994), serve as biological weed control (Roe, et al. 1993), and reduce soil borne diseases (Hoitink, et al. 1993).

Composts can also serve as an additive source of organic matter and nutrients, particularly in soils that are inherently low in fertility. Compost usage in vegetable planting systems may result in a reduction of commercial inorganic fertilizer rates, reduced nutrient leaching, and increase water holding capacity.

The purpose of this experiment was to evaluate a compost (`Compost') derived from the natural byproduct of sugarcane processing mills on tomato seed emergence, seedling growth, and yields.

Materials and methods

Greenhouse experiment 1. Speedling trays (5 cm x 5 cm x 5 cm invert pyramid cells, 72 cells per tray) were filled with sugarcane filtercake compost (a waste byproduct of sugarcane processing mills) (Grand, 1972), plug-mix (Metro Mix 220, Grace Sierra, Milpitas, CA), field soil (Oldsmar fine sand; sandy, siliceous, hyperthermic Alfic Arenic Haplaquods), or 50% field soil:50% compost (v:v). Tomato `Sunny' seeds were sown in each tray (1 seed/cell) with 1 tray (72 cells) per treatment. A randomized complete block design was used with each treatment replicated four times. Percent emergence was recorded dialy and seedling plant height were measured 9 days after seeding. Mean days to emergence (MDE) was calculated by the formula of Gerson and Honma (1978). Mean days to fully developed cotyledons were recorded. Shoots and roots (washed free from the adhering soil or media) were weighed and shoot:root ratios calculated.

Greenhouse experiment 2. Plastic pots (3.7 L) were filled with compost (described in expt 1.), field soil (described in expt. 1), or 50% compost:50% field soil (v:v). Tomato `Sunny' transplants (4 weeks old) were planted into each pot (1 plant/pot). A randomized complete block design was used with each treatment replicated ten times. Plant height (cotyledonary node to growing tip) and stem diameter (just below the cotyledon node) were measured 25 days after planting. Shoots and roots (washed free from adhering soil or media) were dried at 65°C for 3 days, dry weights recorded, and shoot:root ratios calculated.

Field experiment: A field experiment was established at the Agricultural Research and Education Center on 30 September 1994. Soil classification was the same as described in Greenhouse experiment one. Polyethylene covered raised beds (15.2 cm high, 1.1 m wide), with or without compost (compost) (224 mt·ha^{-1};39% moisture) into the beds, were constructed. Fertilizer was applied at 0N-0P-0K, 76.5N-67P-140.5K, or 153N-134P-280K (mt ha^{-1}) (our normal fertilizer rate) resulting in 6 compost-fertilizer treatment combinations. Sunny' transplants (5–week-old) were planted in the center of each bed, 61 cm apart, with a plant population of 7,689 plants/ha. Each plot consisted of 15 plants with the center 10 used for yield and shoot growth data. A randomized complete block design with each treatment combination replicated four times was used. Shoot height (cotyledonary node to growing tip) was measured on 13 October, 1994. Marketable yields (fruit > than

USDA grade 6X7) were counted and weighed during each of 6 weekly harvests beginning on 15 December 1994 through 19 January 1995. Total marketable yields, early marketable yields (mean of first three harvests), and mean fruit size (g/fruit) were calculated. Shoot diameter and weight (without fruits) were measured on 20 January 1995.

Statistical Analyses: Analyses of variance were conducted on all measured and calculated data for each experiment. Treatment means were separated by Duncan's multiple range test, 5% level.

Results and discussion

The compost was a relatively stable compost with a C:N ratio of 13.5 (Table 1). Calcium content (5.3%) is considered relatively high with a moderate N content (2.52%) with other major and minor elements relatively low (Table 1) compared to other composts..

Percent emergence, MDE, and mean days to fully expanded cotyledons did not differ among treatments (Table 3). Seedling roots and shoots were heavier when grown in plug-mix than in field soil, compost, or compost:soil mixture (Table 3). Shoot:root ratios were higher for plants grown in compost or mixture of compost:soil than plugmix or soil.

Tomato transplants grown in pots for 25 days with compost or a mixture of compost and field soil produced plants with thicker stems,taller plants, and heavier shoots and roots than plants grown in field soil (Table 2). Plants grown in compost or a mixture of compost and soil had higher shoot:root ratios than plants grown in soil (Table 2) suggesting that more photosysthates were proportionally translocated to the developing shoots than the roots for plants grown in compost.

In the field, tomato plants grown with compost (224 mt·ha^{-1}) produced taller plants (30 days after transplants) and thicker stems with heavier shoots (100 days after transplanting) than the control, regardless of fertilizer rate (Table 4). Early (first three harvests) and total marketable fruit yields were higher in compost plots compared to the control (Table 4). The significant compost X fertilizer rate interaction for mean fruit size (g/fruit) (Table 4) resulted from an increase in fruit size in plots with Compost and 0 or higher fertilizer rate but not at the moderate fertilizer rate.

Overall tomato yields were less than half of those commercially produced on a ha in Florida (1992–93 means 41524 kg·ha^{-1}) (Florida Agricultural Statistics Service, 1994). The lower yields may have been attributed to the unusual rainfall (20.9 cm on 29–30 October, 1994) during the season resulting in leaching of nutrients. However, compost may have contributed to reducing soil nutrient leaching, thereby resulting in higher tomato yields, regardless of the fertilizer rate. Further investigations with compost are currently being performed under more optimum growing conditions.

Table 1 Nutrient content, pH, and electrical conductivity (EC) of compost.

pH	6.66	Zn (ppm)	239
EC (millinsho)	2.07	Cu (ppm)	218
N (%)	2.52	Mn (ppm)	324
C (%)	34.0	Fe (ppm)	4108
C/N ratio	13.5	Na (ppm)	309
K (%)	0.08	Cd (ppm)	1.75
P (%)	0.94	Pb (ppm)	23
Ca (%)	5.3	Ni (ppm)	5.75
Mg (%)	0.34		

Table 2 Percent emergence, mean days to emergence (MDE), mean days to fully expanded cotyledons, shoot and root weights as influenced by compost for tomatoes. (Greenhouse experiment 1.)

Treatment	Emergence (%)	Mean Days to MDE (days)	Cotyledon Expansion (days)	Plant ht[z] (mm)	Shoot wt[z] (mg/plant)	Root wt[z] (mg/plant)	Shoot: root ratio
Compost	87.0	4.72	6.47	75.2b[y]	247b	36b	7.0a
50% Soil:50% Compost	85.0	4.49	6.37	78.0b	254b	33b	7.6a
Plug-mix	90.5	4.35	6.17	110.8a	383a	74a	5.3b
Soil	85.8	4.55	6.33	59.3c	110c	40b	2.9c

[z]data measured 9 days after seeding.
[y]Mean separation by Duncan Multiple Range Test, 5% level.

Table 3. Mean tomato dry weights of shoots and roots, plant heights, and shoot:root ratios as influenced by compost. (Greenhouse experiment 2.)

Treatment	Stem diameter (mm)	Plant height (cm)	Shoot weight (g)	Root weight (g)	Shoot: root ratio
Compost	2.57ay	4.86a	1.43a	0.24a	5.99a
50% Soil:50% Compost	2.61a	4.73a	1.53a	0.28a	5.70a
Soil	2.04b	4.12b	0.40b	0.19b	2.20b

z Data measured 25 days after transplanting into 3.7 L pots.
y Mean separation by Duncan Multiple Range Test, 5% level.

Table 4. Marketable tomato yields, shoot heights, diameters, and weights as influenced by compost.

Compost (C) (mt.ha−1)	Fertilization (F) (kg.ha−1)	Marketable yields (kg.ha−1)		Fruit size (g/fruit)	Shoot characteristics:		
		Earlyz	Total		Heighty (cm)	Diameterx (mm)	Weightx (kg/plant)
0	0	566	2,212	150	20.8	10.1	0.22
224	0	3,486	12,188	180	58.0	16.4	1.19
0	76.5N 67.5P 140.5K	524	6,535	195	25.5	14.1	0.75
224	76.5N 67.5P 140.5K	2,658	15,008	194	49.3	17.0	1.57
0	153N 134P 280K	1,972	6,568	164	28.7	14.1	0.77
224	153N 134P 280K	2,920	16,979	192	52.0	17.8	1.85
Significance							
C		**	**	**	**	**	**
F		NS	**	**	NS	**	**
C X F		NS	NS	*	NS	**	NS

**, *, NS Significant at the 1 (**) or 5% (*) levels or nonsignificant (NS).
Z Early marketable yields are the total of the first three harvests.
Y Plant height was measured 30 days after transplanting.
X Shoot diameters and weights were measured just after final harvest.

Literature cited

Florida Agricultural Statistics Service. 1994. Fl. Agr. Stat. Vegetable Summary 1992–1993. Orlando, FL.

Gerson, R. and S. Honma. 1978. Emergence response of the pepper at low temperature. Euphytica 27:151–156.

Grand, F. L. 1972. Production of sugar cane. Agronomy Monograph number 1. Univ. FL. Gainesville, FL.

Hoitink, H. A. J., M. J. Boehm, and Y. Hadar. 1993. Mechanisms of suppression of soil borne plant pathogens in compost-amended substrates. P. 601–621. In: Eds: H. A. H. Hoitink and H. M. Keener. Science and engineering of composting: Design, environmental, microbiological and utilization aspects.

Roe, N. E., P. J. Stoffella, and H. H. Bryan. 1993. Municipal solid waste compost suppress weeds in vegetable crop alleys. HortScience 28:1171–1172.

Roe, N. E., P. J. Stoffella, and H. H. Bryan. 1994. Growth and yields of bell pepper and wintersquash grown with organic and living mulches. J. Amer. Soc. Hort. Sci. 119:1193–1199.

Florida Agricultural Experiment Station Journal Series no. *N-01075*.

Composting of Bioreactor Waste From Softwood Processing

R.A.K. SZMIDT and G. BRYDEN – Horticulture Department, Scottish Agricultural College, Auchincruive, Ayr KA6 5HW Scotland, UK Tel. +44(0)1292 520331 Fax.+44(0)1292 520419

Introduction

Production of Medium Density Fibreboard (MDF) from a mix of softwoods involves steam treatment and physical degradation of raw-materials to produce wood fibre for bonding. While the process is typically energy-efficient, using waste materials for steam generation, there may be an excess of wood wastes. In addition, effluent from washing and wood chips has to be treated. This involves a multistage process of filtration, Dissolved Air Flotation (DAF) and Activated Sludge processing. The mixed waste solids are typically belt-pressed and disposed to land-fill.

At the same time there is potential for organic soil conditioners and alternatives to peat. In the UK a significant market exists for peat-like mulching materials. Bragg (1990) noted that approximately 0.5 million cubic metres of peat are used in the UK by landscape / amenity industries. However, the proportion applied directly to land as mulch has not been defined. Softwoods have long been known for phytotoxicity which has been exploited in mulches or removed by uncontrolled composting (Aaron, 1976). The woodwaste solids produced by the MDF process are brown in colour and of a dense, clay-like, nature. Material is biologically very active and so unsuitable for immediate use as a peat substitute. Trials to assess stability of the material, its compostability and any potential phytotoxicity were carried out.

Materials and Methods

Woodwaste solids were loaded into replicated 57 cm x 57 cm x 117cm bins of nominal 380 litre capacity. Compost chambers were of galvalised steel mesh, loosely insulated with expanded polystyrene. Palletised bins were then placed within an insulated temperature controlled room of approximately 20x volume. Temperature was initially boosted within the room to initiated composting. After day 1 no further external heat was applied. Temperature of composts was monitored by dataloggers (Logit: DCP Microdevelopments, Cambridge, UK). Gas evolution was assessed by monitoring air withdrawn from perforate void chambers buried within compost bins (Gastec Ltd.,Yokohama, Japan). Composts were aerated by switching material from bin to bin on levelling-off of temperature compared to the previous 24 hours or on a fall in temperature compared to ambient. Material was assessed for phytotoxicity by bioassay against a range of commercial and weed seeds under controlled conditions at 18°C + 1°C.

Material was assessed for stability by palletising bagged material and logging temperature and gas evolution over approximately 6 months. The use of woodwaste as a mulch material for amenity areas was assessed in replicated plot trials. Comparison was made with other composted materials typically used for this purpose.

Results and Discussion

Material was initially phytotoxic against a range of species. This was reduced approximately four fold for some species by composting (p<0.05). The composting process completed a normal profile of pasteurisation and conditioning (Figure 1), reaching a peak of 58 – 60°C for approximately 4 days. Chemical analysis showed a change in nutrients over the composting period, with volatalisation of nitrogen as ammonia. Initially nitrogen was considered to be bound as microbial-N, from the DAF and activated sludge processes, which was mineralised to ammonium-N and oxidised to nitrate-N in the stabilised end product (Table 1). Ammonia was detected within compost bins in a similar pattern.

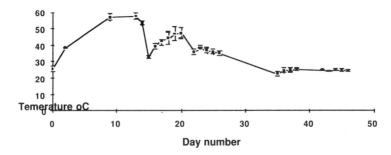

Figure 1 Temperature profile of composted bioreactor woodwaste

Table 1 Typical characteristics of composted bioreactor woodwastes (Mean of 3).

Week	pH	EC (S/cm)	NH3–N mg/l	NO3–N mg/l	P mg/l	K mg/l	Mg mg/l	%AFP
1	7.0	432	<1	<1	4	42	16	NA
2	8.0	556	298	298	1	90	5	14.7
4	8.6	571	374	417	7	90	2	19.1
5	7.4	435	65	231	4	108	12	19.3
6	7.3	352	<1	124	1	90	14	15.3
7	7.3	357	<1	91	1	102	12	17.2

Volume of material reduced during composting, by approximately 35% over a six week period. Use of material in comparison to Spent Mushroom Substrate (SMS) and bark as a soil mulch revealed that phytotoxicity was expressed as weed supression, both when surface applied and incorporated. Weed species were affected to different extents by mulching, supporting evidence of bioasssays for differential species susceptibility ($p<0.05$) (Table 2).

Use of SMS as a mulch has been widely recommended (Rupert, 1995). In these trials it resulted in greatest reduction in weed growth, however this effect was not carried–over to the subsequent season. Weed growth on plots mulched with woodwaste reverted to similar levels of weed growth compared to controls. Plots mulched with bark showed the lowest weed growth, demonstrating natural phytotoxicity of softwoods (Aaron, 1976). Weed numbers were greatest where SMS had previously been applied and the percentage of ground cover by weeds also greatest for this treatment ($p<0.001$), suggesting release of nutrients by the compost.

Table 2 Total weed seedlings per 0.5m2 (mean of five samples per replicate plot)

		TREATMENT				
Species Bark mulch	Control SMS	Woodwaste mulch		Woodwaste incorporated		
Poa annua	19.8	3.0	12.0	0.0	1.2	**
Sonchus aleraceus	12.61	2.40	4.20	4.80	0.00	***
Cerasteum fontanum	40.8	21.6	28.8	13.2	4.2	*
Senecio vulgaris	4.20	0.60	1.80	0.00	0.60	***
Lamium purpureum	12.6	10.8	25.8	26.4	1.8	ns
Urtica urens	53.5	19.8	39.6	16.2	10.2	ns
Capsella bursa-pastoris	84.1	8.4	8.4	6.0	1.2	***
Veronica officinalis	33.6	8.4	19.2	10.8	1.2	*
Chamaenerion angustifolium	5.41	0.00	1.20	0.60	0.00	ns
Cardamine amara	10.8	1.2	3.0	1.8	0.00	ns
Matricaria matricarioides	0.60	0.00	0.00	0.00	0.00	ns
Total weeds	278.0	76.0	144.0	80.0	20.0	**

The use of composting to predict the suitability of materials for use as mulches has been demonstrated. The application of the test material to soil can be safely carried-out in the knowledge that phytotoxicity is biodegradable and that no long-term pollutants would be added to the soil. However, it should be noted that further analysis of product variability and heavy metals is recommended (Jackson,

Merillot and L'Hermite, 1992). Composting served to stabilise the material with reduction in ammonia evolution. Storage of palletised, shrink-wrapped dried material in plastic bags (approximately 30 litre capacity) was satisfactory. Temperature of palletised material remained close to ambient. No hydrogen sulphide, phenolic gases, low molecular weight volatile hydrocarbons nor ammonia was detected. Carbon dioxide reached 1100 ppm within pallets and oxygen (1%) indicated a slow rate of metabolism in stored material.

References

Aaron, J.R. 1976. Conifer bark and its properties and uses. *Forrestry Commission Record*. 110, Pub. HMSO, London (UK).

Bragg, N.C. 1990. Peat and its alternatives. Pub. HDC, Petersfield (UK).

Jackson, D.V., Merillot, J-M. and L'Hermite, P. (Edit.) 1992. Composting and compost quality assurance criteria. Pub. Commission of the European Communities, Brussels. (B).

Rupert, D.R. 1995. Use of Spent Mushroom Substrate in stabilizing disturbed and commercial sites. *Compost Science and Utilization* 3(1). 80 – 83.

Effect of Moisture Content on the Composting of Pig-Manure Sawdust Litter Disposed From the Pig-on-Litter (POL) System

S.M. TIQUIA[1], N.F.Y. TAM[2] and I.J. HODGKISS1

Introduction

The pig-on-litter system, known as in-situ composting, has been developed as one of the recommended methods in Hong Kong to treat pig waste. The system utilizes a mixture of sawdust and a commercial bacterial product as the bedding material on which the pigs are raised, and the pig excreta are decomposed within the bedding material. After 10–13 weeks, the spent pig-manure sawdust litter is removed from the pig pens. This spent litter contains high concentrations of organic matter, nitrogen, phosphorus, potassium and trace elements, and also a significant amount of active microbial biomass, which is similar to an immature compost. In order to improve the quality of the spent litter, further composting to reach maturity is essential. Moisture is one of the most critical factors in controlling the rate of composting and the maturity of the product. Water provides a medium for the transport of dissolved nutrients for the metabolic and physiological activities of microorganisms. Very low initial moisture values would mean early dehydration of the pile which will arrest the biological process giving a physically stable but biologically unstable compost (Bertoldi et al., 1983). On the other hand, high moisture values may produce anaerobic conditions due to water logging. However, the effect of different moisture contents on composting of spent litter and their changes throughout the composting process are not yet understood. Therefore, the study aimed (1) to investigate the changes in the nutrients and organic matter of the spent litter at different stages of composting, and (2) to evaluate the effect of different initial moisture content of the spent litter on this composting process.

Materials and Methods

The spent litter was collected from pig pens employing the pig-on-litter system (Tam and Vrijmoed, 1990) for 12 weeks with 40 piglets raised inside the pig pen.

[1]Department of Ecology and Biodiversity, The University of Hong Kong, Hong Kong.
[2]Department of Biology and Chemistry, City University of Hong Kong, Hong Kong.

The spent litter was mixed homogeneously and piled up in an open shed for further composting and maturation. Three piles of the spent litter with initial moisture content adjusted to 50% (Pile A), 60% (Pile B) and 70% (Pile C) were set up. Each pile was triangular in shape, about 2 m in width at the base and 2 m in height. The piles were turned over twice a week using a front loader tractor. The changes of air temperature and the temperature of each pile at a depth of 60 cm were monitored twice a week (before turning). Composite samples taken from 5 symmetrical locations of each pile were collected immediately after piling (pre-expt.), right after the initial adjustment of the moisture (Day 0), then weekly until the end of the composting process (91 days). The spent litter was analyzed for the content of cation-exchange capacity (CEC) (Inoko and Harada, 1981), ash and total carbon, ratios of C:N humic:fulvic (HA:FA) acid, and different forms of N (Page et al., 1982)

Results and Discussions

The temperature values of piles A, B and C were 50 °C, 44 °C and 30 °C respectively after the moisture content of each pile was adjusted at day 0 (Fig. 1a). Thereafter, the patterns of temperature changes in piles A and B were similar. Temperatures of both piles rose dramatically to about 64–69 °C by day 4 and these readings were maintained until day 21 (thermophilic stage). The temperatures declined slightly and were maintained at a lower level from day 26 to day 57 (cooling stage), but then further dropped slowly to 30 °C (ambient temperature) from day 60 until day 9l (maturing stage). Pile C followed the same trend of changes but the maximum temperature achieved during the thermophilic stage was significantly lower (56–59 °C). Its temperature also dropped more rapidly at day 57 and started to level off by day 67.

The changes in total carbon, ash and total nitrogen of all piles were very similar during the whole composting period. The carbon content decreased as composting proceeded, from an initial value of 52% to about 51% during the first 56 days of composting (Fig. lb). This decrease was due to the disappearance of easily decomposable organic constituents (Tam and Vrijmoed, 1990). From day 63 onwards, the carbon contents of all piles stabilized at around 50% until the end of the composting period. On the other hand, the ash content increased gradually with time from an initial 9.8% to 12.3% during the first 56 days of composting and stabilized at a level of 12.5% until the end of the composting period. It has been reported that the increase in the ash content of composted materials is due to the accumulation of minerals and rapid degradation of organic materials during composting (Tam and Vrijmoed, 1993). The initial total N of all piles was about 1.8–2.0% (Fig. 1d). As composting proceeded, the total N of all piles increased. The total N values increased from an initial 1.8% to about 2.9% by day 60 and were maintained at this level until the end of the composting process. The NH_4^+–N content decreased (Fig. 1e) while the NO_x^-–N increased (Fig. 1f) revealing that the spent

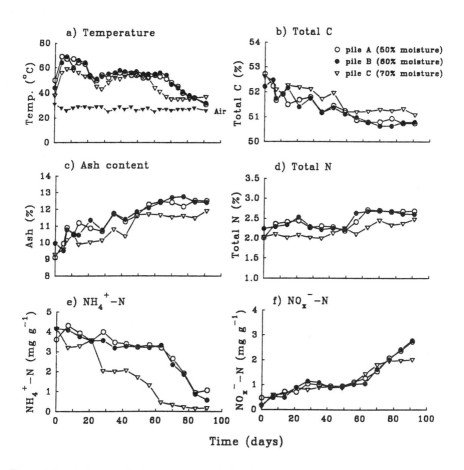

Figure 1 Chemical changes in the spent pig litter during the composting process

litter became more mature. In general, Pile C had lower Total N, ash and NH41–N content throughout the composting process than piles A and B, indicating that the higher moisture content reduced the rate of composting. During the composting, the CEC increased while C:N decreased with time (Table 1). The CEC values of all piles rose to a value greater than 60 meq 100g21 by day 35 indicating that the spent litter became mature. The HA:FA values of all piles were stabilized after day 56 in all piles with values of about 0.36–0.39. Starting from day 60 onwards, the content of CEC, ash and total carbon, and ratios of C:N and HA:FA became stabilized in all piles, suggesting that all 3 piles reached maturity at day 60. Initial moisture content did not significantly affect the quality of the mature product.

Table 1 Changes in CEC, C:N and FA:HA ratios in the spent pig litter during composting

		Day 0	Day 14	Day 21	Day 35	Day 56	Day 77	Day 91
CE C	Pile A	39.19	49.27	53.29	81.44	93.35	98.89	97.14
	Pile B	38.87	47.47	50.06	85.72	95.27	93.13	95.63
	Pile C	39.27	49.24	50.01	77.78	87.78	91.23	88.70
C:N	Pile A	26.19	21.32	21.33	22.18	21.06	19.06	18.93
	Pile B	23.26	22.24	22.24	23.06	19.37	18.88	19.50
	Pile C	25.87	25.91	25.91	25.93	24.30	21.95	20.60
HA:FA	Pile A	0.43	0.28	0.30	0.29	0.39	0.38	0.39
	Pile B	0.37	0.37	0.32	0.35	0.39	0.37	0.40
	Pile C	0.32	0.30	0.25	0.23	0.36	0.37	0.38

Conclusion

In general, all piles had very similar chemical properties during the whole composting period despite of differences in initial moisture content. Two months is sufficient for converting a spent litter from the POL system to a mature compost for land application.

References

Bertoldi, de M. G. Vallini and A. Pera (1983) *Waste Management and Research*. **1**: 157–176.
Harada, Y. And A. Inoko (1980) *Soil Science and Plant Nutrition*. **26**: 123–134
Page, A.L., R.H. Miller and R.D. Keeney (1982) Methods of Soil Analysis Part 2.
Tam, N.F.Y. and L.L.P. Vrijmoed (1990) *Waste Management and Research* **8**: 353–373.
Tam, N.F.Y. and L.L.P. Vrijmoed (1993) City Polytechnic of Hong Kong Research Report.

The Effect of Composted Vegetable, Fruit and Garden Waste on the Incidence of Soilborne Plant Diseases

G. TUITERT & G.J. BOLLEN – Department of Phytopathology, Wageningen Agricultural University, the Netherlands

Abstract

The influence of compost prepared from vegetable, fruit and garden waste (*vfg* compost) on disease incidence caused by *Phytophthora cinnamomi* and *Rhizoctonia solani* from woody ornamentals was assessed in bioassays with lupin and cucumber as test plants, respectively. Peat-perlite mixtures amended with *vfg* composts of different maturity levels and of two compost plants were suppressive to *P. cinnamomi*. Amendment of 20% long matured compost suppressed *R. solani*, whereas amendment of shortly matured compost did not. In a radish bioassay damping-off by *R. solani* was even increased when 20% compost without additional maturation after delivery was added to potting soil. Availability of cellulose affected the suppression of *R. solani* by compost.

Introduction

In the Netherlands an increasing amount of compost is produced from separately collected vegetable, fruit and garden waste (*vfg*) waste of individual households. The compost is used in agriculture and its application in potting mixtures for container-grown woody ornamentals is pending. An important feature for use of compost is its effect on pathogens already present in soil or potting mixture.

We studied the influence of *vfg* compost on two major pathogens of woody ornamentals, *Phytophthora cinnamomi* the cause of foot rot in coniferous trees and *Rhizoctonia solani* causing rot of cuttings in Ericaceae and Rosaceae. As disease symptoms appear sooner on herbaceous plants than on the woody host plants, lupin was used as bioassay host for *P. cinnamomi*, isolated from *Chamaecyparis*, and cucumber for *R. solani*, isolated from *Cotoneaster*.

The *vfg* compost was obtained from two commercial compost plants, coded P and C, with enclosed composting systems. The freshly sieved product was used after different periods of maturation.

In this paper the results of different bioassays to assess the suppressiveness of *vfg* compost will be presented. The significances of differences are based on analysis of variance of data per time of assessment (*P*<0.05).

The effect of vfg compost on development of *Phytophthora cinnamomi* in container medium

Bioassay

Lupinus angustifolius cv Kubesa was used as a test plant in the bioassay. As inoculum an earth-meal culture of *P. cinnamomi* was mixed through the substrates (1% v/v). The substrate mixtures consisted of: a) 0% compost, 85% light peat (Kekkila Finnpeat), 15% perlite, b) 10% compost, 75% peat, 15% perlite, and c) 20% compost, 65% peat, 15% perlite.

One day after preparation of the mixtures, pots of 200 ml were filled with the substrates. Four lupin seeds were added to each pot. Plants with symptoms were counted and stem parts were placed on P_{10}VPH-agar (Tsao and Guy, 1977) to verify infection by *P. cinnamomi*.

Three experiments were performed:

Expt 1, using compost P after 3 weeks additional maturation;

Expt 2, using compost C after 5 weeks additional maturation, and

Expt 3, using compost P immature (3 days after delivery by the compost plant), compost C mature (1 week maturation, but already stable at delivery) and compost P mature (10 weeks additional maturation), with two inoculum levels (0.1% and 1.0%).

Results

Expt 1 and 2. The incidence of infected plants eight days after sowing was reduced by the addition of 20% of both vfg composts tested. The composts used were relatively stable (self-heating to around 20 °C). Addition of 10% compost had no significant effect on disease incidence.

Expt 3. Eleven days after sowing disease incidence was reduced in all substrates amended with 20% compost. With the immature compost also the 10% dose reduced the proportion of diseased plants. Disease incidence was higher in the highest inoculum level, the effect of compost was the same for the two levels (no interaction). In the non-amended peat-mixture 75% of the plants was diseased at the highest inoculum level, the percentage was reduced to 35 (im 99% in the non-amended substrate and was still significantly reduced in the compost-amended substrates to around 80%.

After 24 days the difference between the inoculum levels was not significant anymore. The mean incidence was 99% in the non-amended substrate and was still significantly reduced in the compost-amended substrates to around 80%.

The effect of vfg compost on mycelial growth of Rhizoctonia solani

Bioassay

In commercial nurseries the fungus grows over the soil surface in trays with cuttings covered with polyethylene sheets. Therefore, we developed a bioassay in which mycelial growth from a point source was measured under similar conditions. The isolate of *R. solani* used was pathogenic on *Cotoneaster*, *Pyracantha*, *Erica* and *Juniperus*. It was tested for its pathogenicity on different herbaceous plants to find a cheap and fast-growing test plant. This led to the choice of cucumber as a test plant, showing discoloration and soft rot of the stem base resulting in damping-off.

In shallow trays five rows of 16 cucumber seeds were sown. After emergence of the seedlings the first one in every row was inoculated with a mycelial plug from the edge of a growing colony of the fungus. At different periods after inoculation the number of infected seedlings in each row was counted. Per tray the mean distance of growth and infection by the fungus was determined. Stem pieces of rotted seedlings were placed on water agar with 50 _g ml^{-1} oxytetracycline to verify infection by *Rhizoctonia*.

The substrate mixtures consisted of: a) 0% compost, 85% light peat (Kekkila Finnpeat), 15% perlite and b) 20% compost, 65% peat, 15% perlite. Two batches of long-matured P and C compost (5 and 7 months maturation) were tested, batch 2 of both composts was also tested after only four weeks maturation. In two of the experiments the influence of extra available cellulose was determined.

For an isolate of *R. solani* pathogenic on radish the influence of addition of mature and fresh compost to potting soil was compared with the addition of sand in an assay with radish and inoculum mixed through the soil.

Results

Amendment of the peat-perlite mixture with matured composts (5 and 7 months additional maturation, the same batch) significantly reduced the growth of *R. solani* over the substrate during the 16 days of the experiment, up to 35–75% at day 16 (Fig. 1). With a second batch of 5–months matured P and C compost this effect was repeated. When these second batches had matured for only 32 days, the mean distance of growth of the fungus was larger on the compost-amended substrates than on the non-amended ones, although this difference was not significant.

Addition of low amounts of extra cellulose to the substrates resulted in a different response of *R. solani* for the two compost-amended substrates. With 4-weeks matured compost P the growth of the fungus was increased, whereas with 4-weeks matured compost C it was decreased. When the same composts had matured longer, the addition of cellulose did not have a significant effect.

In the radish bioassay it was found that 20% fresh compost caused an increase

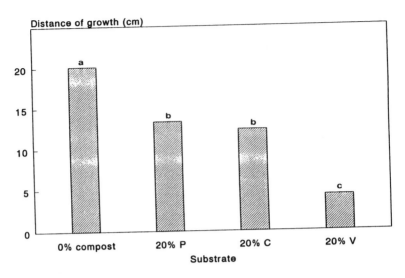

Fig. 1 Effect of vfg compost on growth of Rhizoctonia solani in peat-mixtures planted with cucumber, assessed 15 days after inoculation. Three 7-months matured composts, coded P, C and V, were used. Different letters indicate differences at P<0.05.

in damping-off of radish, whereas 20% of mature compost did not affect disease incidence compared to the 20% sand treatment.

Conclusions

For two major soilborne pathogens of woody ornamentals we found alternative herbaceous test plants for use in bioassays for assessment of the influence of compost on growth and infection of the pathogens. The choice of lupins as an assay plant for *Phytophthora cinnamomi* was derived from Pratt and Heather (1972) and Hoitink et al. (1977). In bioassays both pathogens were suppressed by addition of two types of *vfg* compost to the growth media.

Addition of both fresh and matured *vfg* compost to a common container mixture consisting of light peat and perlite rendered it suppressive to *P. cinnamomi*. In prior experiments with one of the two composts this was also found for *P. nicotianae* on tomato (Coolman, unpublished).

There is substantial evidence that the microbial activity in the substrates is, at least for the greater part, responsible for the effect found; competition for nutrients and/or antibiosis affected mycelial growth in tests for antagonism in the laboratory (Tuitert, unpublished). The mechanism of `general suppression' against fungi like *Phytophthora* and *Pythium* was discussed extensively by Hoitink, Boehm and Hadar (1993). The pH of compost-amended substrates was slightly increased, but not to such an extent that it can explain the results. The problem in estimating the microbial factors of the effects by comparing disease suppression in natural and sterilized compost is that the process of sterilization causes many nutrients to become available from the dead biomass. Addition of the sterile compost, rich in nutrients, to the substrate mixture will lead to rapid recolonization by the resident microflora.

Container mixtures amended with 20% *vfg* compost were suppressive to *R. solani*. Suppression was found when additionally matured compost was used, but not with compost –from the same batch– which had matured only four weeks after leaving the compost plant. In the radish bioassay it was found that fresh compost even increased disease incidence. Therefore, the use of long matured *vfg* compost should be recommended, at least when infestation of the soil with *R. solani* is likely to occur. An increase in suppression of *R. solani* with increasing age of the compost was also found with a completely different type of compost, namely composted hardwood bark (Nelson and Hoitink, 1983).

The exploratory experiments on the effect of cellulose showed that the amount and availability of cellulose influenced the type of effect of compost on *R. solani*. Several factors may be responsible for the difference in response to cellulose, both microbiological (composition of the microflora) and organic (amount of resident cellulose, glucose and other organic compounds). Nitrogen contents in the two composts were at the same level.

The mechanism of suppression of *R. solani* by mature *vfg* compost remains to be elucidated. An adequate indicator for compost maturity in relation to disease suppressiveness is looked for.

References

Hoitink, H.A.J., Van Doren Jr, D.M. & Schmitthenner, 1977. Suppression of *Phytophthora cinnamomi* in a composted hardwood bark medium. Phytopathology 67: 561–565.

Hoitink, H.A.J., Boehm, M.J. & Hadar, Y., 1993. Mechanisms of suppression of soilborne plant pathogens in
compost-amended substrates. In: Hoitink, H.A.J. & Keener, H.M. (Eds), Science and engineering of composting: design, environmental, microbiological and utilization aspects. Wooster, Ohio, p. 601–621.

Nelson, E.B. & Hoitink, H.A.J., 1983. The role of microorganisms in the suppression of *Rhizoctonia solani* in
container media amended with composted hardwood bark. Phytopathology 73: 274–278.

Pratt, B.H. & Heather, W.A., 1972. Method for rapid differentiation of *Phytophthora cinnamomi* from other *Phytophthora* species isolated from soil by lupin baiting. Transactions of the British mycological society 59: 87–96.

Tsao, P.H. & Guy, S.O., 1977. Inhibition of *Mortierella* and *Pythium* in a *Phytophthora*-isolation medium containing
hymexazol. Phytopathology 67: 796–801.

The Experiment of the A.M.I.U. of Modena (Italy) in Composting Sewage Sludges and Source Separated Organic Wastes

L. Valli[1], S. Piccinini[1], P. Bigliardi[2]

Introduction

An important tool in reducing appreciably the flow of disposable refuses and land-filling problems is recycling by composting organic wastes collected by Municipalities, in order to obtain soil conditioners.

A small-scale composting plant built on behalf of the Ministry of the Environment at the organic wastes treatment facility in Soliera (Modena, Italy), was managed in co-operation with the Municipal Wastes Treatment Department of Modena (AMIU) and the Research Centre for Animal Production (CRPA) of Reggio Emilia, with the aim of testing various mixtures of organic by-products.

Materials and Methods

The plant is a reactor pit, closed inside a greenhouse, 3 metres wide and 60 metres long, with a 1 metre maximum height of loaded material, equipped with a turning machine moving on rails. It has a treatment capacity of about 1000 ton/year of fresh material.

The trials campaign started in 1992 and since then several composting runs have been performed with mixtures of municipal dewatered sewage sludges (SS) (35 runs) or bio-wastes source collected (BW) from vegetable markets and restaurants (6 runs), with the addition of different organic agro-industrial by-products as bulking agents.

The bulking agents were: straw cattle litter from the Livestock Market in Modena, woodchips and shredded pieces from trimmings from public garden and parks, chopped barks and grape-stalks.

In table 1 the types of materials treated and the mixing ratios tested are summarized.

[1] Research Centre for Animal Production (C.R.P.A.), Reggio Emilia, Italy
[2] Municipal Wastes Treatment Department (A.M.I.U.), Modena, Italy

Table 1 Mixing ratios of the bulking agents added to sewage sludges and bio-wastes. The ratios are expressed, by volume, dryer/wetter (in brackets the extreme values).

Materials	Cattle litter	Chopped bark	Wood chip trimmings	Shredded trimmings	Grape-stalks
SS	1.5+–2.0 (0.7–2.3)	1.5 (1.0–2.0)		1.2–1.4 (1.0–2.0)	0.5*
BW	0.4–0.6		0.4–0.5	0.4–0.6	

+ in summer; ° in winter; * only together with a dryer bulking agent

The retention time in the reactor (bio-oxidative phase) ranged from 30 to 60 days, according to the seasonal period which influences the decomposition rate. The turning schedule was usually once or twice a week. During summertime it was necessary, in some cases, to moisten the composting mass in order to avoid a slowing down of the process due to an excessive drying of the mass.

Process temperatures were monitored during each trial and samples of the various mixtures were collected at the following stages: the loading of the reactor, the end of the bio-oxidation and the end of the curing period. The main chemical and physical parameters, the heavy metal contents and the humification parameters were determined from these samples.

Results and Discussion

The experimentation made it possible to look into various problems related to the composting process and to produce good quality composts with different analytical characteristics.

Major process difficulties occurred during winter time, because of low outdoor temperatures which slow down the organic matter decomposition rate. In particular, the mixture sewage sludge + cattle litter was problematic, due to the high humidity of both components, the low porosity and the lack of structure of the mix. It seems advisable to use this bulking agent just in more favourable seasons. However with adequate mixing ratios with barks and trimmings (so as to reach at least 35% TS) and reduced turning schedules it was possible to operate satisfactorily during winter too. Moreover, in summer, the process could set out with mix TS content even as low as 25%.

The compost produced shows good characteristics for agricultural use; organic matter and nutrient contents were considerably higher than M.S.W. composts not collected separately (table 2). The thermophilic phase assured the inactivation of pathogens such as Salmonellae, which was never found in the final products.

Table 2 Average composition characteristics of the composts produced.

Parameters		Sewage sludge + straw litter	+ wood wastes	Bio-wastes + straw litter	+ wood wastes
samples	(n°)	16	19	2	4
pH		7.0	7.1	8.9	8.6
TS	(% wb)	41.0	59.9	52.9	67.8
VS	(% TS)	49.8	48.1	57.6	65.2
TKN	(% TS)	2.0	1.6	2.6	1.9
TOC	(% TS)	28.5	28.6	28.0	32.2
C/N		15	18	11	17
Ptot	(% TS)	1.3	0.8	1.0	0.4
Ktot	(% TS)	1.8	0.9	3.9	2.0
Germ. Index	(%)	63	69	68	84
C.E.C.	(mS/cm)	3.4 *	2.0 *	3.6 *	2.2 *

* determined only on 2 samples

The composts produced are suitable as a substitute for various organic soil conditioners in field use, but the relatively high values of pH and salinity, noticed especially in the bio-waste composts, suggest a blending with earth for floriculture and garden use.

The humification parameters (table 3), measured only on last year's trials samples, show the typical trend, from the start to the end of the process, which testify to the increase of the humified organic matter.

Heavy metal concentration in every compost was lower than what is allowed by Italian law for compost use and agricultural utilization of sewage sludge, especially in the case of source collected bio-wastes (table 4).

Table 3 Average humification parameters of the composts.

Parameters		Sewage sludge + straw litter end bio-ox.	end curing	Bio-wastes + wood wastes start	end bio-ox.	end curing	+ wood wastes start	end bio-ox.	end curing
TOC	(% TS)	30.9	26.7	30.2	24.4	21.3	42.0	34.0	32.2
TEC	(% TS)	17.8	15.7	12.9	11.8	10.8	15.5	13.8	14.8
HA + FA	(% TS)	11.1	11.2	8.7	8.6	7.9	9.1	9.6	11.0
NH	(% TS)	6.7	4.5	4.2	3.2	3.0	6.4	4.2	3.8
HI		0.6	0.4	0.5	0.4	0.4	0.7	0.4	0.3
DH	(%)	62.4	71.6	67.6	72.9	72.5	58.2	70.7	75.1
HR	(%)	35.9	42.1	28.8	36.2	36.9	21.5	28.8	35.8

Table 4 Average heavy metals contents of the compost produced.

Parameters		Sewage sludge		Bio-wastes		DPR 915/82 limits	DL 99/92 limits
		+straw litter	+wood waste	+straw litter	+wood waste		
samples	(n°)	15	17	2	4		
Cu	(mg/kg TS)	229	236	95	80	600	1000
Zn	(mg/kg TS)	1325	1046	450	322	2500	2500
Pb	(mg/kg TS)	159	129	5	7	500	750
Ni	(mg/kg TS)	49	47	19	8	200	300
Cd	(mg/kg TS)	2.0	1.5	1.4	0.3	10	20
Cr	(mg/kg TS)	129	128	36	28	510	
Hg	(mg/kg TS)	1.6*	6.6*	0.2*	0.4*	10	10

* determined only on 2 samples

Mass weights were taken to evaluate the loss of the materials during the composting process in the reactor, due to the organic matter decomposition and the water evaporation. Very different results were obtained between cold and hot seasons. The yield of compost related to the starting mixture weight (w.b.) ranged from 0.4 to 0.7, with an average value of 0.5.

In overall esitimated management costs, taking into account the present experimental management which involves scale diseconomy, the production costs ranged from 95 to 115000 Lit/ton of produced compost. The most relevant entry was the labour and machinery operators cost (66%), followed by materials transport and supplies (29%) and energy expenses (5%).

Reference

Valli L., Piccinini S., Cortellini L., (1994) – L'impianto sperimentale di Soliera (MO) per il compostaggio di residui agrozootecnici e fanghi urbani – Rifiuti Solidi, vol VIII, n. 3.

Possibilities for Biological Control of *Pythium* Root Rot in Ornamental Bulb Culture with Composted Organic Household Waste

GERA J. VAN OS and WILMA J.M. VAN GULIK – Bulb Research Centre, PO Box 85, 2160 AB Lisse, The Netherlands

Pythium is a soil-borne fungus which causes root rot in several bulb crops. To reduce the use and dependence on fungicides, non-chemical control methods have to be developed.

Pythium is a potentially fast growing fungus which is relatively susceptible to competition by other microorganisms. A light infestation in non-sterilized soil causes moderate root rot, whereas the same infestation in sterilized soil leads to severe disease development. Pot experiments were performed to study the impact of several cultural practices on *Pythium* root rot in bulbous iris in relation to the condition of the soil microflora. Flooding and soil fumigation are generally applied in ornamental bulb culture to control some diseases and weeds. Flooding, however, does not kill *Pythium*. Flooding of infested field soil resulted in enhanced root rot compared to the non-flooded treatment. A similar effect was found when *Pythium* was introduced after soil fumigation. Infestation of fumigated field soil resulted in enhanced root rot compared to infestation of non-fumigated soil. In absence of other microorganisms (previously heat-sterilized soil) flooding and fumigation treatments had no effect on the disease development, indicating the crucial role of the microflora in the adverse effects of flooding and fumigation.

Apparently *Pythium* benefits from the elimination or disturbance of the soil microflora by sterilization, flooding, or fumigation. To restore the natural disease suppression in treated soils the effect of adding composted organic household waste was investigated. Application of compost in ornamental bulb culture is limited to 6 ton/ha (dry weight) every year or 12 ton/ha once in two years. The equivalence of 12 ton/ha for pot experiments is approximately 1% w/v compost. This very low percentage is not likely to induce additional suppressiveness into naturally colonized field soil. In biologically disturbed soils, however, small amounts of matured compost may effectively serve as inoculum of a broad spectrum of microorganisms which induce general suppressiveness. After the composting process the compost was incubated during five weeks to mature at 20°C and a moisture content of 50% (w/w). The compost was ploughed weekly.

Addition of 1% matured compost to heat-sterilized soil one week prior to infestation with *Pythium*, resulted in a reduction of the pathogen population and the disease development compared to treatments without compost or with sterilized compost. Further experiments are performed to determine the effects of compost application after flooding or fumigation of field soil. Results of these tests are not yet available.

The application of composted organic household waste in ornamental bulb culture may render a solution to the undisirable side effects of flooding and fumigation.

Mineralization of Three Agro-industrial Wastes by an Acid – Producing Strain of *Aspergillus Niger*

VASSILEV N., BACA M.T., VASSILEVA M., FRANCO I., AZCON R.; DE NOBILI, M.[1] – Estación Expermental del Zaidín, CSIC, Prof. Albareda 1, 18008. Granada, Spain

Introduction

The possibility of practical use of rock phosphate as a fertilizer has received significant interest in recent years. Unfortunatelly, rock phosphate is not plant available in soils with pH greater than 5.5–6 and even when conditions are optimal, yields are as a rule lower than those obtained with soluble phosphate (1). It has been repeatedly shown that low-molecular organic acids can strongly increase phosphorus solution concentration by mechanisms involving chelation and exchange reactions. Filamentous fungi are widely used as producers of organic acids and particularly *Aspergillus niger* and some *Penicillium* species have been experimented in fermentation systems or inoculated directly in soil in order to solubilize rock phosphate (2–5).

The objective of this study was to select the best combination between A. niger and three agroindustrial wastes for further application in rock phosphate solubilization.

Materials and methods

Organism. The strain of *Aspergillus niger* NB2 was used in this study, maintained on potato-dextrose agar slants.

Culture media and fermentation conditions. Three agroindustrial lignocellulosic wastes, rice hulls (RH), sugar beet waste (SB) and alperujo (ALP – a waste material obtained from olive oil extraction processes), all ground to 1 mm fragments, at concentrations of 10 % and 20 % were used as substrates for static fermentation in 50 ml Czapek's solution. After sterilization at 120°C/30 min, experiments were

1 Fac. Agraria, Univ. Udine, Via Fagagna, 208. 33100 Udine, Italy.

carried out in 250 ml Erlenmeyer flasks (in triplicate) inoculated with 1.2×10^7 spores/flask. Rock phosphate (12.8% P) at a concentration of 3.0 g/l was added when necessary. Experiments were performed at 30°C for 20 d.

Analytical methods. Mycelial growth was determined by drying the mycelium, cerefully separated from the fermentation medium and washed, in an oven at 100°C, and then weighted. Medium pH was measured with a glass electrode and titratable acidity was determined by titrating each sample to pH 7.0 with 0.1 N NaOH. Weight loss of lignocellulose during the fermentation process was calculated on ash content basis according to Kumar & Sign (6). and presented as a percent of mineralization. Lignin, cellulose and hemicellulose contents were measured according to the method of Goering & Van Soest (7). Phosphorus content was determined by molybdovanado-method (8).

Results and discussion

In the present study *A. niger* grew well on all tested materials but the filamentous fungus showed different level of growth depending both on the type lignocellulosic substrate and its concentration (Table 1). SB at concentrations of 10 % and 20 % appeared to be the best material, which provided a mycelial growth of 1.0 g/fl and 1.24 g/fl respectively, followed by ALP and RH. The amount of mycelial mass grown per day increased as the concentration of each substrate increased. On the other hand, the growth and activity of A. niger were influenced by the composition and degree of complexity of substrates. For this reason, the mineralization percent was lower than 39 % and 21 % for RH and ALP respectively, and higher than 56 % for SB.

Table 1 Growth, titratable acidity and mineralization after 20-day cultivation of Aspergillus niger on RH, SBW and ALP.

Substrate/ concentr (%)	Biomass acidity (g/flask)	Y (g/d)	Titratable (mmol/l)	Mineralization (%)
RH/10	0.63	0.032	11.1	38
RH/20	0.70	0.035	9.3	21
SB/10	1.00	0.051	53.0	69
SB/20	1.24	0.062	42.0	56
ALP/10	0.78	0.039	10.7	21
ALP/20	1.07	0.053	7.3	18

The initial pH value of 6.5–7.0 significantly decreased to about 3 after one week of fermentation in flasks with RH and SB, and 3.5–4.0 when the substrate was ALP but thereafter it increased slightly. The final titratable acidity was detectable in media with RH and ALP and significantly higher when the substrate was SB. The maximum level of mineralization, about 69 %, was achieved by *A. niger* on the 10 % concentration of SB which correlated with the results of more detailed analysis of its compositiion although a part of this material was degraded during the sterilization (data are not presented).

In general, polysaccharides do not give good yields of organic acids without some pretreatment because of the slow rate of hydrolysis and low level of sugars. On the other hand, large amounts of solid substrates resulted in low acid-producing activity due to the inherent problems of this type of fermentation including maintaince of moisture level, aeration and agitation, parameters which are of great importance in organic acid production.

A separate experiment was carried out with 10 % SB supplemented with 3 g/l RP (Table 2). The results indicated a rapid mycelial growth in the beginning of the fermentation followed by a slow growth phase. An increase of the titratable acidity to 72 mmol/l was observed during the first 10 days which resulted in a solubilization of 76 % of the insoluble phosphate but the final amount of acidity decreased later to 48 mmol/l. Although the process was directed towards the biomas growth, the level of acidity obtained by *A. niger* on sugar beet waste was sufficient to overcome the neutralizing effect of rock phosphate. The results also showed that the process of solubilization increased after the active growth phase. However, as the amount of fungal biomass continued to increase slowly, the determined solution phosphate probably corresponded to that amount which was not consumed by the mycelium.. This speculation should not be surprising bearing in mind that when trace metals are not limiting, the additional phosphate results in prolongation of mycelial growth and changes in the fungal metabolism. It was evident that thepresence of rock phosphate added directly to the fermentation medium affected the behaviour of *A. niger,* particularly its growth and citric acid production. The latter accounted for about 2/3 of the titratable acidity at the end of the process and further work should be performed to study the changed fungal metabolism and the nature of acidic metabolites different from the main acid released by the mycelium.

Table 2. Mycelial growth, titratable acidity and RP solubilization by *Aspergillus niger* cultivated on SB.

Time (d)	Biomass (g/flask)	Titratable acidity (mmol/l)	Phosphate concentr. (mg/ml)	% of total phosphate (%)
3	0.57	38	47	12
6	0.68	60	172	44
10	0.77	72	292	76
15	0.89	58	276	71
20	1.20	48	224	58

It is of significance which model of application of *A. niger* would be performed further for improvement of plant growth - inoculation directly into soil–SB-RP mixture in order to ensure a sequential release of soluble phosphate, or to provide preliminary solubilization by a fermentation process as described in this work.

Acknowledgments

N. Vassilev acknowledges a grant from the Spanish Ministry of Education and Science.

References

1) Khasawneh F & Doll E (1978) Adv Agron 30:159–206.
2) Kucey R (1987) Appl Environ Microbiol 55:2699–2703.
3) Asea P, Kucey R Stewart J (1988) Soil Biol Biochem 20:459–464.
4) Cerezine P, Nahas E, Banzatto D (1988) Appl Microbiol Biotechnol 29:501–505.
5) Cunningham J & Kuiack C (1992) Appl Environ Microbiol 52:1451–1458.
6) Kumar N & Sign K (1990) Biol Wastes 33:231–242.
7) Goering H & Van Soest P (1970) Agricul Handbook 379: 1–12.
8) Lachica M, Aguilar A, Yanez J (1973) Anal Edaf Agrobiol 32:103–147.

Organic Waste Treatment – Composting: A Comparative Study on Language usage and Terminology Italian – German – English

CLAUDIA VITTUR – Scuola Superiore di Lingue Moderne per Interpreti e Traduttori, Università di Trieste

Summary

This paper presents the short-term results of a comparative study on language usage and terminology relating to the organic waste treatment method of composting. Approximately 210 terms and their definitions regarding biology, process performance, composting techniques, engineering, starting materials and final product of composting as well as waste management in general are analysed and compared in three language systems.

Introduction

Over the last few years internationally renowned experts have been repeatedly stressing the need for concrete actions that could encourage the wider use of compost in Europe. It has been recognised that promoting the recovery of organic waste through composting requires the setting of appropriate standards on process performance and product quality. Hence, an important task will be the formulation of clear definitons which should reduce approximation and misunderstandings and achieve international consensus. This task demands efficient cooperation between experts and regulatory authorities at national and EC-levels.

Efficiency certainly depends on the usage of clear and homogeneous technical terminology which often seems lacking in this sector. The temporal and substantial differences in the development of composting technology and in the introduction of some important measures within the various nations (i.e. separate collection of the organic fraction from households and commercial/industrial sources), as well as the diversity of classification and specification methods for process and product have been leading to many non-equivalent definitions between the different languages. Along with the lack of equivalence comes the lack of specificity, clarity and consistency of several definitions due to the difficult legislative situation of compost production and marketing which is one of confusion and mistrust.

Objectives

The main objective of this study is to examine and compare the patterns of convergence and divergence among the basic composting-related terminologies and definitions in three languages in order to identify cases of non-equivalence, ambiguity or inconsistency, hence, sets of equivalents and therefore justifiable translations.

The study was started due to the proposal of carrying out a terminological research on the subject 'Waste Treatment', put forward by the 'Interpreter's and Translator's Institut' of the University of Trieste (Italy) and the 'Terminology Section, Directorate of Translation and Terminology of the European Parliament'. Because of its vastness the investigation field was limited to the biological treatment of organic wastes and mainly to composting. For the purpose of the faculty the study's objective is to enable interpreters and translators (or any other lay person) to rapidly achieve all indispensable information on the subject as well as the basic terms and their equivalents in the other languages for the sake of a correct comprehension and reproduction of the message. For experts it may provide a useful contribution to the debate on definitions and specifications for compost production as well as to the development of a clear and homogeneous international language.

Methodological approach, materials and structure

The present study has been evolving through various stages:

- study of the subject from a conceptual point of view
- selection of approximatley 210 concepts
- identification of the existing assignments of terms to concepts in the various languages and investigation on the existing relationships of the concepts concerned
- (present stage) interlingual comparison in all three language directions.

Technical inputs, advice and comments are constantly received from experts belonging to the different language aeras.

The special language of composting is being analysed and compared in its technical and legal aspects. For this reason, terms and definitions were chosen on the basis of the current national laws, guidelines and/or standards and EEC-regulations regarding compost production as well as recent reports, proceedings of symposia and scientific articles in Italian, German and English. As far as the German language is concerned it has been considered necessary to analyse and compare the German legislative system with the Austrian system, since at EC-levels the composting related legislation in Austria, one of the newest EU-member states, is presently considered to be one of the most complete and detailed ones.

The study is structured into three main parts:

a) general introduction to the subject of composting. Following aspects are considered:
- general aspects on waste management and solid wastes (i.e. waste classification)
- biology of composting (i.e. description of biodegradation)
- feedstock (i.e. characteristics and preparation; separate collection)
- process performance (i.e. parameters)
- composting techniques and handling equipment (i.e. windrow vs static pile composting; turning equipment)
- end product 'compost' (i.e. classification of different compost types)
- compost quality (biological and chemical product testing)
- fertilizers (classification; the agronomical use of compost as organic soil conditioner)

b) linguistic analysis:
- description of the legislative situation relating to compost production and its effects on language usage
- introduction to the use of the glossary with special reference to its graphical structure and the methodological criteria

c) glossary – the body of the glossary is formed by the technical terminology, graphically visualized in the introduction which therefore functions as linguistic context. The glossary provides the following data according to the standards DIN and ÷NORM on terminological analysis and terminography as well as the advice given by the terminology section of the European Parliament:
- definitions and explanatory notes – they are given in all three languages; where possible definitions are quoted directly; the explanatory notes give general information on particular legislative, technical and linguistic aspects (i.e. grammatical differences, redundance) as well as eventual cases of ambiguity (due to polysemy or synonymy), inconsistency (due to non-equivalence), obsolescence and improper language usage; these charcteristics are also graphically signalized (i.e. {–} for improper use or obsolescence)
- synonyms – their equivalence degree in relation to the main entry, which may be signed as preferential term, is also graphically signalized (i.e. {<} for a synonym which has a more specific meaning than the concept relating to the main entry)
- the sources, whether direct or indirect, of definitions, synonyms and explanatory notes are always rigorously quoted and listed in the bibliography
- the terminological body of the glossary is structured in several thematic groups introduced by schemes which visualize the interrelations between the various terms. Cross references and an alphabetically ordered index of the terms and their synonyms allow an easy consultation of the glossary.

Short-term results and discussion

The special language of composting in the three analysed language systems shows an elevated degree of heterogeneity and complexity, starting with its multidisciplinary nature which derives from a number of established sciences, such as biology, chemistry, agronomy and engineering. The presence of some neologisms (i.e. the Italian term 'rifiuti verdi', the German term 'kalte Vorbehandlung') and a certain semantic instability (i.e. the terms 'compost', 'closed systems') due to the constant scientific and technological progress as well as the changes induced by the new environmental politics in Europe show a very interesting linguistic evolution which contributes to the terminological complexity.

At present, heterogeneity and complexity are principally fruits of the difficult legislative situation of composting which, mainly in the view of production and marketing of quality compost, is one of confusion and discordance. The main characteristics of this situation are the following:

- lack of clear and detailed guidelines at international levels (as far as the Italian terms 'compost' and 'compostaggio' are concerned, the Italian decree D.P.R. 915/82 on waste in general and the national law L.N. 748/84 on fertilizers are an example for legal and terminological incongruence and obsolecence)
- varying application of different quality criteria and different specification degrees (i.e. Germany and Austria distinguish three classes of compost quality, while Italy distinguishes only one class and Great Britain none at all)
- lack of coordination between the environmental policy of the European Community and the various member states (although EC-directives generally have an obligatory character, the procedures of accomplishment are left to the member states' discretion; since each state has its own particular environmental policy the strength of which differs greatly due to the socio-economic and cultural configuration, the terms of application are quite variable)
- lack of a common strategy in the resarch on compost production and in the formulation of standards and eco-labelling systems at international level (still too many questions are open, the interests involved are so numerous and the points of view are too vexed).

On the linguistic and terminological level this situation explains the origin of a language usage which is often ambiguous or even incorrect (i.e. the use of the terms 'fermentation' or 'aerobic fermentation' in reference to the biooxidative process of composting). The correct usage of technical terms certainly depends on other aspects too, such as the competence and the scientific field of origin of the speaker or author. Clear and homogeneuos terminology, however, can only be achieved if a term, whenever possible, is permanently assigned to a concept or vice versa in an unambiguous way, so as to avoid the phenomena of polysemy and synonymy which are the origin of ambiguities, therefore misunderstandings and incorrect uses.

Conclusion

The results reported here attest the importance and need of clear and homogeneous regulations on compost production and marketing at national and EC-levels (governmental and non-govermental) as a means of creating monoreferentiality and thus terminological homogeneity.

Acknowledgements

A long list would be necessary to thank all of whom have helped with the research of material. The author wishes to express gratitude especially to Mr. G. Zorzi and Mrs. S. Silvestri (Istituto Agrario di S.Michele all'Adige – Italy) and to Mr. R. Rosanelli (Technisches B¸ro f¸r Umweltschutz–TBUGmbH. – Austria) for their valuable advice and their great disposability.

Effects of Application of Municipal Solid Waste Compost on Horticultural Species Yield

VOLTERRANI M., PARDINI G., GAETANI M., GROSSI N., MIELE S. – Dipartimento di Agronomia e Gestione dell'Agro-Ecosistema, Università di Pisa.

Introduction

Composting as a means of waste recycling in order to produce organic fertilizer has long been practised by farmers. Agronomic interest in compost is due to its elevated nutrient and organic matter content. However compost could have an adverse impact on account of its salinity or the presence of pathogens or toxic metals. Numerous studies on this last aspect have shown that heavy metal content in compost treated plant does not differ significantly from the control (Massantini et al., 1988; Stilwell, 1993). The agronomic trials carried out on compost have shown conflicting results for marked variability in composition, type of species grown and pedoclimatic characteristics of the trial area (Del Zan, 1989; Edwards et al., 1993; Paris et al., 1986). The purpose of the present research was to study the direct and residual agronomic effects of compost from treated municipal solid waste on horticultural species.

Materials and methods

Trials were carried out during the years 1990–1991 at Acciaolo (Pisa, Italy). Soil physical-mechanical and chemical characteristics and chemical composition of organic materials applied (referred to dry matter) are listed in the following table 1:

A split plot design was adopted. The main treatment (4 levels) was represented by the control, two doses of compost and manure. The secondary treatment (2 levels) was presence or absence of mineral fertilization. These 8 treatments were replicated 4 times in 40 m² plots.

The trials were carried out as follows, with an identical experimental design for the different crops and in the two years.

Table 1

		Soil	Compost	Manure
Sand	%	68	–	–
Silt	%	23	–	–
Clay	%	9	–	–
Chemical reaction	(pH)	7.9	8.9	8.9
Residue 105°C	%	–	65	21
Organic matter (Lotti met.)	%	1.2	21	66
Total N (Kjeldahl met.)	%	0.2	1.1	2.4
Total P_2O_5	%	0.2	0.7	2.0
Avail. P_2O_5	ppm	63	380	2050
Total K_2O	%	0.3	0.5	3.5
Avail. K_2O (int. met.)	ppm	36	4332	29952

First year:

Spinach (cv 'Melody'), lettuce (cv 'Jory'), tomato (cv 'Sunny') and potato (cv 'Spunta'). The main treatment was compost application at doses of 43.2 and 86.4 t·ha^{-1} and manure at 60 t ·ha^{-1}.

Second year:.

a) lettuce and tomato: crops were repeated on the same plots of the previous year without additional fertilization, in order to assess the residual effect of the previous year's treatments.

b) spinach: organic fertilizer doses were triplicated (129.6 t·ha^{-1} and 259.2 t·ha^{-1} for compost and 180.0 t·ha^{-1} for manure), in order to assess the negative and positive effects of fertilizer.

Organic materials were applied immediately prior to plowing (35 cm). Compost supplied by the ECOSUD S.p.A. factory of Massa was used. Mineral fertilization consisted in 120 kg ·ha^{-1} of N, 150 kg ·ha^{-1}of P_2O_5 and K_2O on all crops except those used for residual effect assessment.

Dry and fresh matter production and the corresponding percent N content of the commercial product were determined on each crop. Results were subjected to analysis of variance separately for each of the 7 experiments.

Results and discussion

None of the characters examined presented a significant interaction between main and secondary treatment in any of the 7 experiments. Therefore only mean effects, where significant, will be discussed.

First year

Spinach. Compost application at doses of 86.4 t·ha^{-1} (table 2) led to greater dry matter production (3.6 t·ha^{-1}) as compared to manure (+ 10.4%) and the control (+ 11.8%). Mean mineral fertilizer effects were a marked increase in fresh matter (80.1 %) and a decrease in percent dry matter content. Consequently, only a slight increase in dry matter was recorded (+10.4 %). In addition, following application of mineral fertilizer, nitrogen concentration increased from 4.1 to 5.5 %.

Lettuce No appreciable differences were observed in this crop following organic fertilizer application. In contrast, mineral fertilizer induced a noteworthy increase in weight of both fresh (from 11.3 to 23.1 t·ha^{-1}) and dry (from 0.6 to 0.9 t·ha^{-1}) biomass, a decrease in percent dry matter content and an increase in whole plant nitrogen concentration (from 2.6 to 3.0 %).

Potato Potato tuber production and composition showed little influence of treatments applied. Only a slight increase in nitrogen concentration following mineral fertilizer administration was observed (from 1.4 to 1.6 %).

Tomato In this species, application of manure resulted in higher fruit production only at the fourth harvest as compared to compost or the control (tab.1). Mineral fertilization led to more abundant production in the first part of the productive cycle. However, in the later harvests, no appreciable effect of these fertilizers could be detected. Dry matter nitrogen concentration was found to be increased as a result of mineral fertilization (from 2.4 to 2.7 %).

Second year

Spinach Application of manure (180 t·ha^{-1}) led to higher fresh matter yield (table 2) as compared to the control (+ 62%) and also as compared to single dose compost (+ 30%). In addition, application of double dose compost resulted in higher production than the control. However, with manure and double dose compost, the lowest values of percent dry matter content were observed. Nitrogen content was found to be decidedly lower in the control than in treatments with organic matter. The mean effect of mineral fertilization showed a marked increase in fresh (from 21.1 to 36.0 t·ha^{-1}) and dry matter production (from 2.6 to 3.1 t·ha^{-1}) and in nitrogen percentage (from 3.1 to 4.2 %), as well as a notable decrease in dry matter content (from 12.3 to 8.6 %).

Lettuce and Tomato In both cases, quanti-qualitative production presented no statistically significant differences attributable to residual effects of the previous year's treatments.

Table 2 First year: mean effect of organic matter treatments on spinach leaves and tomato fruits.

| | YEAR 1990 | | | YEAR 1991 | |
| | Spinach | Tomato 4° harvest | | Spinach | |
	dry weight (t·ha^{-1})	dry weight (t·ha^{-1})	fresh weight (t·ha^{-1})	dry matter %	nitrogen %
Control	3.2 a	13.5 a	21.6 a	11.6 c	3.2 a
Compost dose 1	3.4 ab	16.1 a	26.9 ab	10.4 b	3.7 b
Compost dose 2	3.6 b	14.7 a	30.5 bc	8.5 a	3.9 b
Manure	3.3 a	19.8 b	35.0 c	7.7 a	3.7 b

Conclusions

- Single dose organic fertilization (first year) produced an appreciable effect only on spinach. In particular, compost applied at the higher dose (86.4 t·ha^{-1}) led to an increase in dry matter production as compared to the control.
- Fertilization at triplicated doses (second year) led, on spinach, to an increase in dry matter production (41 %) compared to the control. However, a considerable decrease in percent dry matter was also induced.
- The expected effect of mineral fertilization in inducing an increase in production and nitrogen content was almost always observed. In addition, in winter crops (spinach and lettuce) a marked decrease in dry matter content was also recorded.
- Organic and mineral fertilization applied during the first year had no effect on quanti–qualitative production characteristics of tomato and lettuce in the second year. Residual effects were therefore absent.

In conclusion, it should be emphasized that the type of compost studied showed low short-term nutrient element availability. Its use should therefore be accompanied by mineral fertilization.

References

DEL ZAN F.. 1989. Prove agronomiche di compost diversi per origine e tecnologia produttiva: effetti indotti a breve termine sul terreno e su quattro colture erbacee. Proc. of the Int. Symp. Compost Production and use. S. Michele dell'Adige, 20–30 june.

EDWARDS J.H., WALKER R.H. LU N. BANNON J.S. 1993. Applying organics to agricultural land. Biocycle 34, 10:48–50.

MASSANTINI F., VOLTERRANI M., PARDINI G. 1988. Orientamenti circa l'applicazione dei fanghi conciari in agricoltura sulla base di ricerche lisimetriche e di pieno campo. Atti Conv. Nazionale 'I fanghi e il loro impatto sull'ambiente', Taormina 14–18 marzo.

PARIS P., ROBOTTI A., GAVAZZI C. 1986. Fertilizing value and heavy metal load of some composts from urban refuse. In 'Compost: production, quality and use', de Bertoldi m., Ferranti m.p., L'Hermite P. Zucconi F., Eds. Elsevier Applied Science, London, 643–657.

STILWELL D.E. 1993. Evaluating the suitability of MSW compost as a soil amendment in field grown tomatoes. Part B: elemental analysis. Compost Science & utilization 1, 3:66–72.

Properties of Phosphomonoesterases and ß-glucosidase in Compost Extracts

ARJA H. VUORINEN – Department of Biology, University of Joensuu, P.O.Box 111, SF-80101 Joensuu, Finland

Background

During composting, animal and plant debris is modified by mineralization and humification through a wide variety of biological and biochemical processes. Extracellular enzymes in composts are constituents of the biochemical decomposition, catalyzing reactions in which nutrients are released for later use.

Acid and alkaline phosphomonoesterases are enzymes catalyzing the hydrolyse of organic P esters to orthophosphate, which is a key reaction in the mineralization of organic phosphorus. _-glucosidase is active in cellulose decomposition, and is thus important enzyme when plant depris, like straw, is composted.

Hardly any studies have been made on the enzymatic activity in composts. Godden et al. (1983, 1986) monitored urease, cellulase, invertase and alkaline phosphatase activities during cattle mannure composting, and Garcia et al. (1993) have measured phosphatase activity and kinetics of both fresh and composted urban wastes, but the temperature response of enzymes in composts have not yet been discussed.

Aims

1) to calculate the kinetic and thermal properties of acid and alkaline phosphatase and _-glucosidase enzymes in extracts of manure composts of varying maturity
2) to find out, if these properties can be useful as an index of compost maturity.

Material and methods

Composting of cattle and big manure mixed with barley straw was performed in a continuously working, automatically by computer controlled horizontal drum composting system in several replicates as previously described (Vuorinen & Saharinen, manuscript). The used composts of varying maturity: 1) a material after the digestion of seven days in the composter, called a raw compost, 2) the same material after maturing three or five weeks in an indoor heap 3) after maturing of two months, 4) after maturing of three months

Compost extract was made by shaking 40 g fresh, freezed composts with 90 ml distilled water 1 h in an overhead shaker at 4°C, the homogenate was filtered through Labox 25 laboratory filter system (made by Larox Ltd), the total volume of the filtrate was made up to 150 ml with distilled water.

Acid and Alkaline Phosphomono-esterase activities were assayed according to Vuorinen & Saharinen (in press). Compost extract (0.5 ml) was mixed with 2.5 ml *p*-nitrophenyl phosphate (Sigma) in 1.5 x Modified Universal Buffer, pH 4.8 or 9.0. After incubation 0 and 30 or 60 min in a water bath at 30, 40 and 50°C with shaking, enzyme activity was halted by the addition of ice cold diethyl ether and rapid cooling in an ice bath. Concentrations of the product were calculated based on *p*-nitrophenol standards for each compost extract (Vuorinen 1993). The substrate concentrations for kinetic studies ranged 0.62 to 62.3 mM. All measurements were made in three replicates and substrate controls without enzyme were always included.

The kinetic properties of the enzyme were calculated using reciprocal plots for linear transformation of the basic velocity curve: Eadie-Hofstee plot, a plot of V_0 against $V_0/[S]$ (Segel, 1975). **ß-GLUCOSIDASE** activities were measured as above, but p-nitrophenyl-ß-D-glucopyranoside (Sigma) used as a substrate, concentrations ranged 0.60 to 60 mM. **PROTEIN CONCENTRATIONS** of the extracts were determined by Lowry procedure (Sigma, protein assay kit).

Results

In the extracts of raw composts, the end products of the drum composting system, the activities of both acid and alkaline phosphomonoesterase were low and measurable only with extreme high substrate concentrations. The affinity of the enzymes to the substrate were very low, also at high temperatures. V_{max} values of both acid and alkaline phosphomonoesterase increased during maturing in heaps and remain constant in mature composts. Alkaline phosphatase had higher V_{max} values than acid, but acid phosphatase had a stronger temperature dependence. K_m values of acid phosphatase increased with increasing age of the compost (Tables 1 and 2).

ß-glucosidase activity was also very low in raw manure-straw composts and it was not possible to calculate the kinetic parameters. During thermophilic phase in the maturing heaps V_{max} values were the highest and at the same time the temperature dependence was the strongest (two fold increase in V_{max} per every 10°C increase in temperature). V_{max} values were decreased in the extracts of three months old compost, and increasing incubation temperatures had then only slight effect on V_{max} values of _-glucosidase (Table 3).

CONCLUSIONS

Activities of ß-glucosidase, acid and alkaline phosphomonoesterase were lowest in the extracts of the raw manure-straw composts, digested only seven days in a drum composting system. During maturing in the heaps V_{max} values of all the studied enzymes increased, but the V_{max} of ß-glucosidase was not stable.

The temperature response of all the enzymes varied with varying age of the compost being the strongest in the early stages of the maturing, while the initial temperatures in the experimental heaps were also highest. Acid phosphomonoesterase maintained its temperature dependence also in mature composts.

High phosphatase activity in compost extracts, strong temperature response and high K_m values of acid phosphomonoesterase - can be considered as signs of the maturity of the compost.

References

GARCIA C., HERNANDEZ T., COSTA F., CECCANTI B. & MASCIANDARO G. (1993) Kinetics of phosphatase activity in organic wastes. Soil Biol. Biochem. 25:561-565.

GODDEN B., PENNINCKX M. & CASTILLE C. (1986). On the use of biological and chemical indexes for determining agricultural compost maturity: Extension to the field scale. Agricultural Wastes 15:169-178.

GODDEN B., PENNINCKX M., PIERARD A. & LANNOYE R. (1983) Evolution of enzyme activities and microbial populations during composting cattle manure. Eur. J. Appl. Microbiol. Biochem. 17:306-310.

SEGEL J.H. (1975) Enzyme kinetics, Wiley, New York.

VUORINEN A.H. (1993) Requirement of p–nitrophenol standard for each soil. Soil Biol. Biochem. 25: 295-296.

VUORINEN A.H. & SAHARINEN M. (in press). Effects of soil organic matter extracted from soil on acid phosphomonoesterse. Soil Biol. Biochem.

VUORINEN A.H. & SAHARINEN M. (manuscript). Drum composting of manure-straw compost.

Table 1 Kinetic properties of acid phosphomonoesterase in extracts of two manure-straw composts (standard deviations in paranthesis).

	30°C			40°C			50°C		
	Vmax[a]	Km[b]	r^2	Vmax[a]	Km[b]	r^2	Vmax[a]	Km[b]	r^2
compost 1									
raw	.09 (.01)c	nd		.34 (.00)c	nd		1.27 (.01)c	nd	
5 weeks	.39 (.03)	.92 (.17)	.692	.86 (.08)	1.84(.36)	.664	2.05 (.24)	2.76 (.59)	.630
2 months	.22 (.02)	2.15(.33)	.767	nd	nd	1.55 (.20)	6.32(1.18)	.690	
3 months	.26 (.02)	1.37(.28)	.669	.68 (.06)	2.29(.41)	.704	1.58 (.24)	5.38(1.27)	.579
compost 2									
raw	.09 (.00)c	nd		.38 (.00)c	nd		nd	nd	
3 weeks	.35 (.04)	3.04(.52)	.705	.74 (.08)	2.21(.48)	.618	2.08 (.31)	4.40 (1.05)	.574
2 months	.35 (.02)	9.99(.80)	.957	.96 (.08)	9.74(1.08)	.862	2.34 (.30)	10.43(1.80)	.714
3 months	.28 (.03)	4.59(.99)	.704	.82 (.05)	10.8(.88)	.926	2.65 (.21)	28.2 (2.8)	.899

a) µmol p-nitrophenol/mg protein/60 min
b) mM
c) initial velocities, V0; Vmax not available

Table 2 Kinetic properties of alkaline phosphomonoesterase in extracts of manure-straw compost (standard deviations in paranthesis).

	30°C			40°C			50°C		
	Vmax[a]	Km[b]	r^2	Vmax[a]	Km[b]	r^2	Vmax[a]	Km[b]	r^2
compost 1									
raw	.03 (.00)c	nd		.13 (.02)	14.7(4.0)	.695	.32 (.00)	28.6(.6)	.998
5 weeks	.38 (.03)	1.64(.25)	.769	.79 (.06)	2.23(.31)	.798	1.06(.05)	3.06(.23)	.934
2 months	.59 (.02)	1.89(.15)	.928	1.03(.05)	2.07(.22)	.874	1.45(.09)	1.98(.24)	.843
3 months	.61 (.05)	4.20(.69)	.806	1.21(.04)	1.58(.21)	.822	1.23(.05)	1.95(.23)	.850
compost 2									
raw		nd		nd			nd		
3 weeks	.80 (.07)	3.46(.54)	.760	1.61(.11)	3.58(.40)	.862	2.35 (.11)	3.96(.31)	.925
2 months	.83 (.07)	2.89(.45)	.757	1.36(.06)	2.85(.23)	.921	1.72(.12)	3.20(.38)	.847
3 months	.87 (.06)	2.30(.29)	.830	1.58(.09)	2.75(.28)	.883	1.71(.12)	2.64(.35)	.817

a) μmol p-nitrophenol/mg protein/60 min
b) mM
c) initial velocities, V0; Vmax not available

Table 3.
Kinetic properties of _-glucosidase in extracts of manure-straw compost (standard deviations in paranthesis).

	30°C			40°C			50°C		
	$Vmax^{[a]}$	$Km^{[b]}$	r^2	$Vmax^{[a]}$	$Km^{[b]}$	r^2	$Vmax^{[a]}$	$Km^{[b]}$	r^2
compost 1									
raw	.04 (.00)c	nd		.10 (.01)	2.49(.31)	.835	nd	nd	
5 weeks	.51 (.03)	1.13(.15)	.819	.94 (.05)	1.61(.14)	.906	1.90 (.10)	1.95(.20)	.879
2 months	.30 (.01)	1.74(.17)	.892	.53 (.02)	1.24(.11)	.908	1.00 (.05)	1.29(.18)	.842
3 months	.06 (.01)c	nd		.31 (.01)	4.57(.46)	.933	.42 (.02)	2.17(.18)	.924
compost 2									
raw	.09 (.00)	1.12(.14)	.857	.16 (.01)	.76(.13)	.770	nd	nd	
3 weeks	.68 (.04)	1.04(.15)	.797	1.41 (.05)	1.34(.11)	.915	2.49 (.11)	1.62(.14)	.905
2 months	.46 (.01)	.59 (.05)	.917	.71 (.03)	.63(.08)	.814	1.01 (.05)	.72(.10)	.787
3 months	.30 (.01)	.66 (.05)	.932	.52 (.03)	.85(.12)	.796	.59 (.04)	.73(.19)	.613

[a] _mol p-nitrophenol/mg protein/60 min
[b] _mM
[c] initial velocities, V_0; V_{max} not available

The Effect of Red Mud on Metal Mobility in Anaerobically Digested Primary Sludge During Composting

DANIEL WONG and DR. HARRIE HOFSTEDE –
Institute of Environmental Science, Murdoch University,
Western Australia 6150

Introduction

Anaerobically digested primary sludge (ADPS) refers to the residue obtained after digestion of settled solids from primary wastewater treatment and contains significant levels of pathogens and heavy metals. Disposal of ADPS to land requires composting to reduce pathogens. Land disposal is restricted by total heavy metal content in the sludge. Total metal levels in the sludge can increase during composting as the associated organic matter is decomposed. Heavy metals in the sludge (eg copper), during anaerobic digestion are present in their insoluble reduced forms such as sulphides. During composting, the metals in the sludge become more soluble due to the oxidation of sulphides to sulphates. In addition, decomposition of organic matter and acidification as a result of sulphide oxidation during composting can increase metal mobility in the sludge (Qiao, In Progress). Land application of this compost could thus result in a significant mobilisation of metals into the soil from the compost. The accumulation of metals in soils may result in phytotoxic effects or harm consumers of the contaminated crops (Garcia et al., 1990). In addition, groundwater can also be contaminated by heavy metals leaching into the water table.

Reducing the mobility of heavy metals in ADPS will reduce the risks associated with heavy metal contamination. The addition of bauxite refining residue (red mud) to a mixture of anaerobically digested primary and secondary sludge with sawdust and woodchips prior to composting has shown to reduce the mobility of heavy metals (Qiao et al., 1993). Red mud may also reduce metal mobility in ADPS compost.

The aim of this research was to reduce the mobility of heavy metals in ADPS during composting by the addition of red mud.

Materials and methods

ADPS was mixed with the following materials (in kg) to form four different mixtures: (I) wastewater skimmings (2.12) / ADPS (5.59) / chaff (1.50), (II) sawdust (1.49) / ADPS (4.50) / chaff (1.30), (III) wastewater skimmings (2.23) ADPS (5.59) / green waste (2.18) and (IV) sawdust (1.81) / ADPS (5.03) / green waste (1.96). These materials were used as they were readily available and had suitable C/N ratios as composting amendments for the sludge.

Each mixture was composted with and without red mud. The amount of red mud added was 10 of the solids content of the sludge (20.7%) used in each mixture. Sand was added to the control mixtures in place of red mud to compensate for the metal dilution effect of red mud. The proportions of materials used in mixtures I to IV were based on calculations to obtaln a C/N ratio of 20 and moisture content of 50%.

The mixtures were composted utilising the laboratory scale facility developed by Hofstede (1994). Monitoring of temperature and effluent oxygen and carbon dioxide gas was carried out for each mixture independently on a continuous basis. Composting temperatures were controlled at 55 °C by varying aeration rates between 1 and 3 l/min.

Compost samples were taken at the start, after 5 days and then every 4 days for a 21 day period. Analysis was carried out to determine total metal content, $CaCl_2$ extractable metals (leachable metals) and DTPA extractable metals (plant available metals). The metals were cadmium, copper and zinc. Moisture content, ash content, pH, soluble organic matter and carbon and nitrogen content was also determined (APHA, 1985). Fresh compost samples were used to avoid the effect of drying on metal speciation (Qiao et al., 1993).

Results and discussion

Leachable zinc may exist in forms associated with organic matter or be in free ionic form. Red mud reduced leachable zinc to a greater extent than soluble organic matter in mixture IV (Table 1). The reduction in soluble organic matter on day 21 was 9% for mixture IV. This suggests that zinc not associated with organic matter was immobilised by red mud. This has also been found by Hofstede and Ho (1991). Leachable copper was only reduced in mixtures III and IV. The results in mixtures I and II were too variable between day 0 and 21 to assess the effect of red mud on copper mobility (Wong, 1994).

Leachable cadmium levels were below detection limits in all mixtures. Leachable metals generally formed less than 10% of plant available metals. Plant available metals were generally about 5% of total metals. Hence the metals in the mixtures were mostly immobile which is in agreement to the findings by Garcia et al., 1990.

Red mud reduced plant available metals. Plant available copper was consistently reduced by red mud only in mixtures III and IV (Table 2). This reduction is not due to pH effects as the DTPA extractant solution used was buffered at 7.3. Red mud could have reduced plant available metals by either (i) extracting metals bound to the compost substrate followed by strong adsorption to red mud cation exchange sites or (ii) binding to the compost substrate containing the metals (Wong, 1994).

Table 1 Effect of red mud on reducing leachable zinc

Metal	I		II		III		Iv	
	Day 0	Day 21	Day 0	Day 21	Day 0	Day 21	Day 0	Day 21
Cu	ND	-8	ND	-63	32	24	18	22
Zn	13	46	59	21	42	0	21	53

ND Below detection limit

Table 2 Effect of red mud on reducing plant available metals

Metal	I		II		III		Iv	
	Day 0	Day 21	Day 0	Day 21	Day 0	Day 21	Day 0	Day 21
Cd	33	33	60	5	-6	23	15	5
Cu	96	-21	9	16	93	86	71	39
Zn	54	29	21	46	-1	23	29	22

Table 3 Total metal levels in mixtures with red mud (mpg dry matter)

Metal	I		II		III		Iv	
	Day 0	Day 21	Day 0	Day 21	Day 0	Day 21	Day 0	Day 21
Cd	3.5	3.3	2.7	2.7	2.7	2.9	2.0	2.3
Cu	500	513	29	448	400	55	92	16
Zn	29	29	89	600	655	97	66	09

Table 4 General composting parameters for mixture IV

Parameter	With red mud		Control	
	Day 0	Day 21	Day 0	Day 21
pH	6.8	7.3	6.2	6.6
Soluble organic matter (mg O/l)	1225	645	1225	710
Carbon to nitrogen (C/N) ratio	20	13.5	18	15
Organic matter (%)	49.1	40.8	42.8	39.6

The total metal content increased during composting due to the loss of organic matter during the process (Table 3). Total cadmium, copper and zinc levels in red mud were 3, 81 and 16 mg/kg respectively. Total cadmium, copper, and zinc levels in the sludge were 12, 1855 and 2426 mg/kg respectively (Wong, 1994). Hence addition of red mud to ADPS will not increase total levels of these metals.

The decline in the C/N ratio in mixture IV indicates the loss of carbon in relation to nitrogen, which is common during composting. The reduction in organic matter content is due to the loss as products of respiration in composting (Table 4).

Conclusion

The addition of red mud to ADPS before composting reduced the metal mobility in the sludge. This has the potential to reduce the impact of metals on the environment after land application of compost.

References

APHA (1985), *Standard methods for the examination water and waste water.*, 16th edition. Washington DC.

Garcia C, Hemandez T, and Costa F (1990), The Influence of Composting and Maturation Process on the Heavy-Metal Extractability from Some Organic Wastes, In: *Biological Wastes*, Elsevier Science Publishers Ltd, England, 31, pp 291–301

Hofstede H.T. and Ho G. E. (1991), The effect of the addition of bauxite refining residue (red mud) on the behaviour of heavy metals in compost. In: *Trace metals in the environment, vol. I: Heavy metals in the environment*, J-P Vemet (ed.). Elsevier, Amsterdam, pp. 67–94.

Hofstede H. T. (1994), *Use of Bauxite Refining Residue to Reduce the Mobility of Heavy Metals in Municipal Waste Compost*, PhD thesis, Murdoch University, Western Australia.

Qiao L., Hofstede H., and Ho G. (1993), The Mobility of Heavy Metals in Clay Amended Sewage Sludge and Municipal Solid Waste Compost, In: *Heavy Metals in the Environment, International Conference, Toronto-September*, CEP Consultants Ltd., Volume 2, pp 467–470.

Qiao L. (In Progress), *The Mobility of Heavy Metals in Clay Amended Sewage Sludge and Municipal Solid Waste Compost*, PhD Thesis, Murdoch University, Western Australia.

Wong D. (1994), *Digested sludge composting: Process optimisation and metal Mobility reduction by amendment addition*, Honours Thesis, Murdoch University, Western Australia.

Physical, Chemical and Biological Quality of Four Lombricomposts

JAIRO GOMEZ ZAMBRANO[1] and JUAN CARLOS MENJIVAR FLORES[2]

Abstract

The physical, chemical and biological properties of four lombricomposts, obtained from cow dung, press-filter cake, coffe pulp and green residues, with the aid of the earthworm *Eisenia fetida* were studied.

Except green residues, lombricomposts were produced with pH between 8.4 and 8.9, explained perhaps due to the lack of leaching during the process. The highest earthworm population was reached with coffee pulp and the best bacterial population was found in cow dung. The weight of aggregates retained for sieve No. 10 was high except in the case of green residues. The filter-press cake showed the best content of nutrients, followed by cow dung, coffee pulp and green residues.

Introduction

The aim was to characterize the lombricomposts obtained from different organic residues, to orient its adequate use in agricultural processes.

There are great volumes of cow dung , filter-press cake, coffee pulp and grass residues in the Cauca Valley zone. These four sustrates origin four lombricomposts, that differ in their physical, chemical and biological quality. This research about characterization has been developed in Palmira during the last six years.

Ferruzzi (1986) proposed a table for a categorization of lombricomposts, taking as criteria, any chemical properties as pH, nutrient content, presence of heavy metals and the bacterial charge.

Petrussi (1988) ignores that table, and claimed for a criteria that indicates the quality of lombricompost and then, assure the success of their use. In 1990, Varela and Urueña partially used the Ferruzzi table to categorize lombricomposts that originated from coffee pulp. Martínez (1991) utilized it for lombricompost obtained from flower export residues and validate these categories with a of production trial of pompon flower. In a theoretical paper, Gómez (1993) explored the quality criteria to configure a physical, chemical and biological profile, with the incorporation of physical characterization proposed by Ryser et al (1988) for compost obtained from kitchen and garden wastes.

Experimental procedure

In wood boxes 0.5 x 0.5 x 0.25 m. the four residues refered before were deposited. The boxes were installed in shadow and 200 adult earthworms *Eisenia fetida* were placed in each box. A plastic mesh was placed on top as protection from rodents. Two volumetric irrigations by week were made, so as not to cause leaching. When visual inspection could not distinguish the original waste, the process is over. It occured first in filter-press cake and coffee pulp. Many weeks after it occurs in cow dung and grass residues. The lombricompost were carried to the laboratory for physical, chemical and biolgical analysis. A test for pathogenicity of *Fusarium* in tomato plants for the pulp coffee lombricompost, was realized.

Results

The cuadre 1. resumes the average of four replications.

The cow dung lombricompost has the highest value of both of bulk and particle density. All lombricompost didn't reach the value of 0.3 of bulk density proposed by Ryser et al (1988). The yoder test indicated a high structural stability in water. The pH showed alkalinity except for grass residue, explained by the lack of leaching during the process and doesn't conform to the idea that lombricompost tends to wards neutrality.

The organic matter reached very high values, superior than obtained by Martínez (1991) and Varela and Urueña (1990) the coffee pulp lombricompost was rich in nitrogen and potassium, the filter-pres cake is rich in phosphorus, calcium, iron, copper and manganese; the cow dung in magnesium. For this concept, the best lombricompost originated from filter-press cake and the worst one comes from grass residues.

So far, the highest earthworm population occured in coffee pulp sustrate, material that presents *Fusarium*, but without pathogenicity to tomato seedlings. The bacterial count was high in filter-press cake.

This work is supported by the Contrato 009–93 National of Colombia University and Colciencias.

Table 1 Physical, Chemical and Biological Properties of Four Lombricomposts.

	Sustrate			
	Cow Dung	Filter-Press Cake	Coffee Pulp	Grass Residue
Bulk density (g/cc)	0.99	0.76	0.73	0.72
Particle density (g/cc)	2.11	1.99	1.64	1.78
Porosity (%)	52.90	61.60	55.10	59.50
% retention mesh 10	95.47	99.48	99.25	61.82
Organic matter (%)	65.30	68.30	42.00	47.10
pH	8.42	8.50	8.90	6.67
N (%)	1.82	1.49	3.07	2.22
P	1.15	1.89	0.59	0.44
K	3.59	0.63	3.49	0.98
Ca	1.94	2.96	1.19	1.23
Mg	1.41	0.89	0.34	0.47
Na (ppm)	1925.00	890.00	795.00	1875.00
Cu	64.00	99.00	53.50	43.00
Zn	176.00	143.00	27.50	182.00
Fe	8982.00	16805.00	1053.00	2396.00
Mn	334.00	424.00	167.00	117.00
B	48.00	46.00	20.70	32.70
Earthworms/m2	700.00	1225.00	4550.00	447.00
Bacterial count	4.34	76.10	17.43	43.56
Fusarium presence	0	0	X	0
Fusarium pathogenicity	0	0	0	0

Bibliography

FERRUZZI, U. Manual de Lombricultura. Traducción de Carlos Buxade Mundipiensa Madrid. P. 138. 1986.

GOMEZ, Z. J. Calidad de Lombricompuestos. En Curso de Biología del Suelo. Universidad Nacional de Colombia – Sede Palmira. P. 23. 1993.

MARTINEZ, D. and GOMEZ, Z. J. Utilización de Lombricompuestos en la Producción Comercial de Crisantemo Chrysanthemum morifolium. Acta Agronómica (Colombia). Vol. 45 (1): 79 – 84.

PETRUSSI. Characterization of Organic Matter from Animal Manures after Digestion by Earthworms. Plant and Soil. 105. 41 – 46. 1988.

RUSCHMANN, A. Investigation on Organic Matter from Digestion of Earthworm, Soil Microbiology J. 54: 1316 – 1323. 1983.

RYSER, J. P., T. CANDINAS and C. GYSI. Directives D'utilization et Exigences de Qualite Pour le Compost de Dechets de Jardin et de Cuisine. Revue Suisse Agric. 20 (6): 305 – 312. 1988.

VARELA F. URUEÑA C. Vermihumificación de la Pulpa de Café por medio de la Lombriz Roja de California. Tesis. Universidad Nacional de Colombia, Palmira. P. 3 – 24. 1990.

Integration of 'Biopol' into Biowaste Composting

A Summary of the Composting Studies carried out by PlanCo Tec and the University of Kassel in Germany.

Introduction

Composting may be defined as the 'aerobic biodegradation of organic material to form primarily CO_2 'water and humus'. It is a natural process which has been practised in our agricultural communities for many hundreds of years and by many of us in our own 'backyard' compost piles to generate a clean, valuable soil conditioner.

Up to 60% of municipal solid waste may be organic in nature and a significant proportion of this is suitable for composting. In order to make the best use of this organic material it is essential that the quality of the finished compost is adequate for the intended use.

Although there are as yet no internationally agreed standards, many countries have national quality standards for finished compost. These control issues such as the concentrations of heavy metals and foreign matter content (eg stones, broken glass etc.).

If these high quality standards are to be routinely achieved through the composting of municipal waste it is important that waste is pre-sorted prior to composting. ZENECA BioProducts actively supports the promotion of source separated biowaste composting and the development of composting at all levels of technology from source separated biowaste thorugh yard-waste to back-yard composting. Increases in the amount of composting at these levels will result in a significant diversion of material from landfill. Most countries in Europe are now actively promoting the development of a composting infrastructure and the number of households with access to source-separated biowaste composting is set to rise significantly in the next 5–10 years (see Appendix 2 for more details).

Over a period of two years serveral series of composting studies, using 'BIOPOL', have been carried out by the Ingenieurgemeinschaft Witzenhausen (igw-Kompostverwertung/PlanCoTec) together with the University of Kassel in Germany. The studies involved incorporating 'BIOPOL' into biowaste material

---

1 Profesor Asociado. Universidad Nacional de Colombia, Sede Palmira. A. A. 237 Palmira – Colombia.
2 Ingeniero Agrónomo. M. Sc. Suelos y Aguas. Universidad Nacional de Colombia, Sede Palmira.

for both pilot scale composting trials and for routine compost plant operation. The pilot scale studies were conducted to establish the factors affecting 'BIOPOL' degradation and the full scale composting trials evaluated any interactions arising from the degradation of 'BIOPOL' products in a composting system. The tests covered all the requirements of the test matrix for biodegradable materials stipulated in the draft of LAGA's amended M10 in Germany.

Composting of 'Biopol'

Initial pilot scale work used standard 'BIOPOL' test material (DIN 32 conical chips) and was conducted to establish how 'BIOPOL' degrades under various composting conditions and to what extent the optimum conditions for biodegradation of 'BIOPOL' correspond with standard conditions of organic waste composting. The resulting composts were evaluated in plant tolerance studies according to RAL-GZ 251.

Parameters such as temperature, moisture levels and C/N ratio of the initial compost material were evaluated. The data shows that 'BIOPOL' resin is compostable over a wide range of temperatures and moisture levels. The maximum rates of biodegradation of 'BIOPOL' occurred at a temperature of 60 C and a moisture level of 55%.

The optimum, initial C/N ratio of the organic waste was 18:1. These optimum conditions for 'BIOPOL' correspond well with the optimum composting conditions for traditional biowaste material and are therefore similar to the standard conditions in most larga scala composting facilities.

Parameters such as the extent and frequency of turning did not influence the rate of biodegradation of 'BIOPOL' resin. The various grades of 'BIOPOL' (HV content and the levels of plasticiser) evaluated did not affect the rates of biodegradation during composting. Consequently, using the most favourable composting conditions, corresponding to the optimum composting conditions for biowaste materials, 'BIOPOL' resin was completely degraded within 10 weeks. The 'BIOPOL' had become fully incorporated as part of the compost.

The addiction of 'BIOPOL' resin to compost does not diminish the quality of the final product as judged by seedling growth and chemical analysis. Plant growth studies were compared using control compost and 25% w/dry wt 'BIOPOL' compost. A pass rate is considered to be a plant yield of >90% of that observed with the control material.

Using the respective composts prepared from the above degradation tests and with final mixtures containing 25% of the test compost relative yields were about 125% of that seen with the control compost. Chemical analysis showed no significant changes of organic matter, matrix nutrients or heavy metals in the resulting compost form the pilot scale trials.

Further practical scale studies were then carried out using 'BIOPOL' bottle, both whole and shredded, of varying wall thickness (1000(m – 2000(m). These studies were conducted during routine operation of 3 compost plants employing different composting techniques.

1. Fully enclosed, organic waste windrow composting, with automatic turning.
2. Initial contained, intensive composting with subsequent composting on table windrow with turning.
3. Composting on triangular windrows with manual turning.

These tests cover the requirements of the test matrix for biodegradable materials contained in the draft of the amended M10 of LAGA.

'BIOPOL' was either mixed with the biowaste mateial in bags, or added directly to the compost material in the plants to give a final concentration of 1% w/w. The composting periods in all the tests were between 12 and 19 weeks in accordance with normal operationg procedures. The studies showed that with all the thicknesses and quantities used , 'BIOPOL' resin was rapidly degraded during practical composting providing favourable composting conditions were maintained. This result applied to all the composting systems evaluated.

Under the most favourable composting conditions the 1000(m 'BIOPOL' fragments degraded rapidly and at an equivalent rate to that obtained in the model studies discussed previously. Unfavourable composting conditions, in particular low oxygen contents between 10–12% by volume, delayed 'BIOPOL' degradation. However, complete degradation was still possible during normal composting periods, providing there were no additional restricting conditions such as high temperatures (>70° C).

Since the operation of all compost plants at optimum conditions cannot be guaranteed, the maximum recommended wall thickness for a bottle made solely from 'BIOPOL' (or other moulded item) is 700(m to ensure complete degradation during normal composting periods.

Further chemical analysis on the quality of the 'BIOPOL' composts obtained from the composting trials confirmed the work discussed earlier. It has been concluded that any negative influence of 'BIOPOL' addition to composts can be excluded. A high 'BIOPOL' addition (>15–20%) again improved the quality of 'average' bio-composts in particular with respect to heavy metals such as Pb, Cd and Cr. As with previous results the quality of the compost produced using 'BIOPOL' did not impair plant growth.

Summary

Composting is nature's way of recycling. It is an environmentally beneficial technique for the recovery and recycling of organic bio-wastes. 'BIOPOL' has been demonstrated to rapidly degrade under a wide range of composting conditions, giving rise to a compost of high quality. The compost studies described above and carried out by PlanCoTec in conjunction with the University of Kassel have concluded that 'BIOPOL' meets all the requirements of suitability testing for biodegradable materials, for inclusion in composting, made by LAGA in the draft of the amended M10.

An independent, expert opinion on the 'Processing of the biodegradable material 'BIOPOL' in 'biowaste composting' is included in Appendix 1 in both German and English.

This work was partly funded by the Lower-Saxon Ministry of Economy, Technology and Transport.

A comprehensive report has been compiled by PlanCoTec giving full details of the work carried out and copies can be made available from SAFTA SpA, on request. All enquiries should be addessed to Vittorio Roncoroni, SAFTA SpA, Via Arda, 11, 29100 Piacenza, (Tel. 39.523.5981 – Fax 39.523.597060).

Additional Papers

Biological treatment, the perfect eco-efficient tool in a sustainable integrated waste management

BERT LEMMES, Managing Director, ORCA

The intention behind the session was to avoid the possible overfocussing on the specific composting/biogasification problem, and to avoid the kind of composting euphoria or megalomania that has caused very serious harm to the acceptance of biological treatment in the past.

We only have to think back about the experiments with total MSW-composting, that produced an end-product that one could never call compost. Some people still advocate this strategy as a pure reduction and stabilisation-process, with an end-product called 'STABILAT' (in German) or stabilised matter. (See contribution Paul Bardos.)

But apart from that, the general tendency goes in the direction of source-separated collection, and the biological treatment of the separated organic fraction.

To oversimplify, it boils down to the question of principle whether we should try to compost as much as possible or *impose* a sufficient level of self-discipline on the sector to treat only that part of the organic resources, that will lead ultimately to the production of a valuable product 'compost' and to achieve this result in an **eco-efficient** manner.

We are faced in politics and legislation with the choice between waste and non-waste and a lot will depend on the final decision concerning this issue. But for the sake of this discussion, we will continue to use the definition of recyclables and organic matter as waste for the time being.Nevertheless I want to stress already now that a basic change should be made in the thinking and therefor in the wording and definitions.

We should start calling 'waste' what is definitively 'wasted' and should adopt the positive philosophy of 'RESOURCES-MANAGEMENT' instead of 'waste-management' for all the fractions that can eco-efficiently be recovered.

We are faced with choices in resource-management about how to deal with the organic fraction. Biological processes are one of several means of dealing with this part of the secondary resources, but there are obviously others.

The integration of composting, anaerobic digestion or biogasification and stabilisation into an overall recovery strategy, not only seems the logical way to go but will avail itself the only solution for specific fractions of the waste stream if we want to achieve the recovery targets that will be imposed.

To help the European legislators prepare decisions and mandates to implement biological treatment as a solution, it is necessary to provide a rational framework and a matrix to judge the opportunity of the solution.

This framework will have to take into account:

1) The specific objectives of the waste management (local-regional constraints, etc.)
2) The broad environmental merit via the LCI's and LCA's
3) The economic merit: → impact on the primary production
     → social acceptability and ease of implementation

This should enable the decisionmakers to determine the eco-efficiency of their strategy.

Eco-efficient would mean in this case: *eco*-**logically sound**
     eco-**nomically viable**

Example

Operating Eco-efficiency

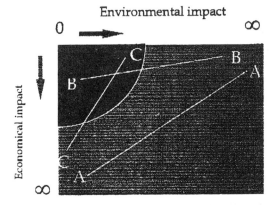

Environmental impact

Legend: **Project A**   **Big ecological impact but very expensive**
     Project B   **Big Ecological impact and low price**
     Project C   **Small ecological impact and very expensive**

Black section is the operating window that can assure eco-efficient sustainable activity

We acknowledge that in the end the decision will remain political, but we can proactively create an *operating widow* to point out to the authorities that any decision outside this operating widow would be unreasonable and certainly not sustainable.

A lot will depend on the definition of the waste and its separation in different fractions.

Depending on the collection system, the source separation system, the fraction of *grey waste could go from 100%* (as in some areas using mass burn-technology) → *to only 10% after source separation and pre-treatment.*

The more efficient the separation, the smaller this grey fraction will become and the more important the **diversion from landfill** will be. But, whatever we do, we admit that there will always be a grey fraction. But this has to be as small as possible.

Dealing with that issue, **Paul Bardos** exposed a possibility to implement the *co-treatment of this grey fraction with industrial wastes in order to reduce the volume and stabilise and sanitise the waste to be landfilled.* At the same time the possibility to recover biogas, and the detoxification of hazardous substances.

Two other aspects of well managing the landfill, where it avails itself unavoidable, were addressed by **Prof. Cossu and Dr. Muntani**. The former addressed the problems with the containment in landfilling and the treatment of the percolate.

This discussion put the finger on every sore spot in the whole waste management strategy.

'How do we estimate the long-term environmental impact of a landfill and the economical implications of its sanitation afterwards.'

How do we reflect this in the actual landfill-levies and do the existing or revised landfill levies such as in the UK reflect the real cost for providing an acceptable solution.'

Dr. Muntani addressed a different issue of landfilling by advocating the use of compost to avoid some of the engineering problems in well-managed landfills. According to him compost can function as a filter, a buffer, an absorbent in landfills to avoid the problems with leachate (among others) by binding toxic ingredients.

From the landfills we turned our attention to the solutions that are provided respectively by recycling of dry matters and the use of R.D.F. and P.D.F. in co-combustion processes.

Julia Hummel presented the pioneering work done by ERRA in setting up a database to a collect data in a dozen pilot cities and to design a number of analyse-tools to interpret those data.

The motivation and participation ratioís for the source separation schemes and *the cost for collection and separation are especially interesting* **for our sector as we are confronted with the same problems.**

For this reason ORCA and ERRA will join forces in the very near future and combine their information and the processing of the data with this data-base to achieve a global view on the recovery-issue.and to put these data at the disposal of the decision-makers. The collaboration with a university and a number of motivated researchers will guarantee the academic and scientific accuracy of these data and make this a tool of high value to the EU, OECD and other authorities,faced with the elaboration of mandates in this complex sector.

Martin Frankenhaeuser of Borealis advocated the reduction of the grey fraction by extracting the recyclables first, *splitting the waste into a dry and a wet fraction.*

This proves once more that the different technologies can be complementary with a common denominator: *good source separation.*

The colours of state of the art composting were defended by: **George Savage**, of California Recovery, who advocated the use of the C/N ratio as a possible tool to judge the efficiency of process design.

At the same time he focused the attention once more on the *importance of the waste characterisation on the quality of the final compost.*

Going deeper into the problem of the definition of biowaste and broadening that definition by adding a paper fraction, **Bruno De Wilde** of Organic Waste Systems N.V. (OWS), Belgium, provided us with a clear insight in the **advantages of paper inclusion in composting**. As there are (among many others):

→ reduction in salts
→ more organic content
→ better compost

The complete text of this research-project is available in the ORCA Technical Publications.

To close this compost section, **Karel Mesuere** commented and analysed the compostability criteria elaborated by **ORCA** as a framework to define the acceptable feedstock for state-of-the-art composting, and in order to make a quality compost.

These criteria create the link between the biodegradability (as tested in the labs) and the practice of composting, by integrating parameters that are relevant to the operations of composting/biogasification-plants. At the same time these criteria are scientifically correct and take into account the technical, biological and economical factors of the different processes. This very important work has been reviewed by a Peer-Review-Committee consisting of the leading authorities in the academic world and has therefore also been accepted by different authorities to form the framework for future legislation.

This publication is also available in the ORCA Technical Publications.

Last but not least, **Peter White**, Procter & Gamble Ltd., UK, author of the book îIntegrated solid waste managementî showed us in his exposÈ, apart from the well-known advantages of biological treatment, the necessity to optimise biological treatment, to optimise source separation and to optimise the whole system, in order to reduce overall environmental impacts and to make integrated waste management economically sustainable.

Through LCI and LCA, we can create tools to model the options, even if questions will always remain concerning the validity of socio-political parameters and their evaluation in the overall LCI. At least we will have a more comprehensive view on the problem that will help us design a possible solution, that will always have to be a **custom made**, local or regional solution.

He warned us once more for the **danger of isolating systems or technologies that will distort the total picture and shift the problems** and advocated *the integrated waste management approach.*

Conclusion

I just want to close this report, by urging all you ladies and gentlemen to do the same. Biological treatment, our technology (including the correct applications of biodegradable materials and products) has so much to offer, that we should not try to pretend it has no limitations

We most certainly have an economical, ecological and socio-psychological advantage over the other waste-treatment options, so why avoid the challenge to situate our technologies in an overall sustainable integrated waste-management system.

Cost-relation Waste-treatment

- ◆ Taking into consideration a similar environmental impact
- ◆ landfill
- ◆ incineration with energy recovery
- ◆ biological treatment

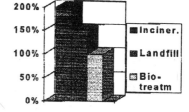

relate as shown here

Lets sit together with the representatives of the other resource-recovery technologies, and discover where we are complementary, where competitive, to ensure an overall better result and the recovery of valuable resources and energy.

ORCA Compostability Criteria: A Framework for the Evaluation of Feedstock for Source-Separated Composting and Biogasification

K. MESUERE, Organic Reclamation and Composting Association (ORCA),Brussels, Belgium

Mailing address: Procter & Gamble GmbH, Sulzbacher Strafle 40, 65818 Schwalbach a/T, Germany

Presented at 'The Science of Composting', international symposium, 30 May–2 June, 1995, Bologna, Italy.
Session: 'Composting as an Integrated System of Waste Management'.

Abstract

Source-separated biowaste composting and biogasification are receiving growing interest as waste recovery technologies because they have significant potential to contribute to waste diversion from landfills while producing valuable end-products such as biogas and/or soil improver with fertilizer value (compost). Organic waste constitutes between 20 and 70% of the municipal solid waste (MSW) stream in EU countries. Vegetable, fruit and garden wastes (VFG) are commonly accepted for composting, but only account for about 3/4 of total organics in the MSW stream. Organic waste other than VFG primarily constitutes paper based products and could include 'biopolymer' waste components in the future as well. Therefore, inclusion of all organic waste fractions into the biowaste definition could further enhance the role of composting and biogasification within integrated waste management systems.

Currently, no technically-based and broadly accepted set of guidelines are available to select paper or other organic waste products for composting or biogasification. To assess the acceptability of waste products for source-separated composting, the Organic Reclamation and Composting Association (ORCA) set out to develop a comprehensive framework of criteria and testing strategy guidelines.

Four criteria were identified as being essential to construct a consistent and adequate framework for the purpose of biowaste definition. The ORCA Compostability Criteria focus on 1) environmental safety, 2) compost quality, 3) plant compatibility and 4) landfill diversion. Testing strategy guidelines were developed for each criterion based on a tiered approach, ranging from lab-scale to full-scale or field testing. These criteria and testing strategy guidelines are intended to assist waste management authorities in searching for an optimised balance between 1) maximising the diversion of organic material from landfills, 2) minimising the overall solid waste management costs and 3) producing a high-quality, marketable end-product.

Introduction

Modern integrated waste management (IWM) include four major interrelated avenues for waste handling ranging from material recycling, biological and thermal treatment to landfilling (Figure 1). Because of legislative, environmental, and practical considerations, the direct landfilling of waste in EU countries will increasingly be phased out and landfill practice will be used only in integration with biological and thermal treatment methods as final depository for sorting residues, nuisance materials or incinerator ashes. Therefore, material recycling, as well as biological and thermal treatment are starting to play an increasingly important role in European solid waste management practices as direct landfilling of waste is phased out and landfill capacity is decreasing.

Maximising landfill diversion and optimising the role of the other IWM elements requires two essential strategies: 1) optimizing the handling and processing of the various fractions of the solid waste stream according to their unique characteristics at minimum environmental and financial cost and 2) setting up appropriate and effective sorting and collection systems for these fractions either at the household level or at a central facility. In this context, source-separated biowaste composting and biogasification take on a unique position because these technologies can handle a relatively large fraction of MSW at competitive cost while producing a high-quality, marketable endproduct.

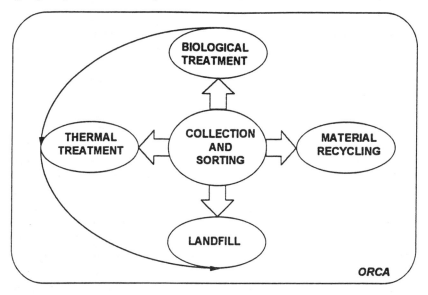

Figure 1 Biological treatment (composting and biogasification) in integrated solid waste management.

Organic waste typically constitutes between 20 and 70% of the municipal solid waste stream in EU countries [1,2]. The weighted average composition of MSW in EU countries indicates that nearly 40% of MSW consists of food and garden waste (Figure 2). It is this so-called vegetable, fruit and garden waste (VFG) fraction that is commonly accepted for state-of-the-art source-separated biowaste composting and biogasification and its seperate collection is increasingly embedded as a mandatory requirement in waste laws of European countries. Nevertheless, the VFG fraction only accounts for about 70-80% of the total organic fraction in the MSW stream. Organic waste other than VFG primarily consists of paper-based products and packaging which are part of the 'paper/board' fraction as wells as the 'rest' fraction. It also could increasingly include 'biopolymer'-based waste components in the future.

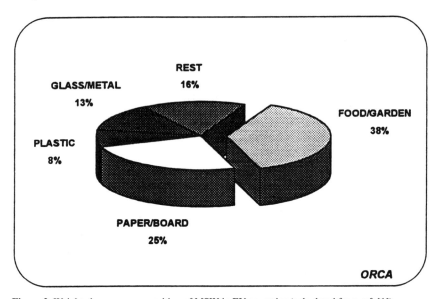

Figure 2 Weighted average composition of MSW in EU countries (calculated from ref. [1]).

Most of the 'paper/cardboard' fraction should not be considered for composting because of an existing and still growing infrastructure for material recycling. Nevertheless, a significant part of it is often soiled with moisture and food residues and therefore not necessarily fit for cost-effective and high-quality material recycling. This non-recyclable paper fraction, accounting for about 5-10% of the waste stream, has been proven to be a suitable feedstock for composting and biogasification. In fact, its value and benefits for the composting process and quality are increasingly well-supported [e.g., 3,4,5]. In addition, there are a number of paper-based products in the 'rest' fraction, such as tissues, towels and hygiene paper products, accounting for about 1-5% of the waste fraction. These products are largely organic in nature and therefore not a-priori incompatible with composting/biogasification. Finally, we may see an increasing number of biopolymer-based products on the market in the future, formulated specifically with composting and biogasification as disposal options in mind. Considering these additional fractions as potential feedstock components would therefore enable waste authorities to divert an additional 5-15% of the MSW stream from landfilling through composting and biogasification technologies (Table 1).

FRACTION	VFG FEEDSTOCK	TOTAL POTENTIAL FEEDSTOCK
FOOD/GARDEN	38%	38%
NON-RECYCLABLE PAPER (paper fraction)	–	5-10%
NON-RECYCLABLE PAPER AND OTHER ORGANICS (rest fraction)	–	1-5%
	35-40%	45-55%
		ORCA

Table 1 Potential waste fractions for composting/biogasification feedstock.

As a bottomline for these introductory remarks, it can be conclude that, in defining biowaste, it may be worthwhile as a general rule to search for a balance between 1) maximising the diversion of organic material from landfills by taking into account the intrinsic and unique characteristics of waste fractions other than food and garden waste, 2) minimising overall solid waste management costs and 3) producing a high-quality, marketable endproduct.

However, with the recognition of a potential growing role of composting/biogasification within IWM systems also comes a need. The need to answer the question: which criteria should be adopted to assess the acceptability of waste products other than food and garden waste? To address this need, ORCA's Technical Committee set out to develop a basic set of criteria and testing strategy guidelines.

ORCA Compostability criteria

Four sets of criteria and assessment requirements are proposed for evaluating the acceptability of waste products for biowaste recovery (Figure 3) [6].

The *first* and foremost criterion requires that the environmental safety of the waste product be demonstrated in the context of composting/biogasification. This means that the waste product should not adversely affect 1) biological activities during composting/biogasification, 2) biota in compost-amended environments, and 3) the physical and chemical properties of compost-amended soil [7]. This criterion requires both detailed fate data (in particular in relation to biodegradability) as well as effect data at anticipated short- and longterm exposure levels. *Secondly*, addition of a waste product to the feedstock for composting/biogasification should not adversely affect the quality of the compost produced as specified by different national or international compost quality standards and parameters or local market-driven demands [8]. *Thirdly*, processing of a waste product must be compatible with the physical operations in a composting facility or newly planned

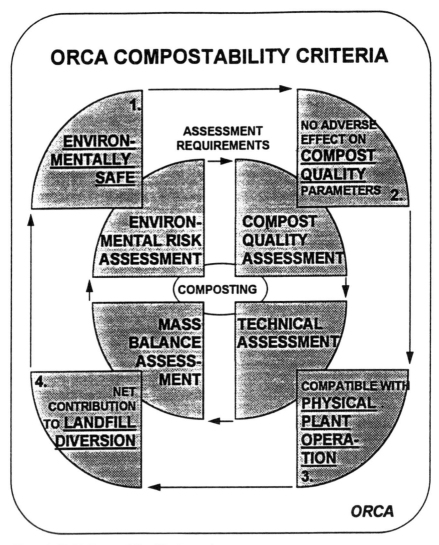

Figure 3 ORCA Compostability Criteria overview.

operations should take into account technical requirements of the envisaged feed-stock. This criterion ensures that addition of a waste product to the feedstock does not adversely affect the operational procedures or maintenance needs of the composting/biogasification facility. *Finally*, inclusion of a waste product in the biowaste definition must result in a positive contribution to waste diversion from landfill at an overall cost locally justifiable versus alternative waste management options.

Taken together, this loop of criteria is thought to form a consistent and adequate framework within which one can start to assess the acceptability of waste products for biowaste recovery. However, in order to actually evaluate a waste product against these criteria, criteria-specific assessment requirements are needed.

Putting the ORCA criteria to use

The evaluation of a waste product for composting is based on criteria-specific assessments (Figure 1). The overall outcome of the assessment procedures must take into account accumulative weight of evidence that a waste product complies with the ORCA criteria, rather than it meeting a number of simplistic pass/fail limits.

For the assessment of the above criteria, it is recommended to adopt an overall testing strategy which includes both laboratory and pilot/full-scale investigations with field studies (Tables 2-5). In test execution it is recommended to follow stan-dardised tests (OECD, ASTM, CEN, DIN, etc.) if available, but allowance should be made to complement such tests with customised test designs, if needed. The *laboratory-scale testing* should be aimed primarily at developing basic, high-quality data for a given waste product and/or its material components with respects to 1) fate, including biodegradability, 2) toxicity to biota in compost and compost-amended soil matrices, 3) physical disintegration, and 4) effects on regulated compost quality parameters. The *pilot/full-scale investigations* should be aimed at confirming data generated at the laboratory-scale. In addition, such tests should evaluate the practical processibility under larger scale operating conditions by assessing the impact of the addition of a given waste product to the feedstock on key process parameters (both biological and mechanical). Finally, the pilot/full-scale investigation should produce a compost sample which can be used in field phyto-toxicity testing. In the following, the assessment guidelines for each criterion are discussed in further detail.

Environmental Safety Criterion

Environmental safety in the context of composting/biogasification means that materials or waste products should not adversely affect the environment in which they are handled, processed and ultimately applied, i.e. during waste collection or sorting, compost processing and final compost application to soil. Specifically, this means that the waste product should not adversely affect 1) worker's safety and health at a facility, 2) biological activities during the active composting or biogasification stage, and 3) biota and physico/chemical properties of compost-amended environments.

To assess this criterion a tiered testing strategy is needed to generate, on the one hand, detailed fate data, in particular in relation to the biodegradation of the product materials, and on the other hand effect data for representative target organisms at estimated exposure levels. Biodegradability should of course form a focal point in this assessment, because it is generally expected to be the primary fate mechanism for those waste product materials which will enter the final compost. On the other hand, it must be understood that biodegradability does not, in and of itself, provide full-proof reassurance for environmental safety. Mostly, it reduces the time of maximum exposure and minimizes the potential for accumulation. So, biodegradability must be assessed as a fate mechanism in this broader context of environmental safety considerations.

The type of questions to be addressed in an environmental safety assessment for this purpose are summarized in Figure 4. *Fate assessment* requires developing an understanding of what ultimately happens to a substance when it is released into the environment and predicting the concentration in relevant environmental compartments [9]. In the context of waste product evaluation for composting, it means the following questions must be addressed (Figure 4):

- What is the predicted environmental concentration (PEC) of the waste product in the feedstock, i.e. at the point of entry of the composting facility?
- Does the product physically disintegrate and does it become part of the final compost? Or are certain waste product components removed during pre-treatment?
- For those product materials entering the final compost: does biodegradation occur and to what extent?

Once compost is applied to soil, an additional set of questions have to be answered (Figure 4):

- Does continued biodegradation occur and at what rate is the biodegradation completed?
- Are persisten intermediates or residues formed?
- What is the predicted environmental concentration (PEC) of waste product materials and compounds under conditions of repeated compost applications?

The effect assessment, on the other hand, requires developing an understanding of the impact of the waste product or materials on selected representative organisms and environmental parameters at the predicted environmental concentration [9]. For the purpose of environmental safety assessment in the context of composting/biogasification, it means one needs to evaluate effects in a composting facility on 1) worker's health during collection and handling and 2) the health of the microbial community in the active composting stage (Figure 4). With regard to

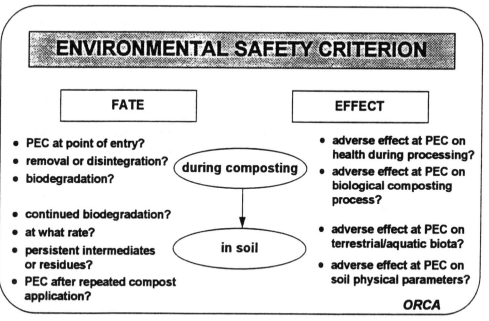

Figure 4 Key aspects of environmental safety assessment of waste product materials in context of composting/biogasification.

the soil environment, it requires measuring effects and on representative terrestrial and possibly aquatic biota as well as impacts on basic soil physical parameters (Figure 4).

The data sources and testing levels required for this purpose will range from relatively simple to more sophisticated: from simply gathering literature data to more elaborate and cost-intensive full-scale facility and field trials (Table 2).

At the very initial level, data gathering will be limited to 1) company data on maximum predicted market volumes and product composition, and 2) available literature data on toxicity and biodegradation for the product materials or ingredients (Table 2). These data can then be used to quantify initial estimated exposure levels for the waste product or materials in the feedstock and the compost-amended soil.

As a second tier, laboratory testing will virtually always be needed and involve both screening and confirmatory testing, in particular for documenting biodegradbility as the primary fate mechanisms (Table 2). At the screening level, one needs to understand 1) the degree of physical/chemical disintegration of the product materials under simulated composting conditions (as measured for example by weight loss and loss of tensile strength), and 2) the inherent biodegradability of the waste product materials, i.e., whether a material can be ultimately biodegraded under a certain set of conditions (as measured by simple in-vitro batch respirometric methods such as Sturm and OECD inherent biodegradability type tests using powdered samples in aqueous medium). Confirmation of these screening data requires additional respirometric methods which use solid matrices and are designed to simulate composting, biogasification or soil conditions. At this confirmatory stage we are interested to answer the question whether the material is practically biodegraded in the compost and terrestrial habitats in which it is discarded and whether the rate of degradation is similar to the projected rate of loading. One example of a useful tool in such an approach is the ASTM D5338 test. Occasionaly, confirmation will require the use of isotopically labeled materials to conclusively document the mineralization of the material to CO_2 or CH_4, or as assimilation of decomposition residues into microbial biomass or stable humus substances. Finally, confirmatory testing at the lab-scale requires an effect assessment and quantification of terrestrial and possibly aquatic toxicity.

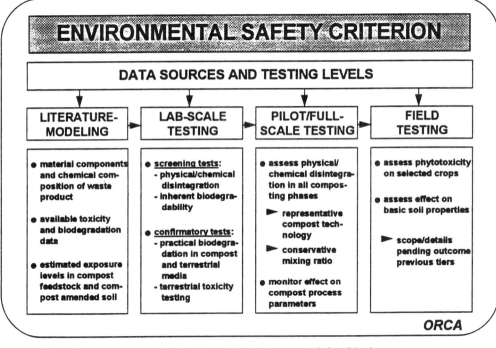

Table 2 Data sources and testing levels for ORCA Environmental Safety Criterion assessment.

Because of the wide diversity in compost technologies, pilot- or full-scale trials using technologies representative for the geography of interest are needed (Table 2). They are essentially needed to confirm that physical/chemical disintegration indeed occurs as the waste product moves through the successive composting phases and to ascertain that the waste product becomes an indistinguishable part of the finished compost under practical operation conditions. A second objective in this higher tier testing level is to assess the effect on the microbial community in the active composting stage, as measured indirectly by unusual variability in process parameters such as temperature, pH, and oxygen uptake or CO_2

evolution. In some cases, where high exposure levels in compost feedstock are expected and lab-scale toxicity testing is not conclusive, some form of field testing will be required, involving, for example, phytotoxicity trials on selected crops. An excellent overview of biodegradability testing methods for synthetic materials in the context of composting/biogasification is provided in reference [10].

Compost Quality Criterion

The compost quality criterion stipulates that including a waste product in the feed-stock definition should not adversely affect regulated compost quality parameters nor the marketability of the compost. The lack of consistent and uniform compost quality regulations across Europe will require to take into account local or national regulations or guidelines according to the area of interest for the feedstock defin-ition.

The testing strategy will again involve a range of tiers and can be done in conjunction with the environmental safety test set-ups by adding the quantification of effects on applicable compost standards as an additional endpoint. Seperate field testing for certain waste product categories may however be required because changing sorting and selection rules at the household level may adversely affect nuisance levels, and therefore possibly compost quality. Therefore, field testing in this context requires co-collection trials which can also provide the input feed-stock for pilot or full-scale testing in the context of other criteria assessments.

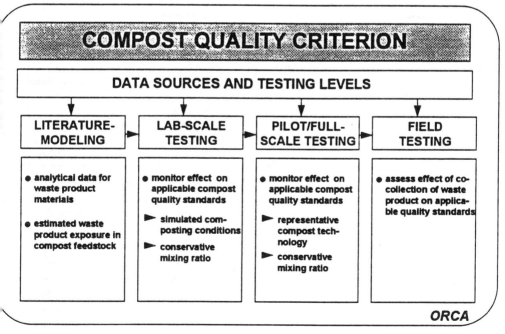

Table 3 Data sources and testing levels for ORCA Compost Quality Criterion assessment.

Plant Compatibility Criterion

Waste products will often differ significantly in their physical form, composition and morfology from vegetable fruit and garden type waste. This will certainly affect their compatibility with the physical operation of composting or biogasification facilities and requires an assessment of the impacts on the sustainability of facility operations and maintenance needs.

From a testing point of view, only a very limited understanding can be achieved on these questions at the laboratory scale. For example, laboratory-scale screening tests could include the assessment of the physical/chemical degradation of the waste product in terms of weight loss, particle size, loss of integrity and changes in tensile strength. However, such tests typically do not provide conclusive and reliable results because the complexity of physical plant operation is not adequately taken into account.

Realistically speaking, one will have to review the technologies in the geography of interest, select representative pilot or full-scale models for the key process phases (pretreatment, active composting, and refining), and monitor or quantify the effect on typical operational and maintenance parameters, ranging from water and energy requirements to possible downtime of machinery. This type of testing will be of high importance for more complex waste products or waste products which differ significantly in their physical form and size from regular VFG waste.

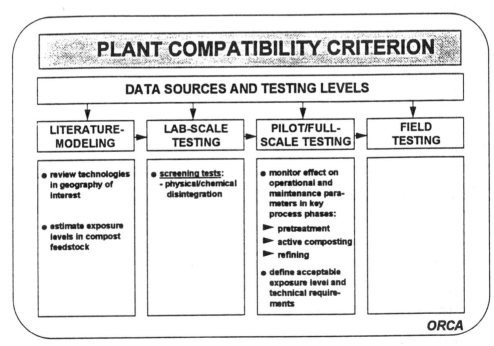

Table 4 Data sources and testing levels for ORCA Plant Compatibility Criterion assessment.

Landfill Diversion Criterion

Finally, it must be ascertained that adding the waste product to the feedstock contributes to an overall improved landfill diversion rate. While such a requirement may appear odd at first glance, it is certainly very relevant when one considers that a waste product may never enter the final compost because it is screened off during pretreatment due its physical form or size, or because a waste product either contains or induces during collection an excessive amount of nuisance material which has to be removed. Since experience teaches that each percent of nuisance material causes about an equal amount of compost to be lost, assessing the overall landfill diversion rate is an important and very relevant requirement.

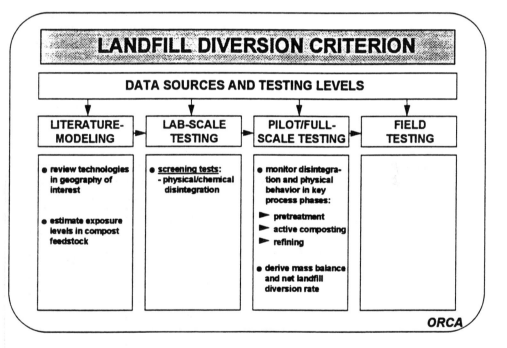

Table 5 Data sources and testing levels for ORCA Landfill Diversion Criterion assessment.

From a testing point of view, reliable mass balance data must be acquired, primarily at the pilot or full-scale testing level using representative technology models for each of the key process phases. Such data can certainly be acquired in conjunction with pilot or full-scale testing for the other criteria assessments. An important piece of the puzzle in this context also can be derived by examining the nuisance level in feedstock material from co-collection field trials conducted in the context of the compost quality criterion.

Summary

When defining biowaste feedstock for composting, one must generally search for a balance between maximising the diversion of organic material from landfills, minimising overall solid waste management costs and producing a marketable compost. In this context it may be desirable to expand the VFG biowaste feedstock to include paper and biopolymer-based waste components. The criteria and testing strategies described here are intended to provide a broad conceptual framework on key factors to assess for the composting/biogasification feedstock definition.

References

[1] White, P.R., M. Franke, and P. Hindle (1995). Integrated Solid Waste Management. A Lifecycle Inventory. Blackie Academic & Professional.

[2] Rijpkema L.P.M., G.W. Krajenbrink, P.W.A. Stijnman, and J.L.B. de Groot (1993). Survey of Municipal Solid Waste Combustion in Europe. TNO Report. Apeldoorn, The Netherlands.

[3] Verstraete W., L. De Baere, and R.-G. Seeboth (1993). Getrenntsammlung und Kompostierung von Bioabfall in Europa. Entsorgungs-Praxis, M‰rz 1993.

[4] Anderson, G. and K. Smith 1994. Mixed Paper Teams Up with Biosolids. Biocycle, March 1994, 61-65.

[5] ORCA (1994). The Impact of Paper Products on Biowaste Composting. ORCA Tech. Publ. Nr. 4 (Ed. B. Lemmes).

[6] ORCA (1994). ORCA Compostability Criteria (Ed. B. Lemmes).

[7] Shimp, R.J. (1993). Assessing the Environmental Safety of Synthetic Materials in Municipal Solid Waste Derived Compost. In: Science and Engineering of Composting (Eds. Hoitink, H.A.J. and H.M. Keener). The Ohio State University.

[8] ORCA (1992). A Review of Compost Standards in Europe. ORCA Tech. Publ. Nr. 2 (Ed. B. Lemmes).

[9] Cowan, C.E., D.J. Versteeg, R.J. Larson, P.J. Kloepper-Sams (1995). Integrated Approach for Environmental Assessment of New and Existing Substances. Regulatory Toxicology and Pharmacology, 21, 3-31.

[10] Pettigrew, C.A., and B.N. Johnson (1995). Testing the biodegradability of synthetic polymeric materials in solid waste. In: A.C. Palmisano, and M. Barlaz (ed.), Microbiology of Solid Waste. CRC Press, Boca Raton, FL. CRC Press, Inc., Boca Raton, Florida. In press.

Acknowledgements

This work represents the joint effort of all ORCA Technical Committee members. Special thanks to Dr. R.J. Shimp and Dr. R.G. Seeboth for their extensive input while chairing the Technical Committee. Special thanks also to the members of the peer review committee whose valuable input and comments were taken into account in the finalisation of the ORCA Compostability Criteria document.

ORCA TECHNICAL COMMITTEE

• Dr. C. Bastioli	Novamont, Italy	• Mr. B.. Lemmes	ORCA, Belgium
• Mr. G. Blackburn	Courtaulds, UK	• Dr. K. Mesuere	Procter & Gamble, Germany
• Dipl.-Ing. A. Canovai	GESENU, Italy	• Dipl.-Ing. T. Obermeier	ITU GmbH, Germany
• Dr. P. Carrera	Hutec/Sorain Cecchini, Italy	• Mr. M.. Schnorr	Herhoff, Germany
• Dr. D.T. Carrick	Zeneca, UK	• Dr. R.-G. Seeboth	Procter & Gamble, Germany
• Dipl.-Ing. B. De Wilde	OWS, Belgium	• Dr. R.J. Shimp	Procter & Gamble, U.S.A.
• Mr. S. Facco	Novamont, Germany	• Mr. A.. Ruckdeschel	Thyssen, Germany
• Dipl.-Ing. H. Hofer	Bühler, Switzerland	• Dipl.-Ing. W.A. Van Belle	Procter & Gamble, Belgium
• Dr. T. Jopski	Zeneca, Germany		

PEER REVIEW COMMITTEE

• Dr. L De Baere	OWS, Belgium	• Dr. E.I. Stentiford	Univ. Leeds, UK
• Prof. Dr. W. Bidlingmaier	Univ. Essen, Germany	• Prof. Dr. ir. W. Verstraete	Univ. Gent, Belgium
• Prof. Dr. M. de Bertoldi	Univ. Udine, Italy	• Prof. Dr. H. Vogtmann	Univ. Kassel, Germany
• Prof. Dr. J. Jager	Univ. Darmstadt, Germany		

The Compostability Criteria document is available from the ORCA secretariat, Avenue E. Mounier 83, Box 1, 1200 Brussels, Belgium. Tel. +32 2 772 90 80.

Index of posters

Baca, M.T. Thermophilic pilot scale composting of olive cake 1057

Baca, M.T. Changes in the amino-acid composition of grass cuttings during turned pile composting 1063

Barberis, R. Composition and chemical characteristics of the main compostable organic wastes 1067

Ben Ammar, S. Impact of separation on compost quality 1071

Bernal, M.P. Nitrogen in composting: relevance of the starting material and the system used 1074

Berner, A. Estimation of N-release and N-mineralization of garden waste composts by the means of easily analysed parameters 1078

Bhurtun, C. Performance prediction of composting processes using Fuzzy Cognitive Maps 1083

Blanc, M. Biodiversity of thermophilic bacteria isolated from hot compost piles 1087

Brun, G. Effect of humic matters extracted from compost and from leonardite on P nutrition of rye-grass 1091

Bruns, C. The suppressive effects of composted separately collected organic waste and yard waste compost on two important soilborne plant pathogens 1094

Ceccati, B. Legislative and scientific aspects of the production and use of vermicompost from biological sludges 1096

Cegarra, J. Composting of fresh olive-mill waste water added to plant residues 1100

Chefetz, B. Municipal solid waste composting: chemical and biological analysis of the process 1105

Ciavatta, C. Chemical parameters to evaluate the stabilization level of the organic matter during composting 1109

D'Angelo, G. Response of three compost-based substrates to different irrigation and fertilization regimes in Poinsettia (*Euphorbia pulcherrima*Willd.) 1113

Degli Innocenti, F. Method for the evaluation of biodegradability of packaging in composting condition proposed by CEN (European Committee for standardization): a technical approach 1118

Deiana, S. Effects of the spreading into the soil of olive mill wastes on the physico-chemical properties of the humic acid. Interaction of natural and synthetic humic acids with caffeic acid 1122

de Toledo, V.C, The use of dairy manure compost for maize production and its effects on soil nutrients, maize maturity and maize nutrition 1126

Diaz-Burgos, M.A. Sewage sludge composting: study of nitrogen mineralization using electroultrafiltration 1130

Diaz, M.J. Co-composting process of (sugarbeet) vinasse and grape marc 1134

Diaz, M.J. Chemical characterization of three composts of (sugarbeet) vinasse with other agroindustrial residues 1138

Domeizel, M. Monitoring of organic matter during composting 1142

Dominguez, J. Effects of bulking agents in composting of pig slurries 1146

Durillon, M. Sewage sludge composting with thermopostage™ process: the platform of Arenthon (Haute Savoie, France) 1150

Fantoni, A. Biodegradability of Thiram in composting process 1152

Ferrara, A.M. First experiments of compost suppressiveness to some phytopathogens 1157

Franco, I. Reduction in phytotoxicity of olive mill waste by incubation with compost and its influence on the soil-plant system 1161

Fuller, M. An assessment of the agronomic value of co-composted MSW and sewage sludge 1166

Garcia, M. New bulking agents for composting sewage sludge (*Pteridium* sp. and *Ulex* sp.). A laboratory scale evaluation 1170

Gronauer, A. Emissions of greenhouse and environmental relevant gases by the decomposition of organic waste from the households 1174

Hollyer, J.R. The compost information kit: business tools for success 1177

Horwath, J.R. The development of low-input, on farm composting of high C:N ratio residues 1181

Kolb, A.   Decentralized compost –
management: case-study of a district of
77 000 inhabitants – the Kulmbach model
1188
Kubocz, T.   Microbial succesion in a
technical composting process   1193
Kuhner, M.   Assessment of compost maturity
1199

Liendeberg, C.   Accelerated composting in
tunnels   1205
Lott, Fisher J.   Composting of organic garden
and kitchen waste in open-air windrows:
influence of turning frequency on the
development of *Aspergillus fumigatus*   1207

Madejon, E.   Sugarbeet fertilization with
three (sugarbeet) vinasse composts   1211
Magagni, A.   Waste collection utility
development, environment project 2000,
projects F.U.S. 20–F.U.S. 21   1215
Makela-Kurtto, R.   Suitable of composted
household waste of Helsinki metropolitan
area for agriculture   1281
Marinari, S.   Effects of compost on soil
biological fertility and maize (*Zea mays*)
production   1221
Meekings, H.   The effect of municipal waste
compost on the development and viability of
Ascaridia galli egge   1224
Michel, F. C.   Fate of lawn care pesticides
during the composting of yard trimmings
1230
Michel, F. C.   Characterization and
composting of source-separated food store
organics   1233
Morselli, L.   Heavy metals determination in
MSW merceological glasses and derived
compost   1236
Mort, N.   The disinfection of anaerobically
digest sewage sludge through composting
1242

Negro, M.J.   Effect of sweet Sorghum
bagasse compost on sweet Sorghum
productivity in pots   1247
Nielsen, L.K.   The Campodan composting
process features rapid process set-off,
uniform process conditions and a built-in
biofilter   1251
Nuntagij, A.   Demostration of the influence
of Mg vermiculite on the activity of
cellulosic agents and diazotrophs during
composting of lignocellulosic residues
1254

Otero, D.   A comparison between chemical
and biological index for measuring compost
quality   1258

Ottabong, E.   Sewage sludge: soil
conditioner and nutrient source   1261

Parades, C.   Composting of fresh and pond-
stored olive-mill waste water by the Rutger
system   1266
Piccinini, S.   Behaviour of biodegradable
Mater- BI ZI01U plastic layers in a
composting pilot plant   1271
Piccinini, S.   The Emilia Romagna
experience in animal manure composting
1275
Pullin, C.J.   Composting wastewater sludges
without the addition of bulking agents   1281

Radanov, S.   The potential use of composted
waste materials for cuttings   1285
Rossi, L.   INALCA experiences in
composting of slaughter house organic
wastes   1289
Ryckeboer, J.   The influence of GFT-compost
extracts on the motility of juveniles of
Heterodera schachtii in vitro   1294
Ryckeboer, J.   The suppression of
Penicillium digitatum by extracts of
GFT-compost   1301

Saharinen, M.H.   Cation exchange capacity
of manure-straw compost – does sample
preparation modify the results?   1309
Sampedro, L.   Vermicomposting solid paper
pulp mill sludge: a three stages biodegradative
process   1312
Sanchez, M.A   Evaluation of turned and
static pile systems on toxicity of water
extracts of biowastes   1316
Schauner, P.   Maximizing both compost
quality and yield   1320
Seck, M.A.   Fresh and composted pea-nut
shells microflora   1322
Shaw, C.M.   Heat transfer in composting
system   1331
Siebert, S.   A microcosm system to
determine the gas production of arable soils
amended with different composts   1335
Silveira, A.   Improved of the composting
process   1339
Silvestri, S.   Use of Mater-Bi ZF03U
biodegradable bags in source separated
collection and composting of organic waste
1343
Solano, M.L.   Oil-mill wastewater sludge
composting   1346
Stoffella, P.J.   Sugarcane filtercake compost
influence on tomato (*Lycoperscion
esculentum*) emergence, seedling growth, and
yields   1351
Szmidt, R.A.K.   Composting of bioreactor
waste from softwood processing   1357

Tiquia, S.M. Effect of moisture content on the composting of pig-manure sawdust litter disposed from the pig-on-litter (POL) system 1361

Tuitert G. The effect of composted vegetable, fruit and garden waste on the incidence of soil-borne plant diseases 1365

Valli L. The experience of the A.M.I.U. of Modena (Italy) in composting sewage sludges and source separated organic wastes 1370

van Os G.J. Possibilities for biological control of Phytium root rot in ornamental bulb culture with composted organic household waste 1374

Vassilev N. Mineralization of three agroindustrial wastes by an acid producing strain of *Aspergillus niger* 1376

Vittur C. Organic waste treatment – composting: a comparative study on language usage and terminology (Italian–German–English) 1380

Volterrani M Effects of application of municipal solid waste compost on horticultural species yield 1385

Vuorinen A.H. Properties of Phosphomonoesterases and B-Glucosidase in compost extract 1389

Wong D. The effect of red mud on metal mobility in anaerobically digested primary sludge during composting 1395

Zambrano J.G. Physical, chemical and biological quality of four lombricomposts 1399

ZENECA Integration of "Biopol" into biowaste composting 1402

Index of contributors

Note: Those references with an 'A' prefix refer to the Additional Papers at the end of volumes 1 and 2.

Accotto, E.   507
Adani, F.   567
Adinarayana Reddy, C.   577, 1230, 1233
Ahlers, S.   1094
Aldo, Panzia Oglietti   1067
Alföldi, T.   1078
Amadon Seck, Mamadon   1322
Amlinger, F.   314
Aragno, Michel   149, 1087, 1207
Atterwall, Stefan   1261
Avataneo, M.   1157
Azcon, R.   1376

Baca, M.T.   255, 1057, 1063, 1161, 1376
Badalucco, L.   1221
Baldoni, G.   431, 457
Barberis, Renzo   175
Bardos, R.P.   767
Barriuso, Enrique   262
Barth, J.   1011
Bastioli, Catia   863, 1118, 1271, 1343
Bazan, Hugues   1150
Beck, R.W.   939
Beffa, Trello   149, 1087, 1207
Bellver, R.   1057
Ben Ammar, Samira   1071
Bernal, M.P.   663, 1074, 1100, 1316
Berner, A.   1078
Beyea, Jan   743
Bhurtun, C.   1083
Bidegain, R.   1254
Bidlingmaier, W.   71
Bigliardi, P.   1370
Blanc, Michel   149, 1087
Bloxham, P.F.   593
Boelens, J.   803
Bollen, G.J.   233, 1365
Bonazzi, Giuseppe   1275
Border, David   983
Bours, G.   346
Brethouwer, Vam   612
Brodie, H.L.   603
Brun, G.   137, 1091, 1254
Brunetti, Antonio   329
Brunetti, G.   195
Bruns, C.   1094
Bryden, G.   1357

Cabrera, F.   394, 1134, 1138, 1211
Carcedo, González   286
Carr, L.E.   603

Carrasco, J.E.   1247
Cato, James C.   557
Ceccanti, B.   1096
Cegarra, I.   663, 1100, 1266, 1316
Centemero, M.   507
Cessa, C.   1109
Chefetz, Benny   382, 1105
Chen, Yona   382, 1105
Cherry, Robert S.   973
Christiana, G.A.   603
Churchill, D.B.   627, 1181
Ciavatta, Claudio   1109, 1289
Ciria, P.   1247
Civilini, M.   870, 884
Claassen, N.   306
Colclough, I.L.   593
Cole, Michael A.   903
Collins, Alan   495
Consiglio, M.   507
Conti, F.   958
Coosemans, J.   1301
Cortellini, L.   431, 457
Cossu, Raffaello   831
Cremonini, Ferdinando   1289
Cristoforetti, Andrea   698

Dall'Orso, Giulio   1275
Dame, Alexandra   495
D'Angelo, G.   1113
Das, K.C.   116, 1020
de Baere, L.   803
de Bertoldi, M.   162, 884
De Nobili, M.   255, 1057, 1063, 1161, 1376
de Toledo, V.C.   1126
De Wilde, B.   803
Degarmo, Richard   983
Degli Innocenti, Francesco   863, 1118, 1271, 1343
Deiana, S.   1122
Dell'Orfanello, C.   1096
Diaz-Burgos, M.A.   1130
Diaz, Luis F.   3, 849, 1037
Díaz, M.J.   1134, 1138, 1211
Dibke, T.   1281
Diener, Robert G.   495
Díez, J.A.   1130
Domeizel, M.   185, 1142
Domenis, C.   884
Domínguez, J.   1146, 1312
Donati, Alfredo   1236

Drew, Susan 1233
Durillon, M. 1150

Ebertseder, Thomas 306
Eggerth, Linda L. 989, 1039
Einzmann, Ursula 329
Elliott, L.F. 627, 1181
Elvira, C. 1146, 1312
Elwell, David L. 1020
Ezelin, K. 137, 1091

Fantoni, A. 1152
Favoino, E. 507
Felipó, M.T. 402
Fernandez-Figares, I. 1063
Ferrara, A.M. 1157
Ferrari, G. 439
Ferrero, G.L. 15
Fieldler, Heidelore 329
Fiorelli, F. 1266
Fischer, Johanna Lott 149, 1207
Fischer, Klaus 81
Fischer, P. 294
Folliet-Hoyte, Nicole 247
Fontaine, Christine 1150
Formigoni, Daniele 1236
Fornasier, F. 255
Forney, J. 577
Forney, Larry J. 1230, 1233
Forsythe, S. 767
Franco, I. 1161, 1376
Frankenhaeuser, Martin 813
Fricke, Klaus 329
Fuchshofen, W. 346
Fuller, M. 469, 1166

Gaetani, M. 1385
Gallardo-Lara, F. 1161
Galli, C. 439
Galli, E. 637, 1266
Ganho, R. 1339
Garcia, D. 1074
García, M. 1096, 1146, 1170, 1258
Gattinger, A. 1094
Genevini, P.L. 567
Gessa, C. 1122
Gianchi, T. 748
Gibbs, R.A. 1242
Goldstein, Jerome 714, 1041
Golueke, C.G. 3, 849
Gottschall, R. 477
Govi, G. 439, 1109
Graeber, Dan 1230
Graetz, Donald A. 1351
Grebus, M.E. 373
Grego, S. 1221
Groenhof, Andy 469, 1166
Gronauer, Andreas 1174
Grossi, N. 1385

Grüneklee, C.E. 1193
Gutser, R. 306

Haak, Enok 1261
Hadar, Yitzhak 382, 1105
Hänninen, K. 673, 1218
Hansen, Robert C. 1020
Haug, Roger T. 60, 1044
Hauke, H. 477
Hedegaard, Mogens 691
Heerenklage, J. 913
Hellman, Bettina 1174
Helm, M. 1174
Hodgkiss, I.J. 1361
Hofstede, Harrie 585, 1242, 1395
Hoitink, H.A.J. 373
Hollyer, James R. 1177
Horwath, William R. 627, 1181
Hout, Sabine 262
Hupe, K. 913
Hummel, Julia 822
Hurtarte, M. 1161

Innocenti, G. 439

Jeng Fang Huang 577
Jury, Sam 469, 1166

Kaemmerer, M. 137, 1091, 1254
Kapetanios, E.G. 924
Kashmanian, Richard M. 648
Keener, Harold M. 116, 1020
Kellogg Johnson, Kathryn 551
Kilbertus, Gérard 1322
Kögel-Knabner, Ingrid 1335
Kolb, Axel 1188
Krüger, I. 631
Kubocz, T. 1193
Kühner, Michael 1199

Lasaridi, K.E. 274
Lawrence, J.L. 1281
Lee, D.L. 1224
Lee, H.C. 447, 1126
Leege, Philip B. 126
Leifeld, Jens 1335
Leita, L. 1161
Lemmes, Bert 1047, A1 (vol. 2)
Lilja, Raimo 892
Lindberg, Charlotta 1205
Liu, Xianzhong 903
López, R. 394, 1134, 1138, 1211
Lopez-Real, J.M. 447, 542, 1052, 1126
Lüth, J.C. 913
Lynch, J.M. 531
Lynch, Nancy J. 973
Lyon, Pierre-François 149, 1207

Madejón, E.   1134, 1138, 1211
Magagni, A.   1215
Mäkelä-Kurtto, R.   1218
Manninen, Helena   813
Mannironi, R.   162
Manunza, B.   1109, 1122
Marilley, L.   149
Marani, G.   162
Marinari, S.   1221
Masciandaro, G.   1096
Massiani, C.   185
Massiani, P.   1142
Mato, S.   1146, 1170, 1312
Mayer, W.   1281
Meekings, H.J.   1224
Menjivar Flores, Juan Carlos   1399
Meneses, Robert   495
Merillot, J.-M.   684, 1052
Mesuere, K.   A8 (vol. 2)
Michel, Consiglio   1067
Michel, Frederick C.   577, 1230, 1233
Miele, S.   1385
Miller, Frederick C.   106
Minshew, H.F.   627
Mohee, R.   1083
Mondini, C.   255, 1063
Montecchio, D.   1109
Morris, Margretta   495
Morselli, Luciano   1236
Mort, N.   1242
Muntoni, Aldo   831
Muraro, P.   1152
Murillo, J.M.   394, 1211

Nakasaki, Kiyohiko   87
Nappi, P.   175, 1157
Nassisi, A.   457
Navarro, A.F.   1100
Negro, M.J.   1247, 1346
Neri, G.   507
Nielsen, Lars Krogsgaard   1251
Nogales, R.   1312
Nuntagij, A.   1254

Otabbong, E.   1261
Otero, D.   1170, 1258

Paavilainen, J.   1218
Panter, Keith   983
Panzia Oglietti, A.   507
Papi, Tiziano   162
Parades, C.   663, 1074, 1100, 1266
Pardini, G.   1385
Parkinson, Rob   469, 1166
Pasetti, L.   637
Pereira Neto, J.T.   729
Persson, Jan   1261
Petruzzelli, Gianniantonio   213
Piavaux, André   A6 (vol. 1)

Piccinini, Sergio   1271, 1275, 1289, 1370
Picco, C.   1152
Pinamonti, Flavio   517
Pistidda, C.   1122
Polo, A.   1130
Popp, L.   294
Prevot, Claude   1150
Pullin, C.J.   1281

Rad Moradillo, J.C.   286
Radanov, Srdjan   1285
Raninger, Bernhard   948
Rausa, R.   1122
Reddy, Adinarayana   577, 1230, 1233
Renzo, Barberis   1067
Revel, J.C.   137, 1091
Rodrigues, M.S.   447
Roig, A.   663, 1074, 1100, 1316
Ron Vaz, M.D.   1134, 1138
Rossi, Lorella   1271, 1275, 1289
Ryckeboer, J.   1301

Sacchini, G.   439
Saez, F.   1247
Saharinen, M.H.   1309
Sampedro, L.   1146, 1312
Sánchez-Monedero, M.A.   663, 1074, 1266, 1316
Sanchez-Raya, A.J.   1057
Savage, G.M.   784, 849
Schattner-Schmidt, Silvia   1174
Schönafinger, Dieter   736
Schüler, C.   1094
Sebastianutto, N.   870, 884
Senesi, N.   195
Sequi, Paolo   23
Serra-Wittling, Claire   262
Shauner, Philippe   1320
Shaw, C.M.   1331
Shoda, Makoto   722
Siebert, S.   1335
Sikora, Lawrence J.   423
Silveira, A.   1339
Silvestri, Silvia   698, 1122, 1343
Sippola, J.   1218
Skinner, John   30
Smith, Wayne H.   413
Solano, M.L.   1247, 1346
Solinas, V.   1122
Sorlini, Giovanni   1289
Stegmann, R.   913
Stentiford, Edward I.   49, 274, 1053, 1224, 1331
Stoffella, Peter J.   1351
Stone, A.G.   373
Stöppler-Zimmer, H.   477
Strauch, Dieter   224
Szmidt, R.A.K.   1357

Tam, N.F.Y.   96, 1361
Tardy, Robert J.   939
Tiquia, S.M.   96, 1361
Tno-me, T.D.   612
Toderi, G.   457
Toledo, V.C.   1221
Tomati, U.   637
Tosetti, A.   1215
Tosin, Maurizio   1118, 1271
Troncoso de Arce, Antonio   394
Tuitert, G.   1365
Tyler, Rodney W.   999, 1177

Udinskey, J.A.   603
Urbini, G.   958

Valentin, N.   1142
Valli, L.   1370
Van Belle, W.A.   41, A1 (vol. 1)
van Gulik, Wilma J.M.   1374
van Os, Gera J.   1374
Vassilev, N.   1376
Vassileva, M.   1376

Verschut, C.   612
Villa, C.   567
Vittur, Claudia   1380
Vogtmann, H.   329, 346, 1094
Volker, D.   233
Volterra, E.   637
Volterrani, M.   1385
Vrijmoed, L.L.P.   96
Vuorinen, Arja H.   1389

Walker, John   357
Watt, T.A.   1126
Westergård, Rune   758
Westlake, K.   767
White, P.R.   792
Whitney, P.J.   531
Wolf, G.   1094
Wong, D.   585, 1395
Wullschleger, I.   1078

Zambrano, Jairo Gomez   1399
Zorzi, Gianni   517, 698, 958, 1152, 1343